Handbook of Computational Approaches to Counterterrorism

V.S. Subrahmanian
Editor

Handbook of Computational Approaches to Counterterrorism

Springer

Editor
V.S. Subrahmanian
Department of Computer Science
University of Maryland
College Park, MD, USA

ISBN 978-1-4899-8766-2 ISBN 978-1-4614-5311-6 (eBook)
DOI 10.1007/978-1-4614-5311-6
Springer New York Heidelberg Dordrecht London

Printed on acid-free paper

Springer is part of Springer Science+Business Media (www.springer.com)

Preface

The last 20 years have witnessed an unprecedented change in information and communications technologies, leading to the ability not only to disseminate information in seconds from one part of the world to another, but also the ability to organize, analyze, understand and predict phenomena on the basis of that information. This book studies how computational methods can substantially improve the collection of data about terrorist groups, the analysis of that data through the construction of behavioral models, the ability to forecast what such groups might do, and how one might respond to a group's behavior. In fact, the first book that uses computational methods to fully analyze a terrorist group's behavior and suggest strategies against the group has just been published[1]—we expect many more detailed analyses of terrorist groups in forthcoming years that use the techniques presented in this book or indeed, use new methods that are emerging both in the social science and computer science literature.

According, the book is divided into several parts.

Part I: Data and Data Acquisition

This section of the book describes methods to automatically collect data in close to real-time

- The time required to gather data about a terrorist group, its actions, and contextual variables surrounding those actions so that the data gathered is real-time data, not data that is manually collected and is several years out of date by the time the collection is completed; Fine temporal granularity so that the data gathered can

[1]Subrahmanian, V.S., Mannes, A., Sliva, A., Shakarian, J., Dickerson, J. Computational Analysis of Terrorist Groups: Lashkar-e-Taiba, Springer, August 2012.

be at as fine a temporal resolution as desired (day, week or month) rather than being aggregated to yearly data as in many past studies;

- Fine-grained quantitative granularity so that instead of merely coding events as having happened (1) or not (0) during a given time frame, we can accurately state how many of a given type of event occurred (e.g. estimated number of bombings during a given time frame, estimated number of fatalities during a given time frame, etc.).

This section consists of 4 chapters. LaFree and Dugan (Chap. 1) describe the Global Terrorism Database—a database of terror events spanning 40 years (1970–2010). This data set was collected manually, not computationally, and provides a baseline of how data collection has been done in terrorism research to date.

The next chapter by Schrodt and Van Brackle (Chap. 2) describes methods to automatically extract "codes" for political events and briefly explains the workings of the authors' freely available TABARI system.

Chapter 3, by Boschee, Natarajan, and Weischedel describes event data extraction by a proprietary system called BBN SERIF™ which focuses on event data extraction. This paper describes experiments carried out by the authors comparing multiple event extraction systems from the point of view of accuracy.

The next chapter by Albanese, Fayzullin, Shakarian and Subrahmanian (Chap. 4) recognizes that not all data about terror groups is event data. To understand the behavior of a terror group, one needs to not only identify the activities the group engaged in (exemplified by the events reported about their actions), but also the context in which those behaviors were carried out. Understanding the context requires understanding a very large number of variables relating to their behavior that is not captured by event data. The chapter describes a prototype system called ACE (Automated Coding Engine) that calculates quantitative values on a fine-grained (monthly or smaller) granularity and was shown to have 82% precision and 77% recall. In addition, the chapter reports experiments showing that if human coders work in conjunction with ACE, then the coding rate of the users increases dramatically with no compromise in accuracy.

Part II: Behavioral Models and Forecasting

Part II of the book consists of 8 papers that study how behavioral models can be learned from the data collected using techniques such as those described in Part I. Learning such behavioral models and explaining them to policy makers and counter-terrorism analysts is key to forecasting—without an explanation, even forecasts that may eventually turn out to be correct may not be believed.

Chapter 5 by Aaron Mannes, a long-time counter-terrorism analyst and author of an excellent book on the topic, focuses on how traditional analysis of terrorism groups is done by policy makers. He summarizes past work on qualitative counter-terrorism analysis and provides a broad perspective on how qualitative analysis methods and computational methods can seamlessly benefit from one another in counter-terrorism analysis.

Chapter 6 by Sliva, Simari, Martinez, and Subrahmanian, presents the well-known SOMA (Stochastic Opponent Modeling Agents) framework. In SOMA, a data set about a terrorist group can be analyzed using sophisticated data mining algorithms to automatically learn probabilistic rules of the form "When the environment in which a given terrorist group operates satisfies a condition C, then the probability that the group will take action A at intensity level I is P". The chapter presents one algorithm to automatically learn SOMA rules from data (such as the data collected using the methods of Part I) and then presents both sequential and parallel algorithms to forecast what the group might do—either in a real or hypothetical situation.

Chapter 7 by Schrodt, Yonamine and Bagozzi provides a broad overview of statistical methods used extensively in political forecasting. The methods, each described briefly, include regression methods, classical time series methods, vector auto-regression models, hazard models, and rare event models.

Chapter 8 by Petroff, Bond and Bond provides algorithms that build upon the well-known Hidden Markov Model (HMM) paradigm in order to come up with forecasts about monthly violence in conflict zones such as Iraq and Afghanistan, irrespective of the group that carries out the attacks. The paper presents the algorithms as well as the results of detailed experiments.

Chapter 9 by Simari, Earp, Martinez, Sliva, and Subrahmanian presents a different approach. Rather than use probabilities directly, this method uses similarities between a given situation and previous situations in order to forecast what actions a group will take in a given situation. The resulting system called CONVEX was tested on 118 groups and was found to be over 95% accurate in the predictions it made.

Chapter 10 by Choi, Asal, Wilkenfeld and Pattipati looks at the problem of forecasting behavior of terrorist groups using methods that impute missing values and then used well-known Support Vector Machines (SVM) to forecast the behavior of a group. As in the case of CONVEX in Chap. 9 above, they report accuracy of over 90%.

A major problem with most forecasting analyses (including those in Chaps. 9 and 10) is that they measure accuracy by evaluating how well a predicted forecast matches reality on a data set where the ground truth of what was done by the group is kept blind from the forecasting algorithm. Though this sounds reasonable, it is the case that many groups just keep doing—in a blind time period—what they did in the training data. As a consequence, a simple algorithm that merely repeats what was done in the most recent training data would have high accuracy while providing no meaningful insight.

Chapter 11 by Martinez, Sliva, Simari and Subrahmanian squarely takes on this issue and studies the problem—not of predicting what the group will do during a given time frame—but when and how the group will *change* its behavior. This time, accuracy is measures solely by looking at the cases when the group changed its behavior and seeing how well those behavioral changes were predicted. The chapter presents the CAPE (Change Analysis Prediction Engine) algorithm and prototype system which predicts both the direction of change (not taking action A to taking action A or vice versa) as well as the intensity of change. The chapter reports experiments showing that CAPE has an overall change forecast accuracy of over 80%.

Chapter 12 by Dickerson, Simari and Subrahmanian focuses on something none of the previous chapters do—it focuses not only on what actions a terrorist group will take and with what intensity—it also specifies *when* the group will take that action. It introduces TP-rules that have the form "If the environment in which terrorist group G operates satisfies a condition C at time T, then the group will take action A at intensity level I at time (T+ΔT) with probability P." The authors describe a patented algorithm to automatically learn such TP-rules from data and briefly describe how they have specifically applied these TP-rules to learn the behavior of the terrorist group, Lashkar-e-Taiba, make forecasts about their behavior, and suggest policies toward them.

Part III: Terrorist Network Analysis

Part III of the book focuses on the structure of terrorist networks, and techniques to analyze those networks using social network analysis techniques. Prior to the emergence of social networks such as Facebook and Twitter, most social network researchers in the social sciences studied relatively small networks. However, criminals, terrorists, and others of their ilk try to blend in, hiding in enormous networks. With the emergence of Facebook, Twitter, and large mobile phone networks, techniques have been invented to massively scale up social network analysis.

Chapter 13 by Hausslage, Lindelauf and Hamers studies a relatively new phenomenon reported by Marc Sageman—that of "leaderless" terrorist networks in which classical centrality measures may not work well because the networks themselves do not have leaders. However, even in a leaderless network, there is a trade-off between the desire of the network's members to stay "secret" and their need to communicate. The chapter describes the concept of game-theoretic centrality of nodes and shows that in the Jemaah Islamiyah 2002 Bali bombings, the leaderless network could be nicely modeled using game theoretic centrality.

Chapter 14 by Fire, Puzis and Elovici recognizes that open source (and even classified) data on terrorism networks may be highly incomplete as the members of an organization and the links between them are normally not clear. Much work has been done on "link prediction" (identifying missing links). The chapter presents methods to identify such missing links and reports experiments on both the "Profiles in Terror" data set, as well as in various online social networks.

Chapter 15 by Dawoud et al. uses a different approach to infer the structure of a network. It uses co-occurrences of specific keywords for link prediction and iteratively constructs the structure of a terror network through repeated applications of this link prediction operation. Both positive link prediction and negative link prediction algorithms are provided. The chapter also studies the evolution (appearance and disappearance) of links in such networks and reports on experiments showing high accuracy.

Chapter 16 by Petersen and Wiil describes a system called CrimeFighter Investigator in which the authors explain how a computational system can support investigators' needs in creating hypotheses, adaptively modeling the expected structure of a terror network, prediction, alias detection, and exploring various perspectives (to reduce cognitive bias). Like the two preceding chapters, the CrimeFighter Investigator system also predicts missing links and neatly brings together social network centrality measures to predict covert network structure. The authors apply this to a synthetic scenario linking a Danish network of radical Islamists and Al Qaeda (and affiliated movements).

Part IV: Systems, Frameworks, and Case Studies

This part of the book consists of six chapters that describe systems, frameworks, and projects related to computational approaches to supporting counter-terrorism efforts. In addition, there are two detailed cases studies.

Chapter 17 by Salerno et al. describes the National Operational Environment Model (NOEM). The authors present the architecture of NOEM, describe how NOEM can be used to generate models of an environment including demographic and infrastructure information and make forecasts. The chapter also suggests policies using space filling Latin Hypercube Sampling in conjunction with a simulation approach.

Chapter 18 by O'Brien describes the DARPA Integrated Crisis Early Warning System (ICEWS) project which seeks to identify instability in countries as early as possible. Though this may seem unrelated to terrorism, it is clear that instability at least in some countries is related to terrorism.

Chapter 19 by Mirza and Memon looks at how terrorists use video on the Internet. The authors use different definitions of violence that are sensitive to the visual content of videos and then develop methods for semantic extraction of violence from raw video. They propose methods to analyze videos for violence and develop methods to quantify the degree of violence in a video.

Chapter 20 by Shieh et al. describes a system called PROTECT that describes how ports in the US can be patrolled using Stackelberg games to define optimal strategies for the adversary and optimal counter-adversary strategies for the port to use. The PROTECT system has been successfully deployed by the US Coast Guard in Boston and future deployments are under consideration.

Chapter 21 by Dugan and Chenoweth does not solely study terrorism—rather, it studies how governments behave towards terrorists and the response that governmental strategies might cause. The authors describe the Government Actions in Terror Environments (GATE) data set which tracks government action in 5 countries over a 17 year period. They conduct specific case studies assessing the impact of Israeli actions on Palestinian terrorist actions and Turkish actions on Kurdish violence.

Chapter 22 by Haken, Taft and Jaeger studies conflict with a special focus on the conflict-prone Niger Delta region of Nigeria. They describe their Conflict Assessment System Tool (CAST). They present CAST scores for Nigeria based on a total of 84 variables and then use these scores to assess risk and map conflict in the Niger Delta.

Part V: New Directions

Chapter 23 by Simari et al. describes the first principled approach to generating policies against a terror group, taking into account a behavioral model of the group that may have been learned using an automated data mining algorithm such as those presented in Chap. 6. Using a small sample of rules about the terrorist group Lashkar-e-Taiba, the chapter shows how analysts can specify constraints on what actions a counter-terrorism organization can or cannot take, and then how to generate policies that have minimal cost. One section of Chap. 12 describes how specific policies against Lashkar-e-Taiba were automatically derived by a different, but related policy generation engine.

Chapter 24 by Fokkink and Lindelauf uses a game-theoretic approach via a widely studied class of games called "search games" in which the goal is to find a "hider" who is savvy enough to realize that intelligent searchers may be looking for him. The chapter describes different classes of search games on networks including patrolling games, games involving decisions about whether to operate jointly or individually (disperse or unite?), and finding moving fugitives. It is clear there is much operational value that could be derived in the future via such game theoretic models of spatial search operations.

Chapter 25 by Du and Yang focuses on the fact that criminals and states have started using cyber-attacks to achieve their ends, and we can expect incidents of cyber-terrorism to increase in the coming years. The chapter briefly explains how attacks can be described and predicted via Bayesian networks. It describes mechanisms to detect botnet attacks and coordinated attacks. Using Attack Social Graphs, the chapter presents mechanisms to conduct a spatio-temporal analysis of coordinated attacks.

Counter-terrorism cannot be solved easily by a single discipline. Till the beginning of the twenty-first century, most counter-terrorism research involved only social scientists. Since the beginning of this century, the study of terrorism has become truly multidisciplinary, involving researchers from a variety of disciplines including computer science, criminology, political science, public policy, business and economics, mathematics, diverse branches of engineering, sociology, psychology, anthropology—and many other disciplines. This Handbook makes a first effort to bring together world-class researchers with distinguished careers in their fields who are making important contributions to understanding, analyzing, and influencing terrorist behavior. Readers who are interested in following up on the material in this book will find auxiliary related material at www.umiacs.umd.edu/research/LCCD/cac.

College Park, MD V.S. Subrahmanian

Acknowledgements

Most of the authors of chapters in this book are world-experts in their field. I would like to thank them for taking time out of their busy schedules to write chapters that would be of great interest to researchers studying how computational approaches can help improve the fight against terrorism.

I would also like to thank several current and former members of my research group at the Lab for Computational Cultural Dynamics at the University of Maryland for helping shape my own ideas and work on this topic: Massimiliano Albanese, Damon Earp, Samir Khuller, Dan LaRocque, Aaron Mannes, Vanina Martinez, Cristian Molinaro, LTG (Ret.) Charley Otstott, Andrea Pugliese, Rami Puzis, Jana Shakarian, Paulo Shakarian, Gerardo Simari, Amy Sliva and Jonathan Wilkenfeld. This work would not have been possible without their strong contributions over the years. In addition, Barbara Lewis provided incredible administrative support and formatting help with the manuscript.

Some of the research reported in this Handbook was funded by AFOSR grant FA95500610405, ARO grant W911NF0910206 and ONR grant N000140910685.

I'd also like to thank Courtney Clark and Susan Lagerstrom-Fife of Springer for continuously helping with the preparation of the manuscript.

Last but not least, I would like to thank my wife, Mary, for her strong support and patience during the editing process.

Contents

Part I
Data and Data Acquisition

The Global Terrorism Database, 1970–2010

Gary LaFree and Laura Dugan

1 Introduction

Over the past four decades there have been a dozen or more major efforts to collect systematic data on the world-wide occurrence of terrorism. In general, terrorism event databases provide systematized descriptive information about terrorist attacks, most often from the unclassified electronic and print media, where the attack is the unit of analysis. These databases generally follow the classic journalistic format of providing information on who is responsible for an attack, what happened, where it happened, when it happened and to a lesser extent, how it happened. At present, the longest running of these event data bases that includes both domestic and international attacks is the Global Terrorism Database (GTD) maintained by the National Consortium for the Study of Terrorism and Responses to Terrorism (START), headquartered at the University of Maryland. In this chapter we provide an introduction to the GTD, consider its strengths and weaknesses and then use the most currently available version of the GTD to provide baseline characteristics of terrorism around the world from 1970 to 2010.

Compared to collecting data on other types of crime, collecting data on terrorism is especially challenging. In criminology, data on illegal behavior come traditionally from three sources, corresponding to the major social roles connected to criminal events: "official" data collected by legal agents, especially the police; "victimization" data collected from the general population of victims and non-victims; and "self-report" data collected from offenders [1]. However, all three of these traditional sources of crime data are problematic in the case of terrorism. Victimization surveys are of little use in the study of terrorism. Despite the attention it gets in the media, terrorism is much rarer than more ordinary violent crime. This

G. LaFree (✉) • L. Dugan
University of Maryland, College Park, MD 20742, USA
e-mail: garylafree@gmail.com; ldugan@umd.edu

V.S. Subrahmanian (ed.), *Handbook of Computational Approaches to Counterterrorism*,
DOI 10.1007/978-1-4614-5311-6_1,

means that even with extremely large sample sizes, few individuals in most countries will have been victimized by terrorists. Furthermore, oftentimes victims of terrorism have no direct contact with perpetrators, making the victim an inadequate source of information on the details of the event. And finally, in many cases, terrorism victims are killed by their attackers, making the subset of survivors a biased source of information.

Self-report data on terrorists have been more informative than victimization data, but they also face serious limitations. Most active terrorists are unwilling to participate in interviews, compromising the generalizability of any findings. And even if willing to participate, getting access to known terrorists for research purposes raises obvious challenges. As Merari has put it, "The clandestine nature of terrorist organizations and the ways and means by which intelligence can be obtained will rarely enable data collection which meets commonly accepted academic standards [2]." Although there has been important research based on self-reported behavior of terrorists or ex-terrorists [3], this methodology is unlikely to ever be robust enough to provide global or even national estimates.

Finally, thus far efforts to compile comprehensive terrorism data from official sources have been problematic. Unlike data on common crimes, which are collected by police departments and other criminal justice agencies in most countries, this is rarely done with terrorism data and when it is done the resulting protocols are inconsistent across countries. Part of the difficulty is that countries rely on different definitions of terrorism; in fact in many cases agencies in the same country have adopted unique definitions. Moreover, terrorist acts often cut across several more common types of criminal categories: assassinations may be included in police data as homicides, destruction of buildings might be listed as arson. To further complicate things, many terrorist attacks have no identified offender and even in those cases where an offender is identified, many of those suspected of terrorism are not legally processed for terrorism, but rather for other related offenses, such as weapons violations and money laundering [4]. And finally, much of the primary data is collected by intelligence agents as classified and thus is unavailable to researchers working in an unclassified environment.

For all these reasons, it has been impossible to rely on traditional sources of data to construct a comprehensive terrorism database, leading researchers and policy makers to instead look to media sources for reliable event data on terrorism. But before open source media outlets could evolve as a major resource for information about terrorism two related technological innovations were necessary. Starting in the late 1960s, the availability of satellite technology and portable video equipment made it possible for the first time in human history to send instantaneously images of conflict and violence from any one place on the planet to any other place. These related developments were not missed by terrorist organizations. On July 22, 1968 three armed members of the Front for the Liberation of Palestine-General Command (PLFP-GC) hijacked an El Al commercial flight scheduled to fly from Rome to Tel Aviv. The hijackers diverted the El Al plane and its 48 occupants to Algeria, releasing some passengers but holding five Israeli passengers and seven crew members hostage. The PFLP-GC subsequently demanded the release of Palestinian guerillas being held in Israeli prisons in exchange for these hostages. The resulting negotiations were broadcast live around the world. In many ways, this event marked the birth of contemporary terrorism event databases.

1.1 Terrorism Data from Open Sources

Since the end of World War II, there have been around a dozen major attempts to construct terrorism event databases [5, 6]. Among the most influential of these are ITERATE, RAND and RAND-MIPT, the US State Department data, WITS, and the GTD, the source we rely on for this chapter. The International Terrorism: Attributes of Terrorist Events (ITERATE) database began coverage in 1968 and has been periodically updated through 2009 [7, 8]. ITERATE contains two different types of files: quantitatively coded data on international terrorist incidents and a qualitative description of each incident included in the quantitative files. The quantitative data are arranged into four files, containing: (1) basic information on the type of terrorist attack, including location, name of group taking responsibility, and number of deaths and injuries; (2) detailed information on the fate of the terrorists or terrorist group claiming responsibility; (3) detailed information on terrorist events involving hostages; and (4) detailed information on terrorist events involving skyjackings.

The RAND Corporation was an early pioneer in developing terrorism event databases and with the support of the Department of State and the Defense Advanced Research Projects Agency (DARPA), in 1972, Brian Jenkins at RAND began to develop a "Chronology of International Terrorism" dating back to 1968. The original RAND data were generally limited to international attacks[1] and with varying levels of support, RAND maintained the Chronology through 1997. In the wake of the Oklahoma City bombings in the late 1990s, RAND received new data collection support from the National Memorial Institute for the Prevention of Terrorism (MIPT). With considerably more resources devoted to the database, RAND staff verified much of the earlier data and also began collecting (going back to 1998) terrorism data on domestic attacks. Funding for the RAND-MIPT data collection ended in 2008. However, shortly after, RAND received additional support and continues collecting terrorism event data, now referred to as the RAND Database of Worldwide Terrorism Incidents.

The U.S. State Department began publishing an annual report on international terrorism in 1982 (reporting 1981 incidents), and in 1983, began calling the report "Patterns of Global Terrorism." The Patterns Report reviews international terrorist events by year, date, region, and terrorist group and includes background information on terrorist organizations, U.S. policy on terrorism, and progress on counterterrorism. The Patterns of Global Terrorism report for 2003 was issued

---

[1] Although RAND did include some cases that were arguably domestic, such as cases in Israel and the Palestine territories.

on 30 April 2004. Starting in 2004, the State Department turned to the National Counterterrorism Center (NCTC) to produce its annual report on terrorism.

The NCTC began collecting its Worldwide Incidents Tracking System (WITS) data in 2004 but did not provide comprehensive annual coverage until 2005. WITS data are collected from open sources manually using commercial subscription news services, the US Government's Open Source Center, local news websites reported in English, and as permitted by the linguistic capabilities of their employees, local news websites in foreign languages [9]. Like GTD and RAND-MIPT, WITS collects both international and domestic data. From its inception, a major goal of those administering WITS was to "cast a wider net on what may be considered terrorism" (p. 5). The NCTC officially stopped collecting WITS data as this chapter was being prepared, in March 2012 (https://wits.nctc.gov).

The Global Terrorism Database (GTD) now includes information on over 98,000 attacks from around the world from 1970 through 2010 and can be accessed directly from the START web site (www.start.umd.edu/gtd). For its first 30 years, the GTD was collected by trained researchers at the Pinkerton Global Intelligence Service (mostly retired military personnel) who identified and recorded terrorism attacks from wire services (including Reuters and the Foreign Broadcast Information Service [FBIS]), U.S. State Department reports, other U.S. and foreign government reporting, and U.S. and foreign newspapers (including the New York Times, the British Financial Times, the Christian Science Monitor, the Washington Post, the Washington Times, and the Wall Street Journal). Today, nearly all data collection is based on sources available on the Internet. For GTD collection between 1998 and 2008, START partnered with a team led by Gary Ackerman and Charles Blair at the Center for Terrorism and Intelligence Studies (CETIS) and from 2008 to November 2011 we partnered with the Institute for the Study of Violent Groups (ISVG), headquartered at New Haven University. Beginning in late 2011 the START Consortium headquartered at the University of Maryland began collecting the original data for the GTD.

The GTD currently includes data on approximately 125 variables. The data incorporate the time and location of the event, the group claiming responsibility, weapons used, tactics used (e.g. bombing, assassination), the target of the attack (e.g., transportation, government), the number of fatalities and injuries, the nationality of the target, as well as more specialized information on aerial hijackings, kidnappings, hostage situations, and ransoms demanded. The GTD includes three additional criteria to help researchers decide how to analyze cases. These are whether the case includes evidence of (1) a political, economic, religious or social goal; (2) the intention to coerce, intimidate or publicize to a larger audience; and (3) behavior that is outside of international humanitarian law. The START Consortium releases annual updates to the GTD each year as they become available. As this chapter was being prepared START was negotiating with the US State Department to supply terrorism data from the GTD for the statistical annex of the 2012 *Country Reports on Terrorism.*

1.1.1 Limitations of Event Databases

Event databases compiled from open sources have serious limitations. On balance, perhaps the most important weaknesses of event databases are: (1) lack of a generally accepted definition of terrorism; (2) biases and inaccuracies in open source data; and (3) lack of consistency over time. The fact that there is no internationally-agreed upon definition of terrorism clearly complicates the process of generating accurate counts. In this regard, relatively little has changed since Schmid and Jongman's review [10] identified 109 different research definitions of terrorism. However, despite the considerable complexity, most commentators and experts agree on several key elements, captured in the operational definition we use here: *"the threatened or actual use of illegal force and violence by non-state actors to attain a political, economic, religious or social goal through fear, coercion or intimidation"* [1]. Note that this operational definition excludes state terrorism and genocide, topics that are important and complex enough to warrant separate treatment.

The media may of course report inaccuracies or outright lies; there may be conflicting information or false, multiple or no claims of responsibility. Government censorship and disinformation may also lead to biased results. When closed societies like North Korea, Sudan, or Myanmar report extremely low terrorism rates, we can never say for sure whether it is because of actual low reporting or the ability of these societies to suppress coverage by the print or electronic media. Because the GTD relies on news sources, it is generally impossible to know the extent to which reported events reflect actual outcomes or the freedom of the press in a particular country or region. It seems incontrovertible that news sources will be more likely to report more serious than less serious attacks. The extent to which countries are covered by the international press also varies by region and over time. Further, data collection efforts have been strongly biased toward coverage of English language sources. We endeavor to monitor non-English sources but in truth resources limit the extent to which this is possible. Moreover, our non-English coverage has varied over time depending on the size and skills of the available staff.

Beyond these obvious issues there are more subtle biases related to the media itself. For example, there is a well-known tendency for news sources to fit individual stories into a particular news frame over time so that events that fit into preselected themes (e.g., improvised explosive devices or suicide attacks) may be more likely to receive coverage [11]. Also, to the extent that media sources must rely on information based on claims made by individuals and groups with strong subjective opinions claims may be unintentionally inaccurate or intentionally wrong.

And finally, given the desirability of developing longitudinal analyses of terrorist attacks, the complexities of maintaining a consistent record of events over time becomes even more challenging. These challenges can be both substantive and financial. In substantive terms, the more time elapsed between real events and data collection the greater the chances that some data are no longer available. Thus, by the time we had computerized the original PGIS data and secured funding for new data collection on the GTD, our data collection was 8 years behind real time. This 8 year gap between the oldest and newest missing data meant that as the analysis came closer to real time, we were likely to have access to a more complete record of actual events. To the extent that print and electronic media are not archived, availability of original sources erodes over time, increasing the extent of missing data. This is likely to be especially problematic for small, regional and local newspapers.

Maintaining longitudinal data on terrorism also raises financial challenges. Collecting global terrorism databases is a relatively expensive enterprise. For governments going through budget crisis and cost cutting pressures, data collection can no doubt look like an attractive target.

1.1.2 Strengths of Event Databases

Despite these limitations, compared to more traditional data options, or even compared to crime data in general, event databases have some important advantages. In particular, because of the compelling interest that terrorist groups have in media attention, open source information may be uniquely useful in the study of terrorism. Terrorists, unlike most common criminals, actively seek media attention. Terrorism expert Brian Jenkins [12] has observed that "terrorism is theatre" and explains how "terrorist attacks are often carefully choreographed to attract the attention of the electronic media and the international press." As discussed above, the media are so central to contemporary terrorist groups that some researchers and policy makers have argued that the birth of modern terrorism is directly linked to the launch by the United States of the first television satellite in 1968. The fact that terrorists are specifically seeking to attract attention through the media means that compared to media coverage of more common crimes, coverage of terrorism can tell us far more. Thus, while no responsible researcher would seriously suggest tracking burglary or auto theft rates by studying electronic and print media, it is much more defensible to track terrorist attacks in this way. For example, it is hard to imagine that it is possible today for an aerial hijacking or politically motivated assassination—even in remote parts of the world—to elude the attention of the global media.

Event databases on terrorism also have another important advantage. One of the most serious limitations of cross-national crime research is that it has been focused overwhelmingly on a small number of highly industrialized western-style democracies. For example, reviews of cross-national research on homicide [13] show that most prior research has been based on fewer than 40 of the world's countries. And of course these countries are not a random sample of the nations of the world but strongly over represent Europe and North America while almost entirely excluding countries of Africa, the Middle East and Asia. By contrast, open source terrorism databases offer at least some coverage for all countries. While it is the case that traditional media under-report news coming from industrializing countries or relatively closed, highly autocratic states, the salience of terrorism as a phenomenon today makes it more likely than ever that media will capture such incidents as information becomes available.

In sum, open source event databases have important limitations. But to be clear, so do all crime databases. For example, official data sources like the FBI's Uniform Crime Reports have long been criticized for many of the same issues as those outlined above for event databases [14, 15]. The bottom line is that despite their drawbacks, at present there are no obvious alternatives to event databases for those interested in tracking terrorism.

1.2 World-Wide Terrorism

In this section we provide an overview of the characteristics of world-wide terrorism based on the most recent version of the Global Terrorism Database (GTD).[2] We begin by examining worldwide trends in total and fatal terrorist attacks for the 41 years covered by the GTD. We also examine more detailed characteristics of these attacks by examining their regional distributions, fatalities, targets, tactics used and weapons. We then consider the countries that have been the location for the largest number of total attacks and fatalities since 1970. Finally, we examine the most active and lethal terrorist groups by examining those that perpetrated the most attacks and killed the most people.

In Fig. 1 we show total and fatal terrorist attacks for the world from 1970 to 2010. Total terrorist attacks increase steeply from a low point in the early 1970s to a peak in 1992—just after the collapse of the Soviet Union. Altogether, we see a more than six-fold increase from the series low point in 1971 with fewer than 500 attacks to the series high point in 1992, with more than 5,000 attacks. Declines in total attacks after the early 1990s were also steep. From their high point in 1992 attacks fell by 82% reaching a low point of 931 attacks in 1998—just before the 9/11 attacks. After a few years of relatively flat rates—from 1999 to 2004— total attacks once again rose sharply to another peak in 2008 with 4,776 attacks. The final 2 years of the series are lower by about 100 attacks with 4,625 in 2009 and 4,674 in 2010. In fact, the total number of attacks in 2010 was at about the same level as it was in 1991 (4,680)—the year before the peak.

In general, the number of fatal attacks clearly follows the pattern of total attacks ($r = 0.94$), but at a substantially lower magnitude (averaging 1,025 fatal attacks per year compared to 2,453 total attacks per year worldwide). Fatal attacks rose above 1,000 per year for the first time in 1980. After hovering close to one-thousand attacks annually for most of the 1980s, fatal attacks more than doubled between 1987 and 1992. Like total attacks, fatal attacks declined somewhat after the peak in 1992, bottoming out in 1998 with 447 attacks and then rising again to a second peak of more than 2,100 fatal attacks in 2008. The 2008 peak had just three fewer fatal attacks than that of the peak year in 1992 (2,176 versus 2,179). As with total attacks, during the last 2 years of the series, the number of fatal attacks declined, but only slightly.

---

[2]These data were downloaded on November 12, 2011.

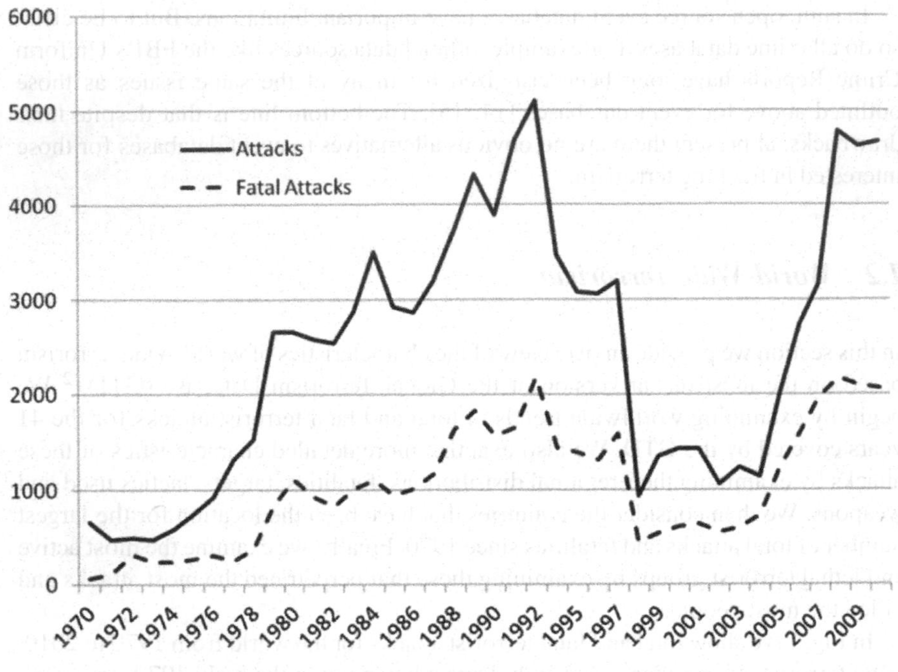

Fig. 1 Total and fatal terrorist attacks, 1970–2010

How confident can we be in these trends? Two issues should be emphasized. First, most of the 1993 data in the GTD were lost by the original data collectors and we have never been able to recover them fully [1]. For Fig. 1 we simply excluded 1993. It turns out that between 1992 (the peak year) and 1994, there were steep declines in total terrorist attacks in the GTD (they dropped from 5,099 to 3,467). So, it might be that had we been able to include the 1993 data, the shape of the drop in the early 1990s would look a bit different. According to an original PGIS report [16], total attacks in 1993 were 4,954, suggesting only a slight decrease since 1992. Unfortunately, we cannot compare the 1993 figure directly to the results in Fig. 1 because when we computerized the original PGIS data we excluded some cases because they differed from the PGIS definition of terrorism or because additional, more recent, information on incidents made them ineligible. Moreover, we also added cases from other sources that were missed by the original data collectors.

Second, by the time we received funding to extend the GTD data collection beyond 1997, it was already 2006. This meant that we were collecting data on events that were in some cases as much as 8 years old. As discussed above, the extensiveness of open source data on terrorism declines over time. Some smaller, local newspapers are not archived; some electronic sources are not indefinitely maintained. In short, our data likely undercount total attacks after 1997 and this undercount is likely most serious for 1998—the most recent low point in both series—and becomes less serious as our data collection process got closer to real time.

Based on these considerations, we would offer the following conclusions about world- wide terrorist attacks over the past four decades based on the GTD. First, both total and fatal attacks increased steadily until reaching a peak in the early 1990s and then declined substantially after the collapse of the Soviet Union in 1991. Second, even though we likely undercount total attacks in 1998, the trajectory of attacks was already steeply down before that year and our ability to identify attacks likely increases as we get closer to the new data collection point in 2006. And finally, both total and fatal attacks have increased considerably since 9/11 so that in 2010 both were back to the historically high levels that they had been at in the mid-1990s.

To develop geospatial comparisons for terrorist attacks, we next divided the countries of the world into thirteen regions (for a list of countries in each region, see Appendix A).[3] In Fig. 2, we compare the total distribution of attacks and fatalities by region. Figure 2 shows that the largest percent of terrorist attacks in the GTD (19.0%) now come from the Middle East and North Africa. South America (17.9%) and South Asia (17.7%) were next highest and had a similar number of attacks. Western Europe ranks fourth with 15.1% of the world's attacks; more than twice as many as Sub-Saharan Africa (6.5%) and Southeast Asia (6.1%). In general, the remaining regions account for a relatively small proportion of terrorist attacks: North America (2.9%), Russia and the Newly Independent States (2.0%), Eastern Europe (0.9%), East Asia (0.7%), Australia and Oceania (0.2%) and Central Asia (0.2%). Taken together, the first three regions to the left of Fig. 2 account for a total of 54.6% of all attacks and the first seven accounts for a total of 93.1% of all attacks. It is worth pointing out that conclusions in Fig. 2 depend a great deal on how regions of the world are defined. For example, if instead of counting South America and Central America/the Caribbean as separate regions we subsume them as "Latin America," this region would have a higher percentage of attacks (28.7%) than any other region in the world.

Figure 2 also shows substantial variation across regions in terms of the relationship of attacks to fatalities. In particular, Sub-Saharan Africa, South Asia, the Middle East-North Africa, and Central America/Caribbean stand out for having a larger percentage of total fatalities than attacks; while South America, Western Europe, Southeast Asia, North America, Eastern Europe and East Asia stand out for having a larger percent of attacks than fatalities. The difference between attacks and fatalities is especially stark for Western Europe: Western Europe has lots of attacks compared to fatalities.

When we looked more closely at the fatality rates for Sub-Saharan Africa, we found that five countries averaged 10 or more fatalities per attack: Rwanda (22.93),

[3]For this classification we treat the country or territory as the target. Thus, an attack on the U.S. embassy in Switzerland is treated here as a Swiss attack. Similarly, an attack on a Swiss ambassador living in the U.S. is counted here as a U.S. attack. Although the vast majority of cases in the GTD involve attacks where the location of the target and the nationality of the target are the same, there are some interesting variations across attacks depending on the geographical country attacked, the nationality of the perpetrators, and the nationality of the target. We are exploring these issues in greater detail in ongoing research.

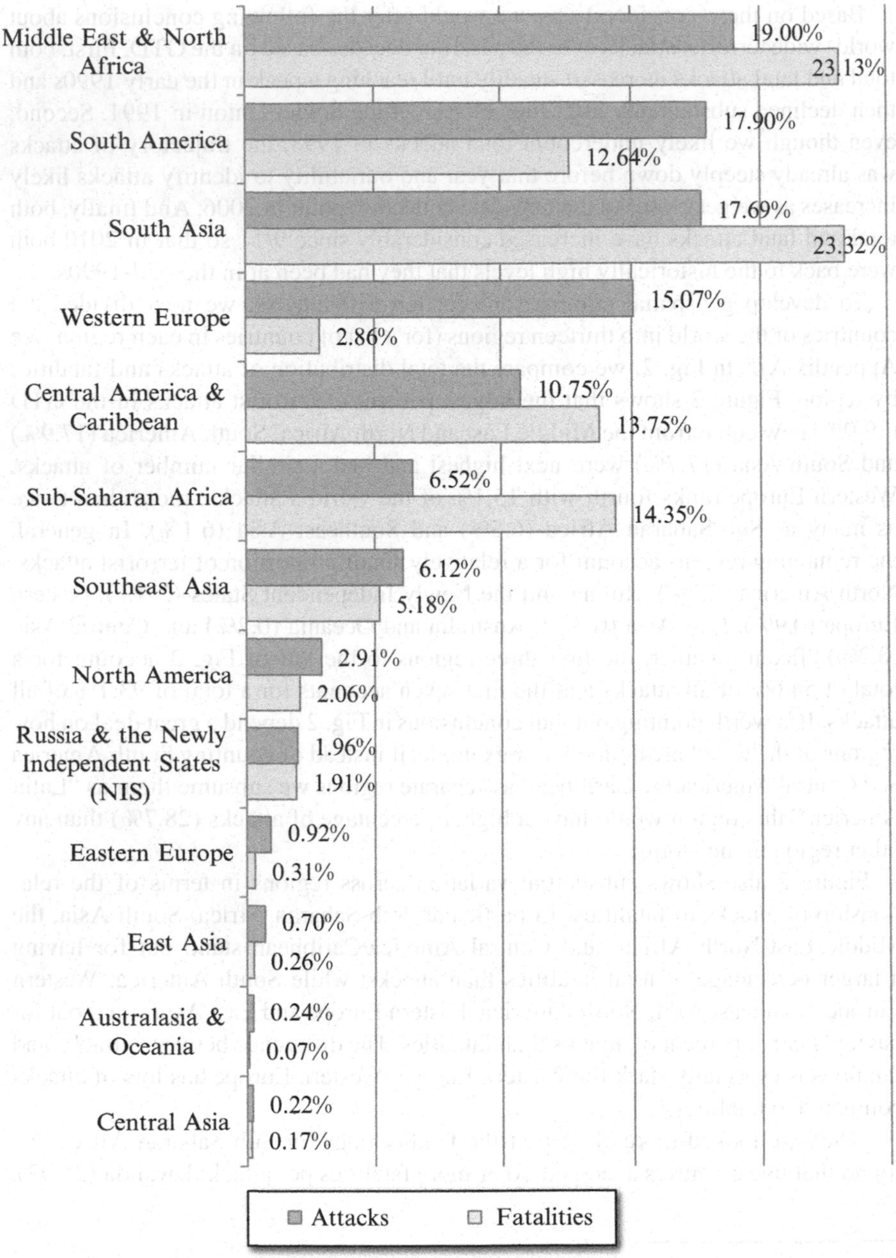

Fig. 2 Worldwide distribution of terrorist attacks and fatalities by region, 1970–2010

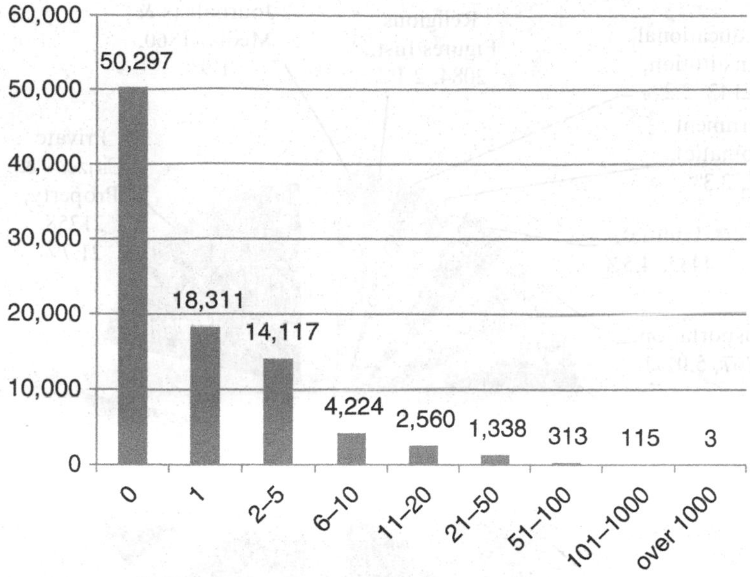

Fig. 3 Number of fatalities per attack, 1970–2010

Djibouti (14.16), Chad (12.98), Mozambique (12.30), and Burundi (11.18). More than 30% (44) of the 141 attacks in Rwanda where a perpetrator was identified were attributed to Hutus. In fact, attacks by Hutus led to nearly 2,500 fatalities, averaging more than 56 deaths per attack. Djibouti only had 19 attacks recorded in the GTD and only eight were attributed to a specific group. Five of these attacks were attributed to the Front for the Restoration of Unity and Democracy, which resulted in 236 fatalities. While 20 of the attacks in Chad were not attributed to a specific group, the remaining 27 attacks were attributed to 19 different organizations. The deadliest organization was the Dissident Military Members of Northern Tribal Group, which killed 150 during a machine gun attack in Southern Chad on September 2, 1992. Mozambique differs from these other countries because most (80.3%) of its terrorist violence was perpetrated by one organization, the Mozambique National Resistance Movement, which was responsible for 179 attacks in that country over more than 20 years (1979 to 1999), leading to 2,355 recorded deaths. Finally, Burundi was driven by the conflict between the Tutsis and the Hutus which resulted on average of more than 70 Tutsi and 13 Hutu deaths per attack.

Figure 3 shows the number of fatalities per attack over the years covered by the GTD. Some may be surprised by the fact that in 55% of the attacks there were no known fatalities. In many cases terrorist groups target property without intending to cause casualties. In other cases, their deadly plans fail. Moreover, some well-known terrorist groups such as the IRA and ETA frequently provide warnings before attacks to minimize casualties. Thirty-five years ago these considerations led Jenkins to suggest that "terrorists want a lot of people watching, not a lot of people dead [12]."

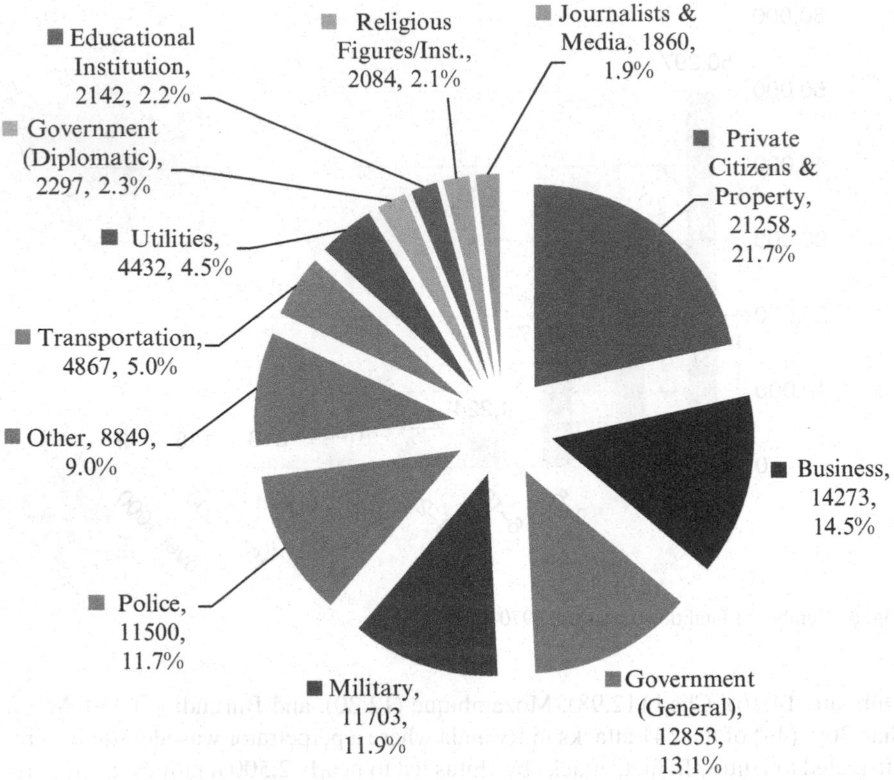

Fig. 4 Distribution of terrorist targets globally, 1970–2010

Of course, it is still the case that 45% of the attacks in the GTD (or more than 41,000 attacks) involved at least one fatality. Attacks that are especially worrisome are the 118 attacks that produced more than 100 fatalities. And in fact, Jenkins has revisited his earlier statement and after reviewing the stated plans of terrorist groups operating in the early twenty-first century concluded that indeed "many of today's terrorists want a lot of people watching and a lot of people dead [17]." Nevertheless, the majority of terrorist attacks in the GTD since 1970 produced no fatalities.

Figure 4 presents the distribution of terrorist targets worldwide. We can see that there is considerable variation in terrorist targeting with the most common target (private citizens) representing 21.7% of the total. Together, private citizens and businesses account for more than 36% of all terrorist attacks. The next most common targets are the government, military, and the police. Utilities and transportation were each targeted about 5% of the time. The remaining targets are attacked even less frequently, and include diplomats, religious figures or institutions, journalists and other media, and educational institutions. Important target types in the "other" category include other terrorists and criminals.

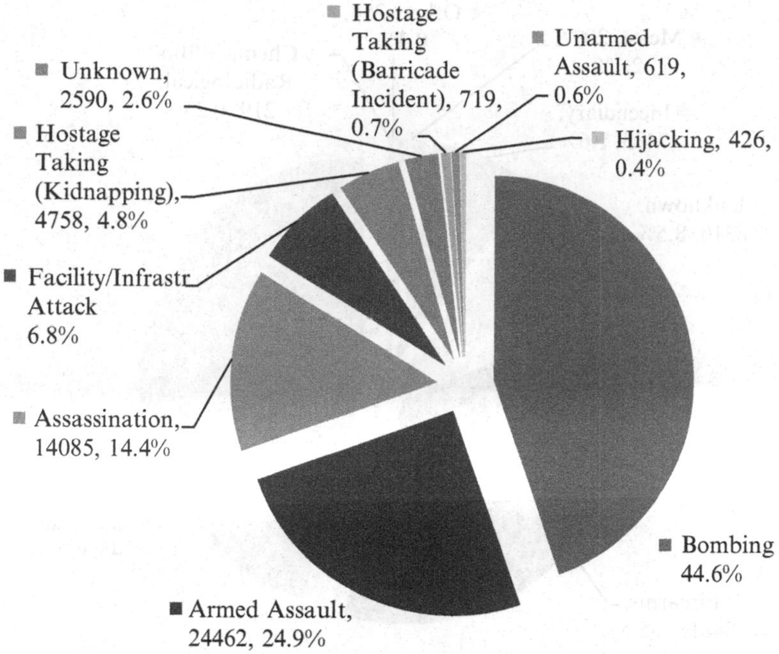

Fig. 5 Distribution of terrorism tactics, 1970–2010

Figure 5 shows the total distribution of terrorist tactics. Perhaps the most striking result here is that more than two-fifths of all terrorist attacks are bombings and nearly a quarter of the attacks are armed assaults. Taken together, bombings and armed assaults account for nearly 70% of the total. Because armed assaults require hand-held weapons rather than explosives, terrorists in these cases are more likely to have direct contact with targets. The next most common tactic is assassination followed by facility attack and kidnapping.

Assassinations are more than twice as common as facility attacks and three times as common as kidnappings. Barricade/hostage situations, unarmed assaults and hijackings each account for less than 1% of the total.

Figure 6 shows the types of weapons that were used in the terrorist attacks included in the GTD. Explosives and firearms were the dominant weapons used by terrorists, jointly accounting for over 80% of all attacks. The most common explosives used were dynamite, car bombs, grenades, and mortars. The most common firearms used were shot guns, pistols and automatic weapons. Incendiaries (including fire and firebombs) contribute nearly 8% to the total for weapons. Melee attacks, where the perpetrator comes in direct contact with the target, account for just over 2% of the total. These attacks usually depend on low technology weapons such as knives, or even stones or fists. In short, most of the weapons used in these cases were conventional, simple and readily available. Fortunately, sophisticated weaponry, especially chemical, biological and nuclear weapons are quite rare, accounting for less than three-tenths of 1% of all attacks.

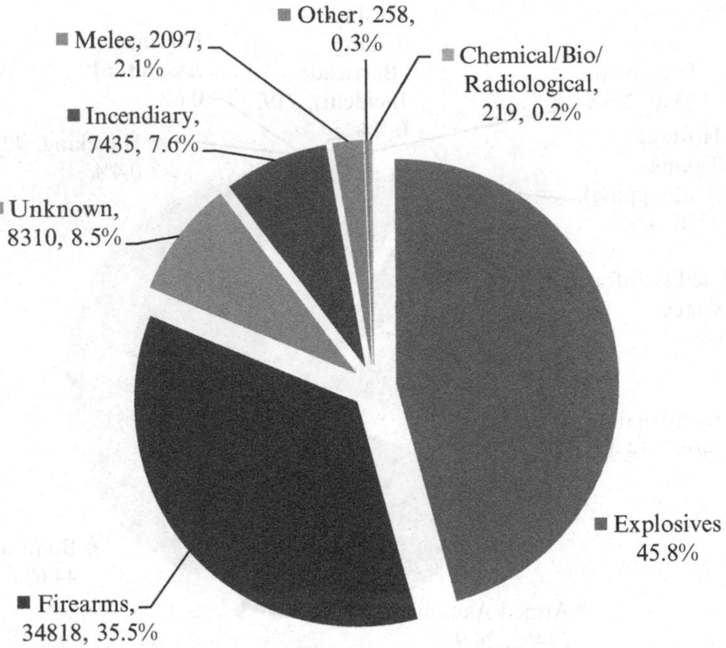

Fig. 6 Weapons used in terrorist attacks, 1970–2010

In Table 1 we present the top 20 countries ranked according to total terrorist attacks and total fatalities in the GTD from 1970 to 2010. The top 20 countries and territories account for more than 70% of all terrorist attacks and nearly 80% of all terrorist fatalities in the world. The top ten countries account for nearly 50% of all terrorist attacks and more than 60% of all terrorist fatalities; and the top five countries account for more than 30% of the world's terrorist attacks and nearly 40% of the world's terrorism fatalities. Comparing total attacks and fatalities in 1 show a good deal of overlap between the two top twenty lists. In fact, the top five countries in both lists are nearly the same: Colombia, Iraq, India, Peru, and El Salvador, the top five in number of attacks, are in the top six for number of fatalities. Sri Lanka rounds out the list as number three for fatalities.

Altogether, 16 countries appear on both lists—although there is sometimes considerable variation in their ranking on the two lists. For example, while Northern Ireland is ranked seventh in the world for total attacks in the GTD, it ranks 20th in the GTD for total fatalities. By contrast, Nicaragua ranks 18th in terms of total attacks but 7th in terms of total fatalities. Most of the countries that appear on one top twenty list and not the other are near the bottom of one or the other list. For total attacks, this applies to Chile (number 15), South Africa (number 19), and Thailand (number 20); for total fatalities, it applies to Burundi (number 14), Russia (number 15), Rwanda (number18) and Angola (number 19). The biggest exception to this general pattern is Spain—which ranks number 8 in terms of total attacks but

Table 1 The twenty top ranking countries in terms of total terrorist attacks and fatalities, 1970–2010

Total terrorist attacks			Total terrorist fatalities		
Rank	Country	Attacks	Rank	Country	Fatalities
1	Colombia	7,180	1	Iraq	25,026
2	Iraq	6,475	2	India	15,877
3	India	6,114	3	Sri Lanka	15,481
4	Peru	6,045	4	Colombia	13,271
5	El Salvador	5,327	5	Peru	12,850
6	Pakistan	4,438	6	El Salvador	12,496
7	Northern Ireland	3,885	7	Nicaragua	11,324
8	Spain	3,205	8	Pakistan	10,107
9	Philippines	3,191	9	Algeria	10,047
10	Sri Lanka	2,924	10	Philippines	6,881
11	Turkey	2,745	11	Afghanistan	6,273
12	Afghanistan	2,585	12	Guatemala	5,150
13	Algeria	2,357	13	Turkey	4,725
14	United States	2,353	14	Burundi	4,114
15	Chile	2,305	15	Russia	3,507
16	Guatemala	2,035	16	United States	3,470
17	Lebanon	2,016	17	Lebanon	3,272
18	Nicaragua	1,986	18	Rwanda	3,210
19	South Africa	1,921	19	Angola	2,853
20	Thailand	1,697	20	Northern Ireland	2,849

fails to make it into the top 20 for total fatalities (it ranks number 32). Of attacks where responsibility is attributed by the GTD to a specific group in Spain, ETA is responsible for the largest percentage of attacks (61.1%). Thus, ETA is noteworthy for carrying out large numbers of attacks with relatively few fatalities per attack.

In Table 2 we again compare total attacks to total fatalities, but this time instead of examining countries we list the top twenty terrorist organizations in the GTD in each category. While the specific rank orderings of the most active and deadliest vary, 14 of the top 20 most active organizations are also in the top twenty for deadliest. The six organizations that are only in the top 20 for most active but not for deadliest are ETA, the Manuel Rodriguez Patriotic Front (Chile), the African National Congress (South Africa), the Corsican National Liberation Front (France), the Tupac Amaru Revolutionary Movement (Peru), and the Movement of the Revolutionary Left (Chile). And the six organizations that are among the 20 deadliest but not among the top 20 most active include Al-Qaida and Al-Qaida in Iraq (counted here as separate organizations), the Mozambique National Resistance Movement, the Democratic Revolutionary Alliance (Nicaragua), the Lord's Resistance Army (Uganda), and the Armed Islamic Group (Algeria).

The range of attacks among the top twenty most active terrorist organizations is large: number one Shining Path has over 4,500 known attacks while number 20 Hezbollah has less than 400 known attacks. Similarly, the GTD attributes more

Table 2 The twenty most active terrorist organizations in terms of attack frequency and fatalities, 1970–2010

Total attacks			Total fatalities		
Rank	Organization	Attacks	Rank	Organization	Fatalities
1	Shining Path (SL)	4,518	1	Shining Path (SL)	11,666
2	Farabundo Marti National Liberation Front (FMLN)	3,357	2	Liberation Tigers of Tamil Eelam (LTTE)	10,841
3	Irish Republican Army (IRA)	2,673	3	Farabundo Marti National Liberation Front (FMLN)	8,508
4	Basque Fatherland and Freedom (ETA)	2,005	4	Nicaraguan Democratic Force (FDN)	7,268
5	Revolutionary Armed Forces of Colombia (FARC)	1,907	5	Revolutionary Armed Forces of Colombia (FARC)	5,004
6	Taliban	1,798	6	Taliban	4,737
7	Liberation Tigers of Tamil Eelam (LTTE)	1,600	7	Al-Qa'ida	4,637
8	National Liberation Army of Colombia (ELN)	1,308	8	Kurdistan Workers' Party (PKK)	3,571
9	New People's Army (NPA)	1,283	9	New People's Army (NPA)	3,385
10	Kurdistan Workers' Party (PKK)	1,191	10	National Union for the Total Independence of Angola (UNITA)	2,562
11	Communist Party of India—Maoist (CPI-M)	1,039	11	Mozambique National Resistance Movement (MNR)	2,443
12	Nicaraguan Democratic Force (FDN)	899	12	Lord's Resistance Army (LRA)	2,029
13	Manuel Rodriguez Patriotic Front (FPMR)	830	13	Al-Qa'ida in Iraq	1,858
14	African National Congress (South Africa)	606	14	Islamic State of Iraq (ISI)	1,839
15	Corsican National Liberation Front (FLNC)	572	15	Irish Republican Army (IRA)	1,807
16	M-19 (Movement of April 19)	564	16	Democratic Revolutionary Alliance (ARDE)	1,803
17	Tupac Amaru Revolutionary Movement (MRTA)	561	17	National Liberation Army of Colombia (ELN)	1,761
18	People's Liberation Front (JVP)	434	18	Communist Party of India—Maoist (CPI-M)	1,553
19	National Union for the Total Independence of Angola (UNITA)	421	19	M-19 (Movement of April 19)	1,402
20	Hezballah	366	20	Armed Islamic Group (GIA)	1,366

Note: The organizations that appear in both lists are shaded than 11,000 fatalities to Shining Path compared to fewer than 1,400 fatalities to the Armed Islamic Group (GIA). Altogether, eleven terrorist groups in the GTD have over 1,000 attacks and 18 groups claimed more than 1,500 lives during the 41 years spanned by the data.

2 Conclusions

Contemporary terrorism event databases became feasible in the late 1960s, along with the application of satellite technology and portable cameras. The nature of terrorism makes it difficult to track through traditional criminological sources such as victimization or self-report surveys or official data. The scope of open source databases on terrorist attacks has greatly expanded since the early 1970s. Open source databases are generated from print and electronic media and face limitations associated with this fact. In particular, the media may report inaccuracies and falsehoods; there may be conflicting information or false, multiple or no claims of responsibility; and despite improvements, coverage no doubt still relies more on western than nonwestern sources. Of course, it is worth pointing out that all of these problems are also well-known drawbacks of international crime data. Certainly government censorship and disinformation affects not only media sources but official government sources. It is especially challenging to disentangle terrorism from acts of war, insurrection or massive civil unrest. Of all the complexities of distinguishing terrorist violence from war-related violence perhaps the most vexing is distinguishing between armed targets with civilian collateral damage and civilian attack with non-civilian casualties. This is an especially big problem in interpreting terrorism in situations of armed conflict such as Iraq following the U.S.-led invasion in 2003. And even though the media now seemingly peer into every corner of the world, obviously media coverage still varies across time and geographic space. On the other hand, event databases have the great strength of tracking a type of behavior whose success is in large part a function of its ability to be publicized.

From the origins of event data bases in the late 1960s to the present there have been truly remarkable technological changes in information technology. The biggest challenge for the original collectors of the GTD in the 1970s and 1980s was to find open source evidence of terrorist attacks in remote parts of the world based on the print media. As we move farther into the Internet age the most pressing challenge in collecting terrorism event data is instead finding the relevant events in an ever increasing mountain of global electronic information. Thus far, all of the event data bases we have reviewed here, including the GTD, are based on identifying a string of relevant print and electronic news sources and then using human coders to enter these sources into a database. In several ways the kinds of computational methods discussed in this Handbook may play an increasingly important role in the collection of open source data in the future. We will conclude by briefly mentioning just two of these ways. First, computational methods can offer assistance in the preparation of event data bases by narrowing the field of relevant news events that must be sourced by human coders. As the sheer volume of information on the Internet continues to grow, this contribution becomes increasingly important. Several of the data bases reviewed above, including WITS and GTD, have experimented with the use of computational methods to narrow the field of potential cases that are entered in their respective data bases. However, there is clearly much room for growth and improvement in this area.

And second, to this point in time all of the major event data bases have relied on identifying a set of relevant sources for terrorism cases and then examining more carefully these sources for valid events to be included. We could consider this a "bottom up" approach to the problem. However, as our ability to synthesize vast quantities of electronic information from around the world continues to grow, it is possible to realistically consider the advantages of a "top down" approach to producing these data bases. With the help of computational approaches it may be possible in the future to produce event data bases by starting with the whole universe of possible electronic news sources and finding the stories most likely to be relevant from this universe. Developments such as these would clearly make event data even more useful in the future.

A.1 Appendix A Countries Listed Under Each Region According to GTD

Region	Countries/Territories
Australasia and Oceana	Australia, Fiji, French Polynesia, New Caledonia, New Hebrides, New Zealand, Papua New Guinea, Samoa (Western Samoa), Solomon Islands, Vanuatu, and Wallis and Futuna, Tonga, and Vanuatu
Central America and Caribbean	Antigua and Barbuda, Bahamas, Barbados, Belize, Bermuda, Cayman Islands, Costa Rica, Cuba, Dominica, Dominican Republic, El Salvador, Grenada, Guadeloupe, Guatemala, Haiti, Honduras, Jamaica, Martinique, Nicaragua, Panama, Puerto Rico, St. Kitts and Nevis, Trinidad and Tobago, and the Virgin Islands (U.S.)
Central Asia	Kazakhstan, Kyrgyzstan, Tajikistan, Turkmenistan, and Uzbekistan
East Asia	China, Hong Kong, Japan, Macau, North Korea, South Korea, and Taiwan
Eastern Europe	Albania, Bosnia-Herzegovina, Bulgaria, Croatia, Czechoslovakia, Czech Republic, Hungary, Kosovo, Macedonia, Moldova, Montenegro, Poland, Romania, Serbia, Serbia-Montenegro, Slovak Republic, Slovenia, and Yugoslavia

(continued)

(continued)

Region	Countries/Territories
Middle East and North Africa	Algeria, Bahrain, Cyprus, Egypt, Iran, Iraq, Israel, Jordan, Kuwait, Lebanon, Libya, Morocco, North Yemen, Oman, Qatar, Saudi Arabia, South Yemen, Syria, Tunisia, Turkey, United Arab Emirates, West Bank and Gaza Strip, Western Sahara, and Yemen
North America	Canada, Mexico, and the United States
Russia and the Newly Independent States (NIS)	Armenia, Azerbaijan, Belarus, Estonia, Georgia, Latvia, Lithuania, Russia, Soviet Union, and Ukraine
South Asia	Afghanistan, Bangladesh, Bhutan, India, Kashmir, Maldives, Mauritius, Nepal, Pakistan, Seychelles, and Sri Lanka
Southeast Asia	Brunei, Cambodia, Guam, Indonesia, Laos, Malaysia, Myanmar, Philippines, Singapore, South Vietnam, Thailand, Timor-Leste, and Vietnam
South America	Argentina, Bolivia, Brazil, Chile, Colombia, Ecuador, Falkland Islands, French Guiana, Guyana, Paraguay, Peru, Suriname, Uruguay, and Venezuela
Sub-Saharan Africa	Angola, Benin, Botswana, Burkina Faso, Burundi, Cameroon, Central African Republic, Chad, Comoros, Congo (Brazzaville), Congo (Kinshasa), Djibouti, Equatorial Guinea, Eritrea, Ethiopia, Gabon, Gambia, Ghana, Guinea, Guinea-Bissau, Ivory Coast, Kenya, Lesotho, Liberia, Madagascar, Malawi, Mali, Mauritania, Mauritius, Mozambique, Namibia, Niger, Nigeria, Rwanda, Senegal, Sierra Leone, Somalia, South Africa, Sudan, Swaziland, Tanzania, Togo, Uganda, Zaire, Zambia, and Zimbabwe
Western Europe	Andorra, Austria, Belgium, Corsica, Denmark, East Germany, Finland, France, Germany, Gibraltar, Great Britain, Greece, Iceland, Ireland, Italy, Luxembourg, Malta, Isle of Man, Netherlands, Northern Ireland, Norway, Portugal, San Marino, Spain, Sweden, Switzerland, Vatican City, and West Germany

References

1. LaFree G, Dugan L (2007) Introducing the global terrorism database. Terrorism and Political Violence 19: 181–204
2. Merari A (1991) Academic research and government policy on terrorism. Terrorism and Political Violence 3:88–102
3. Horgan J (2008) From profiles to pathways and roots to routes: perspectives from psychology on radicalization into terrorism. The ANNALS of the American Academy of Political and Social Science 618:80–94
4. Smith BL, Damphousse KR, Jackson F, Sellers A (2002) The prosecution and punishment of international terrorists in federal courts: 1980–1998. Criminology and Public Policy 1: 311–338
5. LaFree G (2010) The global terrorism database: accomplishments and challenges. Perspectives on Terrorism 4:24–46
6. LaFree G (2012) Generating terrorism event databases: results from the global terrorism database, 1970 to 2008. In: Lum C, Kennedy L (eds) Evidence-based counter terrorism. Springer, New York, pp. 41–64
7. Mickolus EF (2002) How do we know we're winning the war against terrorism? Issues in measurement. Studies in Conflict and Terrorism 25:151–160
8. Mickolus EF, Sandler T, Murdock JM, Flemming P (2010) International terrorism: attributes of terrorist events *(ITERATE)*. Vinyard Software, Dunn Loring
9. Wigle J (2010) Introducing the Worldwide Incidents Tracking System (WITS). Perspectives on Terrorism 4:3–23
10. Schmid A, Jongman AJ (1988) Political terrorism: a new guide to actors, authors, concepts, databases, theories and literature. Amsterdam: North-Holland
11. Fishman M (1980) Manufacturing the news. University of TX Press, Austin
12. Jenkins BM (1975) International terrorism: a new model of conflict. In: Carlton D, Schaerf C (eds) International terrorism and world security. Croom Helm, London
13. LaFree G (1999) A summary and review of cross-national comparative studies of homicide. In: Smith MD, Zahn MA (eds) Homicide: a sourcebook of social research. Sage, Thousand Oaks
14. Gove WR, Hughes M, Geerken M (1985) An affirmative answer with minor qualifications. Criminology 23:451–501
15. O'Brien R (1985) Crime and victimization. Sage, Beverly Hills
16. Pinkerton Global Assessment Services (1995) Annual risk assessment 1994. Pinkerton's Inc., Washington, DC
17. Jenkins BM (2006) The New Age of Terrorism. In DG Kamien (ed), Homeland Security Handbook. Mcgraw-Hill, New York

Automated Coding of Political Event Data

Philip A. Schrodt and David Van Brackle

1 Introduction and Overview

Political event data have long been used in the quantitative study of international politics, dating back to the early efforts of Edward Azar's COPDAB [1] and Charles McClelland's WEIS [18] as well as a variety of more specialized efforts such as Leng's BCOW [16]. By the late 1980s, the NSF-funded *Data Development in International Relations* project [20] had identified event data as the second most common form of data—behind the various Correlates of War data sets—used in quantitative studies. The 1990s saw the development of two practical automated event data coding systems, the NSF-funded KEDS (http://eventdata.psu.edu; [9, 31, 33]) and the proprietary VRA-Reader (http://vranet.com; [15, 27]) and in the 2000s, the development of two new political event coding ontologies—CAMEO [34] and IDEA [4, 27]—designed for implementation in automated coding systems. A summary of the current status of political event projects, as well as detailed discussions of some of these, can be found in [10, 32].

While these efforts had built a substantial foundation for event data—by the mid-2000s, virtually all refereed articles in political science journal used machine-coded, rather than human-coded, event data—the overall development of new technology remained relatively small. This situation changed with the DARPA-funded Integrated Conflict Early Warning System (ICEWS; [25, 26]) which utilized event data development coded with automated methods. The key difference between

P.A. Schrodt (✉)
Political Science, Pennsylvania State University, University Park, PA 16801, USA
e-mail: schrodt@psu.edu

D. Van Brackle
Lockheed Martin Advanced Technology Laboratories, Lockheed Martin Advanced Technology Laboratories 3550 George Busbee Parkway, Kennesaw, GA 30144, USA
e-mail: dvanbrac@atl.lmco.com

V.S. Subrahmanian (ed.), *Handbook of Computational Approaches to Counterterrorism*, DOI 10.1007/978-1-4614-5311-6_2,

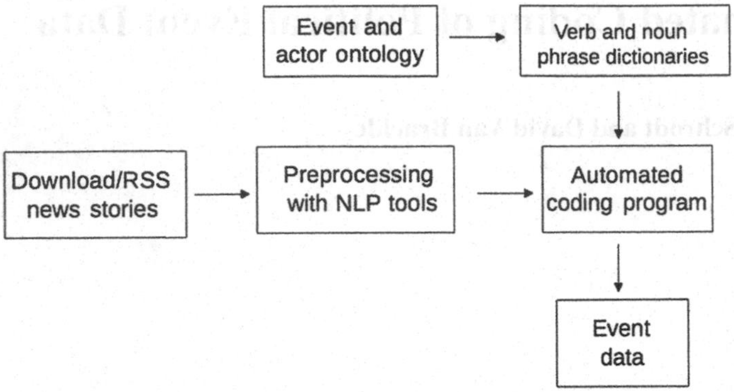

Fig. 1 Process of generating event data by automated methods

the ICEWS event data coding efforts and those of earlier NSF-funded efforts was the scale. As O'Brien—the ICEWS project director—notes,

> . . . the ICEWS performers used input data from a variety of sources. Notably, they collected 6.5 million news stories about countries in the Pacific Command (PACOM) AOR [area of responsibility] for the period 1998-2006. This resulted in a dataset about two orders of magnitude greater than any other with which we are aware. These stories comprise 253 million lines of text and came from over 75 international sources (AP, UPI, and BBC Monitor) as well as regional sources (*India Today, Jakarta Post, Pakistan Newswire,* and *Saigon Times*).

The later phases of ICEWS [25] moved to near-real-time global event data production and the scale of this coding effort increased even further, covering 175 countries and nearly 20-million stories [36].

This chapter will describe a number of incremental improvements and lessons-learned in the recent experience of both our open-source work at Kansas and Penn State, which supported both ICEWS and National Science Foundation-funded basic research, and the proprietary work of the Lockheed Martin Advanced Technology Laboratories, which made several important extensions to that work in conjunction with ICEWS. This chapter is a "how-to" exercise—albeit at a rather high level of generality in places—rather than a theoretical one, and the objective is to provide some guideposts for others who might be interested in undertaking similar efforts, whether as basic research or for applied policy purposes. The chapter essentially goes through the various phases of a machine-coding project, outlined schematically in Fig. 1, starting with the decision on whether to use human coding at all, and discusses both the issues we encountered, the choices we made for resolving these, and thoughts on further developments that might be relevant in the future.

From the outset, we would emphasize that automated coding is a work in progress. It has clearly crossed the threshold into the realm of practical utility—ICEWS models which use event data perform much better than human forecasters—but we do not view it as fully developed. In addition, we are making increasing use of pre-processing software from the much larger field of computational natural language processing, and advances in that area will undoubtedly substantially increase the accuracy of our methods, and quite possibly open avenues for additional coding in areas such as geolocating events, sentiment analysis, coding texts in languages other than English, and resolution of long-standing NLP issues such as noun-verb disambiguation in English, and pronoun co-referencing. Finally, this discussion deals with the field from the perspective of a specific line of related coding programs—KEDS, TABARI and JABARI-NLP—and some of these issues will differ for coding systems using alternative approaches.

1.1 Human Versus Machine Coding

As discussed in some detail in [25], in some circles, automated coding and statistical forecasting can be a very hard sell: many people simply cannot believe that a purely statistical model, generated with well-understood formal methods that are 100 % transparent, and using data generating by automated coding techniques that are also 100 % transparent, can do better than their anything-but-transparent intuition. This is not a problem unique to event data analysis: Nobel Prize-winning psychologist Daniel Kahneman [14, Part III, "Overconfidence"] provides numerous examples from a diverse set of behavioral domains where humans believe they can outperform statistical methods (or dart-throwing chimpanzees) despite overwhelming evidence to the contrary.

Still, before embarking on a coding exercise, you will probably first need to convince skeptical humans. Who will not be impressed by comparisons to chimpanzees, and who usually demonstrate the inferiority of automated methods by pointing to an incorrectly coded sentence—and any event data system, human or machine, will have plenty of those. Meanwhile ignoring the fact that the *total* amount of information in the system is vastly greater than that which can be processed by an individual, and while the intuitive analysis may be better in an individual case (and certainly for an individual news report), the *composite* has better performance. A subject-matter-expert (SME) may perform better on their area of expertise in a particular time frame (though Tetlock's research [37] would suggest not even this is true), but there is little evidence that they can perform broadly. In contrast, using event data, the ICEWS forecasting models predicted five indicators for 29 countries at a monthly granularity for almost 15 years, and effort this is now being scaled to cover the entire world.

As noted in [30], if one is using event data in forecasting models—the objective of ICEWS and most other applications of event data—coding error is only one potential source of error that lies between "events on the ground" and the predictions of the forecasting model. These include

- News reports are only a tiny, tiny fraction of all of the events that occur daily, and are non-randomly selected by reporters and editors;

- Event ontologies such as WEIS, CAMEO and IDEA are very generic and bin together events that may not always belong together in all contexts;
- Forecasting models always contain specification error and cannot consider everything; for example few if any political forecasting models contain a full economic forecasting component;
- Political systems have a degree of intrinsic randomness due to their inherent complexity, chaotic factors even in the deterministic components of those systems, the impact of effectively random natural phenomena such as earthquakes and weather, and finally the effects of free will, so the error intrinsic to a forecasting model will never reduce to zero.

In this chain of events, the impact of coding error in automated systems, while still relevant, is not necessarily dominant. The first and fourth factors also affect SME evaluations; the second and third affect statistical models based on human coding. And the bottom line is that in gold-standard, out-of-sample predictive tests, models using event data consistently show a higher level of predictive accuracy than is typical of SMEs subjected to systematic tests.

When assessing the alternative of human coding for generating event data, there are two additional problems. The first is simple impossibility. In the early phases of the ICEWS project, TABARI repeatedly coded 26-million records in 6 min, resulting in about 3-million events. Sustained human coding projects, once one takes in the issues of training, retraining, replacement, cross-coding, re-coding due to effects of coding drift and/or slacker-coders and so forth, usually ends up coding about six events per hour.[1] The arithmetic is obvious: 6 min of automated coding, or 500,000 labor-hours of manual coding, probably costing on the order of $10-million when labor and administrative costs are taken into effect. And for the manual coding, that amount will code the texts once.

For this reason, human-machine comparisons are of little practical consequence, since human coding is not an option. Multiple published tests [15, 33] have shown that machine coding is comparable in accuracy to human coding. But the human coding accuracy in some of those tests is quite low: King and Lowe [15] use an assortment of measures (and a fairly specific sampling method) but the accuracy on the individual VRA codes alone (Table 2, pg 631)—not the complete record with source and target identification, another major potential source of error—is in the range 25 % (!) to 50% for the detailed codes and 55–70 % for the cue categories. Similarly, [21] show that the reliability of the human coding in the widely-used Comparative Manifestos Project is less than half what is commonly reported, and for some indicators drops as low as 25 %; [28] show similar problems in the coding of governance events in UN peacekeeping. Human coding is anything but flawless.

[1]Individual coders, particularly working for short periods of time, can of course reliably code much faster than this. But for the *overall* labor requirements—that is, the total time invested in the enterprise divided by the resulting useable events—the six events per hour is a pretty good rule of thumb and—like the labor requirements of a string quartet—has changed little over time.

On a supplementary web site (http://eventdata.psu.edu/papers.dir/automated.html) Schrodt has provided an extended rebuttal of the claims in [6] for abysmally low coding accuracy for TABARI. Briefly, while [6] provide almost no information on what combination of software they actually tested, it appears that they attempted to evaluate the system using inappropriate dictionary files. Under any circumstances, it is simply impossible to reconcile their results with the independent assessment of ICEWS Phase I [26] which used the supposedly highly inaccurate data produced by TABARI and yet surpassed the ICEWS 80 % accuracy levels at the same time two competing projects using alternative sets of event data failed to meet those criteria.

Lockheed's internal assessments of the accuracy of TABARI on the initial Asian data evaluated in ICEWS Phase I was around 58 % [36]. This is likely lower than the TABARI accuracy in the Levant and Balkans data sets produced by NSF research—probably closer to 70 %—because during the ICEWS Phase I work relatively few changes were made to the verb-phrase dictionaries, which had been developed on those two regions (see Sect. 3.1). Subsequent work on the JABARI-NLP system during the second two phases of ICEWS brought the accuracy first to 71 % by the incorporation of open-source parsing into JABARI-NLP. Additional enhancement to dictionaries and the processing of various contingencies such as agents and the coding of actions without a clear target led to the current (October 2011) level of "an overall precision of 75.42% with a 3.10% confidence interval." [36]

As noted in greater detail in the web supplement, we don't have a contemporary large, randomly sampled human coded comparison data set—given the futility of human coding as an alternative to automated coding, no one has invested the very substantial amounts of time and money that would be required to do this. The *major* problem with such an exercise is reaching convergence among the human coders: about 10 years ago VRA undertook a substantial, well-designed exercise to do this but no results ever came of it, apparently because the coding never came close to a consensus. Based on our experience and anecdotal reports from various other event data coding projects (Maryland's GEDS, the CACI project for the NSC 1981–1985, Third Point Systems for the Saudis in the 1980s, Russ Leng's BCOW at Middlebury) over the years, that sustained accuracy will be in the range of 70 % at best. The human-coded COPDAB data set somehow manages to miss the Korean War [12], the human-coded GEDS project, which consumed the bulk of the event data expenditures of the NSF-funded "Data Development in International Relations" project has not been used in a single refereed article.

This is not to say that continued efforts should not be made to improve the quality of event coding, and Table 1 provides some general guidelines for situations where human coding is preferable to automated coding. Furthermore, event data provides a "best possible case" for automated coding, since it extracts relatively simple information that usually corresponds to the basic subject-verb-object structure of a typical English-language sentence that is describing an interaction.

Finally, automated coding tools—as well as some of the other NLP software described below—can be effectively used in *machine-assisted coding*. The Chenoweth and Dugan project [7, 8] has used TABARI as a sophisticated pre-filter for coding

Table 1 Tradeoffs between human and automated coding

Advantage to human coding	Advantage to machine coding
Small data sets	Large data sets
Data coded only one time at a single site	Data coded over a period of time or across institutions
No relevant dictionaries	Existing dictionaries can be modified
Complex sentence structure	Simple sentence structures
Metaphorical, idiomatic, or time-dependent text	Literal, present-tense text
Money available to fund coders and supervisors	Money is limited

incidents of terrorism, with a substantial reduction in the required labor costs, and the SPEED event data project [23, 24] uses a variety of customized NLP tools for this purpose.

2 Text Acquisition and Formatting

The first step in generating event data is the acquisition of news reports to code. Following the lead of most event data projects, we initially relied primarily on the Lexis-Nexis (LN) data service; in some of the initial phases of the project these were downloaded; in later phases they were acquired in bulk directly from LN by Lockheed, though this apparently involved the use of the same search engine that is available to ordinary users.

The two key differences between this project and most earlier event data projects was the sheer magnitude of the downloads, and the fact that we were using multiple sources. The eventual text corpus for 1997–2009—after initial filtering—involved about 30 GB of text, which reduced to about eight-million stories.[2] Second, unlike most earlier projects that used a small number of sources—typically the international newswires Agence France Press, BBC, Associated Press and United Press International—we used about 30 different regional sources.[3]

LN, unfortunately, proved problematic, as we also found in [35]. In all likelihood, this is due to LN using a legacy system that was designed to do very narrow searches, rather than providing a large-scale data dumps. In the later phases of the project, we switched to the newer Factiva service [36]. This does not appear to have these search engine problems, presumably because it is working with a relatively new system, and also provided stories from the Reuters news agency. Data providers

[2] The count of "stories" has varied continually as we've updated the downloads, modified the filters and so forth, and so an exact count is both unavailable and irrelevant. But starts around around eight to nine-million.

[3] We've actually identified about 75 distinct sources in the stories, presumably the result of quirks in the LN search engine. However, these additional sources generate only a small number of stories, and by far the bulk of the stories come from the sources we had deliberately identified.

appear to be gradually becoming accustomed to bulk requests that will be used for data-mining, and it is quite possible that these resources will become more available in the future.

The use of multiple sources provides a challenge in extracting the required information—the date, source and individual sentences—from the original download. Following the earlier work in the KEDS project, we were largely using source-specific filters, generally in `perl`. While LN and Factiva are generally consistently formatted, the diverse set of sources—and the sheer size of the files—proved a challenge, particularly since the local sources are more likely to contain minor quirks that will throw off a filter.

As we had discovered in earlier projects, in many sources the task of sentence delineation is a major challenge, both due to the presence of abbreviations, the occasional formatting errors that will cause sentences or entire paragraphs to run together, and the presence of a very large amount of non-sentence material such as tables of sports scores, exchange rates and commodity prices, chronologies, news summaries, weather reports and other such material. In principle, a suitably complex Boolean search term should exclude these; in practice one can't depend on this, particularly for the regional sources. These exceptions are sufficiently widely varied that it is nearly impossible to eliminate all of this using rules on the story itself—though we did have about 30 or so simple rules based on the headline of the story—and instead one needs to use more general rules such as the length of the "sentence." Most news sentences are around 150–300 characters in length, and anything below about 40 characters is almost certainly not codeable. There are also a few patterns easily written as regular expressions that will identify non-material: For example something of the form `\d+\-\d+` is almost always a sports score.

2.1 *Filtering: Irrelevant Stories*

Irrelevant stories have been the bane of the event data source texts from the beginning of our experience. For example, the search string for the now-30-year KEDS "Levant" data set primarily looks for stories containing the names or synonyms of the six actors tracked in the data set: Egypt, Israel, Jordan, Lebanon, the Palestinians, and Syria. However, our early downloads covered the peak of the career of basketball player Michael Jordan and we ended up with quite a number of basketball stories. These are relatively harmless and easily discarded by TABARI or Boolean search exclusions, but they do present problems when downloading—we originally did this using a phone modem [31]—or when one is paying by the story.

However, other types of stories are much more problematic. The most important are chronologies and retrospectives, which describe political events that occurred in the sometimes distant past, yet the dateline of the story is in the present. A good example would be various World War II commemorations, which typically receive extensive coverage and could be miscoded as conflict behavior between the US, Germany and Japan. Recent enhancements to JABARI-NLP specifically address these contingencies.

Another longstanding problem are international sports competitions that use military metaphors. World Cup reports, for example, always use the simple national names—Netherlands versus Spain—and not infrequently use terms such as "battle," "fought," "standoff" and the like. These can usually be solved by discard phrases—a TABARI discard phrase causes the story to be skipped if the phrase occurs anywhere in the text—involving every imaginable form of competition, sporting and others. But even this will fail when the sports context is implicit, such as a [hypothetical] report on the World Cup final on 11 July 2010 that might begin, with little concern that it will be misinterpreted, "Fans eagerly await tonight's battle between the Netherlands and Spain." Furthermore the sheer volume of such stories—as much as a third of the stories in areas where little seems to be happening except sports—decidedly increases download times and costs.

2.2 Filtering: Duplicates

The news downloads contain a very large number of stories that are either literally duplicates, or else are effectively duplicates. These generally come from five sources

- Exact duplicates, where a local source simply reprints the contents of an international newswire story. This is what newswires are for, so it happens a lot;
- Multiple reports of the same event—for example a suicide bombing—as it develops; AFP does this frequently;
- Stories repeated to correct minor errors such as incorrect dates or spelling;
- Lead sentences that occur in general news summaries—which may occur multiple times during a day—as well as in the story itself;
- Multiple independent reports of the event from different news sources: this was a major issue because of the large number of stories we were coding.

Duplicate detection is a very difficult problem, particularly when multiple sources are involved. We dealt with exact and near duplicates by simply seeing whether the first 48 characters of the story matched—this obviously will catch all duplicates and tends to catch minor duplicates such as corrections of spelling errors much of the time.[4] Cross-source duplicates are dealt with using the *One-A-Day* filter discussed below.

When used in a predictive mode, as we are doing with ICEWS, duplicates are not necessarily a bad thing, since they generally will amplify politically-relevant

[4]This will not, however, catching spelling corrections in the first 48 characters. In the Reuters-based filtering for the KEDS project, we did a count of the frequency of letters in the lead sentence, and identified a duplicate if the absolute distance between that vector for two stories, $\sum |x_i - y_i| > \eta$, where the threshold η was usually around 10. This catches spelling and date corrections, the most common source of duplicates in Reuters, but failed on AFP, which tends to expand the details in a sentence as more information becomes available.

signals. In other words, if reporters or editors think that something is important, it is more likely to be repeated, both within sources and across sources, than something that is mundane.

However, when trying to measure changes of "ground-truth" behavior against a baseline over a long period time, duplicates are a serious problem, both across sources and within sources. Cross-source duplication has probably changed considerably over the past 15 years due to local sources putting increasing amounts of material on the Web, and more generally the globalization of the news economy, so that events in once-obscure places are potentially of international interest.[5] In-source duplication can change due both to changes in the resources available to an organization—while not part of the ICEWS source set, Reuters went through something close to an organizational near-death experience during the period 1998–2002 [22] and the frequency of its reporting dropped dramatically during that time—and policies on updating, corrections and the broadcasting of summaries.

As discussed above, duplicate detection is a major challenge in the current environment. Improved story classification to identify, for example, sports stories, historical chronologies and movie reviews, also would simplify the dictionaries by eliminating the need for a number of discard and null-coded phrases that are present only to avoid coding stories that shouldn't be in the data stream in the first place.

Duplicate detection is a fairly specialized application, and one where we've yet to find much in the way of open source software. However, our sense is that algorithms considerably more sophisticated than those we are using exist in various proprietary aggregation systems, notably Google News, European Media Monitor (http://emm.newsbrief.eu/overview.html), and the non-open-source academic project NewsBlaster (http://newsblaster.cs.columbia.edu/). A more thorough review of the computer science literature might produce some guidance on these issues.

In addition, there is a rich literature with well-documented and robust methods—notably support vector machines—for document classification, and these may work considerably better than our current keyword-based methods of detecting sports and business stories in particular. There are no technological barriers preventing this, merely the issue of time and money.

3 Coding Ontologies

For several decades, two coding frameworks dominated event data research: Charles McClelland's WEIS [17, 18] and the Conflict and Peace Data Bank (COPDAB) developed by Edward Azar [1–3]. Both were created during the Cold War and

[5]Notably to traders—carbon-based and silicon-based—in the financial sector, which drives much if not most of the international reporting. The likelihood of an event being reported is very much proportional to the possibility that someone can make or lose money on it.

assumed a "Westphalian-Clausewitzian" political world in which sovereign states reacted to each other primarily through official diplomacy and military threats. While innovative when first created, these coding systems are not optimal for dealing with contemporary issues such as ethnic conflict, low-intensity violence, organized criminal activity, and multilateral intervention. McClelland [19, pg. 177] viewed WEIS as only a "first phase"; he certainly did not anticipate that it would continue to be used, with only minor modifications, for four decades.

3.1 Events

Event categories present in WEIS and COPDAB have both conceptual and practical shortcomings. For instance, WEIS has only a single subcategory for "Military engagement" that must encompass everything from a shot fired at a border patrol to the strategic bombing of cities. COPDAB contains just 16 event categories, spanning a conflict-cooperation continuum that many researchers consider inappropriate. Although there have been efforts to create alternative coding systems—most notably Leng's Behavioral Correlates of War (BCOW) [16]—WEIS and COPDAB remain the predominant frameworks in the published literature.

The lock-in of these early coding systems is readily explained by the time consuming nature of human event coding from paper and microfilm sources. Because human coders typically produce between five and ten events per hour, and a large data set contains tens of thousands of events, experimental recoding is simply not feasible. Automated coding, in contrast, allows researchers to experiment with alternative coding rules that reflect a particular theoretical perspective or interest in a specific set of issues. The effort involved in implementing a new or modified coding system, once it has been developed, is relatively small because most of the work can be done within the dictionary of verb phrases. In most cases verb phrases can be unambiguously assigned to appropriate new categories, while obscure phrases are either removed or modified. Since even a long series of texts spanning multiple decades can then be recoded in a few minutes, this allows researchers to focus on maximizing the validity of the coding scheme for their particular research program since the automated coding process itself guarantees the reliability of the system.

In the early stages of the KEDS research, we felt it was important to work with an existing framework so that we could directly compare human-coded and machine-coded data [33]. For a variety of reasons, we selected WEIS, which despite some obvious drawbacks was good enough for our initial analyses. However, we eventually decided to abandon WEIS and developed CAMEO, much as the VRA group [5, 13, 27] shifted from WEIS to the development of IDEA.

Several considerations motivated this choice. First and foremost was our long-standing concern regarding numerous ambiguities, overlaps, and gaps within the WEIS framework. In addition, the distribution of events in WEIS is quite irregular and several of the two-digit cue categories[6] generate almost no events; we hoped we could improve on this. Third, we wanted to eliminate distinctions among actions that, while analytically discrete, could not be consistently and reliably differentiated using existing news source materials. Finally, as indicated above, the Cold War perspective that permeates WEIS makes it an inappropriate tool for studying contemporary international interactions.

Problems encountered with WEIS are exacerbated due to the lack of a fully specified standard codebook. We based our development of coding dictionaries on the version of the WEIS codebook available through the Inter-university Consortium for Political and Social Research (ICPSR) [18]. The section of the codebook dealing with event categories is quite short—about five pages—and provides only limited guidance. Since McClelland never intended that WEIS would become a de facto coding standard, the ICPSR WEIS codebook was meant to be primarily a proof-of-concept.

We initially intended CAMEO to be an extension of WEIS. Consequently, the first phase of the development of CAMEO involved adding cue and subcategories that we found theoretically necessary for the study of mediation and conflict, while keeping most of the WEIS framework intact. The next phase involved looking for examples of each category and writing definitions for the codebook. This process led to the realization that some of the distinctions we wanted to make for theoretical reasons were simply not possible given the nature of the news leads. For instance, *Promise* (WEIS 07) is almost indistinguishable from *Agree* (WEIS 08) unless the word "promise" is used in the sentence. Therefore, we eventually ended up merging the two into a single cue category—*Agree* (CAMEO 06)—that includes codes representing all forms of future positive commitment. Similarly, because verbs such as *call for, ask for, propose, appeal, petition, suggest, offer*, and *urge* are used interchangeably in news leads to refer to closely related activities, we combined *Request* and *Propose* into a single cue category—*Request/Propose* (CAMEO 05). We made similar decisions with respect to other WEIS categories such as *Grant* and *Reward*, and *Warn* and *Threaten*. We also rearranged the WEIS subcategories, both to reflect these changes and to create more coherent cue categories. As a result, *Nonmilitary demonstration* (WEIS 181) is now part of cue category *Protest* (CAMEO 14) as *Demonstrate* (CAMEO 141) while *Armed force mobilization, exercise and/or displays* (WEIS 182) is modified and falls under the new cue category *Exhibit Military Power* (CAMEO 15).

While developing CAMEO, we paid significant attention to creating a conceptually coherent and complete coding scheme. Having the cue category of *Approve* (CAMEO 03), therefore, necessitated the addition of *Disapprove* (CAMEO 11), which incorporated *Accuse* (WEIS 12) and our new addition *Protest officially* (CAMEO 113). Maintaining the cue category of *Reduce Relations* from WEIS, albeit in a modified fashion, directed us to create a parallel category that captures

[6]The phrase "cue category" refers to the broad two-digit codes, as opposed to the more specific three and four digit subcategories.

improvements in relations: *Cooperate* (CAMEO 04). In other words, we tried to insure that conceptual opposites of each cue and subcategory exist within the coding scheme, although they might not be represented by exact antonyms. We also revised or eliminated all actor-specific event codes.

In addition, we made CAMEO consistent with respect to the order of its main cue categories. Unlike WEIS and IDEA, we start with the most neutral events and move gradually from cooperation to conflict categories. While the initial coding category in WEIS and IDEA is *Yield*, CAMEO starts with *Comment* and locates *Yield* between *Provide Aid* (CAMEO 07) and *Investigate* (CAMEO 09). Technically, all three of these systems use nominal categories so that the placement of each category is irrelevant; in reality, however, the categories are often treated as ordinal or even interval variables. Therefore, CAMEO categories have an ordinal increase in cooperation as one goes from category 01 to 09, and an ordinal increase in conflict as one goes from 10 to 20.

Finally, we developed a formal codebook for CAMEO with descriptions and extensive examples for each category. We have also followed the lead of IDEA in introducing four-digit tertiary subcategories that focus on very specific types of behavior, differentiating, for instance, between agreement to, or rejection of, cease-fire, peacekeeping, and conflict settlements. These tertiary categories have been used only rarely but are available if a researcher wants to examine some very specific behaviors that might be useful in defining patterns.

Despite CAMEO originally being intended specifically to code events dealing with international mediation, it has worked well as a general coding scheme for studying political conflict. This is probably due to the fact that while CAMEO was originally going to involve a minor, 6-month revision of WEIS for a single NSF grant, we ended up spending almost 3 years on the project, with several complete reviews of the dictionaries, and hence effectively created a more comprehensive ontology.

Somewhat to our surprise, the *.verbs* dictionaries—which involved about 15,000 phrases—also needed relatively little work to produce useable data for the first phase of ICEWS. Those dictionaries had been developed for an entirely different part of the world than was coded for ICEWS, but this result was consistent with our earlier experiments in extending the data sets, which have always used a shared *.verbs* dictionary despite using specialized *.actors* dictionaries. We did one experiment where we looked at a sample of sentences where TABARI had *not* identified a verb phrase, and this produced a few new candidate phrases, but only a few. We did considerable work on cleaning up those dictionaries from the accumulated idiosyncracies of two decades of different coders, but they remained largely unchanged.

Under NSF funding, the Penn State project has made extensive efforts to re-define and generalize the entire CAMEO coding ontology using the standardized *WordNet* synsets, rather than using the current categories that were developed inductively, and these dictionaries will be available in the near future. This should help align the event coding with the larger NLP community, and probably simplify its use in languages other than English.

3.2 Actors

One of the major changes in the post-Cold War environment has been the emergence of sub-state actors as major forces in both domestic and international politics. Many commentators have argued that the proliferation of sub-state, non-state, multi-state, and trans-state actors has blurred almost completely the traditional separation of "international" and "comparative" politics. At times these groups exercise coercive force equal to or greater than that of states, whether from within, as in the case of "failed states", or across borders, as with Israel's attempts to control Hizbollah in Lebanon and Hamas in Gaza, or the near irrelevance of borders in many of the conflicts in central and western Africa. Irrespective of the effectiveness of their coercive power, these non-state actors may also be a source of identity that is more important than that of an individual's state-affiliation—the ability of al-Qaeda to attract adherents from across the Islamic world is a good example—or provide examples of strategies that are imitated across borders, as has been seen in the numerous popular revolutions in Eastern Europe or the more recent "Arab Spring."

Because they were state-centered, WEIS and COPDAB paid relatively little attention to non-state actors. A small number of long-lived opposition groups that were active in the 1960s such as the Irish Republican Army, the Palestine Liberation Organization, and the National Liberation Front of Vietnam (Viet Cong) were given state-like codes, as were major international organizations such as the United Nations and the International Committee of the Red Cross/Red Crescent. From the perspective of coding, these actors were treated as honorary states. Beyond this small number of special cases, sub- and non-state actors were ignored.

A major breakthrough in the systematic coding of sub-state actors came with the Protocol for the Analysis of Nonviolent Direct Action (PANDA) project in the early 1990s—the academic precursor to VRA and IDEA—which introduced the concept of sub-state "agents"—e.g. media, politicians, labor unions—as part of their standard actor coding. PANDA's primary focus was on contentious politics within states, and consequently needed to distinguish, for example, between police and demonstrators, or between government and opposition political parties.

Unlike PANDA, which coded the entire world, the KEDS project focused specifically on regions that have experienced protracted conflicts. As a consequence, rather than using the PANDA/IDEA approach of introducing new agent fields, we initially maintained the WEIS/COPDAB convention of using a single "source" and "target" field. However, because the areas we were coding involved quite a few sub-state actors, we eventually developed a series of standard codes that were initially a composite of the WEIS nation-state codes concatenated with PANDA agent codes. Under this system, for example, ISRMIL would be "Israel military", "LIBOPP" would be Liberian opposition parties, "SIEGOV" would be Sierra Leone government and so forth. After realizing that the simple actor-agent model did not accommodate all of the actors we wished to code, we extended this to a more general hierarchical system that was adopted, with modifications, by ICEWS.

Three principles underlie the CAMEO actor coding system. First, codes are composed of one or more three-character elements: In the present system a code can consist of one, two or three of these elements (and therefore three, six, or nine character codes), although this may be extended later. These code elements are classified into a number of broad categories, such as state actors, sub-state actor roles, regions, and ethnic groups.

Second, the codes are interpreted hierarchically: The allowable code in the second element depends on the content of the first element, and the third element depends on the second. This is in contrast to a rectangular coding system, where the second and third elements would always have the same content. The most familiar analogy to a hierarchical coding system is the Library of Congress cataloguing system, where the elements of the catalog number vary—systematically—depending on the nature of the item being catalogued, and consequently may contain very different information despite being part of a single system. The event coding system used in BCOW [16] is another example of a hierarchical scheme in the event data literature.

Third, we are basing our work on standardized codes whenever these are available. This is most obvious in our use of the United Nations nation-state codes (ISO-3166-1 ALPHA 3) (http://unstats.un.org/unsd/methods/m49/m49alpha.htm). This contrasts to the Russett-Singer-Small codes [29] used in WEIS, which are specific to the North American quantitative international relations community. We have generally adopted the IDEA agent codes for sub-state actors. We originally used the HURIDOCS (http://www.huridocs.org/) classifications for world religions, but subsequently expanded this to the much more comprehensive and systematic list found in the CAMEO "Religious Classification System." (http://eventdata.psu.edu/cameo.dir/CAMEO.0.10b2.pdf; this same source also provides a standard set of ethnic codes which we developed by comparing a number of existing sets of ethnicity and languages codes, though we primarily based this on the Joshua Project (http://www.joshuaproject.net/) and Ethnic Power Relations (http://www.epr.ucla.edu/) typologies.

In the later phases of the ICEWS project, Lockheed also developed substate agent typologies which provided considerably more detail than that provided in the classical coding schemes; details on this system and the various proprietary software developed to support it can be found in [36]. Lockheed's system integrates the coding scheme with a large database of group characteristics and allows for the rapid customization of coding schemes.

Unfortunately, standard codes are generally not available. For example, most IGOs are known by acronyms of varying lengths, so we need to decide how to truncate these to three characters. We spent considerable time trying to determine whether the U.S. government had a standard list of militarized non-state actors; as best we can tell, this does not exist (or at least not in a form we can access), and the situation for ethnic groups is similar.

4 Actor Dictionaries and Named Entity Recognition

By far the greatest challenge of scaling-up the KEDS/TABARI system has been in the area of actor dictionary development. The KEDS project had focused on a small number of geographical areas, primarily the Levant, with 10-year data sets on the Balkans and West Africa. We had done some experimental work under small government contracts to code individual countries in other areas of interest, in all parts of the world, for short—typically 2-year—time periods, and graduate student research by Ömür Yilmaz and Baris Kesgin had produced very detailed dictionaries for Turkey, but that was it. ICEWS, in contrast, initially involved coding 29 states that encompass more than half the world's population, and in the final stages was expanded to coding the entire world.

The earlier KEDS data sets were initially developed by individuals—largely undergraduate honors students—who went through sentences item by item and added new patterns to the actor and verb dictionaries as they encountered incorrectly coded sentences.[7] This was later supplemented by a relatively simple named-entity-recognition (NER) program called *ActorFilter* that would locate potential new names based on capitalization patterns, compare these to entries in the existing dictionaries, and then produce a keyword-in-context (KWIC) listing of entities which appeared to be new, listed in reverse order of frequency. This was particularly useful in making sure that any major new actors were not missed, and was our first step in developing dictionaries for new countries.

Neither of these techniques scaled, particularly in the relatively short time frame of the first phase of the ICEWS work. While we did some spot-checking of individual stories, our ability to do this with any meaningful proportion of the 26-million sentences in the ICEWS corpus was limited. *ActorFilter*, unfortunately, had not been designed for a project of this magnitude and while it could be used on a sample, it slowed to an unusable crawl on very large files.

Consequently, three approaches were used.

First, rather than deriving the actors from the texts, we tried to locate lists of actors and incorporate these into both international and nation-specific dictionaries. Various national sources provided lists of parliamentarians and other local leaders, and we've also been expanding the list of NGOs and IGOs. As a consequence, the Asian actors dictionaries now have around 20,000 entries, compared to the 1,000 or so entries typical in earlier KEDS work.

We also augmented a reference file used in earlier NSF-funded work on the Militarized Interstate Disputes dataset [35] with information in the *CIA World Factbook* and rulers.org to a comprehensive list of state names, major cities, regions

[7] To date, all of the successful automated event data coding systems are dictionary and rule based, rather than using statistical-methods: see [36]. While statistical methods would certainly be attractive, and seem to work on highly simplified "toy problems" such as those in [6], all of the successfully-deployed systems to date are dictionary-based, and numerous efforts to scale initially-promising statistical methods have failed.

and geographical features, adjectival forms, and date-delimited lists of heads of state and other members of government. This has developed into the roughly 32,000-entry *CountryInfo* (http://eventdata.psu.edu/software.dir/dictionaries.html) which has a systematic format fairly close to that of XML, and can easily be converted into TABARI dictionary format with a utility program.

Second, we improved the ability of TABARI to automatically assemble codes from combinations of a named actor and an generic agent; this facility is also part of JABARI-NLP. For example "Philippine soldiers" will automatically generate the code `PHLMIL`, whereas "The Philippine Secretary of Agriculture" will automatically generate the code `PHLGOV`. Earlier dictionaries had done this directly, with separate dictionary entries for, say, "Australian police," "Cambodian police," "Chinese police" and so forth. The new system is both faster in terms of the dictionary size and much more efficient. This allows the coding of both generic agents such as "police", "soldiers", "demonstrators" and the like, as well as named individuals where we have the title in the dictionary but not the individual person. For most of our coding, at least for the forecasting efforts in ICEWS, individual identities are not used, so this gets quite a bit of information we were previously missing. In support of this new facility, we also increased the size of the *.agents* dictionary considerably, based on *WordNet* and sampling from the source texts.

Finally, *ActorFilter* was replaced with a new open-source `Python` program, `PoliNER`, which had a similar function but was adapted to the much larger dictionaries and source text files. The sorted output of this program can be combined with a program named `CodeCatcher` for machine-assisted development of dictionaries: `CodeCatcher` guesses the likely code based on known entities in a sentence, and allows rapid combination of codes based on that other information.

These efforts were a major step forward, but dictionary development—and maintenance, as dictionaries need to be updated as political figures change—remains a considerable challenge. Fortunately there is a considerable literature—much of it DARPA-funded—on NER, and some of these methods are very sophisticated—for example using conditional random fields and hidden Markov models—and are certainly far more sophisticated than what we are currently using, and these methods might provide significant additional advances in efficiency.

5 Pre-processing Using NLP Tools

A major shift in automated coding that has been shown to dramatically increase accuracy has been the incorporation of open-source natural language processing (NLP) tools to correctly identify the elements of the sentence required for coding. When KEDS was being developed in the early 1990s, or even in the early 2000s, the development period of TABARI, open-source code was still a relative novelty. As a consequence, these programs handled all of their own linguistic processing with an internal shallow parser written into the code. Parser code written by a political scientist. This obviously *worked*, in the sense of producing useable data, but the internal structure of the program is quite complex and difficult to modify. In the environment of the 2010s, it makes far more sense to leave NLP software development to the computational linguists, and focus only on those remaining tasks that are needed to get convert these structures to events.

This is the approach that was taken with JABARI-NLP. The original JABARI simply duplicated TABARI in a Java environment [38].[8] However, after several key weaknesses were identified in the shallow-parsing approach—most importantly, a tendency to match words in verb phrases that were not actually part of the phrase—the JABARI effort, rather than attempting to deal with these in the program itself, explored a number of open-source options that could provide the NLP processing, then was modified to handle that information. TABARI is gradually being modified in a similar fashion.

For purposes of illustration, consider the following initial sentences for a news story:

> US Supreme Court Justice Stephen Breyer was robbed by a machete-wielding man at his Caribbean vacation home, a Supreme Court spokeswoman said.
> The robber broke into Judge Breyer's home on the island of Nevis around 21:00 EST (02:00 GMT) on Thursday.
> The Supreme Court justice was at home with his wife and guests, but no one was hurt, the spokeswoman said.

Software for the following tasks can be found at open-source NLP software site such as Open-NLP and various other academic sites; we are going to discuss these generally by function rather than making specific recommendations, since this is still very much an evolving field.

- Sentence delineation. As noted in Sect. 2, this is a surprisingly difficult task given the presence of abbreviations, punctuation occurring inside sentences, and the occurrence of character strings that are not actually part of the sentence, particularly across multiple story formats. Linguists have systems that are more robust than our `perl` filters.
- Disambiguation by parts-of-speech markup. One of the major tasks of the TABARI dictionaries is noun-verb disambiguation: this issue accounts for much of their size. Parts-of-speech (POS) marking—or in the example below, a system that makes noun-verb distinctions and also classifies these into general categories—would eliminate this problem.

```
US/noun.group Supreme/noun.group Court/noun.group
    Justice/noun.group
Stephen/noun.person Breyer/noun.person was
    robbed/verb.possession by
```

[8] Including, at the request of the sponsor, some bugs in TABARI, though after the equivalence of the two systems was demonstrated, these were corrected in both systems.

```
a machete-wielding man/noun.person at his/pronoun
     Caribbean
vacation/noun.artifact home/noun.artifact,
     a Supreme/noun.group
Court/noun.group spokeswoman/noun.person
     said/verb.communication.

The robber/noun.person broke/verb.communication
     into/verb.communication
Judge/noun.person Breyer/noun.person's
     home/noun.location on the
island/noun.object of Nevis/noun.location around
     21:00 EST/noun.time on
Thursday/noun.time.
```

- Stemming. TABARI has only recently added capabilities of automatically recognizing the regular forms of nouns and verbs. Many NLP systems use stemming—most frequently the Porter stemming algorithm for English (http://tartarus.org/martin/PorterStemmer/). This should both simplify and generalize the dictionaries.

- Full parsing. An assortment of full-parsers—as distinct from the shallow parsers used in KEDS/TABARI—are available, and the *TreeBank* parse format appears to be a fairly stable and standard output format. This allows a researcher to use the parser of his or her choice (notably some parser developed in the future) so long as these could produce *TreeBank*-formatted output. The most important contribution of the full parsing is insuring that the words associated identified as belonging to a verb phrase are in fact associated with that verb, and not with a subordinate clause or some other part of the sentence.

```
(ROOT (S (S (NP (NNP US) (NNP Supreme) (NNP Court)
     (NNP Justice)
(NNP Stephen) (NNP Breyer)) (VP (VBD was)
     (VP (VBN robbed) (PP (IN by)
(NP (NP (DT a) (JJ machete-wielding) (NN man))
     (PP (IN at) (NP (PRP$ his)
(JJ Caribbean) (NN vacation) (NN home))))))) (, ,)
     (NP (DT a)
(NNP Supreme) (NNP Court) (NN spokeswoman))
     (VP (VBD said)) (. .)))
```

- Pronoun and entity coreferencing. Some of the full-parsing systems provide pronoun and entity coreferencing, another feature coded into TABARI. Alternatively, this can be provided in stand-around coreferencing systems such as the ARK noun phrase coreferencer. (http://www.ark.cs.cmu.edu/ARKref/)

```
<ref id="1" ent="1_4_8">US Supreme Court Justice
     Stephen Breyer</ref> was
robbed by <ref id="2" ent="2">a machete-wielding
     man at
<ref id="3" ent="1_4_8">his</ref> <ref id="4"
     ent="3_7_46">Caribbean vacation
home</ref>, <ref id="5" ent="5_21">a Supreme Court
     spokeswoman</ref> said.

<ref id="6" ent="6_19">The robber</ref> broke into
     <ref id="8" ent="1_4_8">
Judge Breyer's</ref> <ref id="7" ent="3_7_46">
     home</ref> on
<ref id="9" ent="9">the island of Nevis</ref>
     around 21:00 EST on
<ref id="13" ent="13">Thursday</ref>.

<ref id="17" ent="1_4_8">The Supreme Court
     justice</ref> was at home with
<ref id="19" ent="1_4_8">his</ref> wife and guests,
     but <ref id="20" ent="20">
no one</ref> was hurt, <ref id="21" ent="5_21">the
     spokeswoman</ref> said.
```

The use of these tools accomplishes at least the following improvements:

- It aligns automated event coding—which is fundamentally an NLP problem—with the larger NLP community. As their tools improve, we can incorporate those improvements into event data work immediately.
- It considerably simplifies—though not entirely eliminating the need for—the construction and maintenance of coding programs, and in particular the tasks that can now be done with open-source ancillary programs would eliminate many of the most brittle parts of the original TABARI code.
- It introduces a deep—as distinct from a shallow—parser into the system, and the shallow parsing approach has probably reached its limits.
- The use of standardized NLP tools and dictionaries would probably simplify the development of a system for languages other than English, particularly languages such as Chinese and Arabic where considerable NLP work has been invested;
- Many of these features should simplify the *.verbs* dictionaries, or at the very least gain more robust performance from dictionaries of the same length;

Parsing and other pre-processing—in all likelihood a fairly slow process—needs to be done only once for a given sentence, and the marked-up version can be stored, so unlike systems with in-line deep parsers, the resulting coding (which is likely to be re-done many times) should be as fast or faster than the current system. The pre-processing is also trivially divided across multiple processors in a cluster system, so with suitable hardware or using virtual clusters in a cloud computing environment, the processing requirements can be easily adjusted to near-real-time coding environments.

6 Coding and Post-processing

6.1 Cluster Processing

TABARI is an open-source C++ program—compiled under gcc—that runs on a common code base in both the Macintosh OS-X and various Linux/Unix environments. This has proved useful in deploying it across a combination of desktop, server and cluster environments.[9]

The major innovation in conjunction with the 2009 coding for the second phase of ICEWS was the use of a computer cluster to dramatically increase the coding speed. In the 2008 data development for ICEWS Phase I, coding the 1997–2004 data on personal computers required almost a week. This was also slowed by the existence of some bugs in TABARI that occurred only with extremely rare sentence structures and thus had gone undetected in earlier work with the program: there were initially eight of those out of the 26-million sentences.

In 2009, we gained access to a small, 14-processor cluster computer that was sitting unused (and undocumented) at the University of Kansas. Rather than trying to get TABARI to run in parallel at the micro level, we did "parallelism on the cheap" and simply split the text files to be coded across the processors, which shared a common file space, coded these simultaneously, then re-combined the output files at the end of the run. TABARI ran on the individual nodes at around 5,000 sentences per second; the throughput for the cluster as a whole ended up around 70,000 stories per second, allowing the entire 26-million story corpus to be coded in about 6 min. The initial set-up, of course, took quite a bit longer, but this was particularly useful for weeding out the aforementioned problematic records that would cause the program to crash.

A 14-processor cluster is, of course, tiny—Penn State has multiple clusters available to social scientists that are in the 256-processor range—so effectively the coding speed is unlimited, even for a very large corpus. Furthermore, this can be done by simple file spliting, so the gain is almost linear.

---

[9] In principle these enhancements could also be applied to JABARI-NLP, though it is running in secure military systems rather than open environments and to date has made less use of cluster processing.

6.2 *One-A-Day Filtering*

Following the protocols used in most of the research in the KEDS project, the major post-processing step is the application of a "one-a-day" filter, which eliminates any records that have exactly the same combination of date, source, target and event codes. This is designed to eliminate duplicate reports of events that were not caught by earlier duplicate news report filters. In our work on the Levant data set, this fairly consistently removes about 20 % of the events; the effect on the ICEWS data may be somewhat higher due to the use of a greater number of sources.

In areas of intense conflict—where multiple attacks could occur within a single dyad in a single day—this could eliminate some actual events. However, these instances are rare, and periods of intense conflict are usually obvious from the occurrence of frequent attacks across a month (our typical period of aggregation), and do not require precise measures within a single day. Periods of intense conflict are also likely to be apparent through a variety of measures—for example comments, meetings with allies, offers of aid or mediation—and not exclusively through the attacks themselves.

6.3 *Sophisticated Error Detection/Correction*

Thus far, we have been using only limited error detection and correction. Some LM-ATL experiments have shown that even very simple filters focusing on anomalous high-intensity events can eliminate egregious errors such coding USA/Japanese conflict events based on Pearl Harbor travel and movie reviews or anniversaries of the bombings of Hiroshima and Nagasaki. Eliminating these is particularly important when the output is used for the monitoring of unlikely events—for example pattern recognition of potential conflict "triggers" either by humans or machine-learning algorithms—as distinct from conventional statistical approaches which can readily ignore these as noise. In addition, far more sophisticated filtering methods are available, and many of these are of relatively recent vintage due to the computing power required. A multi-category support vector machine (SVM), for example, could be applied to the full text of a story—or possibly a single sentence, but SVMs tend to work better at the document level than the sentence level—to determine whether the story is likely to have produced events of the type coded, based on previously verified correct codings.

From this point, a variety of different things are done with the data, but these fall into the category of data management and model construction, rather than data generation per se. LM-ATL [36] is developing an increasingly elaborate system for the management of the data that includes a wide variety of visualization tools, as well as interactive "drill-down" capability that allow a user to go from the coded events back to the original text, as well as management and display of the coding dictionaries. On the modeling side, the data can be aggregated in a variety of ways, including event counts for various types of dyads as well interval-level scaled data using a modification of the Goldstein scale [11] for the CAMEO ontology.

7 Open Issues

7.1 Geolocation

A still missing component of the system is the ability to tag the entire story with the location, which will allow the agents to be coded even if they are not preceded by a national identifer. This is particularly important in local sources: unlike an international news report, a Philippine news report on Mindanao, for example, will almost never mention that Mindanao is part of the Philippines. There are several software systems for doing this type of tagging and LM-ATL is experimenting with them [36] with some success, though this is still an open issue. As with NLP processing more generally, this is an open research area with a variety of active open-source and proprietary systems available, and is likely to improve substantially in the near future.

7.2 Machine Translation

With the increasing availability of news items in multiple languages on the web—for example European Media Monitor looks at sources in 43 languages—the possibility of coding in languages other than English is very attractive. There are at least three different approaches that could be used here.

The most basic, but by far the most labor intensive, would be to simply write an equivalent automated coding system for other languages, and come up with equivalent *.verbs* dictionaries. The *.actors* dictionaries would probably require little modification for languages using the Latin alphabet; though they would require extended work for systems such as Arabic, Chinese and Hindi. We did this for German in an early phase of the KEDS project [9], albeit with very simple dictionaries. While some modification of the parser is required in this approach, shallow parsing looks at only the major syntactic elements of a sentence and this would be relatively easy, and the *linguistic* work of Noam Chomsky strongly suggests that this modifications will fall into a relatively small number of categories.

The second possibility would be to use NLP tools to handle the parsing—which we are likely to be doing in the next phase of the development of the English-language coders as well—but still use language-specific *.verbs* dictionaries. The modification of the *.verbs* dictionaries would also allow language-specific idiomatic phrases—which are likely to be quite important and quite unsystematic—but would also involve considerable work. This might, however, be justified in the cases of languages where there is a large set of news sources, particularly on local events, which is not covered well in English: Spanish and Arabic come to mind, as would Chinese if an independent press develops in that country.

The final possibility, which was pursued at an experimental level by Lockheed [36], is to use machine translations into English of the source texts, and then continue to use the English-language coders. The extent to which this works depends both on the quality of the automated translators, and the extent to which the existing dictionaries—generally developed on texts at least edited if not written by fluent writers of English—correspond to the phrases encountered in the automated translations, which are often based on statistical methods intended simply to provide a recognizable sense of the text, not an eloquent rendition of it.

Lockheed's initial experiments with several translation systems working on roughly two-million sentences in Spanish and Portuguese achieved accuracy around 67 %, which is probably comparable to human coding accuracy and would provide useful data for statistical modeling but this is not sufficiently high to satisfy human users working with the data at a highly detailed level [25]. There has been extensive work in machine translation into English from Spanish, Arabic, and Chinese, and as with the other NLP tools, these systems are likely to improve over time given the economic motivations for developing good software.

7.3 Real-Time Coding

At Kansas during the 2009–2010 period we undertook an experiment in true real-time coding using RSS feeds. RSS feeds present a potentially very rich source of real-time data because they are available in actual real time using standard software, and, of course, are free. The downside of RSS feeds is the absence—at least at the present time—of any archival capacity, so they can be used for current monitoring but not for generating a long time series.

A variety of RSS feeds are available. The richest would be two major RSS aggregators, GoogleNews and European Media Monitor, which track several thousand sources each. In some experimental downloads in 2008, we found that these generated about 10 Gb of text per month, and that volume has probably only increased. The two downsides with the aggregators are massive levels of duplication, and the fact that they are not produced in a standard format: instead, each source must be reformatted separately. This is not particularly difficult in terms of simply detecting the natural language text of the news report itself—and in fact all of these feeds consist largely of HTML code, which typically takes up more than 90 % of the characters in a downloaded file—but can be difficult in terms of detecting dates and sources.

Instead of looking at the aggregators, we focused on two high-density individual sources: Reuters and UPI. In addition to providing RSS feeds, these also have archives, back to 2007 for Reuters and back to 2001 for UPI; these could be downloaded from the Web. The focus on individual sources meant that only a small number of formats had to be accommodated—even formats within a single source exhibit some minor changes over time—but these two sources, as international news wires, still provide relatively complete coverage of major events. They do not, however, provide the same level of detail as the commercial sources, Factiva for Reuters and LN for UPI. After some experimentation, it turned out to be easier to access the updates to this information from their web sites rather than through RSS feeds per se, but this still allows fairly rapid updating.

Implementation of a real-time coder was a relatively straightforward task of linking together, on a server, the appropriate reformatting and duplicate detection programs, running TABARI at regular intervals on the output of those programs, and then storing the resulting event data in a form that could be used by other programs: mySQL was used for this purpose. While the basic implementation of this system has been relatively straightforward, our 18-month experiment found at least three characteristics of the data that should be taken into account in the design of any future systems.

First, while *in principle* one could get real-time coding—automated news monitoring services used in support of automated financial trading systems routinely do this—there is little reason to do so for existing event data applications, which generally do not work on data that is less finely grained than a day. Furthermore, the news feeds received during the course of a day are considerably messier—for example with minor corrections and duplications—than those available at the end of a day. Consequently, after initial experiments we updated the data only once a day rather than as soon as the data became available.

Second, these are definitely not "build and forget" systems due to the changing organization of the source web sites. Reuters in particular has gone through three or four major reorganizations of their web site during the period we have been coding data from it, and in one instance was off-line for close to a week. Thus far, the changes in code resulting from these reorganizations have been relatively minor, primarily dealing with the locations of files rather than the file formats, but it has necessitated periodic—and unexpected—maintenance. The RSS feeds may have been more reliable—these presumably did not go off-line for a week—but still probably undergo some changes. It is also possible that as the sites mature, they will be more stable, but this has not occurred yet.

Finally, we have not dealt with the issue of automatically updating actor dictionaries, depending instead on general international dictionaries that contain country-level information but relatively little information on individual leaders. International news feeds generally include national identification—"United States President Obama," not just "Obama"—so the country-level coding should generally be accurate, but the data probably is less detailed at the sub-state level.

8 Conclusion

In a history of the first 15 years of the KEDS/TABARI project [31], the final section—titled "Mama don't let your babies grow up to be event data analysts" lamented the low visibility of event data analysis in the political science literature despite major advances in automated coding and the acceptance of analyses resulting from that data in all of the major refereed political science journals.

The situation at the present is very different, largely due to ICEWS, which emerged about 6 months after that history was written. All three of the teams involved in the first phase of ICEWS used some form of event data in their models. Lockheed, the prime contractor for the only team whose models cleared the out-of-sample benchmarks set by ICEWS, has continued to invest in additional developments, both for ICEWS and potentially for other projects, and as noted in the previous section, there are now a number of proprietary systems in active development, in contrast to the previous 15 years which saw only KEDS/TABARI and VRA-Reader. At the same time, there has been substantial NSF funding of further development of the open-source TABARI and various ancillary utilities, so while the open-source work lags somewhat behind the proprietary—though in other aspects, such as the incorporation of *WordNet* into the dictionaries, it is ahead—reasonably up-to-date software is available as open source, and it is still being actively developed.

In 1962, Deng Xiaoping famously quoted the Sichuan proverb, "No matter if it is a white cat or a black cat; as long as it can catch mice, it is a good cat." Statistical models utilizing event data coded with automated techniques are good cats. Some are white, some are black, but they catch mice. Furthermore, the fact that such models exist is now known [25, 26] and from a policy perspective it is likely that they will be continued to be developed for policy applications seems rather high: the open-access textbook on the results of the KEDS project circa 2000, *Analyzing International Event Data,* reportedly has been translated into Chinese.[10] The cat, so to speak, is out of the bag.

Acknowledgements This research was supported in part by contracts from the Defense Advanced Research Projects Agency under the Integrated Crisis Early Warning System (ICEWS) program (Prime Contract #FA8650-07-C-7749: Lockheed-Martin Advance Technology Laboratories) as well as grants from the National Science Foundation (SES-0096086, SES-0455158, SES-0527564, SES-1004414) and by a Fulbright-Hays Research Fellowship for work by Schrodt at the Peace Research Institute, Oslo (http://www.prio.no). The results and findings in no way represent the views of Lockheed-Martin, the Department of Defense, DARPA, or NSF. It has benefitted from extended discussions and experimentation within the ICEWS team and the KEDS research group at the University of Kansas; we would note in particular contributions from Steve Shellman, Hans Leonard, Brandon Stewart, Jennifer Lautenschlager, Andrew Shilliday, Will Lowe, Steve Purpura, Vladimir Petroff, Baris Kesgin and Matthias Heilke.

[10] Though we've not been able to locate this on the web. Itself interesting.

References

1. Azar EE (1980) The conflict and peace data bank (COPDAB) project. J Confl Resolut 24: 143–152
2. Azar EE (1982) The codebook of the conflict and peace data bank (COPDAB). Center for International Development, University of Maryland, College Park
3. Azar EE, Sloan T (1975) Dimensions of interaction. University Center for International Studies, University of Pittsburgh, Pittsburgh
4. Bond D, Bond J, Oh C, Jenkins JC, Taylor CL (2003) Integrated data for events analysis (IDEA): An event typology for automated events data development. J Peace Res 40(6): 733–745
5. Bond D, Jenkins JC, Taylor CLT, Schock K (1997) Mapping mass political conflict and civil society: Issues and prospects for the automated development of event data. J Confl Resolut 41(4):553–579
6. Boschee E, Natarajan P, Weischedel R (2012) Automatic extraction of events from open source text for predictive forecasting. In: Subrahmanian V (ed) Handbook on computational approaches to counterterrorism. Springer, New York
7. Chenoweth E, Dugan L (2012) Rethinking counterterrorism: evidence from israe. Working Paper, Wesleyan University, Middletown, CT
8. Dugan L, Chenoweth E (2012) Moving beyond deterrence: the effectiveness of raising the expected utility of abstaining from terrorism in israel. Working Paper, University of Maryland, College Park, MD
9. Gerner DJ, Schrodt PA, Francisco RA, Weddle JL (1994) The machine coding of events from regional and international sources. Int Stud Q 38:91–119
10. Gleditsch NP (2012) Special issue: event data in the study of conflict. Int Interact 38(4): 375–569
11. Goldstein JS (1992) A conflict-cooperation scale for WEIS events data. J Confl Resolut 36:369–385
12. Howell LD (1983) A comparative study of the WEIS and COPDAB data sets. Int Stud Q 27:149–159
13. Jenkins CJ, Bond D (2001) Conflict carrying capacity, political crisis, and reconstruction. J Confl Resolut 45(1):3–31
14. Kahneman D (2011) Thinking fast and slow. Farrar, Straus and Giroux, New York
15. King G, Lowe W (2004) An automated information extraction tool for international conflict data with performance as good as human coders: A rare events evaluation design. Int Organ 57(3):617–642
16. Leng RJ (1987) Behavioral correlates of war, 1816–1975. (ICPSR 8606). Inter-University Consortium for Political and Social Research, Ann Arbor
17. McClelland CA (1967) World-event-interaction-survey: a research project on the theory and measurement of international interaction and transaction. University of Southern California, Los Angeles, CA
18. McClelland CA (1976) World event/interaction survey codebook (ICPSR 5211). Inter-University Consortium for Political and Social Research, Ann Arbor
19. McClelland CA (1983) Let the user beware. Int Stud Q 27(2):169–177
20. Merritt RL, Muncaster RG, Zinnes DA (eds) (1993) International event data developments: DDIR phase II. University of Michigan Press, Ann Arbor
21. Mikhaylov S, Laver M, Benoit K Coder reliability and misclassification in the human coding of party manifestos. Political Anal 20(1):78–91 (2012)
22. Mooney B, Simpson B (2003) Breaking News: How the Wheels Came off at Reuters. Capstone, Mankato
23. Nardulli P (2011) The social, political and economic event database project (SPEED). http://www.clinecenter.illinois.edu/research/speed.html

24. Nardulli PF, Leetaru KH, Hayes M Event data, civil unrest and the SPEED project (2011). Presented at the International Studies Association Meetings, Montréal
25. O'Brien S (2012) A multi-method approach for near real time conflict and crisis early warning. In: Subrahmanian V (ed) Handbook on computational approaches to counterterrorism. Springer, New York
26. O'Brien SP (2010) Crisis early warning and decision support: contemporary approaches and thoughts on future research. Int Stud Rev 12(1):87–104
27. Petroff V, Bond J, Bond D (2012) Using hidden Markov models to predict terror before it hits (again). In: Subrahmanian V (ed) Handbook on computational approaches to counterterrorism. Springer, New York
28. Ruggeri A, Gizelis TI, Dorussen H (2011) Events data as bismarck's sausages? intercoder reliability, coders' selection, and data quality. Int Interact 37(1):340–361
29. Russett BM, Singer JD, Small M (1968) National political units in the twentieth century: a standardized list. Am Political Sci Rev 62(3):932–951
30. Schrodt PA (1994) Statistical characteristics of events data. Int Interact 20(1–2):35–53
31. Schrodt PA (2006) Twenty years of the Kansas event data system project. Political Methodol 14(1):2–8
32. Schrodt PA (2012) Precedents, progress and prospects in political event data. Int Interact 38(5):546–569
33. Schrodt PA, Gerner DJ (1994) Validity assessment of a machine-coded event data set for the Middle East, 1982–1992. Am J Political Sci 38:825–854
34. Schrodt PA, Gerner DJ, Yilmaz Ö (2009) Conflict and mediation event observations (CAMEO): an event data framework for a post Cold War world. In: Bercovitch J, Gartner S (eds) International conflict mediation: new approaches and findings. Routledge, New York
35. Schrodt PA, Palmer G, Hatipoglu ME (2008) Automated detection of reports of militarized interstate disputes using the SVM document classification algorithm. Paper presented at American Political Science Association, Chicago, IL
36. Shilliday A, Lautenschlager J (2012) Data for a global icews and ongoing research. In: 2nd international conference on cross-cultural decision making: focus 2012, San Francisco, CA
37. Tetlock PE (2005) Expert political judgment: how good is it? how can we know? Princeton University Press, Princeton
38. Van Brackle D, Wedgwood J (2011) Event coding for hscb modeling: challenges and approaches. In: Human social culture behavior modeling focus 2011, Chantilly, VA

Automatic Extraction of Events from Open Source Text for Predictive Forecasting

Elizabeth Boschee, Premkumar Natarajan, and Ralph Weischedel

1 Introduction

Forecasting political instability has been a central task in computational social science for decades. Effective prediction of global and local events is essential to counter-terrorist planning: more accurate prediction will enable decision makers to allocate limited resources in a manner most likely to prove effective. Most recently, programs like DARPA's Integrated Conflict Early Warning Systems [2] have shown that statistically-driven quantitative models can produce out-of-sample predictive accuracy that exceeds 80% for certain events of interest, e.g. rebellion and insurgency. Research on a wide variety of predictive approaches continues to thrive.

At the center of forecasting research is the assumption that predicting the future requires some understanding (or model) of both the present and the past. Approaches vary widely in the methods used to construct these models. In some cases, the models are informed by human subject matter experts, who manually evaluate tendencies like an individual leader's propensity for repression, or a particular country's susceptibility to political unrest. Models may also rely on specific quantitative economic or political indicators, where available, or many other types of information. However, there is also a wealth of political, social, and economic information contained in open source textual resources available across the globe, ranging from news reports to political propaganda to social media. To make use of this rich data trove for predictive modeling, an automatic process for extracting meaningful information is required—the volume is too great to be processed by hand, especially in real-time. When this is done, however, the

E. Boschee (✉) • P. Natarajan • R. Weischedel
Raytheon BBN Technologies, Cambridge, MA, USA
e-mail: eboschee@bbn.com; pnataraj@bbn.com; weischedel@bbn.com

V.S. Subrahmanian (ed.), *Handbook of Computational Approaches to Counterterrorism*,
DOI 10.1007/978-1-4614-5311-6_3,

predictive models gain access to a data source in which trends and possibilities can be revealed that would otherwise have gone unnoticed

One particular area of research for predictive models using open source text has been the incorporation of events involving actors of political interest; event data was used by several of the successful models developed or advanced in the ICEWS program. These events can cover a range of interactions that span the spectrum from cooperation (e.g. the United States promising aid to Burma) to conflict (e.g. al-Qaeda representatives blowing up an oil pipeline in Yemen). Taken as a whole, the fluctuation in the level and severity of such events (as reported in text) can provide important statistical fodder for predictive models.

These types of events were once coded by hand, but over the past 20 years, many have abandoned that time-consuming, expensive, and often inconsistent process and have moved to automated coding as a method to generate large-scale event data sets. One of the pioneers and leaders of this transition has been the KEDS/TABARI project, begun in the late 1980s. However, these automatically-generated event sets remain noisy: many incorrect events are mistakenly included, and many are missed. Because of this, there is still a wealth of event information not yet available to predictive models using existing methods.

Meanwhile, while computational social scientists have moved towards automatic event coding, new algorithms and processes have been separately developed in machine learning and natural language processing, resulting in effective and efficient algorithms that can extract entities, relations between entities, and events involving entities from text. The full breadth and depth of these techniques has yet to be fully applied to the task of event coding for the computational social sciences. In addition, the speed of computer processing continues to increase so significantly that the application of more sophisticated techniques to this task is now much more feasible than it was 10 years ago.

In this chapter we present the results of studies intended to apply these new, state-of-the-art natural language analysis algorithms to the field of event extraction for computational social science. Our initial results demonstrate that using a state-of-the-art natural language processing system like BBN's SERIF™ engine [3] can yield significant gains in both accuracy and coverage of event coding. It is our hope that these results will (1) lead to greater acceptance of automated coding by creators and consumers of social science models that depend on event data and (2) provide a new way to improve the accuracy of those predictive models.

This chapter is organized as follows. We begin by describing the technical approach underlying the two systems we will compare, one based on existing computational social science techniques for event extraction and the other based on BBN SERIF's state-of-the-art statistical natural language processing techniques. We will then lay out the experimental design and present detailed results for each of five different dimensions of performance:

1. **Accuracy.** When an event is reported by the system, how often is it correct?
2. **Coverage.** How many events are correctly reported by the system?

3. **Historical events.** Can the system correctly filter historical events out of the current event data stream? For instance, a reference to World War II should not give downstream models evidence of new unrest in Europe.
4. **Event/document filtering.** How well do systems filter out "red herring" events based on document topics? For instance, sports documents in particular pose a problem for event extraction systems, since sporting "clashes" often read as violence.
5. **Domain shift.** How well do event extraction models perform on data from sources other than traditional newswire articles?

2 Task Description

The experiment described in this chapter focuses on event coding as performed under the CAMEO (Conflict and Mediation Event Observations) scheme. Details can be found in [4] and [5] as well as at http://web.ku.edu/~keds/data.dir/cameo.html. In CAMEO, each event is represented as a "triple" consisting of an event code, a Source actor, and a Target actor. For instance, in the sentence "*The U.S. Air Force bombed Taliban camps*", the appropriate event triple would be (195, USAMIL, AFGINSTAL) or (*Employ aerial weapons, United States military, Taliban*).

The systems evaluated here used the same manually-generated actor and agent lists specifying that, e.g., Hamid Karzai is affiliated with the government of Afghanistan.

The comparative evaluation for this project covers four of the twenty top-level CAMEO event codes: Provide Aid (07), Disapprove (11), Assault (18), and Fight (19). For purposes of this experiment, codes 18 (Assault) and 19 (Fight) were considered the same top-level event code ("Violence"), since humans could not reliably distinguish between the two. The Violence codes were chosen due to their importance to predictive models in social science. The Disapprove code was chosen due to its frequent confusion with Violence (e.g. "*The president attacked Congress*"). The Provide Aid code was chosen as a representative cooperative code to balance the other three non-cooperative codes.

3 System Descriptions

3.1 *Tabari*

TABARI, the pioneering and widely-used automatic event coder in the computational social science community, generates event data from text using a computational method called "sparse parsing"; the system uses pattern recognition and grammatical parsing to identify events using a large inventory of human-generated rules, e.g. *(X) FIRE* MISSILE* AT (Y)*. The system does not use advanced natural language processing techniques but relies on shallow surface structure of sentences to detect events. A full description of the TABARI system can be found in [1].

3.2 BBN SERIF

BBN SERIF (Statistical Entity and Relation Information Finder) is a state-of-the-art, trainable, language-independent understanding system that produces rich linguistic analyses of natural language. The goal of BBN SERIF is to robustly handle widely varying inputs, ranging from formal text (e.g. newswire) to informal speech (e.g. meeting transcripts), and everything in between (e.g. blogs, newsgroups, and broadcast news). The system produces

- Propositional representations of text that capture "*who did what to whom*",
- Entity classification,
- Reference resolution (to whom does "*he*" refer?),
- The detection and characterization of relations between entities, and
- Event detection according to a pre-specified ontology.

All of these components are based on trained statistical models that have learned correct behavior from example data rather than being powered by manually-generated rules, which are often less accurate as well as more difficult to generate and maintain (requiring the use of linguistic experts to hand-craft rules, rather than simply the use of native speakers to highlight examples in text). BBN SERIF has been extensively applied and repeatedly tested in a wide variety of projects and government-sponsored evaluations, including Automatic Content Extraction[1] (ACE) for information extraction, Global Autonomous Language Exploitation (GALE) for query-response distillation and the *who, what, when, where, and why* of distillation [6], and AQUAINT for question answering [7]. BBN SERIF is also being used under an ongoing ONR-sponsored project that detects bias and sentiment with respect to topics of interest.

As the focus of this study is on the improvements in event coding enabled by BBN SERIF, we describe here in detail its primary operations. Significantly more technical detail on specific statistical models employed can be found in [3].

BBN SERIF's event coder consists of a series of natural language analysis components that operate on both the sentence level and the document level. These components are shown in Fig. 1, below.

As shown in the diagram, BBN SERIF first processes an input document's metadata (e.g. the document date and any text zoning information) and then breaks the document into sentences for processing. A set of components then produce

[1] http://www.nist.gov/speech/tests/ace/

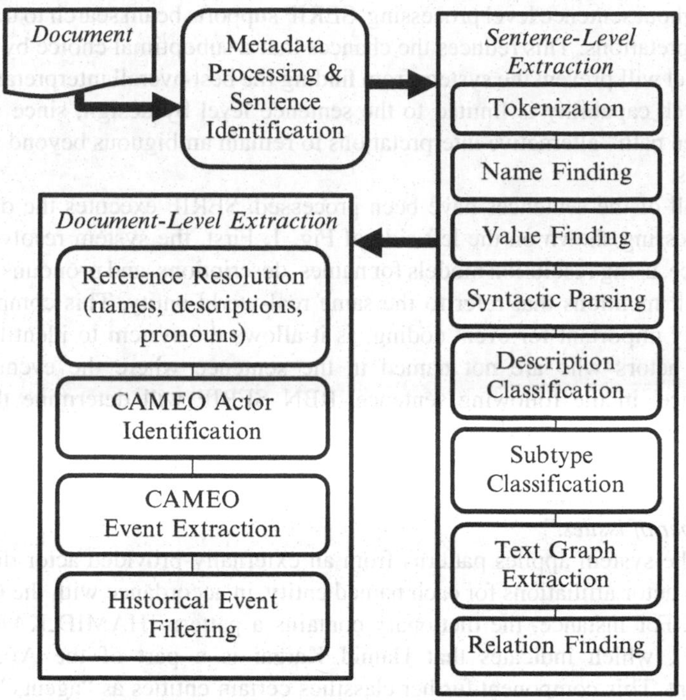

Fig. 1 BBN SERIF system diagram (as used for event coding)

linguistic analyses for each sentence, and then a separate set of document-level components are applied to the document as a whole, using the sentence-level analyses as inputs.

The sentence-level processing components are shown on the right in Fig. 1. Each sentence is first tokenized using a heuristic component that tries to match the tokenization used in downstream components (for English, we use the Penn Treebank tokenization). The tokenized string is then fed to components that use trained models to perform name finding, value finding, and syntactic parsing. ("Values" are abstract concepts such as dates, percentages, and monetary amounts.) Relative dates are resolved with respect to the document's publication date, e.g. by resolving "*Sunday*" to August 8, 2009 or "*last year*" to 2008. The resulting parse tree is then processed by the description classification model to identify nominal mentions of entities, e.g. "the president" or "the missile". "Structured mentions" like appositives and lists are also identified at this stage. A separate component predicts subtype values for each entity mention. The text graph analysis component translates the parse trees into a semantically normalized "propositional" form, which among other things resolves trace information and regularizes the treatment of passive verbs. Finally, another trained model detects and classifies relationships between entities, e.g. *works-for*(A,B) or *located-in*(A,B).

Throughout sentence level processing, SERIF supports beam search to track multiple interpretations. This reduces the chances that a suboptimal choice by an early stage model will prevent the system from finding the best overall interpretation. The beam search capability is limited to the sentence level by design, since it is rare for "garden path" alternative interpretations to remain ambiguous beyond sentence boundaries.

Once all of the sentences have been processed, SERIF executes the document level processing shown on the left side of Fig. 1. First, the system resolves entity coreference, using resolution models for names, descriptions, and pronouns to group together all mentions that refer to the same real-world entity. This component is particularly important for event coding, as it allows the system to identify events involving actors who are not named in the sentence where the event occurs. For instance, in the following sentence, BBN SERIF will determine that "*he*" refers to Ahmadinejad and correctly extract the (*Condemn, Iranian government, United States*) triple stated in the second sentence: "*Iranian President Mahmoud Ahmadinejad called nuclear weapons disgusting. However, he condemned the U.S. on a number of issues.*"

Next, the system applies patterns from an externally-provided actor dictionary to identify actor affiliations for each named entity, in accordance with the CAMEO codebook. For instance, the dictionary contains a pattern [HAMID_KARZAI → AFGGOV], which indicates that Hamid Karzai is a part of the Afghanistan government. This component further classifies certain entities as "agents" of other named actors. For instance, "Iraqi rebels" will be classified as IRQREB. TABARI only classifies agents in cases where the agent word (e.g. "rebels"), is immediately preceded or followed by a known named actor, or when it is followed by the word "in" or "of" and then a known named actor. In contrast, BBN SERIF relies on the linguistic analysis it has already generated to extend the range of its agent matching to include agents detected via *works-for* relations (e.g. "*Jim is employed by Microsoft*") or other syntactic constructions not allowed by TABARI (e.g. "*Jim is a representative for Microsoft*").

As a part of CAMEO actor identification, BBN SERIF propagates all detected actor labels to coreferent entity mentions, as shown above in the example with Mahmoud Ahmadinejad. This obviously significantly improves system actor coverage. It also has the side benefit of increasing system actor accuracy, because it allows the system to be more conservative with named actor patterns. For example, the actor dictionary contains the pattern [HASHIMOTO → JPNGOV]. Although Riyutaro Hashimoto was the prime minister of Japan at one point, there are many other persons named Hashimoto in the world, so this pattern is likely too generic. However, TABARI includes this pattern because it is likely that a sentence might only refer to this important person by only his last name, and without this pattern, any event in that sentence would be missed. In contrast, BBN SERIF can use a more specific pattern like [RIYUTARO_HASHIMOTO → JPNGOV] to label the first instance of Hashimoto mentioned in a document (full names are almost always used at least once in news articles), and then propagate the label JPNGOV to other, abbreviated references to Hashimoto in the same document, using its automatically-generated coreference links.

After actor identification has been performed, BBN SERIF identifies and labels the actual CAMEO events. This component relies heavily on BBN SERIF's generation of "text graphs", which represent the propositional representations of document content. These graphs directly represent who did what to whom, for instance:

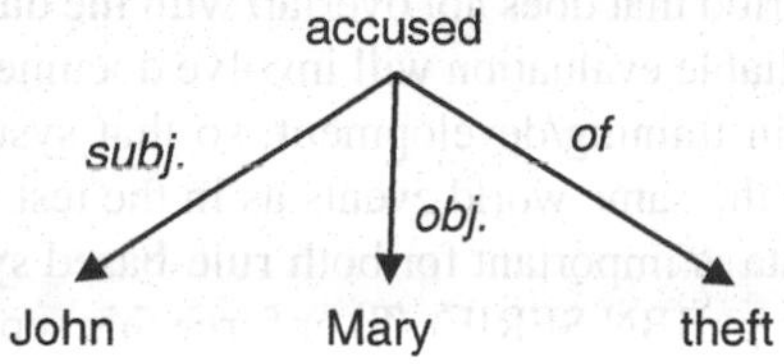

These representations of text are useful for event coding because they are both flexible and accurate. Their flexibility allows a single text graph to represent a wide variety of surface-level phrasings of an event. For instance, the same text graph could represent the core event in "*John accused Mary*", "*John, a friend of Sheila, accused Mary*", or "*Mary was accused by John*". This reduces the set of patterns needed to match various events; one text graph pattern can provide the same coverage as several surface-level patterns.

In addition, a text graph will not match sentences with similar words but a different meaning. For instance, a text graph that matches "*John accused Mary*" will not match the following sentences: "*Bob, a friend of John, accused Mary*", "*Mary was accused by critics of copying John*", or "*John accused someone other than Mary*".

For this study, text graph event patterns were generated semi-automatically from TABARI output on training data: text graphs were automatically extracted from TABARI output and were then pruned and modified by a human developer. In future work, the patterns could be generated by the same process from example data marked by humans, with less or no need for human intervention due to the lower level of noise in the input; the TABARI engine was used as a stand-in for noisy human annotation in this study.

Finally, BBN SERIF identifies and tags "historical" events (defined here as events that occurred more than a month prior to story publication). Historical events can provide misleading information to downstream models and are thus important for the system to identify. As previously described, BBN SERIF already resolves all relative dates with respect to the document's publication date. In this component, BBN SERIF identifies dates connected to extracted events via text graphs, as well as other historical indicators (e.g. "*the anniversary of…*") that modify extracted events. Events that are deemed likely historical are identified and marked for eventual removal, for instance: "*In 1839, the British invaded Afghanistan on the basis of dubious intelligence*" or "*The talks came two months after nine Indians were killed in a suicide attack in Kabul that officials blamed on LeT*".

4 Experiment Design

4.1 Evaluation Corpus

The evaluation corpus consisted of English newswire data collected via LexisNexis. It is standard practice in natural language processing communities that test material be drawn from a time period that does not overlap with the data used to train or tune the system. The most reliable evaluation will involve documents from a newer time period than those used in training/development, so that systems have not already seen (nor been tuned to) the same world events as in the test set. This separation of test and development data is important for both rule-based systems (e.g. TABARI) and statistical systems (e.g. BBN SERIF). To enforce this separation, the evaluation corpus was selected from the most recent data available, namely the 2009–2010 time period.[2] All BBN system development and analysis was performed on documents dated no later than December 31, 2008. TABARI dictionaries were improved by developers looking at events from data through March 2010.

4.2 Evaluation Procedure

For each top-level event-code, we randomly selected 500 triples from each system for evaluation. We then presented these triples in random order to the evaluators (triples produced by both systems were only presented once).

Event triples were selected for evaluation from the first five sentences of each document. This decision was made to ensure the most accurate evaluation: it is important for an evaluator to be able to see and understand the preceding context of any event triple, particularly when actors are referenced using pronouns (e.g. "*he criticized*") or other potentially ambiguous descriptors (e.g. "*the soldiers were attacked*"). Restricting evaluation to the first five sentences of each document ensured that evaluators could easily see and process all necessary context and thereby make accurate assessments about the correctness of system actor labeling.

Each event triple was displayed along with the sentence from which it was derived. The display also included up to four sentences immediately preceding the target sentence, if they existed in the document.

An event triple was considered correct if the sentence it originated in did in fact describe an event of the same top-level type, involving that Source actor and that Target actor. Future, hypothetical, and negated events were not considered correct: the event must actually have been reported as having occurred.

[2]The evaluation corpus included approximately 250,000 documents. Documents judged to be a near duplicate via BBN's semantic de-duplication filter were removed before evaluation.

The evaluator was asked to ignore the event subcode and merely judge the correctness of the top-level event code (e.g. 07). So, if the system said the event was 071 (Provide economic aid) and the correct subcode was actually 073 (Provide humanitarian aid), this would still be acceptable, because the top-level event code was correct. Error rates on subcode detection were separately evaluated and appeared similar across the two systems, but because humans disagreed about 20% of the time on subcode correctness, we looked only at top-level code correctness for our overall evaluation.

Similarly, some flexibility was allowed for actor codes. The evaluator was asked to consider an actor code correct if it was either a "too general" or "too specific" version of the fully correct actor code. An example of a "too general" actor code would be "IRQ" (Iraq) for "IRQMIL" (the Iraqi military). An example of a "too specific" actor code would be "IRQMIL" for "IRQ". However, an actor code could not be considered correct if it referred to a different "branch" of a code than the one suggested by the system. For instance, if the correct code was IRQMIL (the Iraqi military), and the system proposed IRQBUS (an Iraqi business), this is actually contradictory and should have been marked as incorrect.

All triples were seen by at least one evaluator. Most triples were seen by two evaluators, to ensure more stable results.[3] When evaluators disagreed on correctness (i.e. one said "yes" and the other said "no"), the system was given half-credit for that triple.

5 Evaluation Results

5.1 Overview

We evaluated accuracy (precision) directly, based on the number of the 500 triples that were judged to be correct. Table 1 shows these results.

We evaluated coverage, or recall, indirectly, by estimating the number of correct event triples in the corpus from the total number produced multiplied by the estimated precision. Table 2 shows the estimated number of correct triples produced for each top-level event code:

Table 1 Accuracy of extracted events

	Disapprove	Provide Aid	Violence	ALL
TABARI	39%	17%[4]	29%	28%
BBN SERIF	68%	69%	75%	70%

[3]Specifically, the percentages of triples seen by two annotators were 100% for Violence, 100% for Provide Aid, and 69% for Disapprove.

[4]The CAMEO guidelines explicitly prohibit future tense instances of Provide Aid from being marked as correct. This was a significant problem for TABARI; without these errors its performance on Provide Aid would be more comparable to its performance on the other codes.

Table 2 Density of extracted events

	Disapprove	Provide Aid	Violence	ALL
TABARI	1,020	170	620	1,810
BBN SERIF	1,240	270	1,800	3,310

As seen in these tables, the BBN SERIF system appears to both improve accuracy and increase coverage.

5.2 *Comparison to Previous Studies*

Previous studies have also estimated the accuracy of automated event coders in various ways. It is worth briefly addressing the differences in the metrics involved in these studies. In [8], the authors estimate that an automated event coder (in their case Virtual Research Associates Inc. Reader) could place an event in a correct top-level category with 55% accuracy, which rivaled human coders' accuracy at the same task.[5] This number is significantly higher than the accuracy number cited above for TABARI, but it does not require the coder to correctly extract (or code) the Source or Target of the event, so the numbers are not comparable. In fact, using this metric, our evaluation would have assigned TABARI an average accuracy of 75% and BBN SERIF an average accuracy of 93%. However, these numbers should also not be considered comparable to those cited by King and Lowe, since the data set, ontology, and even the task definition differ significantly; we give them as only an approximate frame of reference to demonstrate the important difference between the two metrics.

In addition, we note that the primary goal of this study is to assess the suitability of statistical natural language processing techniques for social science event coding tasks. For this purpose, BBN SERIF is a suitable representative of these techniques, as it has been separately judged to represent the state-of-the-art in event extraction (on different ontologies) in government-sponsored evaluations such as the Automatic Content Extraction (ACE) evaluation. Many other systems do

[5] King and Lowe report a suite of numbers for accuracy; this number assumes a constant weighting across event categories. In addition, King and Lowe report an 85% accuracy number for a very different metric: the probability of a correct event or non-event judgment on a given sentence. We did not compute this number. Given the high percentage of non-event sentences in our data, it would be meaninglessly high—the trivial baseline, where a system never returns an event, would achieve 96% accuracy on our data set. In contrast, the King and Lowe test set is specifically constructed to contain mostly sentences that have a valid event of some kind (and their raw data pool is also more event-heavy than ours), so that number has a very different meaning in their context.

exist which perform event extraction for other ontologies or, more generally, "triple" extraction. Some of these systems rely on manually-generated rules of various sorts, and some also employ statistically trained models. Some encode their output using Semantic Web technology, often in RDF (Resource Description Framework) triple stores; this data representation is agnostic as to the methods of the actual extraction of triples from text. Unfortunately it is difficult to project performance from one event extraction task to another, as seen in the discussion in the previous paragraph. We therefore focus on the specific comparison between TABARI and BBN SERIF as two exemplars of significantly contrasting techniques for event extraction.

5.3 *Error Analysis*

For the purposes of analysis, we identified four primary categories of system error: absolute false alarms (no event of the indicated type exists in the sentence); incorrect actors; actor role reversal; and future, hypothetical, or negated events reported as factual.

In every category, BBN SERIF showed significant reduction in error when compared with TABARI.

The most significant area of error reduction came in the reduction of **absolute false alarms**: cases where no event of the indicated type actually existed in the sentence. Here BBN SERIF reduced TABARI's false alarms from an estimated 1460 to an estimated 320. The constraints placed on the pattern recognition process by text graphs prohibit many spurious matches by requiring the Source and Target actor to be syntactically related to the event verb or noun. For instance, TABARI finds a violent event in the sentence "*Some 150 Japanese soldiers battling piracy are stationed in a US base in Djibouti*", with the United States as the Target. However, since "*US*" is not connected in a meaningful way to "*battling*", BBN SERIF does not find this as a violent event. This 80% reduction in complete noise will provide a significantly more accurate event set to downstream predictive models.

An additional source of error reduction came in sentences where an event of the appropriate type was present in the sentence, but **the actors selected by the system were incorrect**. For instance, the system reported that the perpetrator of a suicide bombing was a U.N. official rather than an Afghani guerrilla. This was the largest category of errors observed for both systems, accounting for an estimated 1740 incorrect TABARI triples and 700 incorrect BBN SERIF triples. Surface-level patterns (as in TABARI) are more vulnerable to this type of error because they have to rely on sentence positioning (rather than logical structure) to hypothesize actors for events. For instance, TABARI finds an event indicating the United States' disapproval of Mexico in the following sentence: "*President Obama, appearing on Wednesday with Felipe Calderon, the president of Mexico, denounced Arizona's new law on illegal immigration.*"

Text graphs help eliminate that guesswork by specifying logical or syntactic roles for both Source and Target actors. Still, text graph patterns can generate errors of this type, as in the following sentence, where BBN SERIF mistakenly believes that the Target of the following Disapprove event is "*Chinese navy*" rather than "*Japan*": "*The Chinese ambassador expressed strong displeasure with the recent monitoring of a Chinese navy fleet by Japan*". This is triggered by the text graph pattern shown below:

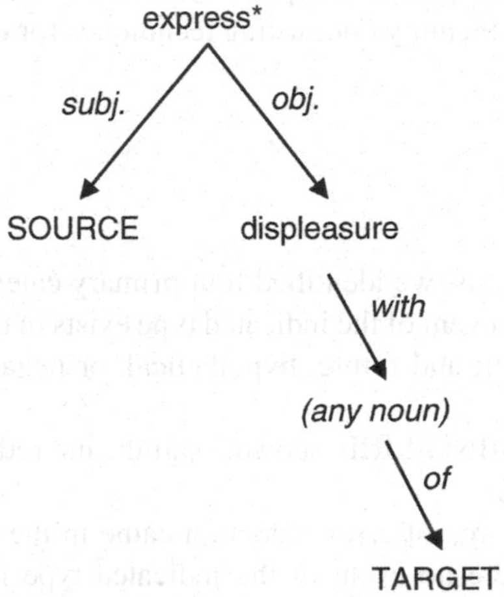

This text graph allows any noun to fill the role played by "monitoring" in this sentence. It could be imagined to fire correctly on a similar but meaningfully different sentence, e.g. "*The Chinese ambassador expressed strong displeasure with the recent movements of the Japanese fleet*".

Another type of error reduction came in the situation where the event type was correct, but **the roles of the actors were reversed**. So, rather than reporting that China criticized the United States, the system reported that the United States criticized China. This category accounted for an estimated 320 TABARI errors and 70 BBN SERIF errors. Again, the constraints of the text graphs (particularly their normalization of predicate-argument structure which accounts for passive verb constructions) allow for better actor role assignment.

Finally, a fourth source of error reduction came in situations where the event triple was correct except for the fact that **the event was future tense, hypothetical, or false**; this accounted for an estimated 240 TABARI errors and 100 BBN SERIF errors. BBN SERIF avoids many of these errors by using "counter-factual" text graph patterns to identify and propagate modality to appropriate events (verbs and nouns). These patterns are not specific to any particular event code but can be applied effectively to any task of this nature as one of BBN SERIF's pre-existing capabilities, allowing BBN SERIF to easily discard false, hypothetical, and future-tense events for any event code.

5.4 System Overlap

We also estimated the overlap in system output by looking at the distribution of triples judged correct by evaluators. We estimate that 21% of correct BBN SERIF triples were also found by TABARI, and 38% of correct TABARI triples were also found by BBN. Stated differently, we estimate that TABARI found approximately 1,100 triples not found by BBN SERIF, and BBN SERIF found 2,600 triples not found by TABARI. The relatively low system overlap is indicative of the very different approaches taken by the two systems.

One focus of continued work on BBN SERIF would be the expansion of the text graph pattern set. This expansion could be done by extracting text graphs from a wider set of training data than was used for this pilot experiment. For instance, TABARI finds a Provide Aid triple in the following sentence: "*Chinese government Thursday added 30 million yuan to the snowstorm disaster relief fund for north China's Inner Mongolia Autonomous Region.*" This triple is triggered by a high-recall/low-precision TABARI pattern that involves simply the word "fund". The version of BBN SERIF used for this experiment does not have a text graph pattern that matches this sentence, but if it had seen this example in training data, it could have automatically generated a simple text graph pattern that would cover this instance and others like it, with relatively low risk of false alarms:

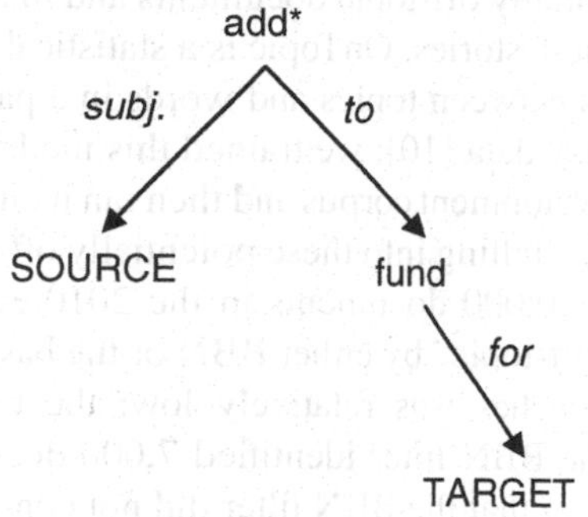

5.5 Historical Events

News articles often discuss the historical context of current events. For instance, an article discussing the controversy over building a mosque in New York City will likely mention the 9/11 attacks. However, this should *not* give a forecasting model evidence of current terrorist attacks in New York City. To avoid this, historical events must be filtered out of the event data provided to models.

The TABARI system does not natively remove historical events. To add this capability, BBN implemented a historical filtering algorithm based on temporal resolution and syntactic parsing (described earlier in this chapter).

We judged system performance on historical event removal as a stand-alone task. Before filtering, 12.3% of the 1,038 correct triples produced by the system were historical. After filtering, only 1.8% were historical. (Of the 910 triples that were *not* historical, only 39 (4.2%) were filtered out incorrectly.)

In summary, BBN's historical filter significantly reduced the number of historical triples provided to downstream models, with only a slight loss in the number of correct non-historical triples.

5.6 Topic Filtering

Avoiding spurious off-topic event triples is crucial for automatic event coding. Both systems used a type of "document filtering" approach to improve performance on this task. The baseline system used for the TABARI event coding used a keyword filter developed by Lockheed Martin to remove documents likely to be about sports, weather, or financial topics. Sports documents in particular are important to remove, lest the system report an act of war in a sentence like "*Japan and Korea clashed during the qualifying World Cup match earlier today*". TABARI did not code any events in documents that were red-flagged by this keyword filter.

BBN SERIF improved significantly on this filter by using BBN's OnTopic™ classifier to identify potentially off-topic documents and to restrict (but not entirely quash) triple output for these stories. OnTopic is a statistical classifier that automatically learns relationships between topics and words in a passage [9] and has been shown to be robust to noisy data [10]; we trained this model using the output from the keyword filter on a development corpus and then ran it on the separate evaluation corpus to detect documents falling into these potentially off-topic categories.

Of the approximately 90,000 documents in the 2010 evaluation corpus, about 15,000 were marked as "off-topic" by either BBN or the baseline filter. The overlap between these two approaches was relatively low: the baseline filter identified 11,000 documents and the BBN filter identified 7,000 documents, but only 3,000 were shared. (Note however that the BBN filter did not consider finance stories off-topic, as the data indicated that events found in these documents were frequently still correct, e.g. the political violence event in "*Police fired water cannons at opposition party activists protesting in New Delhi against the oil price hike, which economists warned would fuel rising inflation.*")

We compared end-to-end system performance on these 15,000 documents. The first system used the baseline keyword filter and the TABARI event engine. The second system used a combination of BBN OnTopic and BBN SERIF to extract events, extracting only high-confidence events in documents considered to be potentially off-topic.

Using simply the keyword filter to remove off-topic documents, the baseline/TABARI system produced 126 incorrect event triples and no correct event triples. In contrast, using the statistical topic classifier combined with the less black-and-white approach to document filtering, the BBN system produced 18

Table 3 Comparison of system performance on different corpora

	TABARI		BBN SERIF	
	Lexis-Nexis	OSC	Lexis-Nexis	OSC
Disapprove	40%	34%	75%	64%
Provide Aid	17%	21%	69%	71%
Violence	28%	21%	66%	67%
TOTAL	31%	28%	69%	66%

incorrect event triples and 13 correct triples. It seems clear that the BBN system identifies and discards many more misleading documents than the baseline keyword filter. (Virtually all incorrect events produced by TABARI were sporting events mistaken for violence.) The BBN system also "recovered" 13 correct triples that were presumably incorrectly discarded by a keyword filter in the baseline system. For instance, the keyword filter discarded a document that contained the following disapproval event, because the document also mentioned the upcoming World Cup: "*Tension between the two Koreas in the wake of the Cheonan's sinking, an act the South blames on the North ...*". The statistical OnTopic engine recognized that although the document mentioned some words relating to sports, the document itself was not fundamentally *about* sports.

5.7 Adapting to New Corpora

A question that often comes up in the context of operational deployment is whether systems will still perform well if the same techniques are applied to new corpora made up of different types of data. Therefore, to test the flexibility of our approach, we performed a secondary evaluation on a corpus made up of data from the Open Source Center. Some of these documents were similar to the newswire available in the LexisNexis corpus, but others were quite different (e.g. radio transcripts). Both systems found approximately 75% as many events in OSC sources as they did in LexisNexis sources. Without human coding of the data, it is impossible to know the reason for this difference: it could be a difference in system performance or simply a difference in the natural event density of the new data source. However, differences in accuracy can be evaluated directly and we did so using the same human evaluators to judge performance on a randomly selected subset of event triples for both systems. Accuracy results were as follows (Table 3).

Both systems degraded only minimally over this new corpus, showing a 15% relative degradation for Disapprove and a slight improvement for Provide Aid. The only significant difference between systems was that the BBN SERIF system sustained its performance on the new corpus for the violence event codes, while the TABARI system suffered a 25% relative degradation in performance. We believe this points further to the robustness of statistical natural language processing when facing problems of domain shift.

6 Conclusion

The BBN SERIF event coding system was designed to build on the "best of both worlds": the knowledge and expertise encoded in the TABARI dictionaries and the most recent statistical learning algorithms for natural language processing. These algorithms perform tasks such as probabilistically assigning a syntactic role for every word and phrase in a sentence, determining who did what to whom in every clause, predicting the real-world entity to which pronouns and other definite descriptions refer, and disambiguating between confusable senses of important event words, such as a verbal *attack* and a military *attack*. Though such algorithms are machine intensive, without any optimization, the BBN SERIF event coding prototype already processes approximately 300,000 sentences an hour on a standard dual quad-core machine. Significant further optimization could be expected as part of a deployed system.

Given that the system capitalizes on the knowledge encoded in TABARI lexicons and sophisticated state-of-the-art statistical learning algorithms, it should not be surprising that the combination substantially improves both in accuracy of output (from 28% to 70%) and also discovers more events (estimated at an 83% increase). The capabilities supplied by BBN SERIF could also improve the scalability and portability of event coding to new genres, new domains, or even new languages, drastically shortening the time needed to generate a dictionary for a new type of event or different type of news source. Initial evidence also suggests that they could significantly automate the currently time-consuming process of actor dictionary maintenance, or actor dictionary creation for a new region.

In addition, the SERIF system

- Significantly reduced reporting of historical events as if current
- Effectively filtered out irrelevant stories, leading to fewer false alarms, and
- Performed robustly on genres outside of news.

The improved precision and coverage of event coding output demonstrated here could significantly help build trust in the output of automated event coding. This has become increasingly important as consumers of the output of forecasting models want to be supplied with evidence convincing them of their validity—to be able to drill down into the details that might support a particular forecast or prediction. In addition, we hope that improving the quality and coverage of the automated event coding data will also improve the quality of statistical models that consume it; we hope to have the opportunity to investigate this in the future.

References

1. Schrodt P (2001) Automated coding of international event data using sparse parsing techniques. Paper presented at the International Studies Association, Chicago
2. O'Brien S (2010) Crisis early warning and decision support: contemporary approaches and thoughts on future research. Int Stud Rev 12:87–104. doi: 10.1111/j.1468-2486.2009.00914.x

3. Ramshaw L, Boschee E, Freedman M, MacBride J, Weischedel R, Zamanian A (2011) SERIF language processing—effective trainable language understanding. Handbook of natural language processing and machine translation: DARPA Global Autonomous Language Exploitation. Springer, New York
4. Gerner D, Schrodt P, Yilmaz O, Abu-Jabr R (2002) Conflict and Mediation Event Observations (CAMEO): a new event data framework for the analysis of foreign policy interactions. Paper presented at the International Studies Association, New Orleans, and American Political Science Association, Boston
5. Schrodt P, Yilmaz O, Gerner D, Hermreck D (2008) The CAMEO (Conflict and Mediation Event Observations) actor coding framework. Paper presented at the International Studies Association, San Francisco
6. Olive J, Christianson C, McCary J (2011) Handbook of natural language proccssing and machine translation: DARPA Global Autonomous Language Exploitation. Springer, New York
7. Maybury M (2004) New directions in question answering. AAAI Press/The MIT Press, Menlo Park
8. King G, Lowe W (2003) An automated information extraction tool for international conflict data with performance as good as human coders: a rare events evaluation design. Int Organ 57:617–642
9. Schwartz R, Imai T, Kubala F, Nguyen L, and Makhoul J (1997) A maximum likelihood model for topic classification of broadcast news. Proceedings of Eurospeech, Greece
10. Prasad R, Natarajan P, Subramanian K, Saleem S, Schwartz R (2007) Finding structure in noisy text: topic classification and unsupervised clustering. Paper presented at IJCAI-2007 Workshop on Analytics for Noisy Unstructured Text Data, Hyderabad, India

Automated Coding of Decision Support Variables

Massimiliano Albanese, Marat Fayzullin, Jana Shakarian, and V.S. Subrahmanian

1 Introduction

With the enormous amount of textual information now available online, there is an increasing demand – especially in the national security community – for tools capable of automatically extracting certain types of information from massive amounts of raw data. In the last several years, ad-hoc Information Extraction (IE) systems have been developed to help address this need [6]. However, there are applications where the types of questions that need to be answered are far more complex than those that traditional IE systems can handle, and require to integrate information from several sources. For instance, political scientists need to monitor political organizations and conflicts, while defense and security analysts need to monitor terrorist groups. Typically, political scientists and analysts define a long list of variables – referred to as "codebook" – that they want to monitor over time for a number of groups. Currently, in most such efforts, the task of finding the right value for each variable – denoted as "coding" – is performed manually by human coders, and is extremely time consuming. Thus, the need for automation is enormous.

In the effort presented in this paper, we leverage our previous work in IE [1] and define a framework for coding terror related variables[1] automatically and in real-time from massive amounts of data. The major contribution of our work consists

[1] Although in this paper we focus on terrorism related variables, the proposed framework is general and can be adapted to many other scenarios.

M. Albanese (✉)
George Mason University, Fairfax, VA 22030, USA
e-mail: malbanes@gmu.edu

M. Fayzullin • J. Shakarian • V.S. Subrahmanian
University of Maryland, College Park, MD 20742, USA
e-mail: marat@umiacs.umd.edu; jana@umiacs.umd.edu; vs@umiacs.umd.edu

V.S. Subrahmanian (ed.), *Handbook of Computational Approaches to Counterterrorism*,
DOI 10.1007/978-1-4614-5311-6_4,

in defining a logic layer to reason about and integrate fine grained information extracted by an IE module. We implemented a prototype of the proposed framework, and preliminary experiments have shown that our approach is promising, both in terms of accuracy and in terms of significantly reducing human intervention in the process. Additionally, we have show that, separating the reasoning component of the process from the low-level processing enables to quickly deploy new variables.

In addition to developing the computational framework, leveraging the interdisciplinary nature of our research group and the experience gained through the collaboration with groups of political scientists, we have developed our own codebook for Computational Monitoring of Terror Groups (CMOT). Codebooks designed by political scientists are primarily conceived for human coders and the definition of many variables leaves enough room to subjective interpretation of facts. In order to avoid this issue, we designed our codebook with the idea of making it an objective recording of evidence and facts, prone to be automated within a computational framework. CMOT is an actor-centered approach to collecting basic aspects of asymmetric conflict anywhere in the world. These aspects include organizational structure, activity profile and social profile of non-state armed groups (NSAGs), as well as a wide range of environmental variables regarding the social, economic, and political situation of the host country. CMOT attempts to cover detailed yet universal ground, and provide analysts with a comprehensive picture of the strategic situation a rational actor faces. To accommodate inaccuracies in the open source media which we are working with, the codebook has been structured into different levels of detail – allowing us to collect general to highly specific facts.

The remainder of the paper is organized as follows. Section 2 discusses related work. Section 3 introduces the Automatic Coding Engine and provides details of its components, whereas Sect. 4 reports preliminary experimental results. Finally, concluding remarks are given in Sect. 5.

2 Related Work

The aim of Information Extraction (IE) is the extraction and structuring of data from unstructured and semi-structured electronic documents, such as news articles from online newspapers [4]. Information Extraction involves a variety of issues and tasks ranging from text segmentation to named entity recognition and anaphora resolution, from ontology-based representations to data integration. There is a large body of work in the IE community addressing single issues among those mentioned, and a variety of approaches and techniques have been proposed. With respect to the issue of named entity recognition, some authors propose knowledge-based approaches [3] while others favor the use of statistical models such as Hidden Markov Models [8]. Amitay et al. [2] introduces Web-a-Where, a system for locating mentions of places and determining the place each name refers to. Several attempts have been made to build a comprehensive IE framework. Unfortunately, most IE tools are domain dependent, as they rely on domain specific knowledge or features, such as page layouts, in order to extract information from text, or fail to answer the type of questions that are typical of the scenarios addressed in this paper. Soderland [12] presents an approach for extracting information from web pages based on a pre-processing stage that takes into account a set of domain-dependent and domain-independent layout features. Similarly [7] focuses on the extraction from web tables, but attempts to be domain independent by taking into account the two-dimensional visual model used by web browsers to display the information on the screen. Other efforts are aimed at developing IE capabilities for specific knowledge domains, such as molecular biology [9], and thus use domain specific knowledge to achieve their goals. More general approaches rely on automatic ontology-based annotation [6].

In conclusion, our approach differs significantly from previous approaches because it relies on domain independent information extraction to perform the extraction task, and adds a logic layer to reason about extracted facts and formulate answers to complex questions. This novel combination of information extraction and logic reasoning has proved to be a key factor in scaling up our approach to coding and enabling quick deployment of new variables.

3 Automatic Coding Engine

In this section, we present the Automatic Coding Engine (ACE), our framework for automated coding of decision support variables. Figure 1 shows the overall architecture of the system, while details of the three major components are provided in the following sections. Documents are either fed into the system by a crawler – which continuously scans online news sources – or imported from any available corpus.[2] The *Document Preprocessor*, using the library of *Linguistic Resources* – which includes dictionaries defined by domain experts and a classification of English verbs [10] – identifies documents, paragraphs, or individual sentences that are good candidates for extraction of relevant information. Candidate text fragments are parsed using the Link Grammar Parser, and *Low-level Linguistic Sensors* extract atomic facts, such as location of events, perpetrators of violent events, types of weapons used.

Finally, the *Rule Engine*, using a library of logic rules, combines such atomic facts with a-priori knowledge encoded in the *Knowledge Base*, and derives group-specific values – also referred to as "codes" – for each variable of interest, at the desired level of temporal granularity. Automatic codes generated by the system are stored in a database and made available to human reviewers through an interface that serves the double purpose of providing a tool to review automatic codes and a mechanism to manually code variables that have not been automated yet. In fact, the user interface was designed to ensure a gradual and smooth transition

[2]A corpus of documents from LexisNexis was used in our experiments.

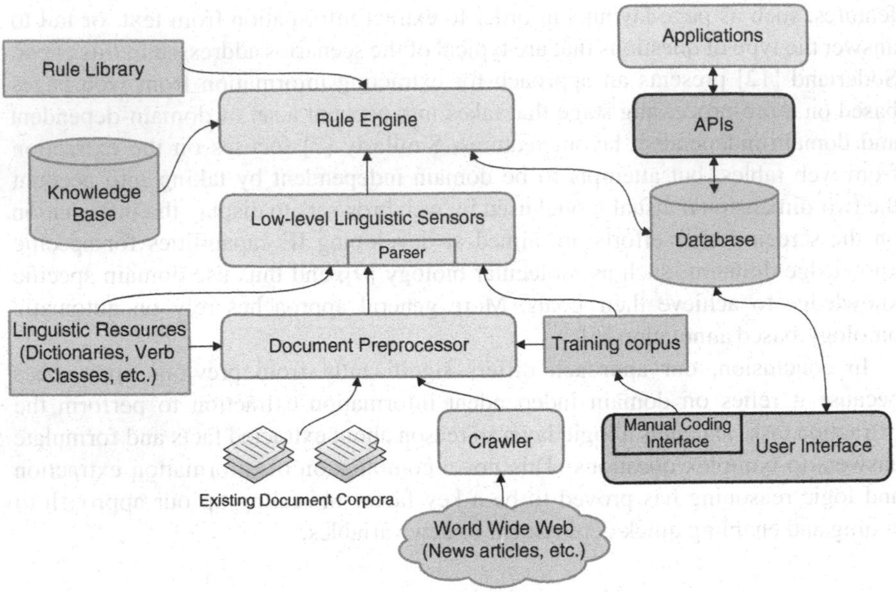

Fig. 1 System architecture

from a fully manual process to a semi-automatic one, where human intervention is limited to validation of automatically generated codes. In Sect. 4.2, we will show that our framework reduces the demand of coder's time by one order of magnitude. Additionally, our framework offers APIs to make data available to other applications.

In the following, we first formalize the coding problem and then describe each major component of the system. Let $\mathscr{V}$ denote the set of variables to be automated, which we will refer to as the *codebook*. Each variable $V \in \mathscr{V}$ has an associated domain $dom(V)$ of possible values that can be assigned to it. Variables can be boolean (i.e., $dom(V) = \{0, 1\}$), enumerative, or numeric. We now give the fundamental definition of *coding function*.

Definition 1. Given a variable $V \in \mathscr{V}$, a set of organizations $\mathscr{O}$, and a set of time intervals $\mathscr{I}$, a *coding function* for V is a function f_V which associates each pair $(I, O) \in \mathscr{I} \times \mathscr{O}$ with a value $v \in dom(V)$.

$$f_V : \mathscr{I} \times \mathscr{O} \rightarrow dom(V) \tag{1}$$

In other words, f_V assigns a value to V for each organization and time interval. In the following, we will denote the function corresponding to the manual coding process as f_V^h, and the one corresponding to the proposed automated process as f_V^a. Our objective is to develop a framework that allows to design coding functions f_V^a for a possibly large number of variables, with average precision and recall figures – as defined in Sect. 4.1 – around or above 70 %. Although we will often refer to individual coding functions, our system is designed to allow concurrent automation and monitoring of multiple variables, a key aspect for achieving scalability of our approach to automation.

3.1 Preprocessing

The first step of the processing pipeline involves selection of text fragments that are good candidates for extraction of atomic facts. The rationale for this stage is that the following steps, namely parsing and information extraction, are computationally intensive, while only a small fraction of the available textual data will actually contain evidence useful for coding. Given a variable V, a time interval I and an organization O, a text fragment T (sentence, paragraph, or entire document) might provide evidence for coding $f_V^a(I, O)$ only if the following necessary conditions are satisfied: (i) T mentions O or any person known to be affiliated with O[3]; (ii) T reports events that occurred during I or facts valid in I; (iii) textual clues for V are found in T. A textual clue for V is any expression (noun phrase or verb) which indicates that the text may possibly contain information useful for coding V. The three conditions above are necessary but not sufficient conditions to code $f_V^a(I, O)$. The information extraction task swill determine whether T enables coding of $f_V^a(I, O)$. In the example of Table 2, the phrase "AK-47 guns" in the text fragment corresponding to 2001 is a strong indicator that the text snippet may support coding of `Equ_F_G_Assault_Rifle`, a binary variable that codes whether a group reportedly utilizes assault rifles or its military arsenal allegedly includes assault rifles of some model and caliber in the period being coded. However, at this stage we cannot yet conclude that $f^a_{\texttt{Equ\_F\_G\_Assault\_Rifle}}$(2001, Lord's Resistance Army) $=$ 1, as the semantic role of relevant entities has not been analyzed. Additionally, intuition also tells us that the closer O and a textual clue for V occur within the text, the higher the likelihood that Linguistic Sensors will extract from T atomic facts useful to code V for organization O. Thus, we pair each occurrence c_V of a textual clue for V, with the closest occurrence o of O and compute the score $sc(o, c_V) = e^{\alpha \cdot d_p(o, c_V)} \cdot e^{\beta \cdot d_s(o, c_V)}$, where $d_p(o, c_V)$ and $d_s(o, c_V)$ are the distances in terms of number of paragraphs and number of sentences respectively between o and c_V. We then consider the top-k co-occurrences, find the text fragments that minimally contain them, and eventually expand the text fragments to include contiguous sentences in order to provide more context for extraction. We remind the reader that this stage does not perform any actual extraction, but rather selects candidate text fragments, thus enabling to scale automatic coding w.r.t. massive amount or raw text documents. An extremely sensitive aspect of automatic coding is identifying the correct time interval in which events occurred or facts are valid. To address this issue, we have developed an algorithm to analyze indirect temporal references like "last year" or "three weeks ago", and infer actual dates.

---

[3] Affiliations can be automatically extracted by Linguistic Sensors.

3.2 Linguistic Sensors

In this stage, text fragments selected by the Document Preprocessor, are first parsed using the Link Grammar Parser [11]. Then Named Entity Recognition and Pronoun Resolution are performed using GATE/ANNIE [5]. The Link Grammar Parser is a syntactic parser of English, based on link grammar, an original theory of English syntax. Given a sentence, the system assigns to it a syntactic structure, which consists of a set of labeled links connecting pairs of words. The parser also produces a constituent representation of a sentence. As in T-REX [1], we then use a library of extraction rules to match constituent trees of previously unseen sentences against a library of templates, and extract pieces of information, in the form of RDF[4] triples, from matching sentences. However, in our modular architecture, the IE component could be replaced by any suitable general purpose IE system. The novelty of our approach consists in combining information extraction with a logic layer to reason about extracted data.

An extraction rule – referred to as *Linguistic Sensor* in this paper – is of type $Head \leftarrow Body$, where the body represents a set of conditions – constraints on subtrees of the constituent tree and on their relative position – and the head represents the set of RDF statements that can be inferred from the sentence if the conditions in the body of the rule are satisfied. A tree matching algorithm is used to match rules against previously unseen sentences, and RDF statements are extracted from matching sentences.

Example 1. Figure 2 shows an example of a Linguistic Sensor aimed at identifying possession of weapons in scenarios where the equipment is surrendered to somebody. Solid edges indicate that there must be a direct link between two nodes, while dotted edges indicate that there must be a path (of any length) between two nodes. The rectangular boxes represent end nodes, i.e., the actual text that will be extracted in case of a match. The sensor in the figure requires that the subject of the sentence, which can be an arbitrarily nested noun phrase, contains a noun phrase identifiable as a named entity (either a person or an organization). It also requires that the main verb belongs to the class `CHANGE_POS_VERBS` of verbs denoting change of possession [10]. Finally, it requires that the verb is followed by a noun phrase – arbitrarily nested within another noun or prepositional phrase – denoting a weapon. If a sentence satisfies such constraints, then two RDF triples can be extracted: (`tdata : EquipPossX`, `trexe : owner`, `Var1`) and (`tdata : EquipPossX`, `trexe : equipment`, `Var3`), where X is a unique identifier assigned to each instance of the `trexe : EquipmentPossession` event. Any other information in the sentence is not relevant to this specific sensor and is ignored.

[4]RDF (Resource Description Framework) is a web standard defined by the World Wide Web Consortium [13], originally created for encoding metadata, but now used for encoding information about and relationships between entities in the real world.

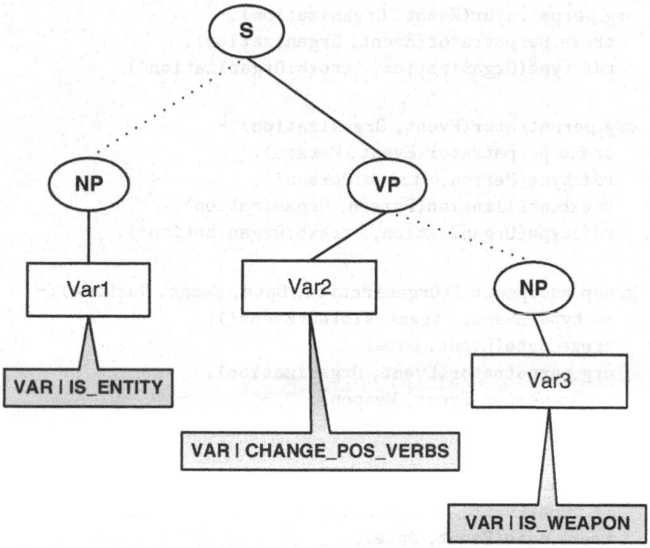

Fig. 2 An example of Linguistic Sensor

3.3 Logic Layer

The last stage in the processing pipeline is the *Rule Engine*, which takes as input the output of the Linguistic Sensors and any a-priori knowledge encoded in the *Knowledge Base* and generates group-specific values for each variable of interest, at the desired level of temporal granularity. The Knowledge Base includes any information that may be available ahead of coding, such as countries in which organizations operate, or names of major known leaders. This information may be provided by the same domain experts who will use the system or automatically extracted from text documents. Each RDF statement extracted by Linguistic Sensors or encoded in the Knowledge Based is converted to an equivalent ground atom and fed into the Rule Engine, a logic engine based on Prolog. The Rule Engine uses rules from the Rule Library, a small fragment of which is shown in Fig. 3. The library of rules includes (i) rules to code individual variables; (ii) rules to code entire classes of variables (e.g. last two rules in Fig. 3); (iii) auxiliary rules to derive intermediate relations (e.g. fist two rules in Fig. 3).

Once a library $\mathscr{L}$ of re-usable linguistic sensors has been developed for extracting a wide range of atomic facts, automation of additional variables only implies writing new logic rules. Thus, the marginal cost of automating additional variables in our framework is negligible w.r.t. the cost of writing ad-hoc algorithms

```
org_perpetrator(Event, Organization):-
  trexe_perpetrator(Event, Organization),
  rdf_type(Organization, 'trexb:Organization').

org_perpetrator(Event, Organization):-
  trexe_perpetrator(Event, Person),
  rdf_type(Person, 'trexb:Person'),
  trexb_affiliation(Person, Organization),
  rdf_type(Organization, 'trexb:Organization').

group_equipment_1(Organization, Date, Event, VarName):-
  is_type(Event, 'trexe:ViolentEvent'),
  trexe_date(Event, Date),
  org_perpetrator(Event, Organization),
  trexe_weapon(Event, Weapon),
  triggers(Weapon, VarName).

group_equipment_1(Organization, Date, Event, VarName):-
  is_type(Event, 'trexe:EquipmentPossession'),
  trexe_date(Event, Date),
  trexe_owner(Event, Organization),
  rdf_type(Organization, 'trexb:Organization'),
  trexe_equipment(Event, Equipment),
  triggers(Equipment, VarName).
```

Fig. 3 A small excerpt of the Rule Library

to automate the same variables. Formally, the average cost to automate a variable in ACE is

$$C = \frac{\sum_{i=1}^{|\mathscr{L}|} C_l(L_i) + \sum_{i=1}^{|\mathscr{V}|} C_r(V_i)}{|\mathscr{V}|} \tag{2}$$

where $C_l(L_i)$ is the cost of developing sensor L_i, and $C_r(V_i)$ is the cost of writing logic rules for variable V_i. Instead, the average cost of developing ad-hoc algorithms is

$$C' = \frac{\sum_{i=1}^{|\mathscr{V}|} C_a(V_i)}{|\mathscr{V}|} \tag{3}$$

where $C_a(V_i)$ is the cost of developing an ad-hoc algorithm for variable V_i. In ACE, atomic facts extracted by linguistic sensors can be re-used in multiple logic rules (i.e., $|\mathscr{L}| \ll |\mathscr{V}|$), and the cost of writing logic rules for variable V_i is clearly much lower than the cost of developing ad-hoc algorithms for V_i (i.e., $C_r(V_i) \ll C_a(V_i)$). Therefore, if $|\mathscr{V}|$ is very large, $C \ll C'$. In other words, our approach scales very well for very large sets of variables. Table 1 reports the number of variables coded each month, and shows how our effort scaled after an initial setup time.

Additionally, dependencies exist between some of the variables in the CMOT codebook. We identified such dependencies and organized subsets of the codebook in a hierarchical fashion. Similarly, we organized textual clues for such variables

Table 1 Progression of automation effort

	Nov. '09	Dec. '09	Jan. '10	Feb. '10	Mar. '10
#. of vars coded	10	30	55	60	55

into dictionaries and analogous hierarchies of concepts. We then mapped such hierarchies of concepts to the hierarchies of variables and encoded all these data structures in the Knowledge Base. Finally, we designed generalized logic rules corresponding to the root node of each hierarchy of variables: such rules (see last two rules in Fig. 3) have the name of the variable to be coded as one of the arguments, thus allowing the logic engine to reason about a whole family of variables. This powerful mechanism allows us to scale up our approach to automated coding even further, as we do not necessarily need to design ad-hoc rules for each individual variable, but rather for a family of variables. Equation 2 then becomes

$$C = \frac{\sum_{i=1}^{|\mathscr{L}|} C_l(L_i) + \sum_{i=1}^{|\mathscr{P}|} C_r(P_i)}{|\mathscr{V}|} \tag{4}$$

where $\mathscr{P}$ is a partition of $\mathscr{V}$ and $C_r(P_i)$ is the cost of writing logic rules to automate variables in $P_i \subseteq \mathscr{V}$.

4 Implementation and Experiments

We implemented a prototype of ACE on top of T-REX [1]. Linguistic resources, logic rules, and additional resources have been developed to enable automation of about 170 variables. We then conducted a number of preliminary experiments to validate our approach. We measured recall and precision with respect to the ground truth provided by human coders (Sect. 4.1). We also tracked the amount of time required to human coders to review and validate the codes generated by the system, compared to the time required to manually code the same variables starting from raw documents (Sect. 4.2).

Table 2 shows a sample of ACE output for variable `Equ_F_G_Assault_Rifle`, coded for the Lord's Resistance Army in Uganda (only cases where the variable was coded as 1 are shown). The first column in the table indicates the time frame to which each code applies (yearly granularity was used in this case). The second column shows fragments of text (possibly from multiple documents) that provided evidence for coding. Finally, the third column is the ground truth generated by human coders.

Table 2 Output for `Equ_F_G_Assault_Rifle`, coded for Lord's Resistance Army

Year	Supporting evidence	GT
1999	Text of report by Ugandan radio on 23rd April; all place names in northern Uganda Over 1,000 AK-47 rifles, nine anti-tank guns with 300 bombs and about 250 anti-tank and anti-personnel mines have been recovered from the Lord's Resistance Army LRA by UPDF Uganda People's Defence Forces in the border areas of northern Uganda over the last 1 year	1
2000	He said that the UPDF recently recovered an assortment of weapons and ammunition from the Lord's Resistance Army LRA rebels, which included 31 AK-47 rifles and 18 guns	1
2001	Four rebels of the Lord's Resistance Army (LRA) under commander Kwo-yelo surrendered to the UPDF Uganda People's Defence Force in Gulu northern Uganda on Sunday evening 18 November with two AK-47 guns with 60 bullets	1
2003	Robert was 16 when guerrillas of the Lord's Resistance Army [LRA] carrying axes, machetes and assault rifles slipped into his village and took him away during a chaotic night attack	0

4.1 Precision and Recall

We conducted experiments on 30 CMOT variables and measured precision and recall of our system. For each variable $V \in \mathscr{V}$, precision and recall can be defined as

$$P_{V,v} = \frac{|RS_{V,v} \cap GT_{V,v}|}{RS_{V,v}}, \forall v \in dom(V) \tag{5}$$

$$R_{V,v} = \frac{|RS_{V,v} \cap GT_{V,v}|}{GT_{V,v}}, \forall v \in dom(V) \tag{6}$$

where $RS_{V,v} = \{(I,O) \in \mathscr{I} \times \mathscr{O} | f_V^a(I,O) = v\}, \forall v \in dom(V)$, and $GT_{V,v} = \{(I,O) \in \mathscr{I} \times \mathscr{O} | f_V^h(I,O) = v\}, \forall v \in dom(V)$, are the result set and the ground truth for $V = v$, i.e., the sets of (I, O) pairs that f_V^a and f_V^h respectively associates with value v. The overall accuracy of f_V^a, i.e., the fraction of (I, O) pairs that human coders and our system associate with the same value, can be computed as

$$A = \frac{|\{(I,O) \in \mathscr{I} \times \mathscr{O} | f_V^a(I,O) = f_V^h(I,O)\}|}{|\mathscr{I} \times \mathscr{O}|} \tag{7}$$

We run our experiments on a corpus of 80,000 LexisNexis documents. We first measured the overall accuracy of our system, obtaining an average accuracy of 90 %. We then observed that, in the ground truth of most binary variables, the cases in which such variables are coded as 1 are quite sparse, therefore one could obtain artificially high accuracy by simply setting every variable to 0. We then focused on the most critical measures $P_{V,1}$ and $R_{V,1}$. On the same set of documents, the system returned 800 (I, O) pairs such that $f_V^a(I, O) = 1$, pertaining to 12 distinct terrorist organizations, between 1990 and 2009. Average precision and recall were found to be 82 and 77 % respectively. It is worth noting that human coders tend to use any previous knowledge they might have about the subject. This means that the actual recall could have been even higher than what we measured, if coders were to base their decisions solely on the information contained in the document corpus. Additionally, we observed that in 35 % of the missed detections, the system was able to correctly code the variable that is the immediate generalization of the one missed. As an example, in several cases the system failed to infer that a group used or possessed AK assault rifles in a certain time frame, but correctly inferred that it was using some sort of assault rifle. In conclusion, our system shows consistently high precision, and recall.

4.2 *Time*

In our experiments, we also evaluated the impact of our framework on the time required to coders for completing the coding task. Since the CMOT coding effort was started, we observed that human coders take between 5 and 8 h to manually code 100 data points, at an average rate of 12–20 codes per hour, and the most time-consuming task is reading or skimming through hundreds of articles in search of relevant information. The large variability in coder's time is due to a number of factors, including the level of expertise of each coder and the ease of finding reliable information for the assigned coding tasks. Instead, a human coder can review automatically generated codes at an average rate of 140 codes per hour, one order of magnitude faster than the fully manual process. As we mentioned earlier, for each code the user is provided with fragments of text that the algorithm used as evidence for coding, along with the list of documents those fragments belong to. If such text fragments do not provide enough context to judge the correctness of a code, reviewers can examine the original documents. We experimentally observed that the text provided as part of the output was sufficient to validate an automatic code in 89 % of the cases, thus reducing the amount of documents to be examined by human coders by a factor of 10. This latest experimental observation confirms that the most expensive task in the manual coding process is reading articles. Reducing the amount of text to be examined by a certain factor, reduces the total time to perform the process by roughly the same factor.

5 Conclusions and Future Work

In this paper, we presented a framework for automatic monitoring of decision support variables and focused on the case of terror related variables. The proposed approach leverages our previous work in Information Extraction and complements it with a number of new features, including a logic layer, in order to address the specific challenges posed by the *coding* task. Although this is still work in progress, we have shown preliminary results, which prove that our approach is promising, and can save analysts an incredible amount of time. Additionally, we showed that our approach to automation scales very well for very large sets of variables. For the future, we plan to complete automation of the CMOT codebook by December 2010, conduct massive experiments and tune the system, with the objective of obtaining recall precision figures around or above 70% for every variable. Future plans also include integrating the system with other systems in the lab to analyze trends and make predictions of future behavior of monitored groups.

References

1. Albanese M, Subrahmanian VS (2007) T-REX: a system for automated cultural information extraction. In: Proceedings of the first international conference on computational cultural dynamics (ICCCD '07). AAAI, Menlo Park, pp 2–8
2. Amitay E, Har'El N, Sivan R, Soffer A (2004) Web-a-where: geotagging web content. In: Proceedings of the 27th annual international ACM SIGIR conference on research and development in information retrieval. ACM, New York, pp 273–280
3. Callan J, Mitamura T (2002) Knowledge-based extraction of named entities. In: Proceedings of the 4th international conference on information and knowledge management. ACM, New York
4. Cowie J, Lehnert W (1996) Information extraction. Commun ACM 39(1):80–91
5. Cunningham H, Maynard D, Bontcheva K, Tablan V GATE: a framework and graphical development environment for robust nlp tools and applications. In: Proceedings of the 40th annual meeting of the Association for Computational Linguistics (2002)
6. Ding Y, Embley DW (2006) Using data-extraction ontologies to foster automating semantic annotation. In: Proceedings of the 22nd international conference on data engineering workshops (ICDEW'06). IEEE Computer Society, Washington, DC, p 138
7. Gatterbauer W, Bohunsky P, Herzog M, Kroepl B, Pollak B (2007) Towards domain-independent information extraction from web tables. In: Proceedings of the 16th international world wide web conference. ACM, New York, pp 71–80
8. GuoDong Z, Jian S (2003) Integrating various features in hidden markov model using constraint relaxation algorithm for recognition of named entities without gazetteers. In: Proceedings of the international conference on natural language processing and knowledge engineering. IEEE Press, pp 465–470.
9. Jensen LJ, Saric J, Bork P (2006) Literature mining for the biologist: from information retrieval to biological discovery. Nat Rev Genet 7(2):119–129
10. Levin B (1993) English verb classes and alternations: a preliminary investigation. University of Chicago Press, Chicago
11. Sleator DD, Temperley D (1993) Parsing english with a link grammar. In: Proceedings of the third international workshop on parsing technologies (IWPT '93). University of Tilburg, The Netherlands
12. Soderland S (1997) Learning to extract text-based information from the world wide web. In: Proceedings of the 3rd international conference on knowledge discovery and data mining. AAAI Press, pp 251–254
13. World Wide Web Consortium (W3C) (2004) Resource description framework (RDF). http://www.w3.org/RDF/

Part II
Behavioral Models and Forecasting

Qualitative Analysis & Computational Techniques for the Counter-Terror Analyst

Aaron Mannes

1 Introduction

Qualitative analysis, with its focus on developing in-depth understanding of specific phenomenon is a useful arrow in the quiver of the counter-terror analyst. Further, advances in computer science have enabled a number of tools that can increase the speed and effectiveness of qualitative counter-terror research.

The plan for this chapter is first to discuss the needs of the counter-terror analyst, followed by a discussion of qualitative research, including a comparison with quantitative research. This is followed by several specific examples of how different modes of qualitative research have been applied to counter-terror analysis, along with cases where the research has been augmented by computational methods. Among the modes of analysis discussed are the strategic perspective, organizational theory, and questions about why and how individuals join and leave terrorist organizations. The analyses cited are not intended to systematically represent the field, but rather serve as a guide to some of the possibilities for counter-terror analysis presented by combining qualitative analysis with computational technology.

1.1 Counter-Terror Research Needs

Counter-terrorism is, essentially, an applied field. Researching terrorism to understand the phenomenon is a useful endeavor that can provide important insights into terrorist behavior. But counter-terror analysis is primarily about preventing terrorist organizations from carrying out their operations. Ideally, this would involve preventing or deterring attacks. But doing so successfully requires extensive detailed

---

A. Mannes (✉)
Laboratory for Computational Cultural Dynamics, University of Maryland,
College Park, MD 20742, USA
e-mail: amannes@umd.edu

V.S. Subrahmanian (ed.), *Handbook of Computational Approaches to Counterterrorism*,
DOI 10.1007/978-1-4614-5311-6_5,

intelligence about the terrorist group. Acquiring this intelligence is a challenge because terrorist groups are clandestine and take great measures to guard their communications. Open societies offer a very large number of potential targets, so that unless intelligence is highly specific, it is impossible to guard all of the possible targets. While certain classes of targets may be hardened in response to past attacks, terrorist groups quickly adapt by identifying new targets. An alternate approach is to view terrorist groups as organizations and take steps that foster dissent and dissatisfaction within a terrorist group [1].

David Kilcullen, an Australian officer who advised the United States government on counter-insurgency in Iraq and Afghanistan, offers a related approach. Kilcullen applies complex systems analysis to developing a strategy against al-Qaeda as a trans-national movement. One crucial component to this model is that these movements thrive on the inputs of energy in the form of new recruits, grievances, weapons, and doctrine, which the organization can then use to fuel new violence and recruit more members and obtain further resources [2]. Denying terrorist organizations these inputs or inserting flawed inputs (double-agents or ideological heresy) that drain the organization's energies can be an effective counter-terror strategy and the examples discussed below focus on that form of counter-terror operation. The particular focus of the examples discussed is on understanding and exploiting organizational vulnerabilities as well as preventing individuals from joining terrorist groups or encouraging individuals to leave. It should be emphasized that this is by no means an exhaustive list of the possible uses of qualitative research for counter-terror analysis. Other possible areas of inquiry include efforts to evaluate the effectiveness of counter-terror policies as well as systems that attempt to predict terror attacks.

1.2 Qualitative Research Overview

The root of qualitative is the Latin *qual*, which means "of what sort." Qualitative research focuses on the specific, distinctive characteristics of a thing. In the social sciences, qualitative research entails in-depth interviews and analysis of historical materials in order to gain a deep understanding of a single or small set of phenomena. As such, qualitative methods, including ethnography, ethnology, unstructured interviewing, and case studies possess:

> Éunrivaled strengths for the elucidation of meanings, the in-depth description of cases, the discovery of new hypotheses, and the description of how treatment interventions are implemented or of possible causal explanations [3].

For counter-terror analysts, the strengths of qualitative analysis, the ability to understand the details of how terrorist organizations function, are essential for identifying workable counter-terror policy options.

Table 1 Contrasts between quantitative and qualitative research [4, 5]

Attributes of quantitative and qualitative research	
Quantitative research	Qualitative research
Deduction: using data to test theories and hypothesis	Induction: developing generalizations and theories from the data
Theory testing	Theory generating
Numerical measurements of specific aspects of phenomenon	Data includes words, pictures, and objects
Measurements and analyses easily replicable by other researchers	Focus is on achieving complete, detailed description of a small number of examples

1.2.1 Contrasting Qualitative and Quantitative Research

Over the past several decades social scientists have debated the merits of qualitative and quantitative research. Some quantitative researchers have caricatured qualitative research as "soft." While improperly conducted research can effectively be little more than reporting, many methodologists note that the dichotomy between the two forms is more a matter of style and that both forms of research fundamentally rely on the logic of inference [4]. Properly conducted, qualitative research can help understand a phenomenon. The following discussion contrasting quantitative and qualitative research is not an effort to resolve the debate or prove the superiority of one form or another, but rather to better understand the essentials of qualitative research and the assets it brings to counter-terror research through a comparison with quantitative analysis techniques (Table 1).

1.2.2 Qualitative vs. Quantitative Research in the Context of Counter-Terrorism

There are many reasons why qualitative methods are essential for understanding and countering terrorism. The emphasis in the descriptions of qualitative analysis cited above on how treatments are implemented and function is particularly relevant because the study of terrorism is an applied field—that is terrorism is studied so that efforts to reduce or eliminate it can be more effective. The focus on specific qualities is important for several linked reasons. First, the *n* of terrorist organizations is relatively small, thus limiting the efficacy of quantitative approaches. Secondly, terrorist groups are unique in the causes and expression of animus. Tolstoy in *Anna Karenina* observed, "Happy families are all alike, every unhappy family is unhappy in its own way." This is not merely a metaphor, in understanding terrorist group behavior and developing counter-strategies, specifics are essential.

These two critical limitations of quantitative methodology and the relative power of qualitative approaches for understanding terrorist group behavior is demonstrated in a study by Seth Jones and Martin Libicki of RAND Corporation on how terrorist groups end. The study included quantitative analysis of 648 terrorist groups active between 1968 and 2006, with a focus on the 268 that came to an end. While these numbers were sufficient to provide statistically significant findings, comparisons between specific types of groups—for example similarly sized terrorist groups with comparable goals yields a much smaller *n*. Even seemingly similar terrorist groups may have vast differences that limits the applicability of comparisons. For example Hezbollah in Lebanon and Lashkar-e-Taiba (LeT) in Pakistan are both large Islamist organizations that have been innovative in using technology, receive state support, and provide education and health care to the public. However, one group is Shia and the other is Sunni. Further, Hezbollah is the dominant political organization among a major sectarian community in a small country. LeT, in contrast, does not dominate its sectarian community in Pakistan and this sectarian community is a small minority within Pakistan. In developing counter-terror strategies these differences could be significant.

Some of the key findings about how terrorist groups end include:

- Most groups that end do so because of police action or because the groups are induced to join the political process.
- Only about 7 % come to an end because of military force.
- There was a strong correlation between the size of terrorist groups and their longevity.
- Police action was particularly effective against smaller groups.

While this kind of information provides useful general guidance to policy-makers, it is of limited practical utility. Knowing that police action can be an effective tool against terrorists provides limited insight into how to actually use police forces to defang a terrorist threat [6].

None of this is meant as a criticism of Jones and Libicki's study. In the mode of the best research, Jones and Libicki combined their quantitative analysis with a series of case studies on how specific terrorist groups end. In describing their use of case studies they cite Alexander George (a leading scholar on case study methodology) and Timothy McKeowan who write about the need to understand:

> What stimuli the actors attend to; the decision process that makes use of these stimuli to arrive at decisions; the actual behavior that then occurs; the effect of various institutional arrangements on attention, processing, and behavior; and the effect of other variables of interest on attention, processing, and behavior [7].

This is an apt statement describing the utility of qualitative approaches for the counter-terror analyst.

1.3 Understanding Terrorist Group Behavior

1.3.1 Employing the Strategic Perspective

One form of qualitative analysis useful for understanding terrorist group behavior is the Strategic Perspective (SP), a mode of analysis that examines the domestic and international coalitions leaders require to hold power and how those coalitions shape their decision-making. There are many modes of analysis that expand on the strict rational actor nation-state model of international affairs that include individuals, sub-national and trans-national organizations, and ideas. This mode of analysis is also applicable to terrorist organizations and other non-state organizations. Even the smallest terrorist groups have internal dynamics that leaders must take into account. Larger terrorist groups such as Hezbollah, FARC, or LeT have substantial internal organizational structures and connections with other international actors. In theory, quantitative methods can be applied to this type of analysis, but, as discussed above, the loss of nuance threatens the utility of the analysis [8].

The RAND monograph cited above includes a case study of the Salvadoran insurgent and terrorist group the FMLN. The FMLN's disarmament in the early 1990s is a case study of a violent non-state armed group choosing to disarm and enter the political process. The key finding is that political solutions to terrorism are heavily reliant on the scale of the organization's goals. Organizations with maximalist goals such as re-making society or the international order (such as al-Qaeda) offer little space for governments to compromise. The FMLN's goals were more limited, focusing on land reform, establishing a democratic political system, and an end to repression. The situation also illustrates the potential of the Strategic Perspective. The FMLN had lost its Soviet patron with the end of the Cold War. The United States was threatening to cut aid to the government of El Salvador if the government did not change its behavior, while also offering incentives to encourage the government to engage in a peace process. Both sides also had internal constituencies that were not motivated to enter the peace process. The Salvadoran military needed to be purged and reformed, while the estates of large landholders needed to be redistributed. The FMLN in turn, had to accept El Salvador as possessing a free enterprise system. While neither side's conditions could be completely satisfied, international pressures (and inducements) effectively shored up the regime of the Salvadoran president so that he could risk the loss of support from other members of his coalition [6]. This example highlights how applying the strategic perspective to understand the needs of the leaders both of the terrorist group and the government dealing with it can help policy-makers develop approaches that influence the leaders of the adversaries to bring a long-running conflict to a close.

A similar analysis of the IRA's decision to abandon armed conflict in favor of participating in elections uses classic case study methodology to test four hypotheses:

1. Participating in the electoral process allows terrorist groups to address their grievances through means other then violence.
2. Participating in the electoral system will lead to cooperation between the elites of the various parties, which will ultimately foster tolerance.
3. Participating in the political process forces terrorist groups to rely less on violence because of its effect on public opinion.
4. Participating in the political process opens the terrorist organization's ranks to non-radicals and thus shifts the organization's priorities.

In studying the targets and quantity of IRA violence as well as the organization's internal dynamics, the study found little support for the first and last hypothesis and only modest support for the second and third hypothesis. Instead it found that an important factor in the IRA's decision to end its long armed struggle was the actions of the British government. The author cites both the government's skilled handling of its engagement with the IRA, but also the successful counter-terror strategy of infiltrating IRA ranks, which effectively removed armed struggle as an effective option. In effect, it was the adroit combination of carrots and sticks—reducing the IRA's military options while providing a credible entry into the constitutional system that gave the advantage in internal debates to the political wing. The author is clear that the circumstances were unique and that this study does not establish a new theory for on electoral participation and terrorism. Nonetheless, it provides a model of analysis that may prove useful to other terrorist conflicts where the non-state-armed group is also participating in an electoral process [9]. This example also illustrates the strategic perspective and how the government's actions can shape the internal dynamics of a terrorist group.

However, these papers—and many like it—were written in hindsight. The great challenge is to provide such analysis in real-time so that it can serve the needs of policy-makers. Computational models can help develop these perspectives in real time by processing large quantities of data systematically and more rapidly than human analysts.

One way computational models can augment this type of analysis is by identifying factors in organizational behavior that might be overlooked by analysts overwhelmed with data. The number of variables shaping an organization's behavior is potentially enormous. Given the huge number of possible factors, analysts, understandably, may discount the importance of the relevance of key variables that relate to a group's behavior. Computational models, however, can include every variable incorporated into the model. The Stochastic Opponent Modeling Agent (SOMA) developed at the University of Maryland's Laboratory for Computational Cultural Dynamics, is one such system. SOMA identifies conditions under which a group will take a given action with a very high probability, based on historical data about a group's behavior. A human analyst can draw similar correlations with a few variables. But when dozens, or hundreds of variables are being tracked there are thousands of possible combinations—far more than a human analyst can realistically track.

The University of Maryland's Minorities at Risk Organizational Behavior (MAROB) database is such a project, systematically collecting data and coding dozens of variables about organizations claiming to represent minority groups across the Middle East and North Africa for several decades. Several of these groups, including Hamas, Hezbollah, and the PKK are major terrorist organizations [10]. When the MAROB data for the Lebanese terrorist group Hezbollah was run through SOMA the most striking finding was that the likelihood of Hezbollah engaging in rocket attacks on Israel decreased dramatically during Lebanese election years [11]. The data on which the model was based was collected from Hezbollah's founding in 1982 through 2004, but the prediction has held since [12].

Besides providing a timely prediction, this finding identified the high level of importance Hezbollah places on its participation in the Lebanese political system. This is a classic illustration of the strategic perspective. While Hezbollah propagates anti-Israeli rhetoric, these findings indicate that Hezbollah's leadership values and benefits from its formal role in the Lebanese political system. The leadership finds the advantages it gains through the Lebanese political process essential to maintaining the support of key constituencies and is important enough to shape the organization's strategic calculations. In shedding light into Hezbollah's priorities and interests, the SOMA findings create potential policy options. Policy-makers may be able to influence Hezbollah's strategic choices by reaching out to communities in Lebanon that would be pleased or unhappy with certain Hezbollah actions. This finding highlights how a computational method can enhance the perspective of a counter-terrorism analyst using the strategic perspective. SOMA can be applied to the study of any terrorist organization for which data has been collected and has also been used to generate findings about the behaviors of Hamas and Lashkar-e-Taiba [13, 14].

1.3.2 Attacking Organizational Weakness

Organizational theory, understanding terrorist groups as organizations with internal procedures, constituencies, and a unique culture is another mode of analysis useful for counter-terror specialists. In a paper written for West Point's Combatting Terrorism Center several analysts examine captured documents from al-Qaida to identify organizational vulnerabilities. Historically, organizing terrorism faces the complex task of balancing the priorities of security, efficiency, and control. Terrorist organizations are clandestine groups that face constraints on their communications. More communications (of any form whether it is electronic or in the form of personal meetings) give greater opportunities for security forces to detect the terrorist group's activities. This means that organization's leaders must delegate tasks. The limited ability of leaders to oversee operations means that agents can diverge from the priorities of their principals. This theory applies to many types of organizations. Efficiency, which refers to the use of the organization's resources, particularly money, is one area in which terrorist groups are vulnerable. The agents who raise money may divert funds or become involved in criminal activities, which then became their primary focus. Other agents, without proper oversight, might misuse funds. Control refers to modulating violence. Terrorist groups seek political impact through violence, but too much violence or violence against the wrong targets can have a negative effect on the group's goals. However, the individuals most committed to violence are likely to be the agents who actually carry out these operations. Thus, they may show initiative in pursuing violence even if the leadership finds this counter-productive. Further, the cells that commit violence live in difficult, clandestine conditions that may lead them to become divorced from reality. Finally, groups compete for prominence, so that violence by one group may drive other groups to also pursue violence. Efforts to mitigate these control problems, such as regular audits of expenditures or punishing transgressors, all exact costs on the organization in the forms of increased communications (which lowers the organization's security) and diverts of the organization's resources from its primary mission [15].

The authors found that these problems were not unique to al-Qaeda but cite an extensive literature going back to Russian revolutionaries in the late 19th century and including the IRA, the PLO, and ETA. The authors then study captured al-Qaida documents, which finds that al-Qaida has also suffered from these difficulties. Examining these documents and seeking al-Qaida's organizational vulnerabilities allows the authors to propose a series of policies that could expand the principal-agent divergence within al-Qaeda and ultimately lower the organization's effectiveness. Among these suggestions were to allow less effective al-Qaeda members to continue to operate, use information operations to expose internal criticisms, not publicize asset freezes so that the individual sanctioned is then held accountable by the group's leadership, interfere with communications between leaders and rank and file, and exploit ideological breaks among jihadis [16].

This paper was written in 2006, but based on the documents captured from al-Qaeda leader Osama Bin Laden's safe house, there is substantial evidence that Bin Laden's ability to exercise control over al-Qaeda had been substantially degraded. Because of security concerns, replies to Bin Laden's messages took months to arrive. Bin Laden regularly asked regional commanders to send him detailed reports of operations, criticized them for linking their operations to local causes rather than al-Qaeda's priorities, and worried that many al-Qaeda operations had damaged the cause by taking innocent lives [17].

As the authors pointed out these theories do not exclusively apply to al-Qaeda. A 2002 paper on the FARC in Colombia discusses how that organization overcame these inherent principal-agent divergence difficulties to become a threat to the state despite the end of the Cold War and the loss of its Soviet patron. The FARC had long been characterized by flexibility and innovation at the strategic, tactical, and operational levels. FARC's engagement in the drug trade ensured that the organization had substantial resources at its disposal to remain equipped and invest in new technology. Colombian geography was also an asset to FARC security. The massive, thinly populated countryside gave FARC operatives ample space for maneuver. However, under the principal-agent divergence theory these assets could have led to factionalization (the huge Colombian country-side demanded that area commanders operate with little oversight) and corruption from participating in the drug trade. The article posits that the FARC leadership remained in control by investing heavily in communications equipment and because the top leaders were very highly regarded throughout the organization and that their reputations limited divergence [18].

A decade later, FARC is a much weaker organization. The Colombian government (with substantial US aid) re-armed and engaged in a classic counter-insurgency that denied FARC some of its operating space as well as encouraging FARC defections. However, the Colombian government specifically targeted the two strengths noted in the article cited above. There were numerous efforts to penetrate FARC's communication system. One FARC commander reported, when she surrendered in 2008, that she had not been in communication with the organization's top leadership for two years [19]. The dramatic rescue of several high-profile FARC hostages, including former Colombian Presidential candidate Ingrid Betancourt, required a successful infiltration of the FARC communications network. Part of the rescue operation involved Colombian intelligence giving instructions to a FARC commander who believed he was receiving instructions from a top FARC leader [20]. Equally importantly, FARC's leadership has been devastated and the top tier of highly regarded leaders have either been killed by the Colombian military or died of natural causes [21, 22]. While the FARC continues to be a threat, it is suffering increasing casualties at the hands of government forces and very high levels of desertion [23].

1.3.3 Applications of Communications Theory

Another approach to exacerbating organizational disorder is applying communications theory concept of noise to counter-terror. Initially the concept of noise applied to technical issues that interfered with communication. But this understanding has expanded to include:

- Semantic noise (due to ambiguities inherent in language),
- Psychological noise (when the receiver's psychological state produces a unpredictable distraction from receiving the intended message), and
- Cultural noise (when the recipient is from a different psychological background from the author and understands the message differently than it was intended.)

Terrorism can be understood as a form of political communication, in which *adding* noise in order to blunt and distort the terrorist's intended message could be a valuable counter-terror strategy. There are two forms of noise that can be introduced. Technological noise interferes with the transmission of the signal. There are a wide range of internet vulnerabilities that could be exploited against terrorist online activities, not only to prevent access but also to monitor traffic and manipulate content and activities. Psychological and social noises come under the rubric of information war or psychological operations (PSYOP). Using a combination of these techniques, noise can be used to undermine the credibility of terrorist messages both to followers, operatives, and potential recruits [24].

Computational tools can dramatically enhance counter-terror strategies based on the theories described above. In developing strategies against the FARC and al-Qaeda, the rapid exploitation of captured documents was essential. In the age of electronic communications, even relatively small organizations can quickly develop enormous files of reports and communiques. However, interfering effectively with terrorist organization communications requires an in-depth knowledge of the organization's communications. However, acquiring that information in a timely fashion is a challenge. Discussing their project monitoring terrorist websites, forums, and chatrooms, Weinmann and von Knop note:

> This monitoring can be done by human analysts and coders (i.e., the manual approach) or by automatic Web crawlers. The manual approach is often used when the relevance and quality of information from websites is of utmost importance. However, this approach is labor-intensive and time-consuming, and often leads to inconclusive results [24].

There have been enormous strides in developing software that can process text and organize text-based materials. Such tools can be used to identify key concepts and how documents sharing concepts relate to one another. This can save time by pointing analysts to the most urgent items. To cite only a few examples of relevant systems:

- Systems that can extract information and place it into a machine readable format creating databases and coding in real-time, allowing the creation of data for modeling systems such as SOMA (described above) [25]
- Programs that can read enormous quantities of text and measure the opinions expressed in them can be used to track the evolution of terrorist group grievances and their response to government actions [26]

Countering online terrorist activity is a circumstance in which qualitative analysis and computational techniques can augment one another. Qualitative analysis is about the nuts and bolts of how phenomenon develops. Ultimately, individuals are radicalized online as part of a search for identity. Undermining the narrative that the radicalized individual finds compelling is ultimately a qualitative endeavor based on identifying how a process occurs and what specific triggers pull people into the jihadi orbit. However, online traffic occurs on a massive scale—far more than can be analyzed. Sophisticated analytics can be used to monitor traffic (who visits a site) so that opinion-makers and trendsetters can be identified and engaged. Systems that seek to maximize or minimize the spread of a phenomenon through a social network by identifying critical nodes can be an enormous asset to counter-terror analysts, allowing them to better concentrate their energies to achieve the maximum result [27].

1.4 Studying the Individual Terrorist

The previous part of this paper focused on organizations. This section focuses on individuals who choose to affiliate with terrorist organizations and those who leave them. The best counter-terror strategy is one in which people are discouraged from joining terrorist organizations in the first place. Persuading individuals to leave terror groups can also rob the organization of vital energy. Qualitative analysis, focusing on the process of radicalization can yield useful insights into how to prevent individuals from joining extremist groups and how to facilitate their leaving.

1.4.1 Counter-Radicalization Strategies

Mariam Abou Zahab conducted one example of a qualitative survey yielding insights into the question of who becomes a terrorist about the martyrs of the Pakistani terrorist group Lashkar-e-Taiba (LeT). Zahab examines the testaments and life stories of over 100 LeT martyrs from LeT publications and taped interviews with relatives. The findings reveal sociological data about class and ethnic background. But the findings also reveal important details about the specific motivations and processes that lead a young Pakistani man to seek martyrdom. One particular detail is the importance of maternal approval for the recruits decision to seek martyrdom. Significantly, the great ideologues that inspired LeT's founders such as Abdullah Azzam stated that an individual does not need anyone's permission to go on jihad. However, in Pakistani society, sons are extremely close to their mothers and unlikely to make a major life decision (such as who to marry or to seek martyrdom) without her approval. The study mentions multiple cases of a boy massaging his mother's feet while seeking this permission. These telling details suggest a potential counter-radicalization strategy focusing on Pakistani women, as the mothers often hold a veto over their son's decisions. Other compelling details discuss how the process of preparing for martyrdom and the martyr's funeral ceremony parallel wedding rituals and the testaments of the martyrs frequently exhort the family to embrace the Salafi Islam espoused by LeT. Finally, Zahab notes that although LeT's interpretation of Salafi Islam is supposed to be universal, it has been extensively adapted to Pakistani norms [28].

This is an area in which a traditional quantitative survey can dramatically augment the qualitative findings. A survey of 141 Pakistani families of slain militants resulted in several studies that helped identify the types of families from which militants are recruited. The survey examined factors such as family size, education, and religious affiliation and wealth. The quantitative findings can be used to target families in which sons are at high risk of joining Islamist jihadi groups. Then, the qualitative findings help show specific mechanisms by which the process can by stopped or stymied [29]. In particular, because support for terrorist organizations in Pakistan is particularly high among uneducated women and mothers, as seen above, play such an important role in the decision-making process of young men, one analyst as specifically proposed education programs for Pakistani women as a counter-terror policy [30].

It is also important to distinguish between the different types of causes into which potential terrorists are recruited. Not all radical militants become terrorists and making these distinctions can help facilitate more carefully targeted counter-recruitment strategies. In an article in *International Security*, Thomas Hegghammer examines the rise of the foreign fighter phenomenon in conflicts in the Muslim world. Hegghammer notes that Western analysts generally group foreign fighters with international terrorists, but that there are significant differences in their motivations, support networks, and their area of operation. However, foreign fighters are a leading source of recruits for terrorist organizations. Disaggregating these phenomena could deny terrorist organizations a major source of personnel.

Hegghammer collects data on the 70 armed conflicts in the Muslim world since 1945 and finds that 18 of them included a foreign fighter component and that 16 of those conflicts were in the past two decades. The paper uses process tracing to test hypotheses about why and how this phenomenon has emerged. The paper determines that simple explanations such as the decrease in travel costs or increase in communications technology are insufficient to explain the rise of the foreign fighters in the Muslim world. Hegghammer finds that the foreign fighters represent a unique movement that is distinct from al-Qaeda and international terrorism. While many al-Qaeda recruits come from the foreign fighter movement, they are distinct. The foreign fighters are not motivated by complex theological motivations, but by a sense of solidarity with fellow Muslims. Unlike al-Qaeda, they do not target nominally Muslim regimes and, unlike terrorists, the broader Arab public holds them in high regard. The author suggests that Western public diplomacy recognize this division and prepare a counter-narrative that targets this particular phenomenon, in order to prevent the foreign fighters from becoming terrorists [31].

1.4.2 Facilitating Desertions

Just as important as the causes for why individuals join terrorist groups is the question on why people leave terrorist organizations. Michael Jacobson, a former staffer on the 9/11 Commission, surveys the literature on terrorist dropouts to identify possible policies that will foster desertions. Although each case was different, a number of themes emerged from interviews with individuals who left al-Qaeda and other Islamist groups. One important motivator for lower level recruits to leave terrorist organizations is the difficult reality of life as a terrorist. Many report being treated as cannon fodder by leaders with little concern for their individual lives. Also, British intelligence reports for example that many British-Pakistani recruits found life in al-Qaeda training camps unpleasant and far less exciting then they had been led believe. The Lackawanna Six, a group of Yemeni-Americans who attended al-Qaeda training camps in Afghanistan found the schedule grueling, the food terrible, and could not share in the overwhelming anti-American vitriol espoused in the camp. Ultimately, five of the six obtained permission to leave and returned to the United States. While a traditional quantitative survey might have revealed some of these factors, the specifics are essential to messaging potential terrorist recruits and discouraging them from seeking to join al-Qaeda or one of its affiliates and encouraging them to leave it. Another important finding was how terrorists estimate the costs of leaving the organization. Here, the knowledge that individuals can leave a terrorist group with minimal repercussions from the group itself was important. Continuing contact with friends and family outside the group was also important in helping to reduce the costs of terrorist recruits leaving the fold. Finally, government jobs programs for terrorists that quit were helpful. Encouraging leaders to defect is particularly fruitful, as the leaders can often inspire more recruits to leave. In addition to all of the other factors, playing both on policy disagreements and petty grievances can be extremely useful in encouraging terrorist leaders to leave the fold and even become vocal opponents of their former group. Each case will have unique attributes and will rely on specific circumstances that help the individual in question form a new narrative to explain his or her actions [32].

In the process of both radicalization and de-radicalization ties of friendship and kinship play a critical role. Analysis of small-world social networks that underpin al-Qaeda have been enormously useful in ascertaining the organization's growth and recruitment strategies [33]. However, these relatively small networks (consisting of a few hundred nodes) are effectively constructed by hand. Computational methods that automate this process and provide indicators of critical nodes and the network's evolution will be invaluable to counter-terror analysts as they seek to identify and prevent potential terrorist recruits from heading down that path.

2 Conclusions

The discussion above is only a modest overview of what will almost certainly be an enormous leap forward in counter-terror analysis. As the proliferation of communications technology accelerates the spread of violent extremist ideas and the movements of terrorist operatives, the ability of the counter-terrorism analyst to track developments will be challenged. The growth of computational tools to augment human analytical capabilities is essential for developing an understanding of terrorist activities as well as designing policies that can effectively counter terrorist efforts to use violence to communicate their message.

References

1. Crenshaw M (1987) Theories of terrorism: instrumental and organizational approaches. J Strateg Stud 10(4):13–31
2. Kilcullen D (2010) Counterinsurgency. Oxford University Press, Oxford
3. Shadish W, Cook T, Campbell D (2002) Experimental and quasi-experimental designs for generalized causal inference. Houghton Mifflin, Boston
4. King G, Keohane R, Verba S (1994) Designing social inquiry: scientific inference in qualitative research. Princeton University Press, Princeton
5. Miller R, Brewer J (2003) The A-Z of social research. SAGE Publications, London
6. Jones S, Libacki M (2008) How terrorist groups end: lessons for countering al Qa'ida. RAND Corporation, Arlington
7. George A, McKeown T (1985) Case studies and theories of organizational decision making. In Coulam R, Smith R (ed) Advances in information processing in organizations: a research annual, vol 2. JAI Press, Greenwich
8. Skinner K (2011) Qualitative analysis for the intelligence community. In Fischhoff B, Chauvin C (eds) Intelligence analysis: behavioral and social scientific foundations. The National Academies Press, Washington
9. Neumann P (2005) The bullet and the ballot box: the case of the IRA. J Strateg Stud 28(6): 941–975

10. Asal V, Pate A, Wilkenfeld J (2008). Minorities at risk organizational behavior data and codebook version. http://www.cidcm.umd.edu/mar/data.asp. Accessed 3 April 2012
11. Mannes A, Michael M, Pate A, Sliva A, Subrahmanian VS, Wilkenfeld J (2008) Stochastic opponent modeling agents: a case study with Hezbollah. First International Workshop on Social Computing, Behavioral Modeling, and Prediction. Arizona State University
12. Mannes A, Subrahmanian VS (2009) Calculated terror. Foreignpolicy.com. http://www.foreignpolicy.com/articles/2009/12/15/calculated_terror. Accessed 2 April 2012
13. Mannes A, Sliva A, Subrahmanian VS, Wilkenfeld J (2008) Stochastic opponent modeling agents: a case study with Hamas. Proceedings of the Second International Conference on Computational Cultural Dynamics. University of Maryland College Park
14. Mannes A, Shakarian J, Sliva A, Subrahmanian VS (2011) A computationally-enabled analysis of Lashkar-e-Taiba Attacks in Jammu & Kashmir. Proceedings of the European Intelligence and Security Informatics Conference, Athens
15. Shapiro J (2007) The terrorist's challenge: security, efficiency, control. Center for International Security and Cooperation, Stanford University. http://igcc3.ucsd.edu/pdf/Shapiro.pdf. Accessed 27 March 2012
16. Felter J, Bramlett J, Perkins B, Brachman J, Fishman B, Forest J, Kennedy L, Shapiro JN, Stocking T (2006) Harmony and disharmony: exploiting al-Qa'ida's organizational vulnerabilities. United States Military Academy, West Point. http://www.ctc.usma.edu/posts/harmony-and-disharmony-exploiting-al-qaidas-organizational-vulnerabilities. Accessed 25 March 2012
17. Ignatius D (2012) The bin Laden plot to kill President Obama. The Washington Post. http://www.washingtonpost.com/opinions/the-bin-laden-plot-to-kill-president-obama/2012/03/16/gIQAwN5RGS_story.html. Accessed 27 March 2012
18. Ortiz R (2002) Insurgent strategies in the post-cold war: the case of the revolutionary armed forces of Colombia. Stud Confl Terror 25(2):127–143
19. Muse T (2008) Colombia rebel urges others to surrender. USA Today. http://www.usatoday.com/news/topstories/2008-05-19-1423133023_x.htm. Accessed 2 April 2012
20. Romero S, Cave D (2008) Carefully planned Colombia rescue exploited FARC weaknesses. The New York Times. http://www.nytimes.com/2008/07/04/world/americas/04iht-colombia.4.14247941.html?_r=1&pagewanted=all. Accessed 2 April 2012
21. Brodzinsky S (2010) Colombia troops kill top FARC rebel leader 'Mono Jojoy.' The christian science monitor http://www.csmonitor.com/World/Americas/2010/0923/Colombia-troops-kill-top-FARC-rebel-leader-Mono-Jojoy. Accessed 2 April 2012
22. Forero J (2011) Alfonso Cano, leader of Colombia's FARC rebels, killed in raid. The Washington Post. http://www.washingtonpost.com/world/americas/alfonso-cano-leader-of-colombias-farc-rebels-killed-in-raid/2011/11/05/gIQAgVxcoM_story.html. Accessed 2 April 2012
23. Mapstone N (2012) Farc frees last government hostages. Financial Times http://www.ft.com/cms/s/0/408eee84-7ce0-11e1-9d8f-00144feab49a.html#axzz1rBLTifrl. Accessed 2 April 2012
24. Weimann G, Von Knop K (2008) Applying the notion of noise to countering online terrorism. Studies in Stud Confl Terror 31(10):883–902
25. Albanese M, Subrahmanian VS (2007) T-Rex: a system for automated cultural information extraction. Proceedings of the First International Conference on Computational Cultural Dynamics, College Park
26. Cesarano C, Picariello A, Reforgiato D, Subrahmanian VS (2007) OASYS 2: An opinion analysis system. Proceedings of the 2007 International Conference on the Web and Social Media, Boulder
27. Shakarian P, Subrahmanian VS, Sapino M (2010) Using generalized annotated programs to solve social network optimization problems. Technical Communications of the International Conference on Logic Programming
28. Abou Zahab M (2007) 'I shall be waiting for you at the door of paradise': the pakistani martyrs of the lashkar-e-taiba (army of the true). In: Rao A, Bollig M, Bock M (eds) The practice of war: production, reproduction and communication of armed violence. Berghahn Books, New York

29. Asal V, Fair CC, Shellman S (2008) Consenting to a child's decision to join a Jihad: insights from a survey of militant families in Pakistan. Stud Confl Terror 31(11):973–994
30. Afzal M (2012) Are the better educated less likely to support militancy and terrorism? women are. Evidence from Public Opinion Polls in Pakistan. Working paper, University of Maryland School of Public Policy
31. Hegghammer T (2010/11) The rise of Muslim Foreign Fighters: Islam and the globalization of Jihad. Int Secur 35(3):53–94
32. Jacobson M (2010) Terrorist dropouts: learning from those who have left. The Washington Institute for Near East Policy. http://www.washingtoninstitute.org/pubPDFs/PolicyFocus101.pdf. Last accessed March 21, 2012
33. Sageman M (2004) Understanding terror networks. University of Pennsylvania Press, Philadelphia

SOMA: Stochastic Opponent Modeling Agents for Forecasting Violent Behavior

Amy Sliva, Gerardo Simari, Vanina Martinez, and V.S. Subrahmanian

1 Introduction

The modern global political environment is growing increasingly complex, characterized by webs of interdependency, interaction, and conflict that are difficult to untangle. Technological expansion has led to an explosion in the information available, as well as the need for more sophisticated analysis methods. In this security and information environment, behaviors in the domain of counterterrorism and conflict can be understood as the confluence of many dynamic factors—cultural, economic, social, political, and historical—in an extremely complex system. Behavioral models and forecasts can be leveraged to manage the analytic complexity of these situations, providing intelligence analysts and policy-makers with decision support for developing security strategies. In this chapter, we develop the Stochastic Opponent Modeling Agents (SOMA) framework as a stochastic model of terror group behavior, presenting several scalable forecasting algorithms and a methodology for creating behavioral models from relational data.

SOMA utilizes action probabilistic logic programs (*ap*-programs), which are a specialized variation of probabilistic logic programs (PLPs) [12] that can be used as a stochastic representation of group behavior. PLPs provide a paradigm for probabilistic logical reasoning with no independence assumptions. In fact, when looking at the behavior of terror groups, we are explicitly trying to model and

A. Sliva (✉)
Northeastern University, Boston, MA 02115, USA
e-mail: asliva@ccs.neu.edu

G.I. Simari • V. Martinez
Department of Computer Science, University of Oxford, Oxford, OX1 3QD, UK
e-mail: gerardo.simari@cs.ox.ac.uk; vanina.martinez@cs.ox.ac.uk

V.S. Subrahmanian
University of Maryland College Park, College Park, MD 20742, USA
e-mail: vs@cs.umd.edu

V.S. Subrahmanian (ed.), *Handbook of Computational Approaches to Counterterrorism*, DOI 10.1007/978-1-4614-5311-6_6,

SOMA behavioral rules

(1) If Hezbollah is involved in electoral politics, it will attack Israel with 0.62 probability.

(2) If Hezbollah is involved in politics only as a minor strategy (i.e., off-election years in Lebanon), probability of targeting Israel increases to 0.78.

(3) If it is an election year in Lebanon, probability of attacks on Israel is negligible.

Fig. 1 Sample from the behavioral model of Hezbollah that can be represented by SOMA rules

understand the dependencies between the contextual setting and the actions taken, and do not want to make any a priori assumptions. This probabilistic logic-based approach to representing terror group behavior lends SOMA not only to automated reasoning over such models, but in producing results that can improve the users' understanding of dependencies or relationships among complex sets of variables.

Using this framework, we can express rules of the form "If conditions $B_1 \ldots B_n$ hold, then terror group G will take the actions A with L to $U\%$ probability," where $B_1 \ldots B_n$ represent the cultural, economic, social, and political factors related to the probability of a group choosing strategy A. Figure 1 provides a small example–in plain English–of such rules extracted from real-world data about the terror group Hezbollah. In the next section we will look at the *ap*-program syntax for these rules. By looking at these SOMA rules, an analyst or policy-maker might see that there is some potential correlation between involvement in politics and Hezbollah's propensity to engage in transnational attacks.

Action probabilistic logic programs used in SOMA—and PLPs in general—use a possible worlds model based on prior work by [3, 4], and [13] to induce a set of probability distributions on a space of possible worlds. The forecasting component of SOMA involves computing the most probable action, or set of actions, that a group might take. This corresponds exactly to the problem of finding a "most probable world" (MPW) from an *ap*-program. As mentioned above, the SOMA formalism is able to solve this problem without making any independence assumptions about the context and strategies of terror groups.

Several applications are currently being developed for counterterrorism and cultural-adversarial reasoning [18] using the SOMA framework to build a model of the behavior of certain groups. Thus far, models have been built of several socio-cultural-economic groups in different parts of the world, including Afghan and Pakistani tribes such as the Afridis, Shinwaris, and Waziris, approximately 50 violent ethnopolitical organizations in the Middle East and Asia Pacific regions–including terror groups Hezbollah [8] and Hamas [10]—various stakeholders in the Afghan drug economy [15], the Pakistani terror organization Lashkar-e-Taiba (LeT) [9], and nation states. Of course, all of these models only capture a limited set of possible actions that these entities might take in any given situation.

The rest of this chapter describes how this behavioral modeling and forecasting has been accomplished with the Stochastic Opponent Modeling Agents (SOMA) framework. Section 2 first recalls the syntax and semantics of the PLPs [11, 12] that are utilized by the SOMA. We then discuss the *most probable world* (MPW) problem for forecasting terror group behavior from a probabilistic logic model. Finding the most probable actions of a group directly uses the linear programming methods of [11, 12]—these methods are exponential because the linear programs are exponential in the number of possible actions a group can take. In the real-world security and counterterror scenarios mentioned above, the complexity of the data is enormous. Real-world data often contain hundreds or thousands of possible actions that a group could take in a given situation. To cope with this complexity, the SOMA framework consists of several efficient exact, heuristic, and distributed algorithms [5, 6] that can forecast behaviors (i.e., find the most probable world) in models containing up to $10^{30,000}$ combinations of actions. In Sect. 3.2, several algorithms for solving the MPW problem and forecasting terror group behavior are reviewed, and distributed algorithms for improving the performance of forecasting in SOMA are presented in Sect. 4.

Behavioral models using the SOMA framework can be constructed automatically from real data using any number of well-known machine learning approaches, such as decision trees. In this chapter, one possible method, the APEX algorithm, is presented in Sect. 5 for extracting such rules from a relational database. However, there are many possible algorithms for finding these rules, and APEX is given merely as a basic proof of concept. Applications of this process to two political science data sets are described for modeling the behavior of terror organizations—specifically Hezbollah and Hamas—in the Middle East and Lashkar-e-Taiba in Pakistan. These behavioral models have produced tangible results of use to U.S. military officers and show promise in applications to other datasets and cultural reasoning domains [1, 17].

2 Representing Terror Group Behavior: Action Probabilistic Logic Programs

The Stochastic Opponent Modeling Agents (SOMA) framework is a language that can be used for modeling and forecasting the behavior of terror organizations. This framework makes use of action probabilistic logic programs (*ap*-programs), a variant of the probabilistic logic programs introduced in [11, 12], to represent the probability of group behaviors. For the SOMA framework, we assume the existence of a logical alphabet that consists of a finite set $\mathscr{L}_{cons}$ of constant symbols, a finite set $\mathscr{L}_{pred}$ of predicate symbols (each with an associated arity), and an infinite set $\mathscr{V}$

SOMA action symbols for Lashkar-e-Taiba

(1) *Attack_Fedayeen(G, R)*: Group G will carry out a fedayeen attack in region R

(2) *Armed_Clash_SF_Casualties(G, R)*: Group G will engage in armed clashes in region R leading to security force (SF) casualties

(3) *Murder(G, R)*: Group G will carry out murders in region R

Fig. 2 Action symbols for a model of Lashkar-e-Taiba

SOMA non-action symbols for Lashkar-e-Taiba

(1) *Religious_Organization(G, R)*: Group G is a known religious organization in region R

(2) *Social_Strife(R)*: There is social strife in region R

(3) *Military_Support_by_Government(G, R)*: Group G receives military support from the government of region R

Fig. 3 Non-action symbols from a SOMA model of Lashkar-e-Taiba (LeT)

of variable symbols. Function symbols are not allowed in this language. Terms and atoms are defined in the usual way [7].

Also, because we are explicitly modeling group behavior, we assume that a subset $\mathscr{L}_{act}$ of $\mathscr{L}_{pred}$ are designated as *action symbols*—these are symbols that denote some strategic-level action or behavior that a group could take.

Example 1. Figure 2 contains some sample action symbols from a SOMA model of the behavior of the terror group Lashkar-e-Taiba.

These action predicates describe particular types of actions LeT can take, and also provide additional information regarding the location, targeting, and consequences (in terms of casualties) of these types of violent behaviors. Here, each predicate is applied to a particular group G taking the action (LeT in our model), and a region where the action occurred, which in this model can be either Jammu and Kashmir, Pakistan, or India.

The remainder of the predicates in $\mathscr{L}_{pred}$ are *non-action symbols* and denote the environmental context—the cultural, economic, social, political, and historical state of the world—in which a group is acting and which may influence its behavior.

Example 2. Some possible environmental, or non-action symbols, for a SOMA model of Lashkar-e-Taiba are given in Fig. 3.

Such context predicates describe characteristics of the group, such as (1) and (3), or aspects of the environment in which they are operating that may affect all groups in a particular region, such as (2). Predicate (1) indicates that the group G is a religious organization in a particular region R, and predicate (3) says that group G is receiving military support from the government in region R. Predicate (2) is describing the general environment, saying that there is currently social strife in the region. As in the previous example, since this is a SOMA model for Lashkar-e-Taiba, LeT is assumed to be the group G in question and the region can be either Jammu and Kashmir, Pakistan, or India.

Thus, an atom $p(t_1, \ldots, t_n)$, where $p \in \mathcal{L}_{act}$, is an *action atom*. Every (resp. action) atom is a (resp. action) wff. If F, G are (resp. action) wffs, then $(F \wedge G)$, $(F \vee G)$ and $\neg G$ are all wffs (resp. action wffs). An *action formula* is any formula in first-order logic that is composed only of action symbols. Using this logical language, we can now define the action probabilistic logic rules at the heart of the SOMA behavioral modeling framework. Without loss of generality, assume that action formula F is in conjunctive normal form (i.e., it is written as a conjunction of disjunctions) in the following definitions.

Definition 1 (*ap*-rule). f F is an action formula, $B_1, \ldots, B_n$ are non-action atoms, and μ is an interval $[\ell, u] \in [0, 1]$ then $F : \mu \leftarrow B_1 \wedge \ldots \wedge B_n$ is called an *ap-rule*. If this rule is named c, then $Head(c)$ denotes $F : \mu$ and $Body(c)$ denotes $B_1 \wedge \ldots \wedge B_n$.

Intuitively, the above *ap*-rule says that an entity (e.g., a terror group g) *will take the actions described in F with probability in the range μ if $B_1, \ldots, B_n$ are true in the current context.*

Definition 2 (*ap*-program). An *action probabilistic logic program* (*ap*-program for short) is a finite set of *ap*-rules.

An *ap*-program provides our model of terror group behavior in the SOMA framework, indicating the relationship between certain contextual or environmental factors—group funding, relations with the state or other rival groups, participation in politics—and the probability that a group will choose a particular strategy.

Example 3. Figure 4 shows a sample SOMA rule base consisting of some automatically derived *ap*-rules about Hezbollah using behavioral data from the Minorities at Risk Organizational Behavior (MAROB) dataset [2, 19].

The behavioral data in MAROB has tracked over 118 ethnopolitical organizations across the Middle East and North Africa for about 25 years from 1980 to 2004. For each year, values have been gathered for about 175 measurable variables for each group in the sample [2, 19]. These variables include strategic conditions such as the tendency to commit bombings and armed attacks, as well as background information about the type of leadership, whether the group is involved in cross-border violence, etc. The automatic derivation of these rules was based on the straightforward data mining algorithm that will be discussed in Sect. 5.

Sample *ap*-program for Hezbollah

(1) *kidnap*:[0.35,0.45] ← *interOrganizationConflicts.*

(2) *kidnap*:[0.60,0.68] ← *unDemocratic* ∧ *internalConflicts.*

(3) *armed_attacks*:[0.42,0.53] ← *typeLeadership(strongSingle)* ∧ *orgPopularity(moderate).*

(4) *armed_attacks*:[0.93,1.0] ← *statusMilitaryWing(standing).*

Fig. 4 Four *ap*-rules for modeling the behavior of a terrorist organization

Figure 4 contains four of these extracted rules for the group Hezbollah, describing some conditions under which it has used a particular strategy, along with a probability range. For example, the third rule indicates that when Hezbollah has a strong, single leader and its popularity is moderate, its propensity to conduct armed attacks is 42–53%. However, when it has had a standing military, its propensity to conduct armed attacks is 93–100%.

Definition 3 (world/state). A *world* is any set of ground action atoms (i.e., action atoms where all variables have been bound to constants). A *state* is any finite set of ground non-action atoms.

A world can be thought of as representing the complete behavior or strategy of a terror group in a certain situation—it is a combination of actions that the group is going to take. In any SOMA model, we are going to have an ***exponential number of possible worlds***. For a model with $|\mathscr{L}_{act}|$ actions, there are $2^{|\mathscr{L}_{act}|}$ possible worlds (i.e., the worlds are the power set of $\mathscr{L}_{act}$), as we want to examine all possible combinations of actions. We will use $\mathscr{W}$ to denote the set of all possible worlds in a SOMA model for a particular terror organization. The state, on the other hand, represents a particular contextual scenario in which the group is acting, describing the environmental factors that are true at a given time.

Example 4. Figures 5 and 6 contain some examples of possible worlds and a sample state for the terror group Lashkar-e-Taiba respectively.

Remember, the full set of possible worlds $\mathscr{W}$ is the powerset of $\mathscr{L}_{act}$. In the example above, we are assuming that there are only three ground versions of the action symbols given in Fig. 2. So, from the three actions available to LeT, we can generate $2^3 = 8$ possible worlds. These eight worlds represent all possible combinations of actions that LeT can take at any given time. For example, world w_3 indicates that LeT is engaging in fedayeen attacks and carrying out armed clashes in Jammu and Kashmir. The last world w_7 is the empty set, indicating that no actions were taken by the group.

Possible Worlds for LeT

Suppose there are only three possible ground actions that the terror group Lashkar-e-Taiba can take: $Attack\_Fedayeen(LeT, J\&K)$, $Armed\_Clash\_SF\_Casualties(LeT, J\&K)$, $Murder(LeT, J\&K)$. Then, there are 8 possible worlds, or combinations of actions, that can be taken by the group in any situation.

(1) $w_0 = \{Attack\_Fedayeen(LeT, J\&K)\}$

(2) $w_1 = \{Armed\_Clash\_SF\_Casualties(LeT, J\&K)\}$

(3) $w_2 = \{Murder(LeT, J\&K)\}$

(4) $w_3 = \{Attack\_Fedayeen(LeT, J\&K), Armed\_Clash\_SF\_Casualties(LeT, J\&K)\}$

(5) $w_4 = \{Attack\_Fedayeen(LeT, J\&K), Murder(LeT, J\&K)\}$

(6) $w_5 = \{Murder(LeT, J\&K), Armed\_Clash\_SF\_Casualties(LeT, J\&K)\}$

(7) $w_6 = \{Attack\_Fedayeen(LeT, J\&K), Armed\_Clash\_SF\_Casualties(LeT, J\&K), Murder(LeT, J\&K)\}$

(8) $w_7 = \{\}$

Fig. 5 Examples of some possible worlds

Example State for LeT

$s = \{Religious\_Organization(LeT, J\&K), Social\_Strife(Pak), Military\_Support\_by\_Government(LeT, Pak), Training\_Camp(LeT, Pak)\}$

Fig. 6 Example state for LeT

The sample state is a small example describing the environmental context in which LeT might be operating. In this state, LeT is a religious organization in Jammu and Kashmir, there is social strife in Pakistan, LeT is receiving military support from Pakistan, and LeT maintains training camps in Pakistan. A full state would include more variables regarding the characteristics of the organization (i.e., leadership, fundraising, relationships with the state) and the region (i.e., elections, economic conditions, etc.).

Reduction of an LeT *ap*-program

Recall the state *s* given in Figure 6 and consider the following *ap*-program Π describing the behavior of LeT:

(1) *Attack_Fedayeen (LeT, J&K)*:[0.53,0.73] $\leftarrow$
Religious_Organization(*LeT*, *J&K*) $\wedge$
Military_Support_by_Government(*LeT*, *Pak*).

(2) *Armed_Clash_SF_Casualties(LeT, J&K)*:[0.6,0.8] $\leftarrow$
Financial_Support_by_Diaspora(*LeT*, *Pak*) $\wedge$
Businesses(*LeT*, *Pak*).

(3) *Armed_Clash_SF_Casualties(LeT, J&K)*:[0.8,1.0] $\leftarrow$
Social_Strife(*Pak*) $\wedge$
Alliance_NSAG(*LeT*, *J&K*).

(4) *Murder(LeT,J&K)*:[0.45,0.65] $\leftarrow$
Religious_Organization(*LeT*, *J&K*) $\wedge$
Training_Camp(*LeT*, *Pak*)

If we take the reduction of the *ap*-program w.r.t. state *s*, we are left with the following set of rules:

(1) *Attack_Fedayeen (LeT, J&K)*:[0.53,0.73] $\leftarrow$.

(2) *Murder(LeT,J&K)*:[0.45,0.65] $\leftarrow$.

Fig. 7 Reduction of an LeT *ap*-program

Given an *ap*-rule $F : \mu \leftarrow B_1 \wedge \ldots \wedge B_n$ named c, we say that a world w *satisfies F*—denoted $w \mapsto F$—iff w is a model for the action formula F. Similarly, we say that a state s *satisfies Body*(c) iff $B_1, \ldots, B_n \in s$. That is, a current state of the world satisfies the body of a rule if all of the body conditions are true in the current context. Using this definition, we can now define the reduction of an *ap*-program, which will form the basis of SOMA's behavioral forecasting.

Definition 4 (Reduction of an *ap*-program). Let Π be an *ap*-program and s a state. The *reduction of* Π *w.r.t.* s, denoted by Π_s is $\{F : \mu \leftarrow \mid s$ satisfies *Body* and $F : \mu \leftarrow$ *Body* is a ground instance of a rule in $\Pi\}$.

Note that Π_s consists only of Heads of ap-rules, and never has any non-action atoms in it.

Example 5. Figure 7 illustrates how to compute the reduction of an *ap*-program for LeT using the sample state given in Fig. 6. We can see that taking the *reduction* of an *ap*-program isolates those elements of a group's behavior that are relevant or applicable in the current state of the world. In this case, only rules (1) and (4) from the original *ap*-program need to be considered in the given state s. Because we know the rules are applicable in the current state, we no longer need to consider their bodies, so we can represent the potential behaviors of LeT with only the probabilistic action formulas in the rule heads. Using this reduced and simplified set of rules, we will now be able to make forecasts regarding what strategy we can expect the group to execute in the current environmental context.

3 Forecasting Terror Group Behavior: Finding the Most Probable World

Suppose we know that the current context can be described by the state in Fig. 6. A policy-maker or security analyst may want determine what LeT is likely to do in this situation and develop the best counterterrorism strategy based on this forecast. How can we use the *ap*-program modeling LeT's behavior to forecast their actions in this situation? In this section we describe the forecasting component of SOMA, presenting several algorithms for making just this kind of "What if?" prediction.

Since the possible worlds for a group describe all possible behaviors they can take at any time, then forecasting the most likely strategy of a terror organization is exactly the problem of finding the most probable world. To do this, we can use the probabilistic relationships described in an *ap*-program, along with information about the state of the world, either the actual current state or some hypothetical scenario to be analyzed. First, we need a way to induce a probability distribution among the possible worlds in the SOMA framework. We can do this by solving a constraint satisfaction problem to assign a lower and upper bound for the probability of each world in a given state.

Definition 5 (Constraints of an *ap*-program). Given an *ap*-program Π and a state s, we can define the *constraints of* Π *w.r.t.* s to be a linear program $\mathsf{CONS}(\Pi, s)$ consisting of the following inequalities:

1. If $F : [\ell, u] \in \Pi_s$ (the reduction of Π w.r.t. s), then $\ell \leq \Sigma_{w_i \in \mathscr{W} \mapsto F}\, p_i \leq u$ is in $\mathsf{CONS}(\Pi, s)$.
2. $\Sigma_{w_i}\, p_i = 1$ is in $\mathsf{CONS}(\Pi, s)$.

Where p_i denotes the probability of each world w_i.

Example 6. Using the reduction Π_s of an LeT *ap*-program from Fig. 7, we can derive the set of linear constraints. As an illustration, we will use the worlds from Fig. 5. For each world w_i we will use a variable p_i in the linear program to represent its probability.

Looking at the first rule in our reduction, *Attack_Fedayeen (LeT, J&K)*:[0.53, 0.73] ←, we see that this is satisfied by worlds w_0, w_3, w_4, and w_6 because each of these

Constraints of an LeT ap-program

Given the reduced ap-program in Figure 7 and the state in Figure 6, $\mathsf{CONS}(\Pi, s)$ will be:

$0.53 \leq p_0 + p_3 + p_4 + p_6 \leq 0.73$

$0.45 \leq p_2 + p_4 + p_5 + p_6 \leq 0.65$

$p_0 + p_1 + p_2 + p_3 + p_4 + p_5 + p_6 + p_7 = 1$

Fig. 8 Linear constraints $\mathsf{CONS}(\Pi, s)$ derived from an ap-program

worlds involves the execution of fedayeen attacks in J&K. So, we can add the following constraint to $\mathsf{CONS}(\Pi, s)$

$$0.53 \leq p_0 + p_3 + p_4 + p_6 \leq 0.73$$

Now taking the second rule in the reduction, *Murder(LeT,J&K)*:[0.45, 0.65] ←, we see that this is satisfied by worlds w_2, w_4, w_5, and w_6 as these are the worlds that include *Murder(LeT,J&K)*. The constraint

$$0.45 \leq p_2 + p_4 + p_5 + p_6 \leq 0.65$$

is also added to $\mathsf{CONS}(\Pi, s)$.

Finally, we need to include a constraint indicating that the probability of all the possible worlds must sum to 1. The final set of constraints $\mathsf{CONS}(\Pi, s)$ is given in Fig. 8.

Intuitively, to find the probability bounds of each world, we first get the reduction of an ap-program with respect to the given state s, isolating those ap-rules in our program that apply to the situation described in s. It is from this reduced set of rules that the linear program $\mathsf{CONS}(\Pi, s)$ is constructed. We can use the linear constraints in $\mathsf{CONS}(\Pi, s)$ to find the probability bounds for each world w_i by minimizing and maximizing its probability p_i w.r.t. the linear program.

Definition 6 (lower/upper probability of a world). Suppose Π is an ap-program and s is a state. The *lower probability*, $\mathsf{low}(w_i)$, of a world w_i is defined as: $\mathsf{low}(w_i) =$ **minimize** p_i **subject to** $\mathsf{CONS}(\Pi, s)$. The *upper probability*, $up(w_i)$, of world w_i is defined as $up(w_i) =$ **maximize** p_i **subject to** $\mathsf{CONS}(\Pi, s)$.

Thus, the low probability of a world w_i is the lowest probability that world can have in any solution to the linear program. Similarly, the upper probability for the same world represents the highest probability that world can have. It is important to note that for any world w_i, a point probability cannot be *exactly* determined. This observation is true because the SOMA framework does not make

Naive Algorithm

1. Compute the reduction of Π w.r.t. s
2. For each world $w_i \in \mathcal{W}$ do:
   a. Compute $\mathsf{low}(w_i)$ by minimizing p_i subject to $\mathsf{CONS}(\Pi, s)$.
   b. If $\mathsf{low}(w_i) > Bestval$ then set $Forecast = w_i$ and $Bestval = \mathsf{low}(w_i)$.
3. If $Forecast = NIL$, then return any world whatsoever, else return $Forecast$.

Fig. 9 Naive algorithm for forecasting terror group behavior in the SOMA framework

any simplifying assumptions (e.g., independence) about the relationship between the variables. Checking if the low (resp. *up*) probability of a world exceeds a given bound is in the class EXPTIME [5].

3.1 A First Approach to Forecasting in SOMA

To forecast terror group behaviors in SOMA, we need to find the most probable world (MPW)—the most probable set of actions—that a group will take in a given situation according to some behavioral model. Essentially, we need a forecasting procedure that will take an *ap*-program describing an organization's behavioral patterns and a state as input, and produce the most probable world. In many cases, it will be more useful for policy-makers or security analysts if we forecast a set of the top-*k* most probable worlds. This will allow experts to make decisions about counterterrorism strategies based on a broader range of possible terror group actions, facilitating contingency planning and the development of more robust policies.

Here, we are going to order the worlds by their low probabilities, that is, the most probable world is one where $\mathsf{low}(w_i)$ is maximal. Such an analysis can be thought of as a conservative forecast—we know that the behaviors taken by a terror group will be *at least* this probable. However, in some cases users may want to forecast the MPW using the up probability or some other calculation, such as an expected value.

Using our definition of the low probability for a world, we can immediately derive an algorithm—albeit a rather naive approach—for forecasting a group's behavior in a given situation. This naive algorithm directly finds $\mathsf{low}(w_i)$ for each world $w_i \in \mathscr{W}$ as follows:

The naive algorithm does a brute force search after computing the reduction of *ap*-program Π w.r.t. the state s, finding the lower bound probability for each world and forecasting the best one (i.e., the most probable according to the lower bound probability) as the expected behavior of the group.

While this approach is straightforward, it has serious shortcomings in terms of scalability. If we think for a moment about the complexity of real-world counterterrorism scenarios, it becomes immediately obvious that the naive algorithm is insufficient for forecasting in these situations due to the exponential number of possible worlds. According to the naive algorithm, we would be solving this linear program with an exponential number of variables an exponential number of times. The following example looks at this issue in more detail.

Example 7. Imagine a relatively small model in the SOMA framework where we are looking at the behavior of a terror organization that can take only 40 possible actions (e.g., *kidnap*, *armed_attack*, *bombing*, etc.). In this case we will have 2^{40}, which is about 10^{12}, possible worlds.

Now, suppose we want to make the model more robust and allow each action predicate to take as an argument a number from 0 to 7 indicating the *intensity* of these violent behaviors. In this case, we will now have $2^{40*8} = 2^{320}$ possible worlds.

What if experts want to make the model more complete, and include geographical information (similar to the Lashkar-e-Taiba model described in previous examples)? If we include 100 possible cities or regions where attacks may occur in this model, the number of possible worlds will now explode to $2^{32,000}$ which is approximately $10^{9,900}$. This figure is orders of magnitude larger than the number of atoms in the universe (estimated to be 10^{80})! Clearly, solving a linear program with $2^{32,000}$ variables (one for the probability of each world) $2^{32,000}$ times (to find the low probability for each world) is an infeasible task.

As the above example makes clear, even for relatively simple models from a counterterror policy standpoint, the number of possible worlds will be inordinately large, making the naive algorithm for forecasting terror group behavior in SOMA intractable. The next section begins to look at how we can improve the efficiency of these forecasts to provide decision support in real-world situations.

3.2 *Scalable Algorithms for Forecasting Terror Group Behavior*

The naive algorithm given in the previous section is guaranteed to find a correct solution to the most probable world problem, forecasting the most likely behavior that a group will take. However, as we increase the size of our model to reflect important aspects of a counterterrorism scenario, this approach quickly becomes utterly useless. In this section, we will review some efficient algorithms developed in [6] that allow for significant improvements in computation time and scalability, making SOMA a feasible approach for real-world decision support and counterterrorism strategy. All of these approaches focus on ways to *reduce the number of worlds* required for forecasting a group's behavior, without requiring additional assumptions in the model.

Two of these algorithms exploit the concept of *head-oriented processing*, grouping worlds (i.e., sets of actions) into equivalence classes according to rules in the *ap*-program. These methods still produce an exact solution—an accurate forecast of the most probable behavior a group will take—but vastly improve the complexity of the forecasting problem. A heuristic approach is also used to further reduce the number of variables in $\mathsf{CONS}(\Pi, s)$, randomly sampling the space of behaviors to decide which worlds to include in the linear program and using a binary search process to adjust the lower-bound constraints accordingly. While this approach does not return a precise solution, it can still be used for forecasting and planning longer-term counterterrorism strategies.

3.2.1 Head-Oriented Processing

The head-oriented processing algorithms are based on the recognition that the linear program $\mathsf{CONS}(\Pi, s)$ is often over parameterized, containing a variable for each world even though several worlds together may actually represent a single probability mass. Using this intuition, worlds can be grouped into equivalence classes, reducing the number of variables required to specify the linear constraints.

Example 8. Recall the possible worlds in our LeT example from Fig. 5. Based on our *ap*-program for LeT, we might find that $world_0$ and $world_1$ always appear in the same constraints in $\mathsf{CONS}(\Pi, s)$. Then, instead of using two variables p_0 and p_1 for these probabilities in the linear program, we can group these worlds together and use a single variable p to represent the combined probability mass. We will make this process more precise in the remainder of this section.

Given a world w, state s, and *ap*-program Π, let $Sat(w) = \{F \mid c$ is a ground instance of a rule in Π_s and $Head(c) = F : \mu$ and $w \mapsto F\}$. Intuitively, $Sat(w)$ is the set of heads of rules in Π_s that are satisfied by w.

Definition 7 ($\sim$-equivalence of worlds). Suppose Π is an *ap*-program, s is a state, and w_1, w_2 are two worlds. Worlds w_1 and w_2 are *equivalent*, denoted $w_1 \sim w_2$, iff $Sat(w_1) = Sat(w_2)$.

In other words, two worlds are equivalent iff the they satisfy the action formulas in the heads of exactly the same rules in Π_s. It is easy to see that $\sim$ is an equivalence relation, and $[w_i]$ denotes the $\sim$-equivalence class to which a world w_i belongs.

Example 9. To demonstrate the equivalence relation between worlds, let's consider the set of possible worlds given in Fig. 5 and the reduction of LeT's *ap*-program from Fig. 7. First, we need to compute $Sat(w)$ for each possible world, indicating which rules they satisfy. For example, $w_0 = \{Attack\_\ Fedayeen(LeT, J\&K)\}$ satisfies rule (1) in the reduction, $w_2 = \{Murder(LeT, J\&K)\}$ satisfies rule (2), etc. Figure 10 contains the set of satisfied rules for each world.

$Sat(w)$ for each possible world for LeT

(1) $Sat(w_0) = \{Attack\_Fedayeen(LeT, J\&K) : [0.53, 0.73]\}$

(2) $Sat(w_1) = \{\}$

(3) $Sat(w_2) = \{Murder(LeT, J\&K) : [0.45, 0.65]\}$

(4) $Sat(w_3) = \{Attack\_Fedayeen(LeT, J\&K) : [0.53, 0.73]\}$

(5) $Sat(w_4) = \{Attack\_Fedayeen(LeT, J\&K) : [0.53, 0.73], Murder(LeT, J\&K) : [0.45, 0.65]\}$

(6) $Sat(w_5) = Murder(LeT, J\&K) : [0.45, 0.65]\}$

(7) $Sat(w_6) = \{Attack\_Fedayeen(LeT, J\&K) : [0.53, 0.73], Murder(LeT, J\&K) : [0.45, 0.65]\}$

(8) $Sat(w_7) = \{\}$

Fig. 10 Rules satisfied by each possible world for LeT

Equivalence classes of worlds for LeT

(1) $c_0 = [w_0] = \{w_0, w_3\}$

(2) $c_1 = [w_1] = \{w_1, w_7\}$

(3) $c_2 = [w_2] = \{w_2, w_5\}$

(4) $c_3 = [w_4] = \{w_4, w_6\}$

Fig. 11 Equivalences classes of worlds for LeT using the $\sim$ relation

Looking at *Sat*(w) for each of these worlds we can immediately see that worlds w_0 and w_3 satisfy the same rules. So, according to the definition given above, we can say $w_0 \sim w_3$. Similarly, because they satisfy exactly the same rules, we have that $w_2 \sim w_5$, $w_4 \sim w_6$ and $w_1 \sim w_7$, giving us a total of four possible equivalence classes for LeT's behavior in the given state. These equivalence classes are listed in Fig. 11.

The Head-Oriented Processing (HOP) approach uses this equivalence relation between behaviors to reduce the number of variables in CONS(Π, s). The key insight is that for any $\sim$-equivalence class $[w_i]$, the summation $\Sigma_{w_j \in [w_i]} p_j$ either appears *in its entirety* in each constraint in the linear program or does not appear at all.

Effectively, the number of variables in the linear program has been reduced from $2^{|\mathcal{L}_{act}|}$—exponential in the number of actions a group can take—to $2^{|\Pi_s|}$—exponential in the number of ap-rules applicable in the current state s; this savings can be significant in some cases (though not always!). The number of constraints in the linear program stays the same, but the number of possible worlds is greatly reduced. This *reduced set of constraints* using equivalence classes is formally defined as follows.

Definition 8 (RedCONS($\mathbf{\Pi}, s$)). For each equivalence class $[w_i]$, there is a variable p'_i in RedCONS(Π, s) denoting the summation of the probability of each of the worlds in $[w_i]$. RedCONS(Π, s) contains the following constraints:

1. If $F : [\ell, u]$ is in Π_s, then $\ell \leq \Sigma_{[w_i] \mapsto F} p'_i \leq u$ is in RedCONS(Π, s).
2. $\Sigma_{[w_i]} p'_i = 1$ is in RedCONS(Π, s).

Here, $[w_i] \mapsto F$ in constraint (1) means that the worlds in $[w_i]$ all satisfy the action formula F.

Example 10. Recall the set of linear constraints CONS(Π, s) from Fig. 8 that were used to forecast LeT's behavior in the previous section:

$$0.53 \leq p_0 + p_3 + p_4 + p_6 \leq 0.73$$

$$0.45 \leq p_2 + p_4 + p_5 + p_6 \leq 0.65$$

$$p_0 + p_1 + p_2 + p_3 + p_4 + p_5 + p_6 + p_7 = 1$$

Using the equivalence classes identified in Fig. 11 and the same state and reduced ap-program from Figs. 6 and 7 respectively, we can derive a new reduced set of constraints RedCONS(Π, s). For each equivalence class c_i we will use p'_i to indicate its probability.

As before, we start by considering the first action formula in the reduction and find that it is satisfied by the equivalence classes c_0 and c_3. So we can add the constraint:

$$0.53 \leq p'_0 + p'_3 \leq 0.73$$

to RedCONS(Π, s).

Next, we look at the second rule and see that it is satisfied by the worlds in classes c_2 and c_3, giving us the constraint:

$$0.45 \leq p'_2 + p'_3 \leq 0.65$$

Finally, we again need to add a constraint that the probabilities of each of the equivalence classes must sum to 1. The final set of constraints is given in Fig. 12. We can see that by using equivalence classes rather than individual worlds, we are able to substantially reduce the size of the linear program from eight variables to

Equivalence class constraints of an LeT ap-program

Given the reduced ap-program in Figure 7, the state in Figure 6, and the equivalence classes of worlds in Figure 11, RedCONS(Π, s) will be:

$0.53 \leq p_0' + p_3' \leq 0.73$

$0.45 \leq p_2' + p_3' \leq 0.65$

$p_0' + p_0 + p_1 + p_2 + p_3 = 1$

Fig. 12 Linear constraints RedCONS(Π, s) derived from an ap-program using world equivalence classes

HOP Algorithm

1. Compute the reduction of Π w.r.t.s. $Forecast = NIL$; $Bestval = 0$.
2. For each equivalence class $[w_i]$ do:
   a. If there is exactly one world in $[w_i]$ do:
      i. **Minimize p_i' subject to** RedCONS(Π, s). Let Val be the result.
      ii. If $Val > Bestval$, then set $Forecast = w_i$ and $Bestval = Val$.
3. If $Bestval = 0$ then return any world whatsoever, otherwise return $Forecast$.

Fig. 13 HOP algorithm for forecasting terror group behavior in the SOMA framework

four. While this savings seems small, in larger real-world problems such a reduction in the size of the constraints will greatly enhance the efficiency and scalability of forecasting terror group behavior.

Before reviewing the HOP algorithm, we can present one additional optimization for forecasting the most probable behaviors of a terror group. Because we are using a conservative approach and finding the low probability of each world, the HOP approach allows us to only solve RedCONS(Π, s) for singleton equivalence classes. In all other cases, we know that the lower bound probability of the class will be zero; we can always assign all of the probability into one world in the class and set the rest to zero, giving the low probability of zero for the whole class. Using our reduced set of linear constraints and this additional optimization, we can now look at the HOP forecasting algorithm:

Though the complexity of HOP is also exponential, it may be preferable to the naive algorithm due to the reduced number of variables in RedCONS$_U$(Π, s). Essentially, it allows us to focus our computation on behaviors that have at least a non-zero probability, making this more efficient, but also more useful for decision-support. However, the required satisfiability checks to compute the equivalence classes can often make HOP intractable in practice.

SemiHOP Algorithm

1. Compute the reduction of Π w.r.t. s. $Forecast = NIL$; $Bestval = 0$.
2. For each sub-partition $[W_i]$ do:
   a. If there is exactly one world in $[W_i]$ do:
      i. **Minimize** $p_i^\star$ **subject to** S_RedCONS(Π, s). Let Val be the result.
      ii. If $Val > Bestval$, then set $Forecast = W_i$ and $Bestval = Val$.
3. If $Bestval = 0$ then return any world whatsoever, otherwise return $Forecast$.

Fig. 14 SemiHOP algorithm for forecasting terror group behavior in the SOMA framework

A variant of the HOP algorithm, called SemiHOP, tries to avoid computing the full equivalence classes. The SemiHOP algorithm finds *sub-partitions* of behaviors rather than finding all pairs of sets that represent the same equivalence class, therefore omitting the exhaustive checks for logical equivalence. This method will still afford us a significant reduction in complexity when trying to use SOMA to forecast terror group behaviors.

Definition 9 (Sub-partition). A *sub-partition* of the set of worlds of Π w.r.t. s is a partition $W_1, \ldots, W_k$ where:

1. $\bigcup_{i=1}^{k} W_i$ is the entire set of worlds.
2. For each W_i, there is an equivalence class $[w_i]$ such that $W_i \subseteq [w_i]$.

A sub-partition can be generated by looking at all possible subsets of $Heads$ in Π_s—rather than looking for distinct equivalence classes of worlds, we pre-define the sub-partitions as the powerset of all rule heads and place each world in all partitions it satisfies. As in the complete version of HOP, SemiHOP uses a single variable to represent each sub-partition, still combining multiple terror group behaviors into a single probability mass. The size of the constraints is still reduced (through the elimination and combination of some variables), but the reduction is not necessarily maximal.

Definition 10 (S_RedCONS$(\mathbf{\Pi}, s)$). Let $W_1, \ldots, W_k$ be a sub-partition of the set of worlds for Π and s. For each W_i, S_RedCONS(Π, s) uses a variable $p_i^\star$ to denote the summation of the probability of each of the worlds in W_i. RedCONS(Π, s) contains the following constraints:

1. If $F : [\ell, u]$ in Π_s, then $\ell \leq \Sigma_{W_i \mapsto F}\, p_i^\star \leq u$ is in S_RedCONS(Π, s).
2. $\Sigma_{W_i}\, p_i^\star = 1$ is in S_RedCONS(Π, s).

Here, $W_i \mapsto F$ in constraint (1) implies that all the worlds in W_i satisfy the action formula F.

The SemiHOP algorithm for forecasting terror group behavior is given in Fig. 14.

The key advantage of SemiHOP over HOP is that the set $[w_i]$ of worlds does not need to be constructed, i.e., finding the equivalence classes $[w_i]$ is not necessary. This advantage comes with a drawback, though, since the size of the set S_RedCONS(Π, s) can be a bit bigger than RedCONS(Π, s). The relative success of these two algorithms will depend heavily on the SOMA model that has been constructed for a particular terror organization, as well as the current state of the world.

3.2.2 Randomized Heuristic Behavioral Forecasts

Even though HOP and SemiHOP try to reduce the number of constraint variables when finding the most probable world to forecast group behavior, the size of the linear program can still be quite large, especially in real-world counterterrorism applications. A randomized heuristic algorithm can be used in conjunction with any of the exact approaches (naive, HOP, SemiHOP) to reduce the number of variables even further. While this heuristic approach is an approximation, these efficient forecasts can be very helpful for long range policy analysis and conflict management strategy development. For example, an approximate forecast of LeT's strategies in Jammu and Kashmir in the event of political turmoil in Pakistan can allow policy-makers to start allocating resources and planning for these contingencies, while allowing for the possibility that the situation on the ground may change.

Let $\mathscr{C}$ be the set of constraints generated for either the naive, HOP, or SemiHOP algorithms. In the randomized algorithm, an *a priori* commitment is made to only look at some set S_k of k variables from the linear program, eliminating all variables not in S_k from any summation in $\mathscr{C}$ and deriving a modified set of constraints $\mathscr{C}'$. This means we are randomly sampling the worlds (resp. equivalence classes or sub-partitions) ignoring those behaviors that were not selected. Of course, this process can be directed in some way (e.g., looking at actions of interest as in [14]), but such targeted sampling procedures are beyond the scope of this paper.

It is immediately apparent that, as all the lower bounds are set to ℓ, a solution to $\mathscr{C}'$ may or may not exist when we randomly remove some variables from the computation. Rather than weakening the lower bound from ℓ to 0 (which would guarantee a solution), a *binary heuristic* is used to *modify the lower bounds* of such constraints as little as possible. If $\mathscr{C}'$ is solvable by itself, then each variable in the random sample S_k is simply minimized subject to $\mathscr{C}'$ and the most probable world (resp. equivalence class or sub-partition) is returned as the behavioral forecast. If not, then the lower bound of one or more constraints in $\mathscr{C}'$ is decreased as follows to derive a new set of constraints $\mathscr{C}^\bullet$.

Suppose $c^\star \in \mathscr{C}'$ is a constraint where $\ell^\star \leq \Sigma_{q_i \in S_k} q_i \leq u$. The heuristic tries to replace $\ell^\star$ by $\frac{\ell^\star}{2}$. If this yields a solvable set of equations, $\frac{\ell^\star}{2}$ is replaced by $\frac{3\times\ell^\star}{4}$. If the resulting system is unsolvable, it is replaced with $\frac{5\times\ell^\star}{8}$, and so forth. This method is called the *binary heuristic* because it resembles a binary search.

Once this process of modifying the lower bounds has completed, each variable in S_k is minimized subject to the new constraints $\mathscr{C}^\bullet$. The variable with the highest minimal value—the approximate most probable terror group behavior w.r.t. the random sample—is forecast.

4 Distributed Computation for Forecasting in SOMA

In the previous sections, several algorithms were reviewed that can be used to efficiently forecast terror group behavior, i.e., find the most probable world given an *ap*-program and a state. However, even with the given constraint simplifications and heuristic approximations, the computation time and resource requirements of these methods will not always permit the desired level of performance. This problem is especially true considering the scale of real-world counterterrorism applications (which, recalling our earlier example, may easily contain 32,000 ground action atoms, or on the order of $2^{32{,}000} \approx 10^{9{,}900}$ possible worlds) where analysts will need to produce multiple iterative "what if?" scenarios and forecasts.

In this section various parallel algorithms are presented that leverage the sequential naive, HOP, SemiHOP, and binary heuristic algorithms. Parallelism will not only reduce the computation time necessary for forecasting a groups most likely actions in a particular scenario, but will also provide increased scalability, allowing a larger number of actions to be modeled and permitting analysis of more robust *ap*-programs. In addition, through parallel sampling in the randomized binary heuristic, a larger proportion of this huge space of possible behaviors can be covered, improving the accuracy of the approximation result and providing a better source for counterterror decision-support.

4.1 *Parallelism for Reducing Computation Time*

All of the forecasting algorithms discussed in Sect. 3 lend themselves to distributed computation in a fairly straightforward, pleasantly parallel way. A new class of algorithms, the Parallel Most Probable World (PMPW) algorithms, operate identically to the serial naive, HOP, SemiHOP, or binary heuristics, except that the computation of $\mathsf{low}(w_i)$ for each world w_i (resp. equivalence class or sub-partition) is distributed among N nodes of a computing cluster. Figure 15 contains the basic PMPW algorithm for forecasting group behavior. Let $\mathscr{C}$ be the set of constraints generated for either the naive, HOP, or SemiHOP algorithms.

PMPW is intentionally designed to be general and applicable to any of the serial forecasting algorithms. The input parameter $\mathscr{C}$ is a set of linear constraints that can be computed as $\mathsf{CONS}(\Pi, s)$, $\mathsf{RedCONS}(\Pi, s)$, $\mathsf{S\_RedCONS}(\Pi, s)$, or the $\mathscr{C}^\bullet$ returned by the binary heuristic. The procedure $divideVariables$ performs the division of the variables in $\mathscr{C}$ into batches to be distributed across the N

PMPWAlgorithm

1. Compute the reduction of Π w.r.t. s. $Forecast = NIL$; $Bestval = 0$.
2. Let $Batches = divideVariables(\mathcal{C})$
3. For each parallel process $n := 0$ to $N - 1$ do:
   a. $Bestval_n = 0$. $Forecast_n = NIL$
   b. For each variable $p_i \in Batches(n)$ do:
      i. **Minimize** p_i **subject to** $\mathcal{C}$. Let Val be the result.
      ii. If $Val > Bestval_n$, then set $Forecast_n = w_i$ and $Bestval_n = Val$.
4. $Bestval = \mathbf{max}_n(Bestval_n)$; $Forecast = \mathbf{argmax}_n(Forecast_n)$
5. If $Bestval = 0$ then return any world whatsoever, otherwise return $Forecast$.

Fig. 15 The general PMPW algorithm for forecasting terror group behavior in the SOMA framework

parallel processes. Like our other forecasting algorithms in the SOMA framework, PMPW finds the lower bound probability for each variable p_i in $Batch(n)$ assigned to parallel process n, storing the maximum of these values in $MaxVals$. After completing all of the distributed computation, the overall most probable world is returned as the behavioral forecast. In the best case, this division of labor in the PMPW algorithm can allow for a computation time improvement of up to a factor of N for N parallel processes.

4.2 *Parallelism for Increasing Computational Capacity*

The computation speedup afforded by the PMPW algorithm can help improve the scalability of the MPW problem. For any given amount of time, a greater number of possible actions or equivalence classes can be processed, allowing us to look at more complicated terror group behaviors. However, rather than simply distributing the serial forecasting algorithms and performing the same computations in parallel, another "pleasantly parallel" algorithm can be designed to better address the issues of scale in finding the most probable world of larger *ap*-programs, i.e., programs with a greater number of possible actions. In the PMPW-LR algorithm presented below, the structure of an *ap*-program Π is exploited to divide the associated linear constraints into distinct components for concurrent behavioral forecasting computations.

Before presenting the PMPW-LR algorithm, a structural representation of the relationships among actions in an *ap*-program must be defined. An *ap*-program Π can be represented as a graph in which the vertices are literals in the program (i.e., individual actions), and the edges indicate co-occurrence of these literals in an *ap*-rule.

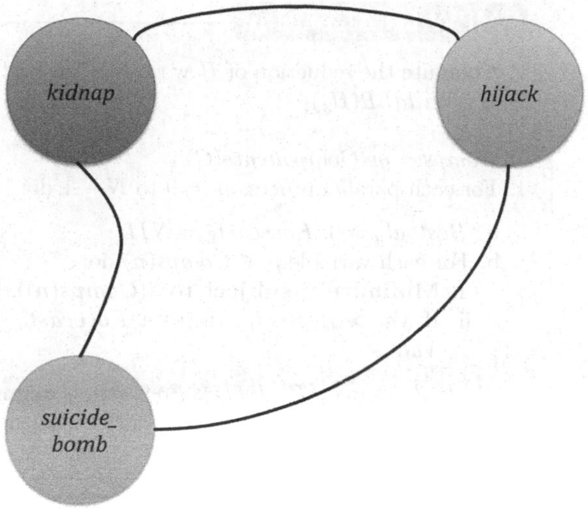

Fig. 16 The literal-relationship graph G_{Π} for a simple *ap*-program Π_s w.r.t state *s*

Definition 11 (Literal Relationship (LR) Graph). Let Π be an *ap*-program. The *literal relationship graph* (LR-graph) $G_{\Pi} = (V, E)$ is an *undirected* graph defined as follows.
$V = \{l \mid l$ is a literal (positive or negative action) appearing in a rule in $\Pi\}$.
$E = \{(l_i, l_j) \mid l_i, l_j \in V$ and l_i and l_j are either complementary literals or they both appear in a rule in $\Pi\}$.

Example 11. Consider a simple *ap*-program for a terror organization. Assume that this is the reduction of the program Π_s w.r.t. some state *s*.

Figure 16 shows the LR-graph associated with this program Π_s. The three action atoms *kidnap*, *suicide_bomb*, and *hijack* are each a vertex in the graph. Then, we draw edges between those actions that appear in a rule together: *kidnap* and *suicide_bomb* appear in rules (1) and (2) together, *kidnap* and *hijack* appear in rule (2) and *suicide_bomb* and *hijack* also appear in rule (2). This means we have three edges in our LR-graph.

Each connected component *c* in an LR-graph G_{Π} represents a subprogram Π_c of Π that utilizes only the literals in that component. Therefore, each connected component comprises a separate set of linear constraints $\mathsf{CONS}(\Pi_c, s)$, $\mathsf{RedCONS}(\Pi_c, s)$, $\mathsf{S\_RedCONS}(\Pi_c, s)$, or $\mathscr{C}^{\bullet}$ that can be used to forecast the group's behavior. By finding the maximally probable world in each component, these individual solutions can be compared to find the most probable actions of a terror organization across all components. The PMPW-LR algorithm given in Fig. 17 uses this methodology to divide a much larger *ap*-program into smaller pieces that can be computed in parallel across *N* nodes. This will allow security analysts to specify larger *ap*-programs—or more detailed models of terror group behavior—for use with the SOMA framework.

PMPW-LR Algorithm

1. Compute the reduction of Π w.r.t. s. $Forecast = NIL$; $Bestval = 0$.
2. $G = buildLR(\Pi_s)$
3. $Comps = getComponents(G)$
4. For each parallel process $n := 0$ to $N - 1$ do:
   a. $Bestval_n = 0$. $Forecast_n = NIL$
   b. For each variable $p_i \in Comps(n)$ do:
      i. **Minimize** p_i **subject to** $\mathcal{C}(Comps(n))$. Let Val be the result.
      ii. If $Val > Bestval_n$, then set $Forecast_n = w_i$ and $Bestval_n = Val$.
5. $Bestval = \mathbf{max}_n(Bestval_n)$; $Forecast = \mathbf{argmax}_n(Forecast_n)$
6. If $Bestval = 0$ then return any world whatsoever, otherwise return $Forecast$.

Fig. 17 The PMPW-LR algorithm for forecasting terror group behavior in the SOMA framework

The PMPW-LR algorithm first calls the procedure $buildLR(\Pi_s)$ to construct the LR-graph for the reduction of the *ap*-program Π. Using this graph, the function $getComponents(G)$ returns the connected components of G. The remainder of this algorithm is similar to the general PMPW algorithm in Fig. 15. First, the components in $Comps$ are divided amongst the N parallel processes. Then, on each node the most probable behavior is found for the assigned components w.r.t. the sub-program CONS($Comps(n)$), the portion of the linear constraints associated with the component $Comps(n)$. These results are combined to return the overall most probable world and forecast the group's behavior for state s.

While the PMPW-LR approach does not provide any explicit savings with regard to the computation time, it does allow the analysis of much larger *ap*-programs that can be divided into computationally feasible parallel components. However, the success of PMPW-LR is highly dependent on the structure of the underlying behavioral model. For example, if an *ap*-program Π for a particular terror organization produces a graph with only a single component (i.e., the original program Π), then PMPW-LR will be identical to the serial algorithms.

4.3 Parallelism for Improving Solution Accuracy

The randomized binary heuristic was introduced in Sect. 3 as an approximation algorithm for forecasting the most probable terror group behavior using a random sample of the set of all possible worlds (resp. equivalence classes or sub-partitions). In addition to improving the running time and scalability of producing a behavioral forecast, distributed algorithms can also be utilized to improve the quality and

PMPW-BH Algorithm

1. Compute the reduction of Π w.r.t.s. $Forecast = NIL$; $Bestval = 0$.
2. For each parallel process $n := 0$ to $N-1$ do:
   a. $Bestval_n = 0$. $Forecast_n = NIL$
   b. For each variable $p_i \in C_n^{\bullet}$ do:
      i. **Minimize** p_i **subject to** $C_n^{\bullet}$. Let Val be the result.
      ii. If $Val > Bestval_n$, then set $Forecast_n = w_i$ and $Bestval_n = Val$.
3. $Bestval = \mathbf{max}_n(Bestval_n)$; $Forecast = \mathbf{argmax}_n(Forecast_n)$
4. If $Bestval = 0$ then return any world whatsoever, otherwise return $Forecast$.

Fig. 18 The PMPW-BH algorithm for forecasting terror group behavior in the SOMA framework

accuracy of these approximation solutions by examining a greater portion of the sample space. In this section, two distributed approximation algorithms are described. The first, PMPW-BH, looks at concurrent random samples of worlds (resp. equivalence classes or sub-partitions) using the binary heuristic independently across N parallel processes. The second algorithm, PMPW-IBH, is an anytime iterative sampling approach that will examine multiple successive sets of parallel random samples to find the most probable group behavior, refining the sample set with each iteration of the parallel computation.

Figure 18 contains the PMPW-BH (Parallel Most Probable World–Binary Heuristic) algorithm, which is able to examine a greater proportion of the possible terrorist behaviors through distributed application of the binary heuristic. With this method, each parallel computation investigates a distinct random sample of possible worlds (resp. equivalence classes or sub-partitions). The procedure $Binary(\mathscr{C}, r)$ returns a set of linear constraints $\mathscr{C}^{\bullet}$, which is the result of applying the binary heuristic to a sample of r worlds (resp. equivalence classes or sub-partitions). To avoid sampling bias, the processes are asynchronous, and each takes its own random sample of size r. The resulting most probable worlds from each sample are then compared to find the most probable actions a terror organization will take in situation s. Using the PMPW-BH algorithm, a greater proportion of the possible worlds may be sampled, thereby improving the chances of finding a more accurate approximate solution with respect to the solutions returned by the exact algorithms.

The PMPW-IBH algorithm is a version of PMPW-BH that incorporates iterative sampling to further increase its ability to examine larger samples of possible worlds and achieve greater solution accuracy. PMPW-IBH maintains a set of the k most probable worlds returned by the current iteration on each of N parallel nodes. This set of known probable worlds is then propagated to the random sample of r worlds in the following iteration, so only $r-k$ new worlds will be sampled and the previous most probable set of k worlds will be retained.

Example 12. Suppose we are running PMPW-IBH where the size of our random sample is $r = 1{,}000$, and $k = 20$ is the number of top-k most probable world we will return on each iteration. In the first iteration the binary heuristic will randomly select 1,000 worlds to use in the forecasting computation of the most probable worlds. From these results, the 20 most probable behaviors will be used as part of the second iteration's sample, choosing only $r - k = 980$ new worlds and generating a new set of constraints $\mathscr{C}^\bullet$ with the binary heuristic. Using these new constraints, a new set of the 20 most probable worlds are again found and propagated to the next iteration.

PMPW-IBH continues this progressive refinement process until $maxIter$ iterations have been completed on each parallel process. Finally, the results from all of the computations are compared and the overall most probable behavior is forecast. As an anytime algorithm, PMPW-IBH allows the user to choose the desired level of solution refinement by setting the number of sample iterations.

5 Applications of *ap*-Programs

Because of its intuitive probabilistic logic representation and the efficient forecasting algorithms presented in the preceding sections, the SOMA framework has been used to develop *ap*-programs that model the cultural, economic, and social dynamics correlated with terror group behavior. Detailed case studies have been conducted to model the violent strategies of Hezbollah [8], Hamas [10], and Lashkar-e-Taiba [9]. In addition, SOMA has been used to model the behavior of other groups, including tribes in the Afghan-Pakistan border region involved in the drug trade (e.g., the Shinwari, Waziri, and Mohmand tribes), about 50 terror organizations from the Middle East and Asia Pacific regions, as well as political parties (e.g., Jamaat-i-Ulema Islami, Pakistan People's Party). SOMA has also been used as the backbone of a decision-support system for intelligence analysts and terrorism experts, the SOMA Terror Organization Portal [16], which makes the SOMA models and various forecasting methods available for analyzing the behavior of terror organizations in the Middle East.

With the exception of the models of the Afghan drug economy, all of the *ap*-programs mentioned above have been automatically extracted from two political science data sets: MAROB [2, 19] and CMOT.

The Minorities at Risk Organizational Behavior (MAROB) dataset [2, 19] from the Center for International Development and Conflict Management at the University of Maryland contains around 175 parameters to monitor about 118 ethnopolitical organizations in the Middle East and Asia Pacific regions that claim to represent the interests of repressed ethnic minorities, often employing violence and terrorism. These 175 attributes describe various aspects of the groups, such as whether or not they engaged in violent attacks, if financial or military support was received from foreign governments, and the type of leadership in each group. Recall from Sect. 2 that the SOMA framework divides the logical symbols into action and non-action components. As an explicitly behavioral dataset, it was easy to divide these attributes into action symbols—or strategic behaviors that could be taken by the group (i.e., bombings, kidnappings, armed attacks, etc.)—and non-action conditions (i.e., the contextual information, such as type of leadership, the kind and amount of foreign support, whether the group has a military wing, etc.). Values for these parameters are available for up to 25 years per group between 1980 and 2004, though there are fewer time points for some groups (e.g., those that have been around for a shorter duration). For each group, MAROB provides a time series as a relational table where the columns correspond to the 175 parameters and the rows correspond to the years. The MAROB data was used in the case studies of Hezbollah and Hamas, as well as providing models of around 50 terror organizations for the STOP decision-support system.

Data from the Computational Modeling of Terrorism (CMOT) project, which is a specialized codebook for developing datasets on terrorist and other violent organizations throughout the world, was used in the case study of Lashkar-e-Taiba. In addition to LeT, CMOT contains data for Jaish-e- Mohammed (JeM), Indian Mujahideen (Mujahideen fi-al Hind), Students Islamic Movement of India, Forces Democratique de Liberation du Rwanda (FDLR), and many others. CMOT is also an explicitly behavioral time series data set, making it ideal for analysis with the SOMA framework. Non-action symbols in CMOT include (i) the group's ideological goals (i.e., religious rule of a region, territorial autonomy, etc.), (ii) the political, economic, cultural, and security context of the state in which the group is active (i.e., diplomatic relations, the inflation rate, ethnic tensions, etc.), (iii) the equipment and weapons possessed or acquired by the group, (iv) basic characteristics of the group (i.e., leadership structure, membership base, etc.), and (v) the relationships of the group with third paries (i.e., financial backing from a diaspora, military support from a foreign state, etc.). Action attributes describe the behaviors and communications engaged in by the group, such as specific types of attacks (e.g., bombings, abductions, assassinations), targets (security personnel, civilians), and propaganda, as well as statistics regarding the outcomes of these events, such as deaths or casualties.

For each time period in the dataset, experts in political science, anthropology, public policy, and terrorism studies, researched the behavior and context of the organization using open source news articles and scholarly publications. The data for LeT has been coded at a monthly granularity from January 1985 to December 2010.

Any well-known algorithm for producing association rules can be used to find *ap*-rules as well. Here, APEX is presented as one possible algorithm for extracting *ap*-rules from standard relational time series data. To extract *ap*-rules automatically, action symbols (i.e., the dependent variables describing a terror group's behavior), which will occur in the *Heads* of the rules, and environmental (non-action) attributes, which will describe the state conditions in the *Bodies* of the *ap*-rules, must be identified in the data.

The significance of an *ap*-rule can be measured using the standard definitions of support and confidence from the literature.

APEX Algorithm

1. Select an action atom (an action attribute with an instantiated value) to be the head of the *ap*-rule.
2. Fix one environmental condition as part of the body of the rule.
3. Add varying combinations of the remaining environmental conditions to the body to determine if significant correlations exist between the body conditions and the head condition.
   a. Compute the support and confidence of the *ap*-rule
   b. If $support > support\_threshold$ and $confidence(Body) - confidence(\neg Body) > confidence\_threshold$ then add the rule to the *ap*-program
   c. The probability range of the rule is given by $[confidence - \sigma, confidence + \sigma]$.

Fig. 19 The APEX algorithm for extracting SOMA models from relational data

Definition 12 (Support). For an action condition AC, an environmental condition EC, and a database DB, the *support* is defined as:

$$S_{AC,EC} = \frac{|\{t \ s.t. \ t \in DB \wedge (AC = true \wedge EC = true)\}|}{|DB|}.$$

Definition 13 (Confidence). For an action condition AC, an environmental condition EC, and a database DB, the *confidence* is defined as:

$$C_{AC,EC} = \frac{|\{t \ s.t. \ t \in DB \wedge (AC = true \wedge EC = true)\}|}{|\{t \ s.t. \ t \in DB \wedge EC = true\}|}.$$

The APEX algorithm calculates the difference between the confidence value produced by an environmental condition and by its negation. If this difference is above a given threshold, then the *ap*-rule is extracted and added to the behavioral model. To obtain the probability range for the extracted rule, the confidence value initially obtained is used, plus/minus the standard deviation σ of the values involved in its calculation. Note that this algorithm is not a novel one and simply performs calculations to capture interesting correlations in the data to construct rules.

The APEX algorithm provides a flexible way to automatically extract behavioral models of terror groups from time series data. It was used to create models of the strategies of violent organizations—many of which are terrorist groups—from the MAROB and CMOT datasets. For instance, approximately 14,000 *ap*-rules have been extracted for Hezbollah, and over 2,000 for LeT. A sample of some of these rules is presented in Fig. 20. These rules indicate environmental conditions under which LeT is likely to attack Indian security forces in Jammu and Kashmir or engage in fedayeen attacks. In fedayeen attacks, the perpetrators assault a target and are willing to (and probably intend to) die in combat, but unlike suicide bombings per

Automatically extracted *ap*-rules for Lashkar-e-Taiba

(1) $Attack\_Indian\_Sec\_Forces(LeT, J\&K)\ :\ [0.71] \rightarrow$
$Financial\_Support\_by\_Diaspora(LeT, Pak) \wedge$
$Businesses(LeT, Pak)$
When LeT is receiving financial support from Pakistan's diaspora and are engaged in businesses in Pakistan, then the group will attack Indian security forces in Jammu and Kashmir with a probability of 0.71.

(2) $Attack\_Indian\_Sec\_Forces(LeT, J\&K)\ :\ [0.89] \rightarrow$
$Social\_Strife(Pak) \wedge$
$Alliance\_NSAG(LeT, J\&K)$
When there is social strife in Pakistan and LeT has entered into an alliance with another non-state armed group (NSAG) in Kashmir, then LeT will attack Indian security forces in Jammu & Kashmir with a probability of 0.89.

(3) $Attack\_Fedayeen(LeT, J\&K)\ :\ [0.7] \rightarrow$
$Military\_Support\_by\_Government(LeT, Pak) \wedge$
$Advocating\_Belief\_System(LeT, Pak).$
LeT will engage in Fedayeen attacks in Jammu and Kashmir with a 0.7 probability when they are receiving military support from Pakistan and are advocating adherence to their belief system in Pakistan.

Fig. 20 *ap*-rules describing the behavior of Lashkar-e-Taiba. These rules were automatically extracted from the CMOT dataset using the APEX algorithm

se, they will not die by their own hand and may in fact survive the attack [9]. Mannes et al. [9] provides further analysis of several rules concerning the behavior of LeT and uses these rules to suggest potential policies for mitigating violent attacks in Jammu and Kashmir.

6 Conclusions

In this chapter, the Stochastic Opponent Modeling Agents (SOMA) framework for modeling, analyzing, and forecasting terror group behavior was presented. SOMA is based on the probabilistic logic programming [12] paradigm, allowing probabilistic reasoning without independence assumptions and leveraging the intuitive representation of logic to provide models that can be interpreted by experts.

We gave several basic algorithms for reasoning in this framework to find the most probable world, which is analogous to forecasting the most likely set of actions a group will take in a particular situation. Much of the focus of this work has been in managing the computational complexity of real world counterterrorism scenarios, where the number of possible actions can be on the order of $10^{30,000}$. With manipulation of the probabilities in the *ap*-program and distributed computation, we are able to achieve levels of scalability and efficiency superior to past efforts with probabilistic logic models. The success of these reasoning algorithms for *ap*-programs makes this framework applicable to real-world counterterrorism situations where experts may need to address a broad range of possible actions or run multiple iterative "what if?" forecasts. Applications of this framework have produced tangible results of use to U.S. military officers and national security experts, showing promise for use with other datasets and expanded international security domains [1, 17].

There are, of course, many problems in *ap*-programs and counterterrorism modeling that remain open. While *ap*-programs provide an effective way to model the behavior of various terror groups, the algorithms for finding the most probable world are not yet efficient enough for real-time computation and behavioral forecasting. Such improvements to this framework would make these models more useful as a tactical decision-support tool.

Acknowledgements Some of the authors of this paper were funded in part by AFOSR grant FA95500610405, ARO grant W911NF0910206 and ONR grant N000140910685.

References

1. Bhattacharjee Y (2007) Pentagon asks academics for help in understanding its enemies. Sci Mag 316(5824):534–535
2. Center for International Development and Conflict Management (2008) Minorities at risk organizational behavior dataset, minorities at risk project. Retrieved from http://www.cidcm.umd.edu/mar.
3. Fagin R, Halpern JY, Megiddo N (1990) A logic for reasoning about probabilities. Inf Comput 87(1/2):78–128
4. Hailperin T (1984) Probability logic. Notre Dame J Form Log 25(3):198–212
5. Khuller S, Martinez V, Nau D, Simari GI, Sliva A, Subrahmanian VS (2007) Computing most probable worlds of action probabilistic logic programs: scalable estimation for $10^{30,000}$ worlds. Ann Math Artif Intell 51(2–4):295–331
6. Khuller S, Martinez V, Nau D, Simari GI, Sliva A, Subrahmanian VS (2007) Finding most probable worlds of probabilistic logic programs. In: Proceedings of the 2007 international conference on scalable uncertainty management. Springer Verlag lecture notes in computer science, vol 4772. Springer, Berlin/New York, p 45–59
7. Lloyd JW (1987) Foundations of logic programming, 2nd edn. Springer, Berlin/New York
8. Mannes A, Michael M, Pate A, Sliva A, Subrahmanian VS, Wilkenfeld J (2008) Stochastic opponent modeling agents: a case study with hezbollah. In: First international workshop on social computing, behavioral modeling, and prediction. Springer, New York/London
9. Mannes A, Shakarian J, Sliva A, Subrahmanian VS (2011) A computationally-enabled analysis of lashkar-e-taiba attacks in jammu & kashmir. In: Proceedings of the European Intelligence and Security Informatics Conference. IEEE. 224–229
10. Mannes A, Sliva A, Subrahmanian VS, Wilkenfeld J (2008) Stochastic opponent modeling agents: a case study with hamas. In: Proceedings of the second international conference on computational cultural dynamics (ICCCD). AAAI

11. Ng RT, Subrahmanian VS (1991) A semantical framework for supporting subjective and conditional probabilities in deductive databases. In: Furukawa K (ed) Proceedings of ICLP '91. MIT, Cambridge, pp 565–580
12. Ng RT, Subrahmanian VS (1992) Probabilistic logic programming. Inf Comput 101(2): 150–201
13. Nilsson N (1986) Probabilistic logic. Artif Intell 28:71–87
14. Simari GI, Martinez MV, Sliva A, Subrahmanian VS (2012) Focused most probable world computations in probabilistic logic programs. Ann Math Artif Intell 64(2–3):113–143
15. Sliva A, Martinez V, Simari GI, Subrahmanian VS (2007) Soma models of the behaviors of stakeholders in the afghan drug economy: a preliminary report. In: First international conference on computational cultural dynamics (ICCCD 2007). ACM, Menlo Park
16. Sliva A, Subrahmanian VS, Martinez V, Simari GI (2008) The SOMA terror organization portal (STOP): social network and analytic tools for the real-time analysis of terror groups. In: Proceedings of the 2008 first international workshop on social computing, behavioral modeling and prediction. Spring Verlag Lecture Notes in Computer Science. Springer, New York/London, pp 9–18
17. Subrahmanian VS (2007) Cultural modeling in real-time. Science 317(5844):1509–1510
18. Subrahmanian VS, Albanese M, Martinez V, Reforgiato D, Simari G, Sliva A, Udrea O, Wilkenfeld J (2007) CARA: a cultural reasoning architecture. IEEE Intell Syst 22(2):12–16
19. Wilkenfeld J, Asal V, Johnson C, Pate A, Michael M (2007) The use of violence by ethnopolitical organizations in the middle east. Technical report, National Consortium for the Study of Terrorism and Responses to Terrorism

Data-based Computational Approaches to Forecasting Political Violence

Philip A. Schrodt, James Yonamine, and Benjamin E. Bagozzi

1 Introduction and Overview

The challenge of terrorism dates back centuries if not millennia. Until recently, the basic approaches to analyzing terrorism—historical analogy and monitoring the contemporary words and deeds of potential perpetrators—have changed little: the Roman authorities warily observing the Zealots in first-century Jerusalem could have easily traded places with the Roman authorities combatting the Red Brigades in twentieth century Italy.

This has changed with the exponential expansion of information processing capability made possible first by the development of the digital computer, followed by the phenomenal growth in the quantity and availability of machine-readable information made possible by the World Wide Web. Information that once circulated furtively on hand-copied sheets of paper (or papyrus) is now instantly available—for good or ill—on web pages which can be accessed at essentially no cost from anywhere in the world. This expansion of the scope and availability of information in all likelihood will change the dynamics of the contest between organizations seeking to engage in terrorism and those seeking to prevent it. It is almost certainly too early to tell which group will benefit more—many of the new media are less than a decade old—but the techniques of processing and evaluating information will most certainly change.

This chapter provides an extensive overview of inductive statistical and computational methodologies used in the analysis and forecasting of political violence, and some of the challenges specific to the issue of analyzing terrorism. It is intended for the non-specialist, but assumes a general familiarity with data and computational methods. Our purpose is not to exhaustively explore any of these methods—each

---

P.A. Schrodt (✉) • J. Yonamine • B.E. Bagozzi
Political Science, Pennsylvania State University, University Park, PA 16801, USA
e-mail: schrodt@psu.edu; jxy190@psu.edu; beb196@psu.edu

V.S. Subrahmanian (ed.), *Handbook of Computational Approaches to Counterterrorism*,
DOI 10.1007/978-1-4614-5311-6_7,

technique would typically require tens or hundreds of pages—but instead to provide a sufficient introduction to the basic concepts and vocabulary, as well as a very extensive sets of references, so that the reader can explore further on his or her own.

The psychologist and philosopher William James, in his Lowell Institute lectures in 1906, subsequently published under the title *Pragmatism: A New Name for Some Old Ways of Thinking* notes that the fundamental split in philosophy, dating to the very origins of the field, is between "rationalists" who seek to find an intellectual structure that will reveal a proper order in the world, and "empiricists," who take the disorder of the observed world as a given and simply try to make as much sense of it as they can. More than a century later, we find exactly the same split in formal approaches in the social sciences: The rationalist position is expressed in deductive approaches such as game theory, expected utility models, systems dynamics and agent-based models, which seek to explain behavior from a set of a priori first-principles and their consequent emergent properties. The empiricist approach is found in inductive statistical and computational data-mining approaches which extract structured information from large sets of observed data. Both approaches are well-represented in this volume [38, 48, 103, 105, 117, 123, 151, 152] but this review will focus only on the empirical approaches.

Throughout this chapter, we will generally be looking at models which focus on forecasting and understanding political violence in general, not just approaches to terrorism per se. This is done for two reasons. First, due to a combination of data limitations and a paucity of interest in the research community prior to the 2000s, the number of quantitative studies of terrorism was quite limited and focused on a relatively small number of approaches [149]. However, the large-scale research efforts did consider various forms of political violence more generally. Second, most of those more general methods are at least potentially applicable to the study of terrorism—in fact, many are generally applicable to almost any behavior for which large amounts of data are available—albeit we will frequently caveat these possibilities with concerns about some of the atypical aspects of terrorism, in particular the distinctions between methods appropriate to rare-events analysis and those dealing with high-frequency behaviors. Finally, in many instances, there is a close relationship between situations of general political violence such as civil war and state failure, and conditions which encourage the formation and maintenance of terrorist groups, so political instability is of considerable interest on its own.

We will first provide a general overview the types of data commonly used in technical forecasting models, then consider the two broad categories of computational models: statistical and algorithmic. Within statistical modeling, we assess the strengths and weaknesses of the widely-used ordinal least squares and logit methods, conventional time series approaches, vector-autoregression models, event-history models, and a variety of techniques involved in rare-events analysis. Within computational modeling, we consider supervised classification methods and unsupervised clustering methods, and then look more specifically at the issue of using event sequences for forecasting. Finally, we will briefly consider some of the recent developments in computational network analysis.

1.1 The Development of Technical Political Forecasting

There is both a push and a pull to the current interest in technical political forecasting. As with many aspects of the Information Revolution of the past several decades, some of this is driven purely by technology and the availability of data, computational power and readily accessible software. We can do things that we couldn't do before, and clearly the impact of intellectual curiosity and technical itches to scratch is a factor.

But if the technological factors were the only driver, we could satisfy that by clicking on virtual cows and broadcasting our prowess on Facebook (http://www.wired.com/magazine/2011/12/ff_cowclicker/all/1). Instead, at the level of serious policy analysis—mess up counter-terrorism, and real people die, with no option to reset—the motivation is different: humans are appallingly bad at predicting human behavior. The pathbreaking studies of Philip Tetlock [161], who assessed the performance of 284 expert political forecasters in the assessments of more than 80,000 predictions over a 20 year period, found that their accuracy was barely better than chance, and the record for some of the most publicly-visible forecasters, who play to a crowd anxious for dramatic predictions, is actually worse than what could be achieved by a dart-throwing chimpanzee. Tetlock's results confirmed for the political realm—albeit in much greater detail—similar findings in economics and psychological assessment [94, 111] but were nonetheless a stunning indictment of the "subject matter expert" as a prognosticator.

Tetlock's work was particularly timely given the intellectual climate for the development of statistical and computational models of political violence. The earliest systematic studies of political violence using modern statistical methods and standards for data collection date from the middle of the twentieth century, in the work of British meteorologist Lewis Richardson [131]. This approach gained considerable momentum in the 1960s with the "behavioral revolution" in political science, which used the emerging technologies of social science statistics, digital computing, and machine-readable data to begin the systematic assessment of various theoretical propositions about political violence.

These found, for the most part, that the patterns in the qualitative "wisdom literature" were far less generalizable than their advocates believed—perhaps unsurprising in a world that had experienced two civilization-shattering wars in only three decades—with the results on the "democratic peace" [12, 15, 81] being the notable exception. This period saw the initial development of many large-scale databases relevant to the study of political violence, including the Correlates of War (http://www.correlatesofwar.org/) on international and civil war, Polity (http://www.systemicpeace.org/polity/polity4.htm) on state political characteristics, and early event data sets [8, 36, 99, 110] on political interactions in general. The data sets collected specifically for the study of political behavior were supplemented by readily-available global data on demographic, development, economic and trade characteristics of states provided by organizations such as the United Nations (UN), International Monetary Fund (IMF) and World Bank.

The U.S. Department of Defense Advanced Research Projects Agency (DARPA) funded a series of projects for the development of statistical forecasting in the 1970s [4, 44, 85], and parallel efforts, using both statistical and computational ("artificial intelligence") methods continued in the political science community under National Science Foundation funding [39, 76, 87, 160]. However, these early efforts were not particularly successful in the production of robust, timely, and accurate forecasts. In retrospect, neither the data nor methods available at the time were sufficient for the task. By the 1980s U.S. government efforts had largely ended, though the basic research in the academic community continued, as did some efforts by international non-governmental organizations [45, 75, 79, 132].

The U.S. government re-engaged with the development of technical political forecasting in the mid-1990s, motivated in part by the considerable progress in the development of methods and data since the earlier efforts, and by dramatic intelligence failures by human analysts on such events as the end of the Cold War [64, 65], state failure in Somalia and the former Yugoslavia, and the Rwandan genocide. The most conspicuous effort was the U.S. multi-agency State Failures Project [60, 61], later renamed the Political Instability Task Force (PITF) [74], a long-term collaboration between government analysts, contractors, and academic researchers. After some initial missteps involving efforts which were overly complex, PITF developed models that predicted various forms of political instability with 70–80% accuracy[1] with a 2-year time horizon and global coverage. In the mid-2000s, DARPA returned to the issue with the Integrated Conflict Early Warning System (ICEWS) [117, 118], which achieved similar levels of accuracy.

As a consequence of these various efforts, technical forecasting of political violence is now quite well developed and the subjects of numerous efforts in both North America and Europe [18, 27, 33, 82–84, 118, 165, 167]. While most of these studies look at violence and irregular political activity such as coups, a quantitative literature has also emerged on the topic of terrorism; extended reviews of this literature can be found in [66, 126, 127, 150, 169].

2 Data Sources

As noted in the previous section, one of the major factors fueling the development of computation methods for the analysis and forecasting of political violence has been the availability of machine-readable data sets. Collection of data sets has now been underway for close to five decades—the Correlates of War Project, for example, began in 1963—and are now the product of thousands of hours of careful research, refinement and coding, and have been used in hundreds of studies. In other cases—automated coding of atomic event data and analysis of the new social media—the

[1] Defined here in terms of the percentage of conflict-onset cases correctly classified.

collections are relatively new and unexplored, but potentially provide very large amounts of data in near-real-time. These data are generally available on the web, either through academic archives such as the Inter-University Consortium for Social and Political Research (http://www.icpsr.umich.edu/icpsrweb/ICPSR/) and Harvard's Dataverse Network (http://thedata.org/), through government and IGO sources such as USA.gov (http://www.usa.gov/Topics/Reference-Shelf/Data.shtml) and the United Nations Statistics Division (http://unstats.un.org/unsd/databases.htm) or—increasingly—through individual web sites established by the projects collecting the data.

In this section, we will briefly describe some general types of data that have been used in models of political violence, with comments on a few of the strengths and weaknesses of each type. As with any typology, not all of the data sets fit clearly into a single category, but most will.

2.1 Structural Data

Structural data refer to characteristics of a single political unit: typically this is a nation-state but it could also be a province or city, or a militarized group. At the nation-state level, classical structural measures include socio-economic variables such as population, literacy and infant mortality rates, gross domestic product (GDP), GDP per capita, percent urban population, ethnic composition, and the Gini index measure of income inequality. Specialized political science data sets such as Polity [104] and Freedom House [62] provide an assessment of a large number of dimensions of governance, often at a very high level of detail: for example the Institutions and Elections Project [130] has coded 127 characteristics of government and electoral institutions for all countries with populations over 500,000 from 1972–2005.

At the (militarized) group level, structural data is available for a wide variety of terrorism-related actors [6, 112, 153]. For instance, the Minorities at Risk (MAR) and Minorities at Risk Organizational Behavior (MAROB) datasets each provide extensive structural information on the social characteristics (e.g., ideology, religion, population) of politically active communal groups throughout the world [6, 112]. Finally, note that structural data can be used either directly or indirectly. GDP and population are good examples of direct measures: they are primarily used to adjust for the economic capacity and number of people in a country. Probably the best known indirect measure is infant mortality rate (IMR), which has consistently emerged as a one of the most important predictors of political instability. This is not due to any tendency of individuals whose children have died to immediately go out and revolt, but rather because IMR has proven, in multiple independent studies, to be a good indirect measure of a combination of poverty and the failures of a government to deliver basic services.

2.2 Dyadic Data

Dyadic data deal with *relations* between two entities. As above, these are usually states but could also be individuals or organizations. Trade data—notably the IMF's *Direction of Trade* [89]—are some of the most commonly used. Again, depending on the application, these can be used as either direct or indirect measures: trade is important in its own right, but measures like bilateral trade can also reflect the extent to which two states have common interests. Data on shared alliances, joint international organization (IGO) membership and military involvement (as allies or antagonists) are also readily available and are among the commonly used concepts used to test traditional international relations theories.

In recent years, the most important development in dyadic measures has been geospatial, which we discuss in Sect. 5.2. While the concept of political geography is quite old, only recently have political-geographic measures reached a level where they can be used systematically. However, political violence has a very strong geospatial component—shared borders are consistently the single most important predictor of whether states will engage in conflict, and overrides issues such as commonalities or differences in religion, language and culture [15], and high levels of terrorism tend to strongly cluster in both time and space—and these methods are gaining increasing attention [34, 35, 37, 129, 162].

2.3 Atomic Event Data

"Atomic" events data—usually called simply "event data" in the literature are basic units of political interaction coded from news sources that provide the date, source, target, event-type of an interaction. These have a long history in the quantitative analysis of conflict, with the early large data sets, the World Event Interaction Survey (WEIS) [109, 110] and the Conflict and Peace Data Bank (COPDAB) [8, 9] dating to the 1960s. While the original data were laboriously coded by (bored) students working with paper and microfilm, starting in the 1990s collection shifted to automated coding from machine-readable news wires [20, 67, 123, 138, 143], which resulted in data sets coded in near-real-time containing millions of events [72, 118, 143]. Along these lines, a number of more specialized event data sets have recently been developed to specifically code terrorism events, and employ a mixture of human and automated coding techniques to do so. Most notably, the National Counterterrorism Center's Worldwide Incidents Tracking System (WITS), the Global Terrorism Database (GTD), and the Terrorism in Western Europe Events Data set (TWEED) all provide wide coverage and detailed codings of terrorist attacks and related events [114, 120, 163].

Most event data sets use systematic typologies to reduce the nearly infinite variety of event descriptions possible in natural language reports to a manageable number of categories that a computational model can be process. All of these typologies have about 20 "macro" categories and around 100 more detailed subcategories. While the original event data collections focused almost exclusively on nation-states, contemporary systems provide higher levels of substrate aggregation, sometimes down to the level of individuals [68,147]. While most studies in the past aggregated the events into a single cooperation-conflict dimension using a scale [73], more recent approaches have created composite events out of patterns of the atomic events [88,140] or looked at sequences of events [50].

2.4 *Composite Event Data*

"Composite" data are those which code the characteristics of an extended set of events such as a war, crisis, or terrorist incident: the Correlates of War project (COW; http://www.correlatesofwar.org/) is the archetype; International Crisis Behavior (http://www.cidcm.umd.edu/icb/) is a more recent example as is the Global Terrorism Database [reference to **The Global Terrorism Database, 1970–2010** Gary LaFree, Laura Dugan, Chap. 1 in this volume]. And multiple characteristics of the incident are coded. Composite event data are typically coded from a number of reports of the event—as distinct from atomic event data, which generally looks at single sentences—and typically code a large number of characteristics of the event. At present, composite event data are usually coded by humans, though machine-assisted coding approaches are becoming increasingly prominent due primarily to automated "data field extraction" methods, able to rapidly locate information from text, such as the number of individuals killed or the amount of aid promised.

In some instances, it may also be possible to define composite events such as "civil war" by using patterns of the atomic events [88]. This would also make the differences between definitions used by various project unambiguous (or at least comparable) and allow the composite events to be easily constructed out of existing data sets rather than starting every new project from the beginning, and dramatically reduce the cost of this type of data.

2.5 *Social Media and Other Unstructured Data Sources*

The new social media—web sites, blogs, chat rooms, Twitter, Facebook, and other information easily available on the web—represent an emerging frontier in data collection. The advantages of these sources are clear: they can provide direct access to information about groups through the materials that they voluntarily make available, they are easily accessible at nearly zero cost, and with automated natural language processing, they can be analyzed in near-real-time. Social communication media such as Twitter and Facebook also provide extremely finely-grained, minute-by-minute data. In addition, a number of analysts have suggested that the social media themselves are changing the character of political mobilization, as demonstrated by the events of the "Arab Spring," particularly in Tunisia and Egypt.

Skeptics, however, have pointed out that social media also have some disadvantages. While access to the Web is increasing rapidly, it is still far from universal, and social media in particular tend to be disproportionately used by individuals who are young, economically secure, and well-educated. In areas with strong authoritarian regimes, notably China, there are substantial (though not uniformly successful) efforts to control the use of these media, and a government agency with even modest resources can easily create a flood of false posts, sites and messages. While there is *some* political content in these media, the vast bulk of the postings are devoid of political content—`OMG, Bieber fever!!!`—and what relevant content does exist may be deliberately or inadvertently encoded in a rapidly-mutating morass of abbreviations and slang almost indecipherably to conventional NLP software. (This contrasts to the news stories used to encode atomic event data, which generally are in syntactically correct English.) Finally, a number of analysts have argued that the critical communications development for political mobilization is the cell phone, both for voice and texting, rather than the Web-based media.

2.6 *The Challenges of Data Aggregation*

In most instances, conflict data comes pre-aggregated: no assembly required. However, researchers interested in analyzing specific events extracted form the text (which is common in event data studies as well as social network analysis), must make critical aggregation decisions in three areas to convert the raw text or 'event triplets' into a format suitable to their models of choice: (1) actors (2) actions, (3) temporal. (In contrast, techniques such as sentiment analysis and unsupervised clustering algorithms (see Sect. 4.2.2), are equipped to analyze text in its raw format.

2.6.1 Actors

Most major event data datasets—including WEIS, CAMEO, ICEWS, and VRA—code the source and target actors for each event in the data set. However, many of these actors may be irrelevant to the specific outcome-of-interest. For example, a study focusing on Israeli-Palestinian conflicts would not want to include events between Aceh rebels and the Indonesian army, as these are not relevant to the conflict of interest. Although excluding Indonesian rebel activity is obvious in this case, more difficult decisions exist, such as whether or not to include events between members of the Lebanese and Syrian armies, or the governments of the United States and Iran in a study of conflict between Israel-Palestine. Yonamine [170] provides a more detailed description of event data aggregation.

2.6.2 Actions

Event datasets use a numerical code to reflect the specific type of event that is occurring between the two actors. Since the numerical codes do not carry intrinsic value, researchers manipulate the code to reflect meaningful information. The majority of extant event data literature either scales all events, assigning them a score on a conflict-cooperation continuum or generates event counts reflecting the number of events that occur within conceptually unique categories. The Goldstein Scale [73], which is the most commonly used scaling technique within the event data literature [73, 77, 125, 136, 138], assigns a value from a −10 to +10 conflict/cooperation scale, with −10 reflecting the most conflictual events and 10 indicating the most cooperative.

Despite the preponderance of the Goldstein scale, a number of other studies [140, 141, 144] utilize count measures. Duvall and Thompson [51] put forth the first event data count model by placing all events into one of four conceptually unique, mutually exclusive categories: verbal cooperation, verbal conflict, material cooperation, material conflict. Although this count approach is more simplistic than scaling methods, [135, 144] find strong empirical results using this count method of action aggregation.

2.6.3 Temporal

Finally, scholars must temporally aggregate data in order to perform empirical analyses at levels appropriate for their theory or empirical models of choice. All of the previously mentioned event data sets code the exact day on which events occur. As the specific time-of-day that events occurred is not reported, events must at the very minimum be aggregated to the daily level [125, 135, 144], though weekly [26, 148], monthly [136, 165], quarterly [93], and annual level aggregations are common within the literature. By aggregated, we mean that the events occurring within the selected temporal length must be jointly interpreted. Common approaches are to calculate the sum or the mean of events that occur within the chosen temporal domain.

3 Statistical Approaches

Most of the work on forecasting political conflict has used statistical modeling, since this has a much longer history in political science than algorithmic and machine learning approaches. While the bulk of these studies have focused on simply interpreting coefficient estimates within the "frequentist" mode of significance testing, a method which has proven to have little utility for predictive modeling [165], more recent work has taken prediction seriously, both using classical time series models and more recently a substantial amount of work using vector autoregression (VAR) models. In addition, recent work has focused on one of the most challenging aspects of forecasting, particularly when applied to counterterrorism: the fact that these events occur very rarely. While this presents a serious challenge to the method, a number of sophisticated methods have been developed to deal with it.

3.1 *Cross-Sectional Regression and Logit*

By far the most common statistical model in contemporary research—probably accounting for at least 80% of the studies—are variants on ordinary least squares regression and the closely-related logit model. Ordinary least squares regression uses equations of the form

$$Y_i = \beta_1 x_{i1} + \cdots + \beta_p x_{ip} + \varepsilon_i = x_i'\beta + \varepsilon_i, \qquad i = 1, \ldots, n, \tag{1}$$

Logistic regression, in contrast, is used to predict a values between 0 and 1—typically interpreted as a probability—and does this by using an equation of the form

$$Y_i = \frac{e^z}{e^z + 1} = \frac{1}{1 + e^{-z}} \tag{2}$$

where the variable "z" is usually defined as

$$z = \beta_0 + \beta_1 x_1 + \beta_2 x_2 + \beta_3 x_3 + \cdots + \beta_k x_k, \tag{3}$$

While these approaches have been used extensively in the academic literature, their practical utility is, unfortunately, quite limited due to a reliance on the "frequentist" significance testing approach. Extended discussions of this issue can be found in [3, 70, 96, 137] but briefly, significance testing was originally developed in the early twentieth century to study problems where the "null hypothesis" of a variable having no effect was meaningful (the archetypical example is whether a new medicine has a different effect than a placebo). In some political science applications, this is still valid—for example, determining whether a forthcoming election is a "statistical tie" based on opinion polling—but in most models a variable is generally not included unless there are theoretical (or common sense) reasons to assume that it will have at least *some* effect. Because this is all that the significance test is assessing—a non-zero effect, not an effect that has some meaningful impact—the information it provides is limited and all but negligible for predictive problems [165].

Although the frequentist approach can be useful in weeding out variables that might seem to be important but in fact are not, in contemporary models that tend to be complex, even this should be interpreted with caution. Linear models are particularly susceptible to problems of "colinearity" when the independent variables x_i are correlated—as is often the case, particularly in models where a researcher is experimenting with several measures of a diffuse concept such as "economic development" or "authoritarianism"—and in the extreme (but not uncommon) case, colinearity can cause the estimated direction of a variable to be the opposite of what it would be in the absence of the colinearity. Specification error—leaving out a variable that in fact has a causal link to the dependent variable Y_i—is another issue, and has the unfortunate impact of inflating the estimated coefficients of the variables which are in the model.

Finally, the existing literature tends to use complex models, often with a dozen or more variables, despite the fact that a large set of literature, going back as early as the 1950s [111], with the lesson repeated by [3] about a decade ago, indicates that simple models are preferable, for at least two reasons. First, complex models tend to "fit the error," providing overly-optimistic assessments of the accuracy of the model based on the existing data, with those estimates *decreasing* the accuracy of the model once new data are available. Second, the nearly inevitable presence of colinearity in non-experimental social science variables tends to increase the variance of the estimated coefficients as the number of independent variables increase.

On the positive side, the strengths and weaknesses of linear models have been studied for decades and are well-understood. The problems noted above are widespread but not inevitable, and models which have confirmed the same result in a wide variety of formulations—for example on the impact of geographical contiguity, joint-democracy and trade on interstate conflict, and the impact of economic development on political instability—are probably robust. There has also been a gradual shift away from the classical frequentist approaches to the "Bayesian" approach [71, 92], which simply uses the data to adjust estimate of the distribution of the coefficient, rather than employing either/or significance tests, as well as methods such as matching cases within a sample [121] which can improve the accuracy of the coefficient estimates.

3.2 Classical Time Series

Classical time series models predict the future value of a continuous variable based on some combination of the past values, usually with a focus on reducing the systematic error in the model to "white noise", that is, errors with a mean of zero and the correlation between any ε_t and ε_s is equal to zero. The classical reference is [22], an open-source introductory is available at http://statistik.mathematik.uni-wuerzburg.de/timeseries/, and an extended treatment can be found in [78]. Examples in the literature on forecasting political conflict include [79, 125, 145, 146]; [124] provides a general reference with respect to applications in political science.

The notation AR("p") refers to the autoregressive model of order "p". The AR("p") model is written

$$X_t = c + \sum_{i=1}^{p} \varphi_i X_{t-i} + \varepsilon_t \tag{4}$$

where $\varphi_1, \ldots, \varphi_p$ are the parameters of the model, c is a constant and ε_t is white noise. In most applications, the error terms ε_t are assumed to be independent, identically-distributed, and sampled from a normal distribution.

The notation $MA(q)$ refers to the moving average model of order q:

$$X_t = \mu + \varepsilon_t + \sum_{i=1}^{q} \theta_i \varepsilon_{t-i} \tag{5}$$

where the $\theta_1, \ldots, \theta_q$ are the parameters of the model, μ is the expectation of X_t (often assumed to equal 0), and the $\varepsilon_t, \varepsilon_{t-1}, \ldots$ are white noise error terms.

The notation $ARMA(p, q)$ refers to the model with p autoregressive terms and q moving-average terms, where a moving average is defined by . This model contains the $AR(p)$ and $MA(q)$ models,

$$X_t = c + \varepsilon_t + \sum_{i=1}^{p} \varphi_i X_{t-i} + \sum_{i=1}^{q} \theta_i \varepsilon_{t-i} \tag{6}$$

In practical terms, classical time series models are useful for phenomena where the current value of the variable is usually highly dependent on past values. This "incremental behavior" is characteristic of a great deal political behaviors, notably public opinion, budgeting, and most conflict behavior: once a civil war or insurgency gets going, it is likely to continue for a while, and once peace is firmly established, that is also likely to the maintained. In such situations, the best predictor of the variable at time t is its value at time $t - 1$ Consequently, these models are very widely used.

Incremental models, however, have a critical weakness: by definition, they cannot explain sudden change, whether isolated incidents or the onset and cessation of protracted behaviors. An autoregressive model can have a very good overall predictive record but still miss the beginning and end of a conflict, and these may be of greatest interest to decision-makers.

The classical time series literature places a great emphasis on the characteristics of the error terms. While these are treated in the literature as random variables, in practical terms, much of the "error" is simply the effects of variables that were not included in the model. Because most social and demographic indicators and many economic indicators are also very strongly autoregressive—for example indicators such as infant mortality rate and literacy rarely change more than a percentage point year to year—these "errors" will strongly correlate with their lagged values, hence the interest in the MA and $ARMA$ models. Unfortunately, disentangling the effects of autoregressive variables and autoregressive errors is extraordinarily difficult in many circumstances, which in turn has led to the development of vector-autoregressive models, discussed in Sect. 3.3.

Finally, one will occasionally encounter questions about "co-integration" and "non-stationarity" in time series models [113]. This issue generally arises in situations where exponential growth is occurring in the variable, as can happen in situations of hyper-inflation, individuals who are really lucky in their choice of stocks or really unlucky in their choice of real estate markets. While conflict data will switch between *modes*—peace, crisis and war—and this is a form of non-stationarity (which refers to a change in the statistical characteristics of the errors), this is not an issue of co-integration. Unfortunately, the tests for co-integrated time series have a very high probability of incorrectly identifying it in situations where the autoregressive coefficient is close to 1.0—that is, the variables change very little from time period to time period—and hence conflict time series can easily be incorrectly identified as co-integrated.

A second common problem encountered in time series with conflict data are "time series cross sections" (TSCS), where data are collected across a number of cases across a period of time (for example, incidents of terrorism by country-year). Because these cases are not independent—in particular, contemporary trans-national terrorist groups operate in multiple countries—this will affect the estimates of the coefficients and methods that do not adjust for this can give misleading results [10, 11].

3.3 Vector Autoregression Models

A relatively recent modification of the classical time series approach has been vector autoregression (VAR) models. Originally developed by Sims [154] and used primarily for financial forecasting, VAR models were also used in some of the earliest applications of sophisticated econometric modeling to assessment of terrorist strategies [52–54, 58] and have been used in a number of policy-oriented studies in the post-9/11 period [28, 29, 55–57, 59], as well as more general applications to conflict early warning [26, 30, 63].

A p-th order VAR, denoted $VAR(p)$, is

$$y_t = c + A_1 y_{t-1} + A_2 y_{t-2} + \cdots + A_p y_{t-p} + e_t \tag{7}$$

where c is a $k \times 1$ vector of constants, A_i is a $k \times k$ matrix (for every $i = 1, \ldots, p$) and e_t is a $k \times 1$ vector of error terms.

The distinctive characteristic of VAR approaches is their emphasis on response functions rather than on the interpretation of individual coefficients. VAR takes as a given the fact that if the specified model is correct, the relations between the variables—or technically speaking, their colinearity—will make the uncertainty in the estimated coefficients very large even when the predictive accuracy of the system as a whole is quite high. Consequently, rather than trying to interpret those coefficients, an analyst looks at the effects—including error bands—of an exogenous change propagating over time through a series of inter-dependent variables. The downside of VAR is that it is very data-intensive and requires data measured consistently across a large number of time points.

A more recent variation on VAR models has been the addition of "switching" components, which allow for the impact of distinct regimes on the behavior of the system [155]. While originally developed for the study of foreign exchange markets, this is clearly applicable to political conflict situations where the system is likely to behave differently in times of peace than during times of war. Bayesian versions have also been developed [26] and applied to cases such as the conflict in the Middle East [25, 27].

3.4 *Event-History and Survival Models*

A different approach to conventional time series and VAR, but somewhat similar to the probabilistic interpretation of logit models, is the event-history or survival model approach [24, 86]. Rather than trying to predict the value of a behavior at a specific time, these models predict *probabilities* of events, and in particular look at the change in probability over time of the next event occurring. The models were thoroughly developed in the bio-medical literature to model disease prognosis—hence the term "survival"—prior to their application in political science. See [23, 41, 42] for applications to political conflict.

The "survival function" is defined as $S(t) = \Pr(T > t)$ where t is time, T is a random variable indicating when the event—for example the next terrorist attack, or the end of a terrorist organization—occurs, and Pr is probability. In many models, the focus of the analysis is the "hazard function"

$$\lambda(t)\,dt = \Pr(t \le T < t + dt \mid T \ge t) = \frac{f(t)\,dt}{S(t)} = -\frac{S'(t)\,dt}{S(t)} \tag{8}$$

which the event rate at time t conditional on survival until time T, $T \ge t$. From these functions, information such as the average number of events in a given period or average lifetime can be derived.

Of particular interest in survival analysis is the overall shape of the hazard function. For example, a survival function which levels off indicates that once an interval has occurred without an event occurring, it is less likely to occur again, whereas a survival function that declines linearly means that the occurrence or nonoccurrence provides no information. Other forms are also possible—for example many biological organisms exhibit a U-shaped hazard function, meaning that mortality is high in the early stages of life, then becomes low for a period of time, then rises again as the organism reaches its natural age limit.

A key downside of survival models is the difficulty that humans have in working with probabilities [94, 161]. In addition, the models are relatively new to the study of political conflict, due to earlier constraints of computational feasibility and the availability of appropriate estimation software. However, with these constraints removed, there are likely to be more such applications in the future.

3.5 Rare-Events Models

In most studies of political violence, the occurrence of violence is rare relative to the instances wherein violence did not occur. This poses a number of challenges to the empirical analysis of terrorism. When the absolute number of observations (or actual terrorism-events) in a sample is low, the rarity of terrorism can bias coefficient estimates and underestimate the probability of terrorist events. Even When one has a large sample of terrorist incidents, the relative rarity of these events and the preponderance of zeroes that this rarity entails can pollute samples with "non-event" observations that effectively have zero likelihood of *ever* experiencing terrorist events. This "zero inflation" contaminates terrorism-event data with multiple (latent) data generating processes, which when unaccounted for, can similarly bias estimates and predicted probabilities—often again by understating the true occurrence of terrorism.

Within logit models of binary terrorism data (see Sect. 3.1 above), scholars have shown that information content deficiencies of these sorts are primarily a function of two dynamics: the total number of observations within one's sample ("N"), and the mixture of 1's and 0's therein [97]. In such instances, estimates will be biased downward due to an underestimation of the variance around $Pr(Z|Y = 1)$ and common methods for computing predicted probabilities will exacerbate this bias through their disregard for estimation uncertainty [97]. While such concerns may not at first appear to be detrimental to 'big data' analyses, they are nevertheless highly relevant to large-N conflict studies given the growing importance of in-sample and out-of-sample forecasting techniques in the discipline [12, 166]. Indeed, even for large-N event data sets, the practices of subdividing data sets into smaller training, test and validation-data sets will frequently necessitate that scholars address these concerns in order to achieve stable and comparable estimates of conflict forecasts across sub-samples.

SOMA behavior models and CAPE models—which we discuss in more detail in Sect. 4.4 below—provide one potential solution to this problem [107, 157]. These models account for behavioral changes, and have been found to be especially useful in modeling the behaviors of terrorist groups under the specific rare-event contexts outlined above. Within the political science literature, an alternative approach is the 'rare-events logit' model, which addresses these challenges by treating rare-events as a sampling design problem. Provided that the dataset has a representative (or complete) sample of actual terrorism events, this approach calls for researchers to then augment their samples of observed terrorism events with a comparable, random sub-sample of all corresponding non-events. When this strategy is consistent with the assumptions of 'case-control' or 'case-cohort' sampling approaches, unbiased coefficient estimates and predictions can be derived by adjusting logit estimates and predicted probabilities for the true ratio of events to non-events within the population (or an estimate thereof) during model estimation [97]. This makes such an approach particularly advantageous for empirical studies of terrorism that (i) employ in-sample and out-of-sample forecasting techniques (ii) focus on a specific region or country over a short period of time, or (iii) require additional variable coding or data collection that, due to time or money constraints, cannot be completed for an entire sample of interest.

A second challenge to the analysis of rare-event data arises from the corresponding preponderance of zero (non-event) observations within such data sets. As recent studies note [126], many zero-observations in terrorism-event data sets correspond to country or region cases that have no probability of experiencing a terrorist-event in any period of interest. In these instances, empirical analyses of conflict-event data risk conflating two distinct types of zero-observations; one for which the probability of terrorism is non-zero but terrorism-events nevertheless didn't occur, and one wherein the probability of terrorist-events is consistently zero. Given that the latter set of zero-cases often arise as the result of covariates that correlate or overlap with one's primary independent variables of interest, ignoring zero-inflation processes of these sorts not only leads to an underestimation of terrorist events generally, but can also bias specific coefficient estimates in indeterminate directions. To correct for these biases, one must conditionally model *both* the zero-inflation process *and* the main outcome of interest.

Zero-inflated mixture-models specifically address these very problems, most notably within the contexts of event-count models such as Poisson or negative binomial estimators. In essence, these models employ a system of two equations to estimate the combined probability of an observation (i) being inflated and (ii) experiencing an event-outcome of interest—usually by including separate but overlapping covariates as predictors for each respective equation. For example, one could expand the logit model presented in Eq. 13 above to the zero-inflated logit approach by incorporating a second, "inflation-stage" logit equation as so,

$$f(z,w) = \left[\frac{1}{1+e^{-w}} + \left(1-\frac{1}{1+e^{-w}}\right)\left(1-\frac{1}{1+e^{-z}}\right)\right]^{1-Y_i} * \left[\left(1-\frac{1}{1+e^{-w}}\right)\left(\frac{1}{1+e^{-z}}\right)\right]^{Y_i} \tag{9}$$

where the variable "w" represents the additional set of inflation-stage covariates;

$$w = \beta_0 + \beta_1 v_1 + \beta_2 v_2 + \beta_3 v_3 + \cdots + \beta_k v_k, \tag{10}$$

which may or may not overlap with z [13]. Zero-inflated models thereby add an additional layer of nuance to the empirical modeling of conflict-events by estimating *both* the propensity of ever experiencing an event of interest *and* the likelihood of experiencing an event of interest conditional on being able to do so. This allows one to use ex-ante observable and theoretically informed covariates to account for the probability that a given zero observation is 'inflated', and to then probabilistically discount these zeroes' leverage within one's primary analysis without dropping these observations entirely. While such zero-inflated modeling approaches have been most extensively applied to political violence count-data [14,40,126], zero-inflated models have also recently been developed and applied by conflict-researchers to a variety of other limited dependent variables [83,159,168].

4 Algorithmic Approaches

Although more traditional "statistical" models still dominate quantitative studies of political conflict, "algorithmic" approaches have proven effective, thus gaining momentum not just within political science [21, 139] but also in other disciplines. For example, computer scientists have developed the CONVEX [157], CAPE [106], and SOMA [107] tools to forecast terrorist group behavior. While we will follow common practice in using the "statistical" vs. "algorithmic" distinction to differentiate between methodologies, there is overlap between the two definitions. For example, linear regression is consider a canonical "statistical" approach, but as we describe in Sect. 4.1.1, it is also a straightforward example of a supervised linear "algorithm".

In general, by algorithmic approaches, we refer to specific models (such as neural networks or random forests) or techniques (like bagging and boosting) that attempt to leverage computational power to specify and train models through iterative resampling techniques and to build and assess out-of-sample predictions, rather than obtaining a single estimate of the model coefficients. Algorithmic approaches can provide a number of benefits over statistical models and are particularly relevant to forecasting terrorism for at least the following four reasons.

First, machine learning algorithms are often better suited than many traditional statistical models at handling 'big data' data sets with large numbers of independent variables that potentially exceed the number of observations. Second, these algorithms are also less dependent on rigid assumptions about the data generating process and underlying distributions. Third, as opposed to some statistical models, many machine learning algorithms were specifically designed to generate accurate predictions, and do this exceedingly well. Finally, a number of the algorithmic approaches approximate the widely used qualitative method "case-based reasoning" [95, 108, 116] which match patterns of events from past cases to the events observed in a current situation, and then use the best historical fit to predict the likely outcome of the current situation; [134, Chap. 6] gives a much more extended discussion of this approach. This similarity to the methods of human analysts accounted for these methods originally being labeled "artificial intelligence" in some of the early studies.

Indeed, major trends in the empirical study of political violence, such as the 'big data' revolution and an increasing interest in predictive models, mean that algorithmic approaches will likely become increasingly popular in the coming years. In the following sections, we address some of the most relevant machine learning algorithms for forecasting political violence. Following standard practices, we divide algorithmic approaches into two general, though not mutually exclusive categories, supervised and unsupervised algorithms, with an additional discussion of sequence analysis techniques.

4.1 *Supervised Cross-Sectional Classification Methods*

Using the language of political science, supervised algorithms require an observed dependent variable value—be it binomial, continuous, or categorical—and at least one independent variable for each observation. The vast majority of quantitative studies of political violence employ data that meet the requirements of supervised algorithms. In the following section, we discuss the two supervised algorithmic approaches that have received the most attention in the political violence literature, linear models and neural networks, and also discuss tree-based approaches that have yet to take hold in the study of political violence but have proven successful at forecasting in other disciplines.

4.1.1 Linear Approaches

Though generally thought of as a statistical approach, liner regression is also one of the most common and straightforward examples of a supervised learning algorithm. The dominance of linear regression among the broader group of existing supervised linear approaches is not fully unwarranted, since linear regression contains nice attributes, such as being "BLUE"—the Best Linear Unbiased Estimator. Additionally, and perhaps more importantly, linear regression approaches are widely taught, straightforward to interpret, and easy to implement with one line of code (or a few clicks of a drop down menu) in virtually every statistical package.

Consistent with the widespread use of the linear method, other supervised approaches similarly utilize a line (in <3 dimensions) or a hyper plane (in >2 dimensions), such as linear discriminant analysis (LDA) and support vector machines (SVM). LDA is a supervised linear classification algorithm that produces nearly identical results to linear regression when Y takes on two classes, but can outperform multinomial logit and probit models as the number of classes exceeds 2 [80, pp. 106–114]. SVMs are an additional linear classification supervised algorithm that is most commonly used in problems such as text classification when the number of covariates (i.e. words in the text example) exceeds the number of observations. [49, 142] demonstrate the strength of SVMs in classifying articles for the Militarized Interstate Disputes 4.0 (MID4) project, while [90] uses SVM to extract events of terrorism from text.

4.1.2 Neural Networks

Aside from traditional linear regression approaches, neural networks have arguably received the most attention in quantitative political violence literature among supervised learning algorithms due almost entirely to Beck, King and Zeng's important article on forecasting interstate conflict [12]. (Neural networks can also be applied in an unsupervised setting: see [98].) Neural networks were originally created to model complex data characterized by a large number of nodes that receive and process input functions and then transfer information to other nodes, such as neurons in the case of the human brain.

Although most work with neural networks seems far removed from terrorism, [12] articulately explain how supervised neural networks are not only an appropriate algorithmic approach to predicting violence but can also be applied as a straightforward extension to logistic regression. Like a logistic regression, neural networks can be applied to a traditional TSCS dataset with a binary dependent variable to generate predicted probabilities that are interoperated identically to the π parameter of logistic regression. However, the primary advantage of a neural network approach is that they are able to account for the potential of 'massive nonlinear interaction effects' that may causally link the independent variables to the outcome of interest without having to directly specify interactive or non-linear terms to the model as required in a logistic. [12] demonstrate that the neural network approach consistently outperforms logistic regression in out-of-sample accuracy measures. Though we are unaware of neural networks being applied in studies of terrorism, it is likely that doing so could yield similar improvements in predictive accuracy.

4.1.3 Tree-Based Algorithms

Though yet to gain traction in the political violence literature, tree-based approaches [80, pp. 305–317, 587–604] are commonly used to generate accurate predictions in a host of other disciplines, including finance, medicine, computer science, and sociology. In brief, tree-based algorithms operate by iteratively partitioning the data into smaller sub-sets based on a break-point in a particular independent variable. This process is repeated with the goal of creating bins of observations with similar Y_i values (for continuous data) or class (for categorical data).

We highlight three important factors that contribute to the strong predictive accuracy of tree-based approaches. First, trees can be used to forecast both continuous (i.e. "regression trees") and binomial (i.e. "classification trees") dependent variables. Second, tree-based approaches, such as random forests, are not sensitive to degrees of freedom and can handle more independent variables than observations. Third, leading tree-based approaches incorporate iterative re-sampling, weighting, and model averaging strategies like bagging and boosting techniques [156], which tend to enhance accuracy and stability vis-à-vis other supervised learning classification and regression algorithms. In social science applications of tree-based algorithms, [16] and [17] demonstrates that random forests can help generate accurate forecasts of violent crime rates in the United States. Despite the scarcity of tree-based approaches in political science, scholars in other disciplines ranging from ecology (see [43]) to transportation studies (see [158]) have generated accurate predictions using this approach. As the quantitative political violence literature continues to progress, it will behoove scholars to continue experimenting with supervised forecasting algorithms that have demonstrated their value in other disciplines. We believe that tree-based approaches, especially boosted regression trees (BRTs) and boosted random forests, are particularly useful for scholars attempting to forecasting terrorism.

4.2 Unsupervised Methods

Unlike supervised learning algorithms that train a model based on relationships between a matrix of covariates and a corresponding vector of observed dependent variables for each observation, unsupervised approaches are applied to datasets for which dependent variables are 'latent' and therefore not directly provided. Though the concept of unsupervised algorithms may seem abstract, useful applications exist in the study of political violence. We describe two types of particularly relevant unsupervised learning algorithms in the following sections dimensionality reduction and clustering approaches. (Hidden Markov Models (HMMs) can also be classified as an unsupervised learning algorithm, but we present it in Sect. 4.3 as a sequence-based approach.)

4.2.1 Dimension Reduction

Factor analysis is a general algorithmic approach to reduce the dimensionality of matrix of potential covariates by identifying latent attributes. Principal components analysis (PCA) is a more restricted form of factor analysis and [91] is among the oldest (circa 1901) and most commonly employed dimension reduction tools in the social science [69]. In brief, the PCA algorithm works by analyzing relationships among the covariates and using these to create a new set of orthogonal variables called "principal components" that reflect the latent attributes and can be used in place of the original set of covariates in a quantitative model. Certain models, like linear regression, are sensitive to degrees of freedom and cannot operate when a dataset contains more covariates than observations. In these instances, a researcher can implement PCA or a similar approach to sufficiently reduce the dimensionality of covariates to enable the linear regression to converge [2].

4.2.2 Clustering

Clustering approaches such as k-means and Latent Dirichlet Allocation are similar to dimension reduction in that they attempt to identify latent classes amongst a set of observations, but differ in that they identify discrete, rather than continuous solutions like PCA and FA [69]. Within the machine learning literature, k-means approaches are among the most commonly used clustering algorithms, though their application to the study of political violence has been scarce [80, pp. 509–520].

However, [139] demonstrate that k-means can successfully identify latent clusters within event data that identify phases of violence in the Middle East.

Latent Dirichlet Allocation [19] is another clustering algorithm that is primarily applied to raw text. In the typical Latent Dirichlet Allocation application to document classification, each document is assumed to be a mixture of multiple, overlapping *latent topics*, each with a characteristic set of words. Classification is done by associating words in a document with a pre-defined number of topics most likely to have generated the observed distribution of words in the documents.

The purpose of LDA is to determine those latent topics from patterns in the data, which are useful for two purposes. First, to the extent that the words associated with a topic suggest a plausible category, they are intrinsically interesting in determining the issues found in the set of documents. Second, the topics can be used with other classification algorithms such as logistic regression, support vector machines or discriminant analysis to classify new documents.

Despite the surface differences between the domains, the application of Latent Dirichlet Analysis to the problem of political forecasting is straightforward: it is reasonable to assume that the stream of events observed between a set of actors is a mixture of a variety political strategies and standard operating procedures (for example escalation of repressive measures against a minority group while simultaneously making efforts to co-opt the elites of that group). This is essentially identical to the process by which a collection of words in a document is a composite of the various themes and topics, the problem Latent Dirichlet Analysis is designed to solve. As before, the objective of Latent Dirichlet Analysis will be to find those latent strategies that are mixed to produce the observed event stream. These latent factors can then be used to convert full event stream to a much simpler set of measures.

The Latent Dirichlet Analysis approach is similar in many ways to the hidden Markov approach (Sect. 4.3). In both models, the observed event stream is produced by a set of events randomly drawn from a mixture of distributions. In an HMM, however, these distributions are determined by the state of a Markov chain, whose transition probabilities must be estimated but which consequently also explicitly provides a formal sequence. An Latent Dirichlet Analysis, in contrast, allows any combination of mixtures, without explicit sequencing except to the extent that sequencing information is provided by the events in the model.

4.3 Sequence Development: Hidden Markov Models

Hidden Markov models (HMM) are a type of stochastic signaling model that has become increasingly popular in computing the probability that a noisy sequence was generated by a known model, most notably in speech recognition and protein sequence comparison. A detailed discussion and application of the method to the problem of forecasting terrorism can be found in [123] in this volume; other applications to political forecasting include [21, 133, 135, 144].

An HMM is a variation on the well-known Markov chain model, one of the most widely studied stochastic models of discrete events. Like a conventional Markov chain, a HMM consists of a set of discrete states and a matrix $A = a_{ij}$ of transition probabilities for going between those states. In addition, however, every state has a vector of observed symbol probabilities, $B = b_j(k)$ that corresponds to the probability that the system will produce a symbol of type k when it is in state j. The states of the HMM cannot be directly observed and can only be inferred from the observed symbols, hence the adjective "hidden."

In empirical applications, the transition matrix and symbol probabilities of an HMM are estimated using an iterative maximum likelihood technique called the Baum-Welch algorithm which finds values for the matrices A and B that locally maximize the probability of observing a set of training sequences. Once a set of models has been estimated, they can be used to classify an unknown sequence by computing the probability that each of the models generated the observed sequence. The model with the highest such probability is chosen as the one which best represents the sequence.

The application of the HMM to the problem of classifying international event sequences is straightforward. The symbol set consists of the event codes taken from an event data set such as IDEA [123] or CAMEO [143]. The states of the model are unobserved, but have a close theoretical analog in the concept of crisis "phase" [101]. Different political phases are distinguished by different distributions of observed events from the event ontology. Using CAMEO coding as an example, a "stable peace" would have a preponderance of cooperative events in the 01–10 range which codes cooperative events; a crisis escalation phase would be characterized by events in the 11–15 range (accusations, protests, denials, and threats), and a phase of active hostilities would show events in the 18–22 range, which codes violent events.

An important advantage of the HMM is that it can be trained by example rather than by the deductive specification of rule. Furthermore, HMMs require no temporal aggregation. This is particularly important for early warning problems, where critical periods in the development of a crisis may occur over a week or even a day. Finally, indeterminate time means that the HMM is relatively insensitive to the delineation of the start of a sequence: It is simple to prefix an HMM with a "background" state that simply gives the distribution of events generated by a particular source (e.g. Reuters/IDEA) when no crisis is occurring and simply cycle in this state until something important happens.

4.4 Sequence Analysis: Sequence Matching

Sequence analysis is a method that assesses the degree of similarity between two or more $1xN$ vectors, commonly populated by numerical or categorical inputs. Scientists first developed the approach as a tool to locate matches between different genomic D.N.A. sequences [31, 46] and it has since been applied to various disciplines including sociology [1] and demographics [102]. Although existing

Table 1 Archetype sequence the precedes an attack against the base

Variable	Week$_{t-4}$	Week$_{t-3}$	Week$_{t-3}$	Week$_{t-1}$	Week$_t$
Level of civilian casualties	1	2	3	4	N/A
Number of ambush attempts against patrol units	3	3	3	0	N/A
Attack on base?	No	no	No	No	Yes

studies using sequence analysis primarily focus on matching rather than forecasting, Martinez et al. [106], Silva et al. [157], and D'orazio et al. [50], do utilize sequence analysis-based algorithms for the explicit goal of prediction.

4.4.1 Archetypal Sequence Matching

D'orazio et al. suggest an archetype-driven approach to sequence analysis with the explicit goal of predicting political violence that uses sequence analysis to build features on which a statistical or algorithmic model produces out-of-sample forecasts. For this archetype approach be succeed, it requires the existence of a distinct pattern of events (i.e. an archetype) that tends to precede a given outcome-of-interest. If such an archetype exists, then a model may be able to forecast the outcome of interest based on the extent to which a sequence whose outcome is unknown (for example, events being observed in real time) is similar to the archetype sequence. The process of using this archetype-driven sequence analysis approach for prediction follows four main steps:

1. Define the archetype(s)
2. Calculate distances
3. Train a model
4. Build out-of-sample forecasts

To more clearly explain this approach, we use a hypothetical example to demonstrate each of the four steps above. Imagine our goal is to predict a binary variable indicating whether or not an American military base will be attacked in the upcoming week. Looking back, officers notice patterns in the level of civilian casualties (scale of 1–4, 4 being most severe) and the number of attempted ambushes on patrol missions that have occurred in peaceful weeks that precede an attack on the base. Table 1 completes <Step 1> as it reflects the pattern of events that tends to precede an attack against the base in terms of archetypical sequences. If officers observe 4 week progressions in real time that are sufficiently similar to the archetypical sequence of the level of civilian casualties and the number of ambush attempts on patrols, then they can confidently predict that an attack against the base will occur in the next week.

Just how similar must the observed sequences be to the archetypical sequences in order to predict that the following week will experience an attack against the base? To address this, a researcher must first complete <Step 2>, calculating the mathematical degree of similarity (i.e. distance) between the archetypes and

$$Distance_{civilian_casualty} = \sqrt{\sum_{i=1}^{4} (Archetype_{civiliancasulaty(i)} - Sequence_{civiliancasualty(i)})^2}$$

$$Distance_{ambush_attempts} = \sqrt{\sum_{i=1}^{4} (Archetype_{ambushattempts(i)} - Sequence_{ambusattemps(i)})^2}$$

Fig. 1 Building co-variates with Euclidean distance

set of training sequence of equal length (i.e. 4). Euclidean distance is one of the most common and robust measures, though many other distance measures exist. Figure 1 demonstrates how Euclidean is applied to calculate distances–which serve as the covariates in `<Step 3>`–between training sequences and the archetypical sequences.

The two distances, $Distance_{civiliancasualty}$ and $Distance_{ambushattempts}$ are calculated for every observation (i.e. 4 week sequence) in the training set. To complete `<Step 3>`, we choose a logistic regression, which is suitable to a parsimoniously specified model with a binary dependent variable. To train the logistic model, estimate the β values that maximize the likelihood function below.

$$L(\beta|y) = \prod_{i=1}^{N} \frac{n_i!}{y_i!(n_i - y_i)!} \pi_i^{y_i} (1 - \pi_i)^{n_i - y_i} \tag{11}$$

To complete `<Step 4>` and build actual forecasts based on the archetype-driven sequence analysis approach, we apply the two β estimates that result from (Eq. 6) to the logistic regression formula in order to calculate $f(z)$, which reflects the likelihood that the week following the 4-week period used to generate the $Distance_{civiliancasualty}$ and $Distance_{ambushattempts}$ distances from the archetypes will experience an attack on the base.

$$z = \beta_0 + \beta_1 Distance_{civiliancasualty} + \beta_2 Distance_{ambushattempts} \tag{12}$$

$$f(z) = \frac{1}{1 + e^{-z}} \tag{13}$$

To the extent that archetypical sequences that tend to precede an outcome-of-interest exist, sequence analysis may be an effective tool of prediction that can provide leverage over other approaches.

4.4.2 Convex Algorithms

In addition to the sequential archetype approach, Martinez et al. [106] put forth two alternative, sequence-based algorithms (Convex$_N^k N$ and ConvexMerge) that have proven successful at predicting terrorist activities. The key difference between these algorithms and the previously discussed archetypical approach is that the former do not assume the presence of an archetype. Thus, rather than calculating the distances between out-of-sample sequences with unknown outcomes and an archetypical sequence, the Convex algorithm calculates the various distance measure between each out-of-sample sequence and *all* in-sample sequences. In order to form out-of-sample predictions, the Convex algorithm determines the K-number of the most similar sequences based on various distance measures. If all of the k-nearest neighbor sequences have the same dependent variable, then the predicted value for that out-of-sample sequence is assigned as the dependent variable value of the single nearest neighbor. When this occurs, the $\text{Convex}_N^k N$ and ConvexMerge approaches generate identical predictions. However, when the dependent variable values of the nearest neighbor sequences are not identical, $\text{Convex}_N^k N$ rounds to the nearest integer while ConvexMerge assigns more value to more similar sequences. In addition to Martinez et al., Sliva et al. [157] apply both variants of the Convex algorithm to terrorist group behavior.

5 Network Models

We label our final category "network models:" these are approaches that specifically consider the relationships between entities, either according to their interactions (SNA models) or location (geospatial models). As with many of the algorithmic applications, network models have only become feasible in the last decade of so, as they require very substantial amounts of data and computational capacity. Consequently the number of applications at present is relatively small, though these are very active research areas.

5.1 *Social Network Analysis Models*

Social network analysis (SNA) in a type of graph theory that models systems in terms of nodes, which are the unit of analysis and can be anything from states in the international system to members of a terrorist cell, and edges that connect the nodes. The edges can reflect any type of interactions that occur between the nodes, such as levels of international trade if states are the nodes or amount of communication if the nodes reflect members of a terror cell. Though still an emerging method, SNA is becoming particularly attractive approach for studying terrorism [122]. For example, [7] use SNA to model the organizational structure of the terrorist groups responsible for the 2008 Mumbai attacks and [47] develop an edge-based network to forecast the source and target locations of transnational terrorist attacks. We speculate three potential explanations for its rapid growth.

First, terrorist scholars are often interested in information regarding within-group dynamics of terrorist organizations—such as chains of command, lines of communication, and policy preferences of leadership—which SNA may reveal more effectively than statistical or algorithmic approaches. Second, the rise of social networking platforms like Facebook, Twitter, less publicized online forums, and any other modes of communication open source or otherwise, has greatly increased the amount of information that can be used to populate a social network. Third, SNA has been almost simultaneously embraced by the intelligence, policy, and academic communities—a rare feat for a new methodology. These factors, as well as its ability to deliver unique insights, is likely to continue driving its growth among quantitative studies of terrorism.

5.2 Geo-spatial Models

Similar to our discussion of temporal dynamics above (Sects. 3.2 and 3.3), the incorporation of spatial dynamics into statistical models of terrorism can often enhance modeling and forecasting capabilities. Political violence frequently clusters or diffuses in geographic space [35, 164, 166], and failure to account for these dependencies can bias inferences [5, 164]. Spatial statistics allow one to model the existence of dependencies or influences across spatial units with a dependency matrix that measures geographic distances between units, which can then be used to specify a spatial autoregressive model with spatially lagged dependent variable or a spatially correlated error structure model [5, 164]. A number of recent terrorism studies employ such approaches [115, 119]. Scholars have also tailored these spatial statistics to the study of binary dependent variables—akin to the binary terrorism events mentioned above—through the use of (e.g.) autologistic models that incorporate distance-based connectivity matrices [164, 166]. Finally, spatial forecast methods employing feature reduction techniques have recently been applied to terrorism events in efforts to incorporate the explanatory power of high dimensional geographic spaces into probabilistic forecasting models, and show great promise in this regard [32, 128].

6 Conclusion

In 1954, the psychologist Paul Meehl [111] published a path-breaking analysis of the relative accuracy of human clinical assessments versus simple statistical models for a variety of prediction problems, such as future school performance and criminal recidivism. Despite using substantially less information, the statistical models either outperformed, or performed as well as, the human assessments in most situations. Meehl's work has been replicated and extended to numerous other domains in the subsequent six decades and the results are always the same: the statistical models win. Meehl, quoted in Kahneman [94, Chap. 21], reflecting on 30 years of studies, said, "There is no controversy in the social sciences which shows such a large body of qualitatively diverse studies coming out in the same direction as this one."

This has not, of course, removed the human analysts from trying—and most certainly, *claiming*—to provide superior performance. These are justified in the face of overwhelming empirical evidence that they are inaccurate using a wide variety of excuses—dare we call them "pathologies"?—that Kahneman [94, Part 3] discusses in great detail in the section of his book titled, appropriately, "Overconfidence." It is not only that simple statistical models are superior to human punditry, but as Tetlock [161] established, the most confident and a well-known human analysts actually tend to make the worst predictions. Furthermore, in most fields, the past performance of analysts—notoriously, the well-compensated stock-pickers of managed mutual funds—provides essentially no guide to their future performance.

At present, there is very little likelihood that human punditry, particularly the opinionated self-assurance so valued in the popular media, will be completely replaced by the unblinking assessments of computer programs, whether on 24-hour news channels or in brainstorming sessions in windowless conference rooms. Humans are social animals with exquisite skills at persuasion and manipulation; computer programs simply are far more likely to provide the correct answer in an inexpensive, consistent and transparent manner.

Yet with the vast increase in the availability of data, computational power, and the resulting refinement of methodological techniques, there is some change in the works. While the covers of investment magazines are adorned with the latest well-coffed guru who by blind luck has managed to have an unusually good year, in fact algorithmic trading now accounts for all but a small fraction of the activity on financial exchanges. Weather forecasts are presented on television by jocular and attractive individuals with the apparent intelligence of tadpoles, but the forecasts themselves are the result of numerical models processing massive quantities of data. In the political realm, sophisticated public opinion polling replaced the intuitive hunches of experts several decades ago, and polls are so rarely incorrect that it is a major news event when they fail. Even that last bastion of intuitive manliness, the assessment of athletes, can be trumped by statistical models, as documented in the surprisingly popular book and movie *Moneyball* [100].

Similar progress in the adoption of models forecasting political violence, particularly terrorism, is likely to be much slower. As we have stressed repeatedly, one of the major challenges of this research is that political violence is a rare event, and acts of terrorism in otherwise peaceful situations—the "bolt out of the blue" of Oklahoma City, 9/11/2001, Madrid 2004 and Utøya Island, Norway in 2011—are among the rarest. Consequently even if the statistical models are more accurate—and there is every reason to believe that they will be—establishing this will take far longer than is required in a field where new predictions can be assessed by the day, or even by the minute. In addition, a long series of psychological studies have shown that human risk assessment is particularly inaccurate—and hence its validity overestimated—in low probability, high risk situations, precisely the domain of counter-terrorism. Getting the technical assessments in the door, to say nothing of getting them used properly, will not be an easy task, and the initial applications will almost certainly be in domains where the events of interest are more frequent, as we are already seeing with the success of the Political Instability Task Force. But as this chapter has illustrated, the challenges are well understood, a plethora of sophisticated techniques await experimentation, and the appropriate data is readily available, all that is needed is the will and skill to apply these.

Acknowledgements This research was supported in part by a grant from the U.S. National Science Foundation, SES-1004414, and by a Fulbright-Hays Research Fellowship for work by Schrodt at the Peace Research Institute, Oslo (http://www.prio.no).

References

1. Abbott A (1995) Sequence analysis: new methods for old ideas. Ann Rev Sociol 21:93–113
2. Abdi H, Williams LJ (2010) Principal components analysis. Wiley Interdiscip Rev 2(4): 433–459
3. Achen C (2002) Toward a new political methodology: microfoundations and ART. Ann Rev Pol Sci 5:423–450
4. Andriole SJ, Hopple GW (1984) The rise and fall of events data: from basic research to applied use in the U.S. Department of Defense. Int Interact 10(3/4):293–309
5. Anselin L (1988) Spatial econometrics: methods and models. Kluwer, Dordrecht
6. Asal V, Pate A, Wilkenfeld J (2008) Minorities at risk organizational behavior data and codebook version 9–2008. http://www.cidcm.umd.edu/mar/data.asp
7. Azad S, Gupta A (2011) A quantitative assessment on 26/11 mumbai attack using social network analysis. J Terror Res 2(2):4–14
8. Azar EE (1980) The conflict and peace data bank (COPDAB) project. J Confl Resolut 24: 143–152
9. Azar EE, Sloan T (1975) Dimensions of interaction. University Center for International Studies, University of Pittsburgh, Pittsburgh
10. Beck N, Katz. JN (1995) What to do (and not to do) with time- series cross-section data. Am Pol Sci Rev 89:634–634
11. Beck N, Katz JN, Tucker R (1998) Taking time seriously: time-series-cross-section analysis with a binary dependent variable. Am J Pol Sci 42(4):1260–1288
12. Beck N, King G, Zeng L (2000) Improving quantitative studies of international conflict: a conjecture. Am Pol Sci Rev 94(1):21–35
13. Beger A, DeMeritt JH, Hwang W, Moore WH (2011) The split population logit (spoplogit): modeling measurement bias in binary data. http://ssrn.com/abstract=1773594. Working paper
14. Benini AA, Moulton LH (2004) Civilian victims in an asymmetrical conflict: operation enduring freedom, afghanistan. J Peace Res 41(4):403–422
15. Bennett DS, Stam AC (2004) The behavioral origins of war. University of Michigan Press, Ann Arbor
16. Berk R (2009) The role of race in forecasts of violent crime. Race Soc Probl 1(4):231–242
17. Berk R, Sherman L, barnes G, Kurtz E, Ahlman L (2007) Forecasting murder within a population of probationers and parolees: A high stakes application of statistical learning. J R Stat Soc 172(1):191–211
18. Besley T, Persson T (2009) Repression or civil war? Am Econ Rev 99(2):292–297
19. Blei DM, Ng AY, Jordan MI (2003) Latent dirichlet allocation. J Mach Learn Res 2:993–1022
20. Bond D, Jenkins JC, Taylor CLT, Schock K (1997) Mapping mass political conflict and civil society: Issues and prospects for the automated development of event data. J Confl Resolut 41(4):553–579
21. Bond J, Petroff V, O'Brien S, Bond D (2004) Forecasting turmoil in Indonesia: an application of hidden Markov models. Presented at the International Studies Association Meetings, Montréal

22. Box GE, Jenkins GM, Reinsel GC (1994) Time series analysis: forecasting and control, 3rd edn. Prentice Hall, Englewood Cliffs
23. Box-Steffensmeier J, Reiter D, Zorn C (2003) Nonproportional hazards and event history analysis in international relations. J Confl Resolut 47(1):33–53
24. Box-Steffensmeier JM, Jones BS (2004) Event history modeling: a guide for social scientists. Analytical methods for social research. Cambridge University Press, Cambridge
25. Brandt PT, Colaresi MP, Freeman JR (2008) The dynamics of reciprocity, accountability and credibility. J Confl Resolut 52(3):343–374
26. Brandt PT, Freeman JR (2006) Advances in Bayesian time series modeling and the study of politics: Theory testing, forecasting, and policy analysis. Pol Anal 14(1):1–36
27. Brandt PT, Freeman JR, Schrodt PA (2011) Real time, time series forecasting of inter- and intra-state political conflict. Confl Manage Peace Sci 28(1):41–64
28. Brandt PT, Sandler T (2009) Hostage taking: understanding terrorism event dynamics. J Policy Model 31(5):758–778
29. Brandt PT, Sandler T (2010) What do transnational terrorists target? Has it changed? Are we safer? J Confl Resolut 54(2):214–236
30. Brandt PT, Williams JT (2007) Multiple time series models. Sage, Thousand Oaks
31. Brochet J, Marie-Paule-Lefrance, Guidicelli V (2008) Imgt/v-quest: the highly customized and integrated system for ig and tr standardized v-j and v-d-j sequence analysis. Nucl Acids Res 36:387–395
32. Brown D, Dalton J, Hoyle H (2004) Spatial forecast methods for terrorist events in urban environments. Intell Secur Inform 3073:426–435
33. Buhaug H (2006) Relative capability and rebel objective in civil war. J Peace Res 43(6): 691–708
34. Buhaug H, Lujala P (2005) Accounting for scale: measuring geography in quantitative studies of civil war. Pol Geogr 24(4):399–418
35. Buhaug H, Rod JK (2006) Local determinants of African civil wars, 1970–2001. Pol Geogr 25(6):315–335
36. Burgess PM, Lawton RW (1972) Indicators of international behavior: an assessment of events data research. Sage, Beverly Hills
37. Cederman LE, Gleditsch KS (2009) Introduction to special issue of "disaggregating civil war". J Confl Res 24(4):590–617
38. Choi K, Asal V, Wilkenfeld J, Pattipati KR (2012) Forecasting the use of violence by ethno-political organizations: middle eastern minorities and the choice of violence. In: Subrahmanian V (ed) Handbook on computational approaches to counterterrorism. Springer, New York
39. Choucri N, Robinson TW (eds) (1979) Forecasting in international relations: theory, methods, problems, prospects. Freeman, San Francisco
40. Clark DH, Regan PM (2003) Opportunities to fight: a statistical technique for modeling unobservable phenomena. J Confl Resolut 47(1):94–115
41. Cunningham DE (2011) Barriers to Peace in Civil War. Cambridge University Press, Cambridge
42. Cunningham DE, Gleditsch KS, Salehyan I (2009) It takes two: a dyadic analysis of civil war duration and outcome. J Confl Resolut 53(4):570–597
43. Cutler DR, Edwards TC Jr, Beard KH, Cutler A, Hess KT, Gibson J, Lawler J (2007) Random forests for classification in ecology. Ecology 88(11):2783–2792
44. Daly JA, Andriole SJ (1980) The use of events/interaction research by the intelligence community. Policy Sci 12:215–236
45. Davies JL, Gurr TR (eds) (1998) Preventive measures: building risk assessment and crisis early warning. Rowman and Littlefield, Lanham
46. Dereux J, Haeberli P, Smithies O (1984) A comprehensive set of sequence analysis programs for the vax. Nucl Acids Res 12(1):387–395

47. Desmarais BA, Cranmar SJ (2011) Forecasting the locational dynamics of transnational terrorism: a network analysis approach. In: Paper presented at the 2011 European Intelligence and Security Conference, Athens. http://ieeexplore.ieee.org/xpls/abs_all.jsp?arnumber=6061174&tag=1
48. Dickerson JP, Simari GI, Subrahmanian VS (2012) Using temporal probabilistic rules to learn group behavior. In: Subrahmanian V (ed) Handbook on computational approaches to counterterrorism. Springer, New York
49. D'Orazio V, Landis ST, Palmer G, Schrodt PA (2011) Separating the wheat from the chaff: application of support vector machines to mid4 text classification. Paper presented at the annual meeting of the midwest political science association, Chicago
50. D'Orazio V, Yonamine J, Schrodt PA (2011) Predicting intra-state conflict onset: an events data approach using euclidean and levenshtein distance measures. Presented at the midwest political science association, Chicago. Available at http://eventdata.psu.edu
51. Duval RD, Thompson WR (1980) Reconsidering the aggregate relationship between size, economic development, and some types of foreign policy behavior. Am J Pol Sci 24(3): 511–525
52. Enders W, Parise GF, Sandler T (1992) A time-series analyssis of transnational terrorism. Def Peace Econ 3(4):305–320
53. Enders W, Sandler T (1993) The effectiveness of anti-terrorism policies: a vector-autoregression-intervention analysis. Am Pol Sci Rev 87(4):829–844
54. Enders W, Sandler T (2000) Is transnational terrorism becoming more threatening? A time-series investigation. J Confl Resolut 44(3):307–332
55. Enders W, Sandler T (2005) After 9/11: is it all different now? J Confl Resolut 49(2):259–277
56. Enders W, Sandler T (2006) Distribution of transnational terrorism among countries by income classes and geography after 9/11. Int Stud Q 50(2):367–393
57. Enders W, Sandler T (2006) The political economy of terrorism. Cambridge University Press, Cambridge
58. Enders W, Sandler T, Cauley J (1990) Assessing the impact of terrorist-thwarting policies: an intervention time series approach. Def Peace Econ 2(1):1–18
59. Enders W, Sandler T, Gaibulloev K (2011) Domestic versus transnational terrorism: Data, decomposition, and dynamics. J Peace Res 48(3):319–337
60. Esty DC, Goldstone JA, Gurr TR, Harff B, Levy M, Dabelko GD, Surko P, Unger AN (1998) State failure task force report: phase II findings. Science Applications International Corporation, McLean
61. Esty DC, Goldstone JA, Gurr TR, Surko P, Unger AN (1995) State failure task force report. Science Applications International Corporation, McLean
62. Freedom House (2009) Freedom in the world. http://www.freedomhouse.org/reports
63. Freeman JR (1989) Systematic sampling, temporal aggregation, and the study of political relationships. Pol Anal 1:61–98
64. Gaddis JL (1992) International relations theory and the end of the cold war. Int Secur 17:5–58
65. Gaddis JL (1992) The United States and the end of the cold war. Oxford University Press, New York
66. Gassebner M, Luechinger S (2011) Lock, stock, and barrel: a comprehensive assessment of the determinants of terror. Public Choice 149:235–261
67. Gerner DJ, Schrodt PA, Francisco RA, Weddle JL (1994) The machine coding of events from regional and international sources. Int Stud Q 38:91–119
68. Gerner DJ, Schrodt PA, Yilmaz Ö (2009) Conflict and mediation event observations (CAMEO) Codebook. http://eventdata.psu.edu/data.dir/cameo.html
69. Ghahramani Z (2004) Unsupervised learning. In: Advanced lectures on machine learning. Springer, Berlin/New York, pp 72–112
70. Gill J (1999) The insignificance of null hypothesis significance testing. Pol Res Q 52(3): 647–674
71. Gill J (2003) Bayesian methods: a social and behavioral sciences approach. Chapman and Hall, Boca Raton

72. Gleditsch NP (2012) Special issue: event data. Int Interact 38(4)
73. Goldstein JS (1992) A conflict-cooperation scale for WEIS events data. J Confl Resolut 36:369–385
74. Goldstone JA, Bates R, Epstein DL, Gurr TR, Lustik M, Marshall, MG, Ulfelder J, Woodward M (2010) A global model for forecasting political instability. Am J Pol Sci 54(1):190–208
75. Gurr TR, Harff B (1994) Conceptual, research and policy issues in early warning research: an overview. J Ethno-Dev 4(1):3–15
76. Gurr TR, Lichbach MI (1986) Forecasting internal conflict: A competitive evaluation of empirical theories. Comp Pol Stud 19:3–38
77. Hämerli A, Gattiker R, Weyermann R (2006) Conflict and cooperation in an actor's network of Chechnya based on event data. J Confl Resolut 50(159):159–175
78. Hamilton J (1994) Time series analysis. Princeton University Press, Princeton
79. Harff B, Gurr TR (2001) Systematic early warning of humanitarian emergencies. J Peace Res 35(5):359–371
80. Hastie T, Tibshirani R, Friedman J (2009) The elements of statistical learning. Springer, New York
81. Hegre H, Ellingson T, Gates S, Gleditsch NP (2001) Toward a democratic civil peace? democracy, political change, and civil war, 1816–1992. J Policy Model 95(1):34–48
82. Hegre H, Ostby G, Raleigh C (2009) Property and civil war: a disaggregated study of liberia. J Confl Resolut 53(4):598–623
83. Hill DW, Bagozzi BE, Moore WH, Mukherjee B (2011) Strategic incentives and modeling bias in ordinal data:the zero-inflated ordered probit (ziop) model in political science. In: Paper presented at the new faces in political methodology meeting, Penn State, 30 April 2011. http://qssi.psu.edu/files/NF4Hill.pdf
84. Holmes JS, Pineres SAGD, Curtina KM (2007) A subnational study of insurgency: farc violence in the 1990s. Stud Confl Terror 30(3):249–265
85. Hopple GW, Andriole SJ, Freedy A (eds) (1984) National security crisis forecasting and management. Westview, Boulder
86. Hosmer DW, Lemeshow S, May S (2008) Applied survival analysis: regression modeling of time to event data. Series in probability and statistics. Wiley, New York
87. Hudson V (ed) (1991) Artificial intelligence and international politics. Westview, Boulder
88. Hudson VM, Schrodt PA, Whitmer RD (2008) Discrete sequence rule models as a social science methodology: an exploratory analysis of foreign policy rule enactment within Palestinian-Israeli event data. Foreign Policy Anal 4(2):105–126
89. International Monetary Fund. Statistics Dept. (2009) Direction of trade statistics yearbook, 2008. http://www.imf.org/external/pubs/cat/longres.cfm?sk=22053.0
90. Inyaem U, Meesad P, Haruechaiyasak C (2009) Named-entity techniques for terrorism event extraction and classification. In: Paper presented at the eighth international symposium on natural language processing. http://ieeexplore.ieee.org/xpls/abs_all.jsp?arnumber=5340924&tag=1
91. Jackman RW, Miller RA (1996) A renaissance of political culture? Am J Pol Sci 40(3): 632–659
92. Jackman S (2009) Bayesian analysis for the social sciences. Wiley, Chichester
93. Jenkins CJ, Bond D (2001) Conflict carrying capacity, political crisis, and reconstruction. J Confl Resolut 45(1):3–31
94. Kahneman D (2011) Thinking fast and slow. Farrar, Straus and Giroux, New York
95. Khong YF (1992) Analogies at war. Princeton University Press, Princeton
96. King G (1986) How not to lie with statistics: avoiding common mistakes in quantitative political science. Am J Pol Sci 30(3):666–687
97. King G, Zeng L (2001) Logistic regression in rare events data. Pol Anal 9(2):12–54
98. Lee K, Booth D, Alam P (2005) A comparison of supervised and unsupervised neural networks in predicting bankruptcy of korean firms. Expert Syst Appl 29:1–16
99. Leng RJ (1987) Behavioral correlates of war, 1816–1975. (ICPSR 8606). Inter-University Consortium for Political and Social Research, Ann Arbor

100. Lewis M (2011) Moneyball. Norton, New York
101. Lund MS (1996) Preventing violent conflicts: a strategy for preventive diplomacy. United States Institute for Peace, Washington, DC
102. Malo MA, Munoz-Bullon F (2003) Employment status mobility from a life-cycle perspective: a sequence analysis of work-histories in the bhps. Demogr Res 9:119–162
103. Mannes A (2012) Qualitative analysis & computational techniques for the counter-terror analyst. In: Subrahmanian V (ed) Handbook on computational approaches to counterterrorism. Springer, New York
104. Marshall MG, Jaggers K, Gurr TR (2010) Polity iv project: political regime characteristics and transitions, 1800–2010. Center for Systemic Peace, Vienna, VA
105. Martinez MV, Silva A, Simari GI, Subrahmanian VS (2012) Forecasting changes in terror group behavior. In: Subrahmanian V (ed) Handbook on computational approaches to counterterrorism. Springer, New York
106. Martinez V, Simari G, Sliva A, Subrahmanian V (2008) Convex: similarity-based algorithms for forecasting group behavior. IEEE Intell Syst 23:51–57
107. Martinez V, Simari G, Sliva A, Subrahmanian V (2008) The soma terror organization portal (stop): Social network and analytic tools for the real-time analysis of terror groups. In: Liu H, Salerno JJ, Young M (eds) Social computing, behavioral modeling, and prediction. Springer/Wien, Norderstedt, pp 9–18
108. May ER (1973) "Lessons" of the past: the use and misuse of history in American foreign policy. Oxford University Press, New York
109. McClelland CA (1967) World-event-interaction-survey: a research project on the theory and measurement of international interaction and transaction. University of Southern California, Los Angeles, CA
110. McClelland CA (1976) World event/interaction survey codebook (ICPSR 5211). Inter-University consortium for political and social research, Ann Arbor
111. Meehl P (1954) Clinical and statistical prediction: a theoretical analysis and a review of the evidence. University of Minnesota Press, Minneapolis
112. Minorities at Risk Project (2009) Minorities at risk dataset. Center for International Development and Conflict Management, College Park. http://www.cidcm.umd.edu/mar/
113. Murray MP (1994) A drunk and her dog: an illustration of cointegration and error correction. http://www-stat.wharton.upenn.edu/~steele/Courses/434/434Context/Co-int%egration/Murray93DrunkAndDog.pdf
114. National Consortium for the Study of Terrorism and Responses to Terrorism (START) (2011) Global terrorism database [data file]. http://www.start.umd.edu/gtd
115. Neumayer E, Plümper T (2010) Galton's problem and contagion in international terrorism along civilizational lines. Confl Manage Peace Sci 27(4):308–325
116. Neustadt RE, May ER (1986) Thinking in time: the uses of history for decision makers. Free Press, New York
117. O'Brien S (2012) A multi-method approach for near real time conflict and crisis early warning. In: Subrahmanian V (ed) Handbook on computational approaches to counterterrorism. Springer, New York
118. O'Brien SP (2010) Crisis early warning and decision support: Contemporary approaches and thoughts on future research. Int Stud Rev 12(1):87–104
119. Öcal N, Yildirim J (2010) Regional effects of terrorism on economic growth in turkey: A geographically weighted regression. J Peace Res 47(4):477–489
120. Oskar Engene J (2007) Five decades of terrorism in europe: the tweed dataset. J Peace Res 44(1):109–121
121. Pearl J (2009) Understanding propensity scores. In: Causality: models, reasoning, and inference, 2nd edn. Cambridge University Press, Cambridge/New York
122. Perliger A, Pedahzur A (2011) Social network analysis in the study of terrorism and political violence. Pol Sci Pol 44(1):45–50

123. Petroff V, Bond J, Bond D (2012) Using hidden markov models to predict terror before it hits (again). In: Subrahmanian V (ed) Handbook on computational approaches to counterterrorism. Springer, New York
124. Pevehouse JC, Brozek J (2010) Time series analysis in political science. In: Box-Steffensmeier J, Brady H, Collier D (eds) Oxford handbook of political Methodology. Oxford University Press, New York
125. Pevehouse JC, Goldstein JS (1999) Serbian compliance or defiance in Kosovo? Statistical analysis and real-time predictions. J Confl Resolut 43(4):538–546
126. Piazza J (2011) Poverty, minority economic discrimination and domestic terrorism. J Peace Res 48(3):339–353
127. Piazza J, Walsh JI (2009) Transnational terrorism and human rights. Int Stud Q 53:125–148
128. Porter MD, Brown DE (2007) Detecting local regions of change in high-dimensional criminal or terrorist point processes. Comput Stat Data Anal 51(5):2753–2768
129. Raleigh C, Linke A, Hegre H, Karlsen J (2010) Introducing ACLED: an armed conflict location and event dataset. J Peace Res 47(5):651–660
130. Regan P, Clark D (2010) The institutions and elections project data collection. http://www2.binghamton.edu/political-science/institutions-and-elections%-project.html
131. Richardson LF (1960) Statistics of deadly quarrels. Quadrangle, Chicago
132. Rupesinghe K, Kuroda M (eds) (1992) Early warning and conflict resolution. St. Martin's, New York
133. Schrodt PA (2000) Pattern recognition of international crises using hidden Markov models. In: Richards D (ed) Political complexity: nonlinear models of politics. University of Michigan Press, Ann Arbor, pp 296–328
134. Schrodt PA (2004) Detecting united states mediation styles in the middle east, 1979–1998. In: Maoz Z, Mintz A, Morgan TC, Palmer G, Stoll RJ (eds) Multiple paths to knowledge in international relations. Lexington Books, Lexington, pp 99–124
135. Schrodt PA (2006) Forecasting conflict in the Balkans using hidden Markov models. In: Trappl R (ed) Programming for peace: computer-aided methods for international conflict resolution and prevention. Kluwer, Dordrecht, pp 161–184
136. Schrodt PA (2007) Inductive event data scaling using item response theory. Presented at the summer meeting of the society of political methodology. Available at http://eventdata.psu.edu
137. Schrodt PA (2010) Seven deadly sins of contemporary quantitative analysis. Presented at the american political science association Meetings, Washington, DC. http://eventdata.psu.edu/papers.dir/Schrodt.7Sins.APSA10.pdf
138. Schrodt PA, Gerner DJ (1994) Validity assessment of a machine-coded event data set for the Middle East, 1982–1992. Am J Pol Sci 38:825–854
139. Schrodt PA, Gerner DJ (1997) Empirical indicators of crisis phase in the Middle East, 1979–1995. J Confl Resolut 25(4):803–817
140. Schrodt PA, Gerner DJ (2004) An event data analysis of third-party mediation. J Confl Resolut 48(3):310–330
141. Schrodt PA, Gerner DJ, Abu-Jabr R, Yilmaz, Ö, Simpson EM (2001) Analyzing the dynamics of international mediation processes in the Middle East and Balkans. Presented at the American Political Science Association meetings, San Francisco
142. Schrodt PA, Palmer G, Hatipoglu ME (2008) Automated detection of reports of militarized interstate disputes using the svm document classification algorithm. Paper presented at American Political Science Association, San Francisco
143. Schrodt PA, Van Brackle D (2012) Automated coding of political event data. In: Subrahmanian V (ed) Handbook on computational approaches to counterterrorism. Springer, New York
144. Shearer R (2006) Forecasting Israeli-Palestinian conflict with hidden Markov models. Available at http://eventdata.psu.edu/papers.dir/Shearer.IP.pdf
145. Shellman S (2000) Process matters: Conflict and cooperation in sequential government-dissident interactions. Journal of Conflict Resolution **15**(4), 563–599
146. Shellman S (2004) Time series intervals and statistical inference: the effects of temporal aggregation on event data analysis. Secur Stud 12(1):97–104

147. Shellman S, Hatfield C, Mills M (2010) Disaggregating actors in intrastate conflict. J Peace Res 47(1):83–90
148. Shellman S, Stewart B (2007) Predicting risk factors associated with forced migration: An early warning model of Haitian flight. Civil Wars 9(2):174–199
149. Silke A (ed) (2004) Research on terrorism: trends, achievements and failures. Frank Cass, London
150. Silke A (2009) Contemporary terrorism studies: issues in research. In: Jackson R, Smyth MB, Gunning J (eds) Critical terrorism studies: a new research agenda. Routledge, London
151. Silva A, Simari G, Martinez V, Subrahmanian VS (2012) SOMA: Stochastic opponent modeling agents for forecasting violent behavior. In: Subrahmanian V (ed) Handbook on computational approaches to counterterrorism. Springer, New York
152. Simari GI, Earp D, Martinez MV, Silva A, Subrahmanian VS (2012) Forecasting group-level actions using similarity measures. In: Subrahmanian V (ed) Handbook on computational approaches to counterterrorism. Springer, New York
153. Simi P (2010) Operation and structure of right-wing extremist groups in the united states, 1980–2007. http://www.icpsr.umich.edu/icpsrweb/NACJD/studies/25722/detail
154. Sims CA (1980) Macroeconomics and reality. Econometrica 1, 634–4
155. Sims CA, Waggoner DF, Zha TA (2008) Methods for inference in large multiple-equation Markov-switching models. J Econom 146(2):255–274
156. Siroky DS (2009) Navigating random forests and related advances in algorithmic modeling. Stat Surv 3:147–163
157. Sliva A, Subrahmanian V, Martinez V, Simari G (2009) Cape: Automatically predicting changes in group behavior. In: Memon N, Farley JD, Hicks DL, Rosenorn T (eds) Mathematical methods in counterterrorism. Springer/Wien, Norderstedt, pp 253–269
158. Sun S, Zhang C (2007) The selective random subspace predictor for traffic flow forecasting. IEEE Trans Intell Transp Syst 8(2):367–373
159. Svolik MW (2008) Authoritarian reversals and democratic consolidation. Am Pol Sci Rev 102(2):153–168
160. Sylvan DA, Chan S (1984) Foreign policy decision making: perception, cognition and artificial intelligence. Praeger, New York
161. Tetlock PE (2005) Expert political judgment: how good is it? How can we know? Princeton University Press, Princeton
162. Urdal H (2008) Population, resources and violent conflict: A sub-national study of india 1956–2002. J Confl Resolut 52(4):590–617
163. U.S. National Counterterrorism Center (2009) Worldwide incidents tracking system (wits). http://wits.nctc.gov/
164. Ward MD, Gleditsch KS (2002) Location, location, location: an mcmc approach to modeling the spatial context of war and peace. Pol Anal 10(3):244–260
165. Ward MD, Greenhill BD, Bakke KM (2010) The perils of policy by p-value: predicting civil conflicts. J Peace Res 47(5):363–375
166. Weidman NB, Ward MD (2010) Predicting conflict in space and time. J Confl Resolut 54(6):883–901
167. Weidmann NB, Toft MD (2010) Promises and pitfalls in the spatial prediction of ethnic violence: a comment. Confl Manage Peace Sci 27(2):159–176
168. Xiang J (2010) Relevance as a latent variable in dyadic analysis of conflict. J Pol 72(2): 484–498
169. Young JK, Findley MG (2011) Promise and pitfalls of terrorism research. Int Stud Rev 13:411–431
170. Yonamine JE (2012) Working with even data: a guide to aggregation. Available at http://jayyonamine.com/wp-content/uploads/2012/06/Working-with-Event-Data-A-Guide-to-Aggregation-Choices.pdf

Using Hidden Markov Models to Predict Terror Before it Hits (Again)

Vladimir B. Petroff, Joe H. Bond, and Doug H. Bond

1 Introduction

The authors used auto-coded events data to forecast indicators associated with country instability using Hidden Markov Models (HMM). Forecasts were based on weekly updates of event data records for selected countries of interest. These events data were developed from open source global news agency reports. Updates built upon an archive of historical data on the same countries since January 2000 (for Reuters' reports) and January 2003 (for AFP reports). The primary focus of the forecasts was the use of force, and more generally, conflict and cooperation processes in the short term.

Forecasting of conflict in the international relations literature has been discussed extensively by both skeptics and non-skeptics. The former category includes scholars like Chapman and Doran who argue that the international system is too complex and non-linear to forecast with adequate precision [10, 15]. Other scholars contend that forecasting conflict falls within the limits of scientific pursuit and should be treated as such. Building on earlier work of Richardson involving systems of differential equations [25, 26]—Rothstein states that if we fail to use sophisticated, emerging methods of forecasting we would not gain understanding of the regularities in the political behavior in the international arena [27]. To this

V.B. Petroff (✉)
Political Science, University of Kansas, The University of Kansas Lawrence, KS 66045
e-mail: vlado77@hotmail.com

J.H. Bond
Harvard Division of Continuing Education, VP, Virtual Research Associates, Inc., Harvard University Cambridge, MA 02138
e-mail: joe.bond@vranet.com

D.H. Bond
Lecturer in Extension, Harvard University Division of Continuing Education, President, Virtual Research Associates, Inc., Harvard University Cambridge, MA 02138
e-mail: doug.bond@comcast.net

V.S. Subrahmanian (ed.), *Handbook of Computational Approaches to Counterterrorism*,
DOI 10.1007/978-1-4614-5311-6_8,

end, Ayers uses morphological analysis, trend extrapolation, and heuristic models [2]. Choucri discusses econometric models of forecasting [11] and time series while Rummel does three-mode factor analysis [28].

Recognizing that the international system is a non-linear, complex, and sometimes chaotic, other scholars employ non-heuristics and non-linear forecasting techniques. The main conceptual difference between these two classes is that heuristics and linear models make a host of simplifying assumptions—normality, constant variance, homogeneity, etc—or are restrictive as King points out in his political methodology book [16]. To tackle non-linear, complex, context contingent, political systems we need unrestrictive models that make less constraining assumptions about the underlying physical mechanism generating the observations. Similarly, Beck, King, and Zheng discuss neural network analysis [4], Schrodt [32] as well as Bond, Petrov, Bond, and O'Brien discuss Hidden Markov Models (HMM) [8].

Numerous authors have contributed to the field via structural, complex, and/or agent based models. For example Clauset, Young, and Gleditsch used Richardson's study of deadly conflicts and found that the center of the distribution oscillates slightly with a period of roughly 13 years [13]. Cioffi-Revilla discuss an agent-based complex systems simulation approach that focuses more or less on "static" variables that can be controlled to determine what the behavior of the system could be in a descriptive manner [12]. Bar-Yam describes multi-scale representations of littoral warfare in more of a networked context in an effort to shed some light on the nature of the system under investigation [3]. In order to avoid the shortcoming of these more traditional methods without losing their advantages we chose to use HMMs in their generative capacity. The next sections describe the HMM model, the data and forecast generation using empirically derived guidelines to steer the overall process.

1.1 Hidden Markov Models

The HMM approach discussed by Rabiner is premised upon the idea that the sequence of social, political and economic events is important in influencing which events occur in the near-term future [23, 24]. Thus, the sequencing of observed events is considered in the HMM algorithms that yield a likely path of "hidden states" or phases in which the events occur. A common analogy may be drawn between our observations of weather conditions to determine the seasons, where the latter are not directly observable.

Figure 1 illustrates a six state, three-symbol HMM where the letters A through F represent the states whereas the arrows emanating from the states represent the observed symbols. The arrows between/among states represent phases through which the conflict goes though. Loops on the top of the states represent the fact that some countries are engulfed in a state of perpetual conflict—Afghanistan and Iraq being two examples. September 11, for example, might be viewed as a jump from state A (peace) to state F (outright war) in a single day. Stated differently, HMMs are theoretically suited to deal with such sudden jumps as well as all other possible transitions between/among phases.

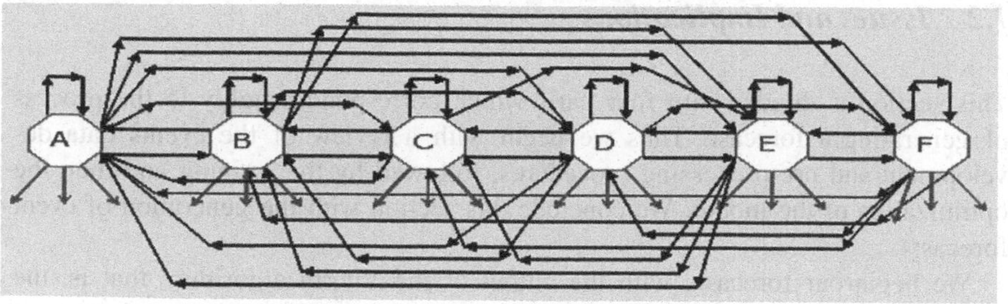

Fig. 1 Six state, 3-symbol fully connected HMM

As Schrodt asserts, the application of the HMM to the problem of generalizing the characteristics of international event sequences is straightforward [33]. "The states of the model are unobserved, but have a close theoretical analog in the concept of crisis "phase" that has been explicitly coded in data sets such as the Butterworth international dispute resolution dataset [9], CASCON [5] and SHERFACS [34], and in work on preventive diplomacy [17]. For example, Lund, in pages 38–39 [17] outlines a series of crisis phases ranging from "durable peace" to "war" and emphasizes the importance of an "unstable peace" phase."

In the HMM, these different phases are distinguished by different distributions of observed events. A stable peace, for example, would generate mainly, if not only, cooperative, non-violent events. The escalation phase of the crisis might be characterized by events like accusations, protests, demonstrations, denials, and threats and a phase of active hostilities would involve physical assaults, suicide bombings, riots, etc. The more often we observe violent events over time the more likely it is that the conflict is in a violent state.

Given the phase theory of conflict, Iraq and Afghanistan are obvious choices not only because the wars there have been dragging on for years—in the case of Iraq more than 6 years as of 2009 and in the case of Afghanistan more than 30 years—but also because the outcomes are indeterminate. Not knowing the outcome of a war ensures that our predictions are not just fitting past data to a known resolution but are instead focused on the future. Finally, both countries remain high in the policy agenda of all top decision makers. The HMM approach is applicable to other countries as well [8] and could be readily applied to Syria, Egypt among other countries.

The forecasts are presented as monthly time series of target activities, with a focus on forceful conflict. This process is guided by the HMM model's optimized output sequence of hidden states but manifest in measurable, observed activities providing a direct means of tracking past performance and offering a clear measure of confidence in the forecasts. The results of this research to date are encouraging. The HMM one-month horizon forecasts are able to predict the correct direction in the trend of forceful conflict with up to 82% accuracy. In the section that follows, we discuss the methods used, the issues encountered, and the implications of automated events data monitoring as input for forecasting events using an HMM approach.

1.2 Issues and Implications

This section is divided into four parts which correspond roughly to the process of generating a forecast. Thus we begin with a review of the events data development and pre-processing procedures, followed by the training and then the optimization of the model. We conclude this section with the generation of event forecasts.

We begin our forecasts with the output of the Viterbi algorithm; that is, the optimized sequence of hidden states that represents the most likely trajectory to a given set of previously observed events. However, we extend this procedure by generating a new set of events based on the Viterbi state trajectory and use the parameter estimates from the Baum-Welch algorithm to find locally maximum solutions. To date there is no known training algorithm used in HMM training to produce global maxima [23, 24]. We use this set of synthetic events to create daily and monthly arrays of the most likely events that build upon the prior event observations that are informed by the optimized state trajectory. In this way extend the prior use of the HMM approach to directly forecast the most likely events into the near-term future. This approach is discussed further below along with the specific issues and implications that arose during our research.

1.3 Data Development and Pre-processing

In this subsection, we address three main sets of related issues and their implications: (1) Event forms, symbol sets, states and mappings, (2) Data assumptions, density, multiple sentence processing and (3) Contingent and null events, and padding.

Our approach to HMM forecasting with events data begins with the automated processing of global news reports by the VRA Reader. VRA's patented technology codes the basic parameters (who does what to/with whom, when and where) of the social, political, environmental and economic events reported in news reports. The coded event forms (that is, the "what" parameter) follow the Integrated Data for Events Analysis (IDEA) [7]. The current version of the IDEA framework includes approximately 249 events forms, including all of the original World Event Interaction Survey (WEIS) events that have been widely used in international relations research for over three decades [18].

Our four symbol set includes (the use of) force and compliance activities (that encompass conflict and cooperation respectively) supplemented with an ambiguous and a null event (the absence of news event reports) category. The five symbol set adds one symbol by distinguishing between the use of force in conflict from non-conflict challenges such as accidents, disasters and other disruptive events. As for the six symbol set, the added distinction here focuses on nonviolent sanctions. We believe this symbol set may be useful in situations where challenges to a

Table 1 Preferred symbol sets

Polarity	N = 4	N = 5	N = 6
Negative	Force	Force	Force
Negative	Force	Challenges	Sanctions
Negative	Force	Challenges	Adversity
Ambiguous	Ambiguity	Ambiguity	Ambiguity
Positive	Compliance	Compliance	Compliance
None	Null	Null	Null

regime's stability are based on coercive but nonviolent sanctions or "people power" movements.

To be clear, for all symbol sets we take all of the IDEA event forms in cue categories 1 through 22, and associate each of them with one of these symbols. For example, the IDEA event form physical force maps to the symbol labeled "force" on our four symbol set. In the minimal (N = 4) symbol set, therefore, accidents and natural disasters are mapped together with force. While parsimonious, we found that such aggregation caused problems with countries that experienced natural disasters and conflict; in other words, we could not differentiate between them. Thus we added the five and six symbol sets to discriminate between the challenges of conflict and other adversity involving physical force as well as nonviolent sanctions that may also be operative (for the six symbol set).

Another issue that we considered was the practical utility of the symbol sets. We asked, is it better to forecast correctly one of fifteen types of events as opposed to one of four types. When larger symbol sets were tested, we found ourselves focusing on one or two symbols of particular interest, typically at the two extremes of conflict and cooperation or force and compliance. Thus our preferred symbol sets retain these foci, and add incrementally where we thought the differentiation would be most useful. The adversity wrought by accidents and natural disasters and nonviolent sanctions were the two categories we chose to highlight as we expanded beyond our minimum set. By using smaller sets the impact of coding errors, both false negatives and false positives, in the underlying events data was mitigated due to the higher levels of aggregation in mapping event forms to symbols. Our three preferred sets are outlined below (Table 1).

Our approach to the specification of states or phases was also anchored in interpretability. Although the states are unobservable, we sought empirical interpretability of the states so that we could better assign the initial symbol pair values for state transitions. Of course the interpretability also meant that we could better understand the output. The process of assigning and counting all possible symbol pair transitions yields the state transition matrix that is used to initiate the training.

In the preferred states table below, one can see that using an odd number of states allows for an ambiguous category. We found this helpful as the observed symbols were often ambiguous in terms of their role or context in the conflict process. Six states or phases do not provide such an ambiguous category, assuming we wish to preserve some sense of symmetry in escalation and de-escalation of conflict. Moving to a seven state model renders the interpretability of many if not most event forms difficult.

Table 2 Preferred states

Polarity	N = 3	N = 4	N = 5
Negative	Conflictive	Escalating	Escalating
Negative	Conflictive	Conflictive	Conflictive
Ambiguous	Ambiguous	Ambiguous	
Positive	Peaceful	De-escalating	De-escalating
Positive	Peaceful	Peaceful	Peaceful

Our three state model is simple; it consists of negative (conflictive), positive (cooperative) and ambiguous phases. We also worked extensively with a four state model. The four phase model includes conflictive and peaceful phases in addition to escalation and de-escalation phases. We found it extremely difficult to interpret the empirical results from higher order models as presented in the literature on conflict; five and six state (conceptual) models, for example, have been widely used [1], but in practice we found the additional phase distinctions to be vague at best (Table 2).

Next we address our primary data assumptions. We assume independence among the observations of discrete events that are ordered chronologically in equal time intervals. This assumption of independence among events is pretty standard and it, along with chronological ordering, are both fairly benign as assumptions. The more challenging assumption is that of equal temporal intervals across the observations. We address this requirement of the HMM algorithms by considering the character of our news report events data as well as other events data that may be used.

One approach in addressing the equal temporal interval assumption is to aggregate events to a daily (or any other time interval such as an hour or week). This approach, however, requires transforming the nominally scaled event data, typically by applying weights. Although this is a widely practiced procedure in the international relations research it embodies the assumption that conflict and cooperation are bi-polar opposites along a one-dimensional continuum. We reject this assumption and thus chose a different approach.

We treat each event form observation (event record) as a data point such that if twelve events per day are reported then their temporal spacing is two hours, and if forty eight events are reported in a day, then their temporal spacing is a half hour. We use this approach for all calculations leading up to the event generation for our forecast. At this point we adjust the forecast horizon by the most recent (week) density. In other words, we assume that the most recent past is the best predictor of the immediate future. We suggest that this is a reasonable and realistic view of conflict and cooperation processes that involves no aggregation or assumptions about their dimensionality. For densely reported countries like Iraq and Afghanistan the number of daily events is far higher than the empirically established minimum of around 5 events per day.

The assumption of equal temporal intervals is intricately linked to the issue of when the event occurred, or in our case, when the event report was published by Reuters. The news reports that we process include date time stamps that reveal the time of their release. We assume that the journalistic process of transforming observations into published reports is fairly constant across events even though we recognize the reality that certain crisis oriented events may be rushed to press ahead of the more mundane reports. In other words, we assume the date time stamp of the news report is an attribute of the event that is reasonably accurate overall, and more accurate for events of interest to us (i.e. crisis events). At a minimum, we assume that the ordering of event reports is a reasonable approximation of the sequence of the actual events. We base this assumption on our knowledge that the overall volume of Reuters' reports in any given country is fairly constant in the medium term—it has trended upwards over the longer term, and it often includes erratic jumps in the shorter term. It is the mix or sequence of events that vary on a day-to-day basis that we seek to exploit in our forecasting. Given all these considerations we used report date since only around 0.03% of Reuters' news reports differed on report date and event date.

We also assume that the absence of news reports on any given day indicates an absence of (internationally) significant events. This assumption is linked to our international news agency source, and its daily and weekly news cycles that are apparent in our data. Thus a day without a news report is padded to represent a null event day. In other words, we impute a null event into the dataset for each day that has no reports. Logically we could carry this padding procedure down to an even smaller time interval, say to an hour or less. However, we do not consider the time stamp on each news reports to be as accurate as the date stamp in terms of providing a temporal attribute of the event observation. If one had better time stamp data, padding on smaller units would likely be more desirable.

The density of our news event input is highly variable across countries and across time. In some countries Reuters reports no events in a given day while in the US, the count is routinely well over a hundred per day with the weekly mean over 10 years being more than 1,700 events. We have built into the HMM application a series of user selectable parameters to handle this range of densities. These are discussed more fully below.

Finally, we have experimented with processing multiple sentences to increase the density of events for countries not sufficiently covered by Reuters. In general the lead sentence is sufficient to track event trends in news reports. We consider multiple sentence coding to be less than optimal because it tends to violate our assumption of ordered events. However in the absence of an alternative, more dense news or other (field-based, direct data entry) feed, we continue to use a two sentence input when needed.

1.4 Training (Baum-Welch estimates)

1.4.1 Approach, Initial Estimates and Alternative Models

The Baum-Welch algorithm begins with a sequence of observations (nominal events represented as discrete symbols), and from this input it calculates the most likely set of state transition probabilities along with their likely symbol distributions and initial state vector. The states, however, are not directly observable; rather, states are deduced from the presence or absence of the observed sequence of events. Appendix A provides a more technical discussion.

We have settled on the following preferred models:

1. Three states with four and six symbols

1.4.2 Sequence Length, Iterations and Losses

Regardless of the model selected, the number of iterations used in the Baum-Welch training algorithm, the length of the sequences, the estimates used in training (i.e., whether we rely on global estimates or estimates generated in the previous forecast) and the number of simulations generated in the simulator all affect the forecast.

In general, longer sequences, defined as 1,500–3,800 symbols, tend to produce superior results in countries with high baseline levels of violence/force (i.e. approaching at least 20% of all reported activity). Applied to countries with low density (i.e. <5 events/symbols per day), long sequences tend to produce relatively flat forecasts, converging on the overall mean of the time series. In low density countries, we are somewhat constrained in our selection of sequence length because as we extend the length of the sequence, we reduce the number of iterations used in the Baum-Welch training algorithm. Because we have established a minimum threshold of five iterations in the Baum-Welch training algorithm (based on our extensive experimentation), we are unable to select long sequences for very low-density countries because the number of iterations generated is inextricably linked to the sequence length.

Because the sequence length and iterations are inextricably linked, an increase in one results in a decrease in the other and vice versa for a constant number of states. Again, we assume that longer sequences are generally preferable, as they appear to capture longer-term patterns of behavior inherent in the data. Density also dictates the sequence length. Lengthy sequences cannot be used on low-density countries. Therefore, sequences between 500 and 700 events are used in order to generate five or more iterations for purpose of training.

Sequence length appears to be a key factor in the accurate prediction of trends. In most cases, we can retrospectively predict or "post-dict" the trend if we know whether the next period's density is increased or decreased from its current level. Increasing the length of the sequence in cases where the most recent density (i.e., the density over the past seven days) is decreasing tends to capture the trend. Similarly, as the recent density increase, we can use shorter sequence lengths.

1.4.3 Global and Prior Estimates

Another parameter selected with each run is the choice to use either the (initial) global estimates or to use the ending or output estimates from the most recent run.

The first estimation, of course, requires the use of the initial estimates, but for each subsequent estimate one can choose either setting. We have experimented with this and have determined that it has the least impact on the forecasting results. In general, for lower density countries, we tend to use global estimates exclusively, and for higher density countries we tend to mix the selections. Again, we find this parameter to have little impact on our forecasts.

1.4.4 Global, Cut (Training Set) and Most Recent Densities

As noted above, the density of the data stream is interrelated with our chosen sequence length and number of possible iterations. The overall density measure is supplemented with a cut density measure; the cut density refers to the density in the training set data only. The cut density can be greater or lesser than overall density. Typically, the cut density indicates the coverage closer to the present time than the overall density, and thus it is useful for gauging recent activity within a country, or at least the international media coverage of activity. We also calculate four additional densities to guide our parameter selection. We calculate the most recent, the second recent, the third most recent and the fourth most recent week densities prior to choosing the parameters. We find these to be useful guides in our parameter choices. For example, if one knows the future densities, parameter selection becomes simply a matter of optimizing the chosen sequence length and number of iterations, decreasing or increasing the lengths with the future density in mind. Given our goal of prediction, however, where the event density within the forecast horizon is unknown, we rely on two guidelines to assist in the parameter choices; the most recent (one week) density and the most recent (one month) volatility of the known densities.

Utilizing the most recent density (i.e. one week) is premised on the idea that the best single predictor of asymmetric conflict in the immediate future is the (ongoing) conflict in the most recent past. Our second guide—one month—is premised on that idea that conflict is volatile and news reporting may be alternately repressed and flood out of a country during agitation and turmoil just before an escalation into a full-blown crisis. By monitoring the most recent density and volatility, we are able to adjust the parameters to better account for rapid escalations into violence.

1.4.5 Optimization (Viterbi State Trajectories)

The most likely set of state transition and symbol distributions calculated by the Baum-Welch algorithm are used in the Viterbi algorithm to produce the most likely sequence of (hidden) states that result in the sequence of previously observed events. We identify the most recent sequence of states as found in the last 10% of the last row. In practice this works out to represent a temporal span of approximately one week.

We consider the modal state from the last section of the Viterbi vector to be the final (hidden) state of the observed sequence. We also use this value to represent the initial (hidden) state for our forecasting. As noted earlier, our use of this vector only as a starting point for our forecasts differs from conventional usage. Our choice to work with symbols instead of states in forecasting was driven by a desire to employ direct measures of success.

2 Forecasting

Our forecasting procedure is outlined below.

1. Retrieve the most recent (modal) state from the Viterbi algorithm output as described above. The modal state is the current state for initial iteration.
2. Go to the row in the symbol distribution [B] matrix representing the current state and select a symbol using a weighted throw of a die based on the estimated distribution for that state
3. Go to the row in the state transition [A] matrix representing the current state and select the next state using a weighted throw of a die based on the estimated transition probabilities for that state. The next state becomes the current one.
4. Repeat steps #2 and #3 until the number of symbols selected equals the most recent week's daily density times the forecast horizon in days. For example, if the most recent daily density is 10 events and the forecast horizon is 30 days, then the number of symbols to be generated is 10*30 = 300.
5. This event generation procedure or scenario may be repeated for a user selectable number of times. We have found 3,000 scenarios to be an upper bound number, beyond which the results are indistinguishable. However, for expediency we often use a very low number of scenarios, for instance, one, ten or a hundred, when we are testing one isolated part of the process, or are running baseline comparisons among different settings. Using the same density (10) and forecast horizon (30) as described in the example above, the total number of symbols generated in this forecast would be 300 symbols times 3,000 scenarios for a set of 900,000 symbols.

The procedure outlined above demonstrates the generative capacity of HMMs [23, 24]. In computer science code generation by compilers roughly corresponds to symbol generation whereas parsing corresponds to pattern matching; thus, the generative and pattern-matching capacities are the two complementary facets. Finally, our approach is novel in the area of international relations thus further prompting our interest.

The symbol array generated in our forecast procedure follows the optimum (Viterbi) state trajectory and are drawn from the (Baum-Welch) estimated symbol distribution within each of those states in succession. We provide a graphic representation of our results using a monthly time series of forecasted versus

Table 3 Forecast direction only results (Summary of our Results, Comparing Observed to Forecasts)

Country-Target activity	Direction only (%)
Iraq-Force	73
Afghanistan-Force	82

observed results. We assess the performance here with a measure of the degree to which the forecast captured the right direction of the trend. This simple measure is also the most visually intuitive as it matches the immediate sense of the general performance that one acquires from a time series.

2.1 Iraq and Afghanistan Results

In this section, we present in this section the results for Iraq—4 symbols, 3 states HMM—while Afghanistan is described by 6 symbols and 3 states. We have generated models for Iraq and Afghanistan with data from 1990 through January 2009. The summary table below presents the bottom line results for our forecasting. Nothing was optimized in this set of results and no cases were added or dropped.

Given the significant differences among the country contexts and reporting practices, these levels of reported activity should be interpreted over time rather than across countries. In other words, these time series present baselines of force that reflect the trends over time within a given country (Table 3).

Figure 2 illustrates the gradual process of withdrawal from Iraq of the "coalition of the willing." In January of 2008 the Iraqi Parliament passed legislation allowing former officials from Saddam Hussein's Baath party to return to public life—a stark reversal from the earlier bans, imprisonment, prosecution of former Baathists. In February of 2008, suicide bombings at markets in Baghdad killed more than 50 people in the deadliest attacks in the capital in months. Concurrently, Turkish forces mounted a ground offensive against Kurdish rebels in northern Iraq. In order to stabilize his country, both internally and externally, Prime Minister Maliki hosted the Iranian president for an unprecedented 3 day visit and also ordered a crackdown on militia in Basra, sparking pitched battles with Moqtada Sadr's Mehdi Army. Hundreds were killed in March and April of 2008; as a result, the percentage of violence as a proportion of all reported activity increased slightly over 30% during these months. In April of 2008, Sadr threatened to scrap the Mehdi Army truce which he declared in August 2007 after the crackdown of his militias. In June of 2008 Australia officially ended its combat operations in Iraq. Realizing that the political solution should be a priority, Prime Minister Maliki—in July of the same year—for the first time raised the prospect of a withdrawal timetable of US troops as part of negotiations over a new security agreement with Washington. The main Sunni Arab bloc, the Iraqi Accordance Front, rejoined the Shia-led government

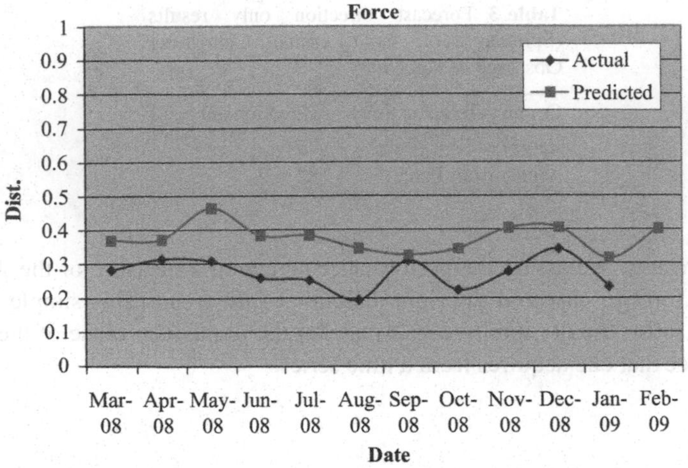

Fig. 2 Iraq time series

almost a year after it pulled out shortly after the withdrawal timetable proposal. As part of a political solution, US forces handed over control of the western province of Anbar to the Iraqi government in September. Once a hotbed of the anti-US insurgency and later an al-Qaeda stronghold, Anbar is the first Sunni province to be returned to Baghdad's Shia-led government. In October of 2008 members of the Baghdad Awakening Council, estimated to number about 54,000, moved to the Iraqi government payroll, with other members of the Sunni militia councils set to follow—such militia cooptation mirrors the return of Baathists to the government. Most importantly, with the election of President Obama, the Iraqi parliament approved a security pact with the United States under which all US troops were due to leave the country by the end of 2011. Therefore, all these substantive political measures contribute to reduction of violence in the summer/autumn of 2008 as evident from Fig. 2. Moreover, Iraq took control of security in Baghdad's fortified Green Zone and assumed more powers over foreign troops based in the country as Prime Minister Maliki welcomed the move as Iraq's "day of sovereignty" in January of 2008.

Commanders of UK forces in southern Iraq also handed power over to a US general, marking the beginning of Britain's official withdrawal. The spike in Fig. 2 for our predictions for February 2009 signifies the fact that there are remaining hostile elements—three suicide attacks in a week kill 33 people at a reconciliation conference in western Baghdad, more than 30 in the east of the city at the main police academy, and ten at a cattle market in Babel.

Table 4 shows that the squared differences between actually observed and predicted proportions of force from all reported events is within 10% error margin of magnitude.

Table 4 Iraq error metrics

Date	Actual force	Predicted force	(Actual-Predicted)^2
Mar-08	0.28205	0.370027	0.007739953
Apr-08	0.31438	0.37189	0.0033074
May-08	0.30822	0.463448	0.024095732
Jun-08	0.25843	0.38338	0.015612503
Jul-08	0.25166	0.382965	0.017241003
Aug-08	0.19403	0.345281	0.022876865
Sep-08	0.30909	0.325789	0.000278857
Oct-08	0.22283	0.344743	0.01486278
Nov-08	0.27692	0.403872	0.01611681
Dec-08	0.34286	0.406054	0.003993482
Jan-09	0.23188	0.316011	0.007078025

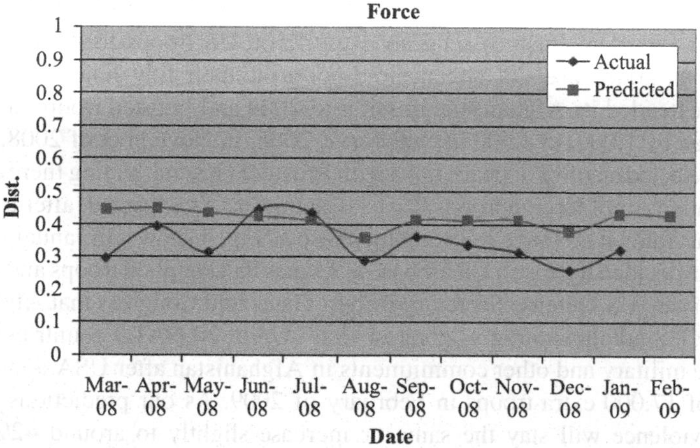

Fig. 3 Afghanistan time series

In April of 2008 NATO leaders met in Bucharest to say the peacekeeping mission in Afghanistan is their top priority. They pledged a "firm and shared long-term commitment" there. Acting, rather than taking pledges, the Taliban engineered a massive jail break freeing at least 350 insurgents in June of 2008 according to cross-checked BBC reports. In the same month, the British Defense Secretary Des Browne announced British troop numbers in Afghanistan to increase by 230 to a new high of more than 8,000 by spring 2009. To make matters worse, in July of 2008 suicide bombers attacked the Indian embassy in Kabul killing more than 50 people. As Fig. 3 illustrates, it is not surprising that force events accounted for about 45% of the total volume of reported events by mid-summer. Although ten French soldiers were killed in an ambush by Taliban fighters, the levels of violence decreased in August of 2008 as both the predicted and the actual graph show. Finally, in the same month, President Karzai accused Afghan and US-led coalition forces of killing at least 89 civilians in an air strike in the western province of Herat. Despite the

Table 5 Afghanistan error metrics

Date	Actual force	Predicted force	(Actual-Predicted)^2
Mar-08	0.29703	0.444696	0.021805248
Apr-08	0.39535	0.448867	0.002864069
May-08	0.31579	0.4344	0.014068332
Jun-08	0.448	0.42457	0.000548965
Jul-08	0.43624	0.41054	0.00066049
Aug-08	0.29032	0.359703	0.004814001
Sep-08	0.36364	0.412707	0.00240757
Oct-08	0.33824	0.412272	0.005480737
Nov-08	0.31618	0.414292	0.009625965
Dec-08	0.2623	0.378746	0.013559671
Jan-09	0.3253	0.432367	0.011463342

decision of President Bush to send an extra 4,500 US troops to Afghanistan, in September of 2008, levels of violence did not go to their July-June peaks even as Germany extended its Afghanistan mission to 2009 and boosted troop numbers in Afghanistan by 1,000, to 4,500 in October of 2008. In November of 2008, Taliban militants rejected an offer of peace talks with President Karzai, stating there could be no negotiations until foreign troops leave Afghanistan. As expected, after a summer combat, the Taliban go into a re-grouping phase as Fig. 3 shows. In January of 2009 Kyrgyzstan decided to close a US air base at Manas that supplied troops and materiel to Afghanistan. US Defense Secretary Robert Gates told Congress that Afghanistan is the new US administration's "greatest test." Up to 20 NATO countries pledged to increase military and other commitments in Afghanistan after USA announced a dispatch of 17,000 extra troops in February of 2009. As our predictions suggest, levels of violence will stay the same or increase slightly to around 42% of the total volume as the weather warms and the Taliban start a new round of guerrilla warfare. The relative decline in activity for September through December of 2008 is influenced not only by winter conditions, but also by concurrent US presidential elections as well as the most serious economic recession since 1929 to date.

Table 5 shows that the squared differences between actually observed and predicted frequencies are within the 10% range or so over the past year.

2.2 *Testing of Results and Technical Discussion*

We did not compare the HMM forecast results against those of linear models for several reasons. Clements and Smith note—based on Monte Carlo simulation—that "Our results indicate that non-linear models have an edge in certain states of nature but not in others, and that this can be highlighted by evaluating forecasts conditional upon the regime" [14]. Stated differently sharp/sudden changes in the states of

Table 6 RMSFE test results for Iraq and Afghanistan

Case	RMSFE
Iraq	0.110042723
Afghanistan	0.089085449

nature cannot be easily captured by linear models but can be potentially done with non-linear ones [22]. Second, events data is discrete, (under/over) dispersed, non-symmetric around 0, lower bounded by 0, skewed to the left with occasional "fat" tails on the right thus easily, but mistakenly, classified as outliers [16]. Pesaran and Timmermann (PT) note that "For example, the test is applicable to non-stationary distributions as long as y, and x, are symmetrically distributed around 0" [21]. Events data is not symmetric around 0—it is bound by 0 since the minimal number of events per day is 0 in cases where no violence was reported. Stated differently, we cannot record negative number of events; moreover, 0 happens to be the extreme value to be recorded over periods of absence of attacks and thus another assumption of the PT test is violated. Given that the assumptions of this test are not satisfied, we chose not to do it as not to skew the results.

Although Kolomogorov-Smirnov test is not that sensitive to the normality, zero-mean, distribution/shape, serially/contemporaneously non-correlation we chose not to run it due to small sample size—11 months of in-sample predictions and one out-of-sample forecasts. For similar normality deviation reasons we chose not to equate forecast errors with a zero test. With all these caveats in mind, root mean squared forecast error tests (RMSFE) were done for the cases of Afghanistan and Iraq and are presented in Table 6 above:

The RMSFE for Afghanistan is closer to 0 as compared to the Iraqi case, thus the RMSE implies that the magnitude of forecast error is smaller for Afghanistan. In both cases the RMSFE appears close to 0, thus the predicted values and the observed values are not that different.

3 Conclusions

Our experience working with the HMM approach to forecast social, political and economic events data leads us to an optimistic conclusion. Conceptually, the approach makes sense and the mathematical tools are well developed; conducted tests indicate that error margins are robust thus warranting potentially more research in this area. Our results suggest that the presented approach can be used by three kinds of end users: military, bureaucratic, and academic. Knowing what the approximate levels of violence in location X allows military commanders better plan accordingly. Bureaucrats can decide on resource allocation based upon violence estimates whereas academics can further explore and apply HMMs to novel problems.

4 Training

Although the hidden states of a Markov system are not directly observable, they "have a close theoretical analog in the concept of crisis 'phase'" [31]. Thus when we observe mounting casualties from continuous fighting over a certain period of time we may consider ourselves to be in a state of conflict. In applying Hidden Markov models to international relations, states correspond roughly to phases in the conflict process. Like daily weather (snow, clouds, sunshine and the like) individual (observable) events are used to infer a particular season or state. In the realm of international relations, events such as visits, agreements, armed raids, bombings and the like correspond to the observation sequence.

A.1 Appendix A: Technical Details

Entries for the 3 State 4 Symbol Model (Table A.1).

After entering the initial paired event values in the cells of the state transition matrix, we count each of the symbol pairs in the input stream for each cell and divide by the total to derive the initial Pi vector of state transition probabilities. The obvious alternative is to treat all state transitions as equally probably (0.33 each on a three state model). However, we consider our initial values—derived from the data—to be more useful in that they are bound to our understanding of the observed events in our data; that is, we consider them as actual pairs of events and from this empirical understanding, infer their most appropriate hidden state context. Similar input data derived estimates were used for the initial state and the symbols matrices. The more traditional initializing K-means family of algorithms was avoided since they rely on computing Euclidean-like distances which are not applicable to our discrete data.

Table A.1 State transition matrix

Cell	Description of state change	Symbol pair (N = 16) initial values
1.1	Conflictive to Conflictive	1 > 1
1.2	Conflictive to Ambiguous	1 > 2
1.3	Conflictive to Peaceful	1 > 3, 1 > 4
2.1	Ambiguous to Conflictive	2 > 1
2.2	Ambiguous to Ambiguous	2 > 2
2.3	Ambiguous to Peaceful	2 > 3, 2 > 4
3.1	Peaceful to Conflictive	3 > 1, 4 > 1
3.2	Peaceful to Ambiguous	3 > 2, 4 > 2
3.3	Peaceful to Peaceful	3 > 3, 3 > 4, 4 > 3, 4 > 4

References

1. Alker HR, Schmalberger T (2001) The double design of the CEWS project. In: Alker HR, Gurr TR, Rupesinghe K (ed) Journeys through conflict: narratives and lessons. Rowman & Littlefield Publishers, Lanham, MD
2. Ayres RU (1969) Technological forecasting and long-range planning. McGraw-Hill, New York
3. Bar-Yam Y (2003) Complexity of military conflict: multiscale complex systems analysis of littoral warfare. New England Complex Systems Institute at http://necsi.edu/research/management/military/. Accessed 26 June 2012
4. Beck N, King G, Zeng L (2000) Improving quantitative studies of international conflict: a conjecture. Am Polit Sci Rev 94(1):21–35
5. Bloomfield LP, Moulton A (1997) Managing international conflict: from theory to policy. St. Martin's Press, New York
6. Bond J, Bond D (1995) Panda codebook. Cambridge, MA: The Program on Nonviolent Sanctions and Cultural Survival, Weatherhead Center for International Affairs, Harvard University
7. Bond D, Bond J, Jenkins JD, Taylor CL (2003) Integrated Data for Events Analysis (IDEA): an events typology for automated events data development. J Peace Res 40(6): 727–745
8. Bond J, Petrov V, Bond D, O'Brien S (2004) Forecasting Turmoil in Indonesia: an application of hidden Markov models. A paper prepared for presentation at the International Studies Association Convention. Montreal, Quebec, May 17–20, 2004
9. Butterworth RL (1976) Managing interstate conflict, 1945–74: data with synopses. University of Pittsburgh University Center for International Studies, Pittsburgh
10. Chapman JW (1971) Political forecasting and strategic planning. Int Stud Quart 15(Sept): 317–357
11. Choucri N (1973) Forecasting in international relations: problems and prospects. Int Interact 1
12. Cioffi-Revilla C (2012) Modeling and simulation at the mason center for social complexity. The voice of technology. Winter 2012:30–31
13. Clauset A, Young M, Gleditsch KS (2007) On the frequency of severe terrorist events. J Conflict Resolut 51(1):58–88
14. Clements M, Smith J (1999) A Monte Carlo study of the forecasting performance of empirical setar models. J Appl Econom 14(2):(Mar.–Apr., 1999)
15. Doran CF (1999) Why forecasts fail: the limits and potential of forecasting in international relations and economics. International Studies Association, Blackwell Publishers, Oxford
16. King G (1989) Unifying political methodology: the likelihood theory of statistical inference. Cambridge University Press
17. Lund MS (1996) Preventing violent conflicts: a strategy for preventive diplomacy. United States Institute for Peace, Washington, DC
18. McClelland CA (1976) World Event/Interaction Survey Codebook. (ICPSR 5211). Inter-University Consortium for Political and Social Research, Ann Arbor
19. McClelland CA (1983) Let the user beware. Int Stud Quart 27(2):169–177
20. O'Brien S (2002) Anticipating the good, the bad, and the ugly: an early warning approach to conflict and instability analysis. J Conflict Resolut 46(6):791–811
21. Pesaran H, Timmermann A, (1992) A simple nonparametric test of predictive performance. J Bus Econ Stat 10(4):461–465
22. Priestley MB (1988) Non-linear and non-stationary time series analysis. Academic Press, Waltham, Massachusetts
23. Rabiner LR (1989) A tutorial on hidden Markov models and selected applications in speech recognition. Proc. IEEE 77(2) Feb 1989
24. Rabiner LR, Juang BH (1985) A probabilistic distance measure for hidden Markov models. AT&T Tech J 64(2) Feb 1985
25. Richardson LF (1960) Arms and insecurity. Boxwood Press, Pittsburgh

26. Richardson LF (1960a) Statistics of deadly quarrels. Boxwood Press, Pittsburgh
27. Rothstein RL (1972) Planning, prediction and policymaking in foreign affairs. Little, Brown, Boston
28. Rummel RJ (1969) Forecasting international relations: a proposed investigation of a three-mode factor analysis. Technol Forecast Soc 197–216
29. Rummel RJ (1975) Conflict watch: a continuous computer monitor of the Asia-Pacific region. Presented at the Conference on Peace and Unification on the Korean Peninsula in a New International Order, September, Seoul
30. Schrodt PA (1997) Early warning of conflict in southern Lebanon using hidden Markov models. Presented at Annual meetings of the American Political Science Association, Washington, DC
31. Schrodt PA (2000a) Forecasting conflict in the Balkans using hidden Markov models. A paper presented at the APSA meeting in Washington, DC
32. Schrodt PA (2000b) Pattern recognition of international crises using hidden Markov models. In: Richards D (ed) Political complexity: nonlinear models of politics. University of Michigan Press, Ann Arbor, pp 296–328
33. Schrodt PA, Gerner D (2001) Event data in international relations. Department of Political Science, University of Kansas, Lawrence, KS
34. Sherman FL, Neack L (1993) Imagining the possibilities: the prospects of isolating the genome of international conflict from the SHERFACS dataset. In: Merritt RL, Muncaster RG, Zinnes DA (eds) International event-data developments: DDIR phase II. University of Michigan Press, Ann Arbor, pp 87–112

Forecasting Group-Level Actions Using Similarity Measures

Gerardo I. Simari, Damon Earp, Maria Vanina Martinez, Amy Sliva, and V.S. Subrahmanian

1 Introduction

In real-world settings, and in particular in counterterrorism efforts, there is a constant need for a given reasoning agent to have the means by which to "stay ahead" of certain other agents, such as organizations or individuals who may carry out actions against its interests. In this work, we focus on one such way in which a reasoning agent can do this: *forecasting group-level actions*. This ability is indeed a useful one to have, and one that is definitely attainable if the right kind of data is available. For example, consider the Minorities at Risk Organizational Behavior (MAROB) dataset [4, 19]. This data tracks the behavior for 118 ethnopolitical organizations in the Middle East and Asia Pacific on a yearly basis from 1980 to 2004. For each year, values have been gathered for about 175 measurable variables for each group in the sample. These variables include strategic conditions such as the

---

G.I. Simari (✉) · M.V. Martinez
Department of Computer Science, University of Oxford, Oxford OX1 3QD, UK
e-mail: gerardo.simari@cs.ox.ac.uk; vanina.martinez@cs.ox.ac.uk

D. Earp
University of Maryland Institute for Advanced Computer Studies (UMIACS), University of Maryland College Park, College Park, MD 20742, USA
e-mail: dearp@umiacs.umd.edu

A. Sliva
College of Computer and Information Science, Northeastern University, Boston, MA 02115, USA,
e-mail: asliva@ccs.neu.edu

V.S. Subrahmanian
Department of Computer Science, University of Maryland College Park, College Park, MD 20742, USA
e-mail: vs@cs.umd.edu

V.S. Subrahmanian (ed.), *Handbook of Computational Approaches to Counterterrorism*, DOI 10.1007/978-1-4614-5311-6_9,

tendency to commit bombings and armed attacks, as well as background information about the type of leadership, whether the group is involved in cross border violence, etc. Only a subset—around 43—of the approximately 175 attributes in the data represent strategic actions taken by the group, while the others represent variables relating to the environment or context in which the group functions. This context includes variables about the degree of military and financial support the group gets from foreign nations or the ethnic diaspora, the degree of state government repression and persecution against the group, and so forth. It also includes variables about the structure of the group and how factionalized it may or may not be, the level of violence and protests in which the group engages, and the amount of participation in the political process.

The ontology generated from this schema is quite shallow in contrast to the deep ontologies seen in semantic web approaches such as RDF or OWL [1]. However, experts still want to use this information on past behavior, particularly data related to the contextual or situational factors, to forecast the actions taken by an agent, even without deep ontological or probabilistic knowledge, such as the ones discussed below. More formally, suppose there is historical data about an agent's behavior over many time periods, i.e., there is a set $\mathscr{PB}$ of past behaviors. Each time period in $\mathscr{PB}$ is a pair (c, a) consisting of two vectors: c is a vector containing the values of the context attributes associated with the agent, and a is a vector containing the values of the action variables. Suppose now that a user wants to identify what the agent might do in the current situation or in a hypothetical scenario. In either of these cases, the situation can also be represented by a vector q describing the context or the environment in which the agent is hypothesized to be, or actually is, functioning. The user is interested in determining what the associated action vector will be for this given situation, providing a forecast of the likely strategy an agent will employ in certain contexts. For example, the action vector might indicate that a terrorist group will engage in bombings with a "high" degree of intensity, while simultaneously showing that they will not resort to kidnappings.

1.1 Related Work

Past work on forecasting terror group actions has mainly focused on model-based techniques. For instance, the Stochastic Opponent Modeling Agents (SOMA) [7, 11–15, 17, 18] framework has been extensively studied as a robust method for behavioral modeling and forecasting, with applications to groups (i.e., terror organizations, drug traffickers, political parties) involved in global security situations. SOMA uses action probabilistic logic programs (*ap*-programs) and scalable distributed reasoning algorithms to compute the most probable actions an agent will take in a given situation. This and other similar approaches are based on *indicators* that have previously been identified, and possibly encoded as probabilistic logic [5, 6, 9] rules, as being correlated with specific agent behaviors being forecasted. However, there are many cases in which these indicators are not known a priori or it may be necessary to make a timely forecast or "what if" analysis without constructing an intermediate behavioral model to learn these indicators in advance.

Our goal in this work is to develop computational methods to cope with the intractability intrinsic to modeling, forecasting, and analyzing the behavior of agents. In this regard, there has been relatively little work on automated forecasting of what an agent (ethnopolitical group, terrorist organization, etc.) might do in the future in a specific situation. All previous efforts to build such predictive models in the socio-cultural-political domain have focused on three phases: a data collection phase, a model construction phase, and a forecasting phase using the model.

Efforts by Schrodt [10], Bond et al. [2], and the previous work on *ap*-programs mentioned above have all tried to address the data collection phase in an automated manner. However, data collection is not addressed here, since it has been extensively covered in other architectures.

When it comes to the model building phase, these previous approaches vary. Schrodt [10] and Bond et al. [2] have developed methods to build hidden Markov models (HMMs) to describe how a particular conflict might evolve over time. However, these HMMs are painstakingly constructed in a very time-consuming process, which must be repeated for each country or group involved and requires a degree of subjectivity. Nevertheless, HMMs and their variants (e.g., stochastic automata, stochastic Petri nets, etc.) can form very valuable modeling tools and mechanisms for behavioral forecasting and deserve continued study.

The APEX algorithm associated with the SOMA framework (described in [8]) for automated extraction of stochastic behavioral rules (*ap*-rules) has an advantage over the HMM models of [2, 10] in that it is fully automated and fast. These behavioral rules are then used to forecast agent behavior using efficient linear programming and scalable distributed algorithms. However, while this method is very robust and the behavioral rules can aid in understanding the forecasted actions, computational efficiency is still a major challenge of this logic-based approach, especially for real-time tactical analysis.

In contrast to all of these methods, the CONVEX framework described in this work does *not* build an intermediate model of an agent's behavior. Rather, time series data is used directly to assess the similarity between a given context (expressed as the query vector) and information about the agent's behavior in the past when confronted with similar situations. These past situations that the agent encountered are then used for forecasts—the "model building phase" present in related works is therefore skipped completely. CONVEX forecasts can be made solely by examining the data without an agent-specific model that takes time to build and may require subjective methods.

1.2 Contributions and Organization of This Work

In this work, a computational theory and algorithms for making behavioral forecasts are developed having the following features:

1. The proposed algorithms are very fast.
2. The algorithms are highly accurate for forecasting agent behavior, as demonstrated through application to the MAROB data [4, 19].
3. The results produced by the algorithms are easily explainable.

The goal here is to develop computational methods to better tame the complexity involved in modeling, forecasting, and analyzing agent behavior, especially in international security or conflict situations.

In Sects. 2 and 3, our behavioral forecasting problem is formalized. Section 4 then describes two classes of algorithms to address this problem. Strictly speaking, these are two general algorithms; however, their accuracy depends upon various parameters. Section 5 describes a prototype implementation showing that these algorithms are scalable, extremely general, and capable of producing highly accurate forecasts from time series behavioral data. Section 6 briefly discusses the SitCAST algorithm, which is similar in spirit to CONVEX but focuses on forecasting the values of environmental variables instead and can be used in conjunction with it in forecasting efforts. Finally, Sect. 7 presents our conclusions.

2 Behavioral Time Series Data

To forecast agent behaviors directly from data, a formal representation of this problem and the underlying time series is required. Such behavioral time series datasets consist of a relational database describing the behavior of an agent g. Assume the existence of some arbitrary universe $\mathscr{A}$ whose elements are called *attribute names* (attributes for short). Each attribute V_i has an associated domain $dom(V_i)$.

In these datasets the attributes fall broadly into two categories—*environmental or contextual (independent) attributes* describing the context in which an agent functioned during a given time frame, and *action (dependent) attributes* describing actions taken by the agent during a given time frame. Note that environmental attributes can also include actions taken by other agents or external actors that may impact the agent or group in question. Using this intuition, assume that any agent g has an associated *context schema* $CS(g) = (C_1, \ldots, C_n)$ of context attributes and *action schema* $AS(g) = (A_1, \ldots, A_m)$ of action attributes where each $C_i, A_j \in \mathscr{A}$ is an attribute and $\{C_1, \ldots, C_n\} \cap \{A_1, \ldots, A_m\} = \emptyset$. The full behavioral time series for a particular agent or group g is denoted by T_g.

Year	LEAD	ORGPOP	DEMORG	FORSTFINSUP	ARMATTACK	HOSTAGE
1993	4	2	1	0	1	0
1994	3	2	1	0	1	0
1995	3	2	1	0	0	0
1996	3	2	1	0	1	0
1997	3	2	1	0	0	0

Fig. 1 Small subset of actual data for a group in the MAROB dataset. LEAD, ORGPOP, DEMORG, and FORSTFINSUP are context attributes, while ARMATTACK and HOSTAGE are action attributes

Example 1. Figure 1 shows a small example of the behavioral time series data T_g for a terrorist organization g from the MAROB data [4, 19]. The table shown represents actual data for a subset of the approximately 175 MAROB attributes. Here, $CS(g) = \{\text{LEAD}, \text{ORGPOP}, \text{DEMORG}, \text{FORSTFINSUP}\}$, where the context in which the group is operating is described by categorical attributes denoting the type of group leadership, the amount of popular support, whether the group is internally democratic, and whether they receive financial support from a foreign state, respectively. The action schema $AS(g) = \{\text{ARMATTACK}, \text{HOSTAGE}\}$ contains two action attributes indicating whether the group uses armed attacks or takes hostages as a strategy at a particular time point. ■

3 A Formal Vector Model of Agent Behaviors

A behavioral time series T_g can also be represented as a collection of pairs of vectors assigning values to the context attributes and action attributes, respectively.

Definition 1 (g-behavior). Suppose g is an agent with context schema $CS(g) = (C_1, \ldots, C_n)$ and action schema $AS(g) = (A_1, \ldots, A_m)$. A *g-behavior* is a pair

$$\langle (c_1, \ldots, c_n), (a_1, \ldots, a_m) \rangle$$

where $c_i \in dom(C_i)$ and $a_j \in dom(A_j)$.

Definition 2 (Past Behavior). A *past behavior* for agent g is a finite set, $\mathscr{PB}(g)$, of g-behaviors.

Note that given the entire behavioral dataset T_g for an agent g, any $T'_g \subseteq T_g$ can be expressed as a past behavior $\mathscr{PB}(g)$.

Example 2. Consider again the behavioral time series for terror organization g given in Fig. 1. The corresponding past behavior in vector form consists of five g-behaviors: ■

$$\mathscr{PB}(g) = \{\langle(4,2,1,0),(1,0)\rangle,\ \langle(3,2,1,0),(1,0)\rangle,\ \langle(3,2,1,0),(0,0)\rangle,\ \langle(3,2,1,0),(1,0)\rangle,\ \langle(3,2,1,0),(0,0)\rangle\ \}$$

Definition 3 (Context/Action Vector). Suppose g is an agent with context schema $CS(g) = (C_1,\ldots,C_n)$ and action schema $AS(g) = (A_1,\ldots,A_m)$. A *context vector* w.r.t. agent g is an expression of the form $(c_1,\ldots,c_n)$ where each $c_i \in dom(C_i)$.

An *action vector* w.r.t. agent g is an expression of the form $(a_1,\ldots,a_m)$ where each $a_j \in dom(A_j)$.

Intuitively, a context vector specifies a value for each context attribute, and thus can be viewed as either a real or hypothetical context within which the agent functions (or is hypothesized to function). An action vector assigns a value to every action attribute and represents the actions taken by an agent.

Given a past behavior $\mathscr{PB}(g)$ and a query context vector q, the goal is to find an appropriate action vector $(a_1,\ldots,a_m)$ describing the agent's behavior given q. Clearly, the past behavior constitutes a set of historical behavioral data from which to forecast what actions the agent g might take when in a situation or context characterized by the query vector q.

4 Algorithms for Forecasting Agent Behavior

Using the vector data representation, this section presents two general algorithms to forecast an action vector $(a_1,\ldots,a_m)$ from a given past behavior $\mathscr{PB}(g)$ of an agent and a query context vector $(c_1,\ldots,c_n)$. Both algorithms use distance functions in metric spaces to compare the query vector to context vectors in the past behavior.

4.1 Distance Functions

It is clear that each context vector (and the query vector) is a point in the n-dimensional vector space $dom(C_1) \times \cdots \times dom(C_n)$. A distance function d can be defined on this vector space. As is commonly the case with distance functions, d must satisfy the following three axioms:

A1. $d(x,x) = 0$
A2. $d(x,y) = d(y,x)$
A3. $d(x,z) \leq d(x,y) + d(y,z)$

In the above formulas, x, y, z are all context vectors, i.e., members of $dom(C_1) \times \cdots \times dom(C_n)$. There are any number of well-known distance functions in the literature that meet these requirements. Six such distance functions are studied in this work and experimentally evaluated, but the forecasting algorithms given below also work with other distance functions, as well as weighted distance functions.

In the following, suppose context vectors $(c_1, \ldots, c_n), (c'_1, \ldots, c'_n)$ are members of $dom(C_1) \times \cdots \times dom(C_n)$.

1. **Euclidean distance.** $d_{EUC}((c_1, \ldots, c_n), (c'_1, \ldots, c'_n)) =$

$$\sqrt{(c_1 - c'_1)^2 + \cdots + (c_n - c'_n)^2}.$$

2. **Canberra distance.** $d_{CAN}((c_1, \ldots, c_n), (c'_1, \ldots, c'_n)) =$

$$\sum_{i=1}^{n} \frac{|c_i - c'_i|}{|c| + |c'|};$$

when divisions by zero occur, $d_{CAN}((c_1, \ldots, c_n), (c'_1, \ldots, c'_n)) = 0$.

3. **Chebyshev distance.** $d_{CHEB}((c_1, \ldots, c_n), (c'_1, \ldots, c'_n)) =$

$$\max_i \left(|c_i - c'_i| \right).$$

4. **Cosine distance.** $d_{COS}((c_1, \ldots, c_n), (c'_1, \ldots, c'_n)) =$

$$\frac{\sum_{i=1}^{n} |c_i * c'_i|}{|c| * |c'|}.$$

5. **Hamming distance.** $d_{HAM}((c_1, \ldots, c_n), (c'_1, \ldots, c'_n)) =$

$$\sum_{i=1}^{n} isDiff\left(c_i, c'_i\right)$$

where $isDiff(c_i, c'_i)$ is 1 if c_i and c'_i are different, and 0 if they are equal.

6. **Manhattan distance.** $d_{MAN}((c_1, \ldots, c_n), (c'_1, \ldots, c'_n)) =$

$$\sum_{i=1}^{n} |c_i - c'_i|.$$

```
Algorithm CONVEX^{k_NN}(PB(g), q, d, AS(g), k)
 1.   NearK = list of k elements initialized to +∞;
 2.   Forecast = ∅;
 3.   for each (cv, av) ∈ PB(g) do
 4.       Dist = d(q, cv);
 5.       if Dist < d(q, NearK[k − 1]) then
 6.           insert_sort((cv, av), NearK);
 7.   for each A_j ∈ AS(g) do
 8.       V = average of NearK[0].A_j, ..., NearK[k − 1].A_j;
 9.       if V is an integer then add V to Forecast;
10.         else add [⌊V⌋, ⌈V⌉] to Forecast;
11.   return Forecast.
```

Fig. 2 The CONVEX^{k_NN} algorithm for forecasting agent behavior

Example 3. Recall the past behavior $\mathcal{PB}(g)$ from Example 2 for terror organization g. Using the Manhattan distance function, the distance between the context vectors in the first two g-behaviors can be computed as:

$$d_{MAN}((4,2,1,0),(3,2,1,0)) = |4-3| + |2-2| + |1-1| + |0-0| = 1.$$

■

4.2 The CONVEX^{k_NN} Algorithm

In order to forecast what an agent will do in a given situation, either real or hypothetical, a query vector can be compared to similar contexts that the agent experienced in the past by utilizing a distance function over $dom(C_1)\times\cdots\times dom(C_n)$. In the CONVEX^{k_NN} algorithm given in Fig. 2, the k contexts from an agent's past data closest to the query vector are identified, and the average over the associated action vectors is computed to forecast actions in the query context.

Given a past behavior $\mathcal{PB}(g)$ for an agent of interest g, a query vector q, a distance function d, a number $k \geq 1$, and an action schema $AS(g)$, CONVEX^{k_NN} begins by initializing a sorted list $NearK$ of k elements. In the loop on Lines 3–6, each g-behavior in the past behavior is examined to find the k past context vectors closest to the query vector according to the distance function d. The function $insert\_sort$ on Line 6 is a procedure that will insert a new element into a list of fixed size, maintaining ascending order w.r.t. distance from the query vector. Then, in the loop beginning on Line 7, for each action attribute A_j, the values $a_{j_1},\ldots,a_{j_k}$ are retrieved from the action vectors associated with the k nearest context vectors in $NearK$. V is then computed to be the average of these values; if V is an integer, then it is simply added to the behavioral forecast in Line 9. Otherwise, the interval

$[\lfloor V \rfloor, \lceil V \rceil]$ is added to the forecast. This interval indicates that CONVEX$^{k\text{-NN}}$ cannot forecast the value of A_j exactly, but it is expected to either be $\lfloor V \rfloor$ or $\lceil V \rceil$ (assuming $dom(A_j)$ consists of integers—otherwise, if $dom(A_j)$ consists of real values, the forecast can be anywhere in the closed interval in question or the value V can be returned, depending on the semantics of the domain).

Example 4. Consider again the behavioral time series table T_g shown in Fig. 1 for a terrorist organization g with the context schema:

$$CS(g) = \{\text{LEAD, ORGPOP, DEMORG, FORSTFINSUP}\}$$

and the action schema:

$$AS(g) = \{\text{ARMATTACK, HOSTAGE}\}.$$

Suppose we want to forecast the behavior of group g in the scenario described by query vector $q = (4, 1, 1, 0)$ using CONVEX$^{k\text{-NN}}$, where d is the Euclidian distance function and $k = 1$ (i.e., only the single closest context vector from the past behavior will be considered). Of the 5 years of data shown in the table, the year that is closest to the query vector is 1993 where $d(q, (4, 2, 1, 0)) = 1$, so the g-behavior $\langle(4, 2, 1, 0), (1, 0)\rangle$ is added to $NearK$. The next step is averaging over the action vectors associated with the k closest context vectors to obtain a forecast. In this case CONVEX$^{k\text{-NN}}$ would forecast the action vector $(1, 0)$, where ARMATTACK $= 1$ and that HOSTAGE $= 0$, by simply looking at the single year 1993.

Now, consider the query vector $q = (3, 2, 1, 1)$ and $k = 4$ with the same distance function. In this case, years 1994, 1995, 1996, and 1997 are all the nearest neighbors with $d(q, (3, 2, 1, 0)) = 1$. To forecast the behaviors for the query context, CONVEX$^{k\text{-NN}}$ will average the values of the two action attributes over these four g-behaviors stored in *NearK*. For ARMATTACK, $V = avg(1, 0, 1, 0) = 0.5$, so the forecast for ARMATTACK will be the interval $[0, 1]$, meaning that the value of ARMATTACK will be either 0 or 1. For HOSTAGE, $V = avg(0, 0, 0, 0) = 0$, so the forecast for HOSTAGE is 0. CONVEX$^{k\text{-NN}}$ then returns the action vector $([0, 1], 0)$ as the forecasted behavior of group g given the query context.

Note that if $k = 3$ in the above case, the years 1994, 1995, 1996, and 1997 would all still be nearest neighbors of the query vector, but the algorithm would select only the first three and return the same answer as above based on the values of the action attributes in these 3 years. ■

The following result shows that CONVEX$^{k\text{-NN}}$ runs very fast. In practice, k is usually small (as will be shown later in Sect. 5), and this proposition states that the algorithm is linear in the size of the past behaviors $\mathcal{PB}(g)$ of the agent g.

Proposition 1. *Suppose $\mathcal{PB}(g)$ is a past behavior for agent g, d is a distance function over $dom(C_1) \times \cdots \times dom(C_n)$, $AS(g)$ is an action schema for g, and $k \geq 1$. If d is computable in $\mathbf{O}(n)$ time, then the* CONVEX$^{k\text{-NN}}$ *algorithm runs in time $\mathbf{O}\big(k \cdot n \cdot |\mathcal{PB}(g)| + k \cdot |AS(g)|\big)$.*

Proof. The loop in Lines 3–6 of the CONVEX^{k_NN} algorithm is executed at most $|\mathcal{PB}(g)|$ times. In each iteration of the loop, an insertion into a sorted list is necessary, which can be done in $\mathbf{O}(k)$ time. The second loop beginning on Line 7 executes at most $|AS(g)|$ times, with each iteration performing a linear retrieval operation on the list, again taking $\mathbf{O}(k)$ time. □

4.3 *The* CONVEXMerge *Algorithm*

The basic CONVEX^{k_NN} algorithm assigns an equal weight to each of the k nearest neighbors regardless of the variances in their distance to the query vector. The CONVEXMerge algorithm addresses this issue by assuming that the importance of the k nearest neighbors is inversely proportional to the distance between those context vectors and the query vector. In other words, suppose the two nearest neighbors in $\mathcal{PB}(g)$ of the query vector q are under consideration. The first neighbor may be at distance 1 away, while the second nearest neighbor may be at distance 10 away. In this case, the value assigned to action attribute A_j in the nearest g-behavior must have greater priority than the value assigned to action attribute A_j by the second nearest g-behavior. This intuition is captured by the following definition, where the set of k nearest neighbors w.r.t. a query vector q and a past behavior $\mathcal{PB}(g)$ is denoted $kNN_{q,\mathcal{PB}(g)} = \{(cv_1, av_1), \ldots, (cv_k, av_k)\}$.

Definition 4 (Conditional Probability of an Action). Let k be a fixed integer s.t. $k \geq 1$, q be a query vector, $\mathcal{PB}(g)$ be a past behavior for agent g, and $AS(g)$ be an action schema. The *probability* $\mathbf{P}(A_j = a|q, \mathcal{PB}(g))$ of action attribute $A_j \in AS(g)$ having value a in the context described by q with past behavior $\mathcal{PB}(g)$ is:

$$\mathbf{P}\left(A_j = a|q, \mathcal{PB}(g)\right) = \frac{|\{(cv_i, av_i) \mid (cv_i, av_i) \in kNN_{q,\mathcal{PB}(g)} \wedge A_j = a \in av_i\}|}{k} \tag{1}$$

if $\Sigma_{(cv_i,av_i)\in kNN_{q,\mathcal{PB}(g)}} d(q, cv_i) = 0$;

$$\mathbf{P}(A_j = a|q, \mathcal{PB}(g)) = 1 \tag{2}$$

if $\{(cv_i, av_i) \mid (cv_i, av_i) \in kNN_{q,\mathcal{PB}(g)} \wedge A_j = a\ in\ av_i\} = kNN_{q,\mathcal{PB}(g)}$;

$$\mathbf{P}(A_j = a|q, \mathcal{PB}(g)) = 0 \tag{3}$$

if $\{(cv_i, av_i) \mid (cv_i, av_i) \in kNN_{q,\mathcal{PB}(g)} \wedge A_j = a\ in\ av_i\} = \emptyset$; otherwise

$$\mathbf{P}(A_j = a|q, \mathcal{PB}(g)) = 1 - \frac{\Sigma_{(cv_i,av_i)\in kNN_{q,\mathcal{PB}(g)} \wedge A_j = a\ in\ av_i} d(cv_i, q)}{\Sigma_{i=1}^{k} d(cv_i, q)} \tag{4}$$

The first case of Definition 4 occurs when all k nearest neighbors are at distance zero from the query vector. The subsequent definitions are the cases when all of the k nearest neighbors agree on the value a for attribute A_j, none of the neighbors have value a for A_j, and the general case when none of these special conditions is true. The denominator of the term:

$$\frac{\Sigma_{(cv_i,av_i)\in kNN_{q,\mathscr{PB}(g)}\wedge A_j=a \text{ in } av_i} d(cv_i,q)}{\Sigma_{i=1}^{k} d(cv_i,q)}$$

in the last definition is the sum of the distances to q from each of its k nearest neighbors. The numerator is the sum of the distances between q and those nearest neighbors whose attribute A_j has a given value a. Thus, the smaller the numerator is, the "closer" (or more similar) the attribute value $A_j = a$ is to the g-behaviors that are most similar to q. This ratio is subtracted from 1 because a small distance must indicate a high probability.

Example 5 Suppose we want to forecast the behavior of agent g given the past behavior $\mathscr{PB}(g)$, the action schema $AS(g) = A_1$, and the query vector $q = (0, 0)$ where $k = 3$. The k nearest neighbors to q are the set

$$\begin{aligned} kNN_{q,\mathscr{PB}(g)} = \{&(c_1 = (0,1),\ a_1 = (0)),\\ &(c_2 = (0,2),\ a_2 = (0)),\\ &(c_3 = (1,1),\ a_3 = (1))\} \end{aligned}$$

In this case,

$$\begin{aligned} d_{EUC}(q, c_1) &= 1,\\ d_{EUC}(q, c_2) &= 2, \text{ and}\\ d_{EUC}(q, c_3) &= \sqrt{2} \end{aligned}$$

For each possible value in $dom(A_1)$ the probability of that action occurring in the context of query vector q can be computed:

$$\begin{aligned} \mathbf{P}(A_j = 0|q, \mathscr{PB}(g)) &= 1 - \frac{1+2}{1+2+\sqrt{2}}\\ &= 0.32\\ \mathbf{P}(A_j = 1|q, \mathscr{PB}(g)) &= 1 - \frac{\sqrt{2}}{1+2+\sqrt{2}}\\ &= 0.68. \end{aligned}$$

The reason that the probability of $A_1 = 1$ is higher than that of $A_1 = 0$ is intuitively because two of the three nearest neighbors of the query vector q have $A_1 = 0$, and

```
Algorithm CONVEXMerge(𝒫ℬ(g), q, d, AS(g), k)
    NearK = list of k elements initialized to +∞;
    Forecast = ∅;
    for each (cv, av) ∈ 𝒫ℬ(g) do
        Dist = d(q, cv);
        if Dist < d(q, NearK[k−1]) then
            insert_sort((cv, av), NearK);
    for each A_j ∈ AS(g) do
        V = argmax_{a∈dom(A_j)} (P(A_j = a | q, 𝒫ℬ(g)));
        add V to Forecast;
    return Forecast.
```

Fig. 3 The CONVEXMerge algorithm

these two nearest neighbors have a distance of 1 and 2 from the query vector. In contrast, while there is only one nearest neighbor having $A_1 = 1$, its distance from the query vector is $\sqrt{2}$. Because it is closer to the query vector, context vector c_3 is more important for predictive purposes than context vector c_2 of distance 2 away, which leads to a reduced probability for A_1 having the value 0. ■

Figure 3 presents the CONVEXMerge algorithm, which finds the probability that $A_j = a$ for each $a \in dom(A_j)$. The values with the highest probability are returned in the forecasted action vector. With slight modifications, this algorithm can return the entire probability distribution over all action values. The CONVEXMerge algorithm is slightly less efficient than the CONVEX$^{k\text{-NN}}$ algorithm; its complexity includes an additional multiplicative factor, $|dom(A_j)|$, because the probability of each $A_j = a$ possibility must be computed.

Proposition 2. *Suppose $\mathscr{PB}(g)$ is a past behavior for agent g, d is a distance function over $dom(C_1)\times\cdots\times dom(C_n)$, $AS(g)$ is an action schema for g, and $k \geq 1$. If d is computable in $\mathbf{O}(n)$ time, then* CONVEXMerge$(\mathscr{PB}(g), q, d, AS(g), k)$ *runs in time $\mathbf{O}(k \cdot n \cdot |\mathscr{PB}(g)| + k \cdot |AS(g)| \cdot |dom(A_j)|)$.*

5 Implementation and Experiments

Both the CONVEX$^{k\text{-NN}}$ and CONVEXMerge algorithms were implemented and experimentally evaluated to test this approach for behavioral forecasting. The implementations required about 1,200 lines of Java code, and all experiments described in this section were run on a computer with a dual-core processor at 2 GHz, with 2 GB of RAM, running the Windows Vista operating system.

The Minorities at Risk Organizational Behavior (MAROB) dataset [4, 19] was used as the behavioral time series for these experiments, consisting of yearly behavioral information about violent ethnopolitical groups (many of them engaged

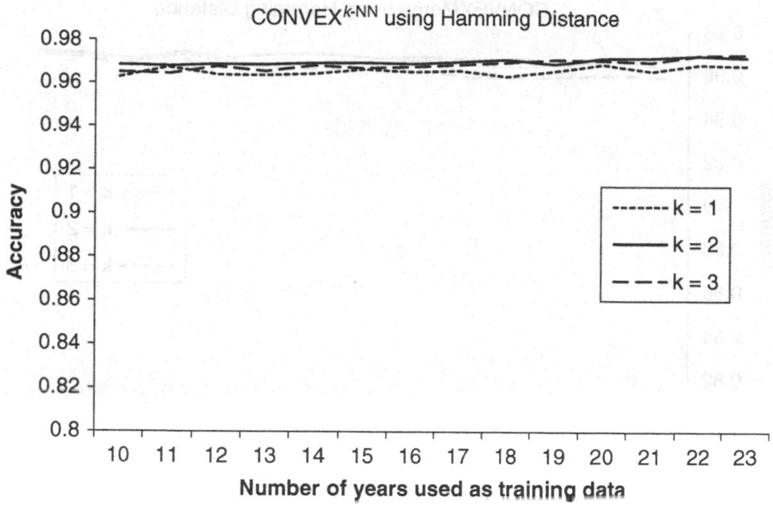

Fig. 4 Average accuracy of the CONVEX$^{k\text{-NN}}$ algorithm

in terrorism) in the Middle East from 1980 to 2004. All of the attributes in MAROB are categorical variables with integer domains that were populated by humans using large volumes of trusted news reports. For each group and each year, the human coders tried to arrive at a judgment of which categorical value should be specified for each attribute during that year. For example, for the variable DOMORGPROT measuring the level of domestic protest by a group in a given year, the coders try to determine from press reports the number of protests staged (and protesters involved) by the group for the year in question and arrive at a reasoned judgment of the appropriate value for this attribute. As described in Example 2, for each group g, the g-behaviors for MAROB contain the coded data for a given year. Thus, for each group g, $\mathcal{PB}(g)$ has at most 25 g-behaviors.

Only groups with more than 10 years of data were considered in the experiments. If a group had y years of data for $y > 10$, training sets of size $t = 10, \ldots, y - 1$ were constructed. Every single year was used as a query vector, and CONVEX$^{k\text{-NN}}$ and CONVEXMerge were used to try and forecast the actual action vector in that year from random training sets of size t chosen from the remaining years of data. This process was repeated for all 118 groups in the MAROB dataset.

The *correctness ratio* of the forecasting algorithms was used as a metric of accuracy and is defined as the number of correct predictions, divided by the total number of predictions. A prediction was considered correct if, for a given query year, the actual value of a given action variable coincided with the *single* value predicted by the algorithm. However, recall that the CONVEX$^{k\text{-NN}}$ algorithm can yield *intervals* rather than single values. To compute the correctness ratio in these cases, a single value was obtained by rounding V off to an integer value.

Figures 4 and 5 show the accuracy results for the CONVEX$^{k\text{-NN}}$ and CONVEXMerge algorithms for a variety of distance functions and $k = 1, 2, 3$. Each

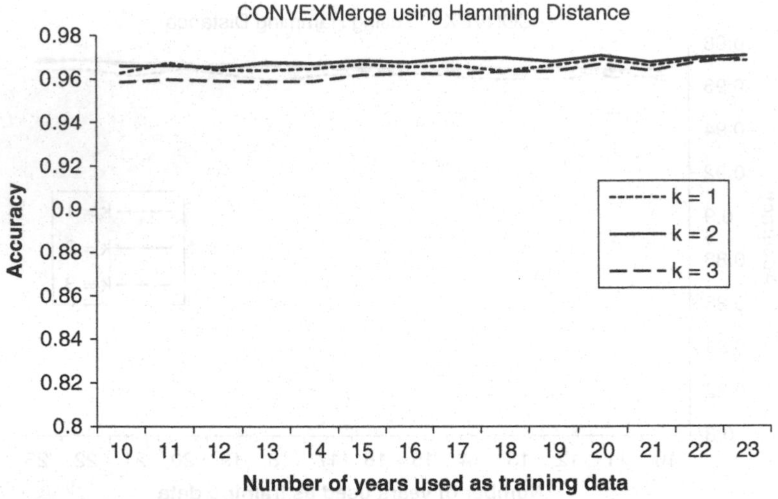

Fig. 5 Average accuracy of the CONVEXMerge algorithm

entry in the tables shows the accuracy of the algorithm, averaged as t (the size of the training data) was varied from 10 to $y - 1$ as indicated above. There are several important observations that can be made from these results:

1. **Always check 2 nearest neighbors.** Choosing $k = 2$ seems to give the best results, irrespective of which algorithm or distance function was used. *Thus, it is best to look at the two nearest neighbors, not one or three.*
2. **CONVEX^{k_NN} consistently outperforms CONVEXMerge.** Regardless of which distance function was used or what value of k was selected, the CONVEX^{k_NN} algorithm was more accurate than CONVEXMerge. Of course, when $k = 1$, both algorithms had identical performance.
3. **The accuracy of k varies with the distance function.** When $k = 2$ or $k = 3$, CONVEX^{k_NN} seems to perform almost equally well. However, for three distance functions (Hamming, Manhattan, and Canberra), the accuracy observed for CONVEX^{k_NN} with $k = 2$ is better than that with $k = 3$, while the situation is reversed for the other three distance functions.
4. **Hamming and Manhattan are the best distance functions.** In terms of the best accuracy, CONVEX^{k_NN} with $k = 2$ using either the Hamming or the Manhattan distance seems to yield the best performance—95.1 % accuracy. Actually, Hamming wins by a marginal amount (0.951355 vs. 0.951263).

The fact that using Hamming and Manhattan distance yielded the best result is not entirely surprising. Intuitively, Hamming distance is simply a count of the number of vector positions that are different, while Manhattan distance is the sum of the differences in each position. For the type of data being used here, this approach to distance seems more applicable than others, such as the Euclidean or Cosine distances.

Distance	CONVEX[1]	CONVEX[2]	CONVEX[3]
Canberra	0.857	0.938	0.942
Chebyshev	0.857	0.949	0.945
Cosine	0.857	0.950	0.947
Euclidean	0.857	0.950	0.948
Hamming	0.857	0.951	0.948
Manhattan	0.857	0.951	0.947

Fig. 6 Hamming distance, CONVEX$^{k\text{-NN}}$ algorithm

Distance	CONVEXMerge[1]	CONVEXMerge[2]	CONVEXMerge[3]
Canberra	0.857	0.851	0.849
Chebyshev	0.857	0.860	0.859
Cosine	0.857	0.859	0.857
Euclidean	0.857	0.861	0.860
Hamming	0.857	0.861	0.861
Manhattan	0.857	0.861	0.861

Fig. 7 Hamming distance, CONVEXMerge algorithm

To drill down a bit deeper and see how the accuracy of the algorithms varies with the amount of training data, the performance of the CONVEX$^{k\text{-NN}}$ and CONVEXMerge algorithms was compared as t was varied for the Hamming and Manhattan distance functions. Figures 6 and 7 show how the accuracy of the CONVEX$^{k\text{-NN}}$ and CONVEXMerge algorithms vary for $k = 1, 2, 3$ using the Hamming distance and a training set size varied from 10 to 23. Both algorithms show virtually no change with t, indicating that $t = 10$ is adequate for making a highly accurate behavioral forecast, at least for the MAROB dataset. Figures 8 and 9 show the corresponding results when the Manhattan distance is used instead. In both cases (as well as for the other four distance functions) the results are similar—choosing a training set greater than 10 offers virtually no improvement in accuracy.

6 Forecasting Situations

In much the same spirit as we are interested in forecasting action vectors, we may also wish to forecast environmental vectors, or *situations*. In [16], an algorithm called SitCAST was designed with this goal in mind; as this algorithm was developed to act as a companion to CONVEX, we will provide here a brief overview of how these two algorithms can be used together.

Using behavioral time series data of the kind we've been considering up to now (with integer domains), a user may want to forecast what the environmental

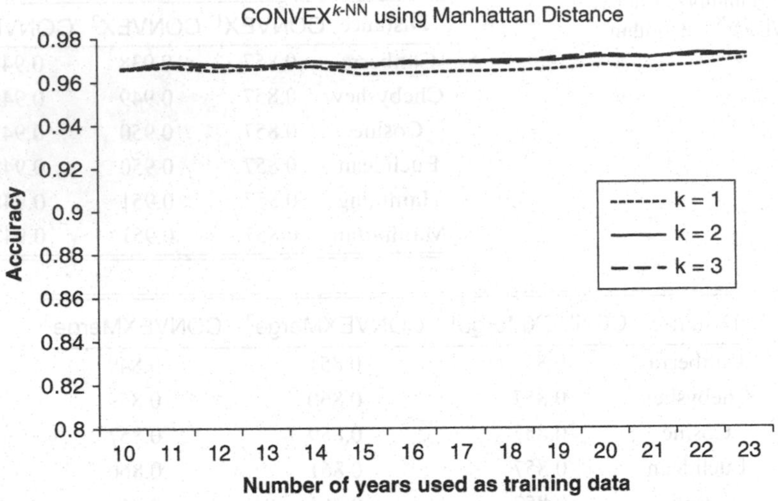

Fig. 8 Manhattan distance, CONVEX$^{k\text{-NN}}$ algorithm

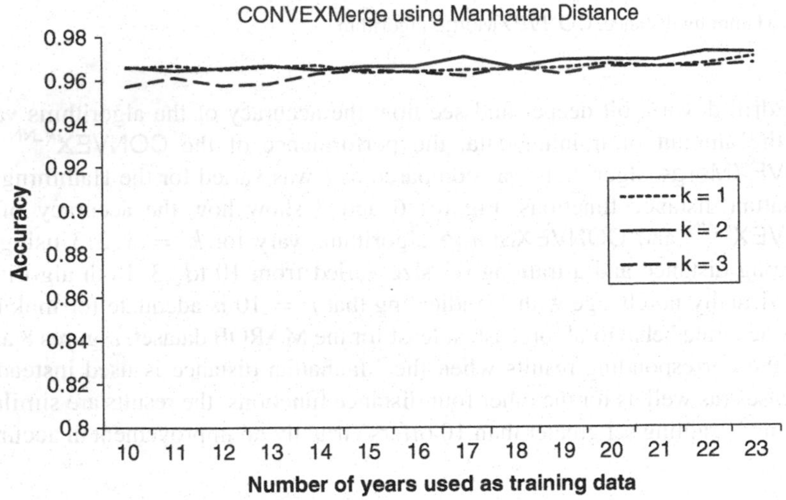

Fig. 9 Manhattan distance, CONVEXMerge algorithm

context might be like, for the entity being modeled, at a certain time period in the future. That is, we want to answer the question: what value will the environmental variables have at time point t? The SitCAST forecasting algorithm was proposed for answering this question. One of the strong points of this algorithm is that it is designed to work with any standard statistical time series forecasting algorithm [3] as part of its input (e.g., linear regression, quadratic regression, logistic regression, etc.). The SitCAST algorithm applies this input in order to come up with a candidate value for each environmental variable; since this value is not guaranteed to be an integer, the algorithm will in general output a *set* of possible context vectors. Given this setup, the main hurdle that needs to be overcome is how to define a probability distribution over all possible situations that are forecast by the algorithm; we refer the reader to [16] for the technical details of how this is done.

SitCAST and CONVEX. In the discussion of CONVEX, we assume that the algorithm receives a query vector with values for each environmental variable; however, rather than relying on users to provide such a possible situation for a "what if" analysis, the future situations forecast by SitCAST can be used as query vectors, and CONVEX can be used to predict the behaviors an agent will take in each case. Thus, SitCAST and CONVEX are able to jointly forecast what actions an agent will take at some future time point, together with a probability.

The efficacy of SitCAST as a means of forecasting the context attributes in a behavioral time series has been empirically shown in [16], as the algorithm achieves a very high accuracy in predicting future situations, as well as its combination with CONVEX to forecast the actions resulting from the most probable situation produced by SitCAST.

Finally, another related effort in this vein aims at adequately addressing settings in which this kind of approach is likely to fail, i.e., when the agent being modeled exhibits a *radical change* in its behavior. Since this is likely to cause regression-based techniques to perform poorly, a different kind of prediction engine was also developed in [16] to contemplate and identify such divergence from the norm.

7 Conclusions

In this work, two general algorithms, CONVEX$^{k\text{-NN}}$ and CONVEXMerge, were presented based on viewing behavioral data as points in a high dimensional metric space and assessing the distance between a query context vector and historical data. By examining the k nearest neighbors with respect to different distance functions, both CONVEX$^{k\text{-NN}}$ and CONVEXMerge are able to take user-specified query vectors, representing a real or hypothetical situation, and produce a forecast of what an agent might do in that query context, allowing for highly accurate "what if" analyses.

The main distinction between these two approaches is their weighting of the k nearest neighbors—one regards all the selected k nearest neighbors the same, while the other looks at the actual distances between the query vector and the k nearest neighbors and tries to normalize the answer based on these distances.

The initial expectation was that the second method, CONVEXMerge, would outperform the first by taking the specific distances into account. Surprisingly, empirical tests indicate that the first algorithm, CONVEX$^{k\text{-NN}}$, is consistently more accurate, independent of the amount of training data used, the choice of k, and the choice of distance function. Moreover, for the CONVEX$^{k\text{-NN}}$ algorithm, $k = 2$

provides consistently better results than $k = 1$, and the best results are obtained with either Hamming or Manhattan distance (though there is not much difference in terms of the distance metric used). Last, but not least, 10 years of training data seems to be more than enough for accurate behavioral forecasts using the MAROB dataset. Experiments on approximately 25 years of real data for 118 ethnopolitical groups in the Middle East show that the best configurations of CONVEX$^{k\text{-NN}}$ and CONVEXMerge are able to consistently produce forecasts with over 95 % accuracy.

It is important to qualify these comments with several caveats. It is impossible to assert that the algorithms studied here will accurately forecast any kind of action vector from any kind of context vector. The MAROB data with which these algorithms were evaluated has characteristics that are symptomatic of a class of problems in the social sciences, but are by no means representative of all problems where forecasts may be desired. For example, the domains of all attributes are small (in MAROB, no variable has more than seven possible values and most have between two and four), which is also true of many other datasets in the social sciences. The accuracy of these methods when the domains of attributes are large is still an open question. Moreover, there are issues related to numerical data that need to be addressed (MAROB only contains categorical data). For example, to study variables such as GDP, percentage of arable land, etc., it is necessary to have a good approach for handling numeric data. This raises interesting questions of scale; if some attributes are on a small scale such as 0 or 1, while others are on a wider scale (e.g., $1,000$–$100,000$), then the wider scale may have a disproportionate impact on the distances computed. Fortunately, there seem to be straightforward ways to normalize them.

Another important problem relates to checking whether certain subsets of context attributes (rather than all) will elicit better results. Because the accuracy of the current algorithms is already over 95 %, this possibility has not yet been investigated, but it is certainly worth examining. It is also important to explore whether placing weights on attributes would lead to improved accuracy of forecasts, or whether a temporal weighting of the nearest neighbors in the CONVEXMerge algorithm (i.e., giving greater weight to more recent contexts) may prove more effective.

Acknowledgements Some of the authors of this paper were funded in part by AFOSR grant FA95500610405, ARO grant W911NF0910206 and ONR grant N000140910685.

References

1. Antoniou G, van Harmelen F (2004) A semantic web primer. MIT, Cambridge, MA
2. Bond J, Petroff V, O'Brien S, Bond D (2004) Forecasting turmoil in indonesia: an application of hidden markov models. International Studies Association, USA. pp 17–21
3. Bowerman B, O'Connell R, Koehler A Forecasting, Time series and regression, 4th edn. Southwestern College Publishers, Cincinnati, OH, USA (2004)

4. CIDCM (2008) Minorities at risk organizational behavior dataset, minorities at risk project. Retrieved from http://www.cidcm.umd.edu/mar
5. Fagin R, Halpern JY, Megiddo N (1990) A logic for reasoning about probabilities. Inf Comput 87(1/2):78–128
6. Hailperin T (1984) Probability logic. Notre Dame J Form Log 25(3):198–212
7. Khuller S, Martinez MV, Nau D, Simari GI, Sliva A, Subrahmanian VS (2007) Finding most probable worlds of probabilistic logic programs. In: Proceedings 2007 international conference on scalable uncertainty management. Lecture notes in computer science vol 4772. Springer, Berlin/Heidelberg, pp 45–59
8. Khuller S, Martinez MV, Nau DS, Sliva A, Simari GI, Subrahmanian VS (2007) Computing most probable worlds of action probabilistic logic programs: scalable estimation for $10^{30,000}$ worlds. Ann Math Artif Intell 51(2–4):295–331
9. Nilsson N (1986) Probabilistic logic. Artif Intell 28:71–87
10. Schrodt P (2000) Forecasting conflict in the balkans using hidden markov models. American Political Science Association, USA
11. Simari GI, Subrahmanian VS (2010) Abductive inference in probabilistic logic programs. Schloss Dagstuhl - Leibniz-Zentrum fuer Informatik 2010, ISBN 978-3-939897-17-0
12. Simari GI, Sliva A, Nau D, Subrahmanian VS (2006) A stochastic language for modelling opponent agents. In: AAMAS '06: proceedings of the fifth international joint conference on autonomous agents and multiagent systems. ACM, New York, pp 244–246
13. Simari GI, Dickerson JP, Subrahmanian VS (2010) Cost-based query answering in action probabilistic logic programs. Lecture Notes in Computer Science 6379 Springer 2010, ISBN 978-3-642-15950-3
14. Simari GI, Martinez MV, Sliva A, Subrahmanian VS (2012) Focused most probable world computations in probabilistic logic programs. Ann Math Artif Intell 64(2–3):113–143
15. Sliva A, Martinez MV, Simari GI, Subrahmanian VS (2007) SOMA models of the behaviors of stakeholders in the afghan drug economy: a preliminary report. In: First international conference on computational cultural dynamics (ICCCD 2007). AAAI Press, USA
16. Sliva A, Subrahmanian VS, Martinez MV, Simari GI (2009) CAPE: automatically predicting changes in group behavior. In: Mathematical methods in counterterrorism. Springer, Wien, pp 247–263
17. Subrahmanian VS (2007) Cultural modeling in real-time. Science 317(5844):1509–1510
18. Subrahmanian VS, Albanese M, Martinez MV, Nau D, Reforgiato D, Simari GI, Sliva A, Wilkenfeld J (2007) Cara: a cultural adversarial reasoning architecture. IEEE Intell Syst 22(2), 12–16
19. Wilkenfeld J, Asal V, Johnson C, Pate A, Michael M (2007) The use of violence by ethnopolitical organizations in the middle east. Technical report, National Consortium for the Study of Terrorism and Responses to Terrorism

Forecasting the Use of Violence by Ethno–Political Organizations: Middle Eastern Minorities and the Choice of Violence

Kihoon Choi, Victor Asal, Jonathan Wilkenfeld, and Krishna R. Pattipati

1 Introduction

Can analytic models, informed by social scientific theories using computational engineering approaches, offer effective forecasting of violent behavior? This chapter discusses a new data set which codes the structure and behavior of ethno-political organizations and the use of a new approach for forecasting political behavior drawn from computer engineering. In the chapter, we build a forecasting model and then test the model against existing data as well as a predictive analysis for the year 2009 (the analysis was done in 2008 and data for 2005–2009 has not yet been collected for this data set). The data used was drawn from the Minorities at Risk Organizational Behavior (MAROB) data set. MAROB was created through collaboration between the National Consortium for the Study of Terrorism and Responses to Terrorism and the Minorities at Risk (MAR) Project. This data focuses on ethno-political organizations in the Middle East to test factors that make it more or less likely that an organization will choose to use violence. While the variables on which data was collected were informed by theories of contentious politics, this chapter focuses primarily on the data itself and the forecasting approach that we used and less on the

K. Choi (✉)
Qualtech Systems, Inc., 99 East River Dr., East Hartford, CT 06108, USA
e-mail: kihoon@teamqsi.com

V. Asal
State University of New York, Milne Hall, 135 Western Ave., Albany, NY 12222, USA
e-mail: vasal@albany.edu

J. Wilkenfeld
University of Maryland, 1122 Holzafel Hall, College Park, MD 20742, USA
e-mail: jwilkenf@umd.edu

K.R. Pattipati
University of Connecticut, ITE Building 350, Storrs, CT 06269, USA
e-mail: krishna@engr.uconn.edu

V.S. Subrahmanian (ed.), *Handbook of Computational Approaches to Counterterrorism*,
DOI 10.1007/978-1-4614-5311-6_10,

social science theoretical models as such. Analytically we use multiple approaches for data massaging, classification and forecasting to achieve high classification accuracies (measured in terms of overall accuracy, recall, precision, false positives, and F-measure). We also strive for parsimony in the number of variables we use to make our forecasting predictions.

2 Efforts at Forecasting in Past

Much of the focus in international relations, comparative politics and political science in general has been on explaining and understanding political outcomes and political behavior [29, 42]. Much of the empirical work in political science has focused on sample prediction but not on out-of-sample analysis, which is critical for doing useful forecasting that allows for a measure of success [37]. Despite this focus on the past and a reluctance to forecast, many in the discipline have a strong interest in forecasting, particularly in the area of political violence and state instability [33, 37]. However, forecasting presents many problems. Despite claims to the contrary, history does not repeat itself and often the most important events in the near future are the ones that are the most unexpected [64].

The policy world has tended to rely on subject matter experts with varying levels of breadth to help formulate forecasts [48]. Many of these experts though "... are not mere objective experts. They develop belief systems, operational codes, theories, and agendas; they are subject to the same cognitive and social psychological and group dynamics that affect decision makers" [48]. The problem is deeper than simply a concern that analysts let their personal views color their analysis. Investigation of the use of subject matter experts indicates that they are often asked not about what they have studied and have empirical support for but are asked to give opinions – what they think. "Evidence from surveys suggests that forecasts of decisions in conflicts are typically based on experts' unaided judgments" [24]. Green and Armstrong [24] have found that asking experts to make predictions based on opinion is particularly dangerous because experts are no more right than non-experts. Tetlock [66] also finds that experts (like leaders) often ignore data that disagree with their views.[1]

Despite the problems that forecasting presents, there have been important efforts to try using approaches to forecast behavior related to political violence that can be tested and assigned confidence intervals about their accuracy. These efforts have produced substantive results. Substantive testable relevant social science forecasting is possible [5, 21, 37, 52, 58]. Despite these positive results, efforts to forecast are rare and are often restricted in the types of methodologies they employ [4]. Part of the problem in terms of exploiting different methodologies is disciplinary because

[1] Whether experts are listened to is an important but different issue that we do not address in this chapter (see [2, 20]).

computational scientists and social scientists rarely collaborate on research projects (or even talk) [3]. In this chapter, we bridge this gap. Using a data set constructed with a political science theoretical model in mind, we attempt to forecast which ethno-political organizations are likely to use violence and which are not.

3 Forecasting Ethnic Violence: MAROB

In addressing the issue of forecasting political events, we have chosen to explore the choice of political violence by ethno-political organizations in the Middle East. While admittedly a very specific question and regional focus, the widespread nature of these events and phenomena in the Middle East make it a good testbed for the development of forecasting models. We draw on new data – the MAROB dataset[2] – in order to identify which factors make it more or less likely that violence will be used in any particular year. The MAROB dataset has data on over 100 ethno-political organizations in the Middle East for the years 1980–2004 with each organization being coded yearly. MAROB builds on the Minorities at Risk (MAR) Project [27, 28], which focuses specifically on non-state communal groups that have an ethnic or ethno-religious identity. A MAR group needs to be at least 1 % of the population of their host country or have at least 100,000 members. In addition, the identity of the group must have political significance because the group either benefits or suffers from discrimination and/or the identity is used to mobilize politically for collective action [27, 28]. Drawing on the MAR Project for our forecasting effort allows us to build on a base that is well established both theoretically and in terms of data.

Our goal though is not to stay at the level of groups but to focus on organizations. The Palestinians did not launch rockets into Israel or not renew the cease fire with Israel-Hamas did both of these things. By disaggregating to the organizational level we hope to get a better handle on the factors that lead to the choice of political violence. By coding all organizations that claim to represent MAR groups for every MAR group in the region, MAROB allows us to get what we believe to be the right level of granularity.

The MAROB project codes every organization claiming to represent a MAR group that has survived for at least 3 years. As coding progresses MAROB will encompass all MAR groups internationally but the project started by focusing on the Middle East because of the active use of violence as a political strategy and wide variance inside groups and over time. The database currently has data on 112 organizations representing 29 MAR groups in the Middle East and North Africa. Branches of the same organization in different countries are coded separately so that, for example, the Palestinian Liberation Organization is coded three times because

[2]The data can be downloaded at http://www.cidcm.umd.edu/mar/data.asp (see [52]).

it has active branches in the West Bank/Gaza, Lebanon, and Jordan. Organizational branches were coded differently because these branches often behaved very differently from one another both domestically and transnationally.

The data set design was driven by a desire to understand what organizational, behavioral and environmental factors drive certain organizations towards using violence while other organizations do not use violence. While our efforts in this chapter focus on our analytical approach the variables that were chosen for inclusion in the data collection effort were motivated primarily by existing literature in the contentious politics research programs. Underlying the data collection effort was a theoretical perspective that certain factors made it more likely that organizations would be violent. Generally the data allow researchers to test how state behavior, organizational capability, and ideology condition some organizations to be more likely to use violence and other organizations to be less likely to use violence.[3]

Specifically MAROB collects data on both violent and non-violent organizational behavior – so, for example, the electoral, protest, and social service behavior of the organizations and whether the groups target civilians or security personnel with violence. The data set also focuses on the external support of organizations and what kind of support they might receive from diasporas, foreign states, international governmental organizations, and international nongovernmental organizations. In addition, MAROB collects information on the organizational characteristics of the group including its type of leadership and popularity. One of the most extensively coded aspects of organizational characteristics is the organization's ideological make up, its type of grievances and strategic goals. Finally, MAROB codes the type of organization – state relations that exist. For example, MAROB has data on the prevalence of government repression and violence directed at the group as well as whether or not the organization and the state are negotiating or if they have reached any kind of negotiated agreement.

4 Forecasting from Engineering to the Social Sciences

In the engineering arena, forecasting, as an add-on capability to fault diagnosis, assesses the current health of a system, detects the precursors of a failure, and anticipates the incipient failures and systems' residual life time based on features that capture the gradual degradation in the functional and structural properties of engineering systems. Effective methods of forecasting in engineering systems cannot only help avoid the undesirable effects of faults, but also significantly reduce maintenance cost and time. Unlike fault diagnosis, forecasting is relatively new and plays an important role in engineering systems. Forecasting methods have been widely studied, and they can be classified into three categories: model-based, knowledge-based, and data-driven [15, 16]. Each of these approaches has

[3]See [9–14, 18, 25–27, 34, 42, 44, 45, 49, 55, 65, 67, 72].

its own advantages and consequently they are often used in combination in many engineering applications. The model-based approach uses a mathematical representation of the system. The approach is applicable to information-rich systems, where satisfactory models and an adequate number of factors are available. The main advantage of a model-based approach is the ability to incorporate a physical understanding of the system for monitoring [46]. The knowledge-based approach uses qualitative models to develop process monitoring measures. The approach is especially suitable for systems in which detailed mathematical models are not available. The data-driven approach is appropriate when monitoring data is available, but a mathematical model (model-based) or cause-effect graph model of system failures and their manifestations (knowledge-based) are not available.

Although it is anticipated that models and simulation tools for most modern engineering system components are available, there is a fair chance that not all of them are obtainable by the users. In this situation, the collected data becomes the sole means to obtain knowledge about the systems. Data-driven techniques might be the only possible way to perform forecasts in such restricted technology systems. The forecasted features can be put through the data-driven based model. This model can run the same tests and reasoning algorithms (as it does for the real-time and archived data) to identify projected faults on the forecasted data. Data-driven forecasting is beneficial in the sense that it can be developed on the fly (just by using archived data) and the forecasting model parameters can be updated dynamically without much effort. In addition, it provides the predicted values of not only the primary system variables, but also of the health and performance metrics (e.g., projected system efficiency) directly. Time series analysis methods have proven to be very accurate for performing data-driven based forecasting [46].

In our view, the data-driven techniques for conflict assessment in social science potentially have close relationship with fault diagnosis/forecasting in the engineering field, wherein one seeks to categorize the input-output data into normal or conflict levels (intensities). Owing to its simplicity and adaptability, customization of a data-driven (inductive) approach does not require an in-depth knowledge of the application context, while model-based (deductive) and knowledge-based (e.g., rule-based) techniques invariably do. Our approaches enable us to develop hybrid models that include both inductive and deductive elements. In this chapter, we propose an end-to-end probabilistic modeling process for the dependent variable classification. Our approach integrates five key steps in a novel way to achieve high accuracies (measured in terms of overall accuracy, recall, precision, false positives, and F-measure). The key process steps are: (1) imputing missing values via time series analysis, (2) mutual information and genetic algorithm based feature extraction, (3) massaging data via a data reduction technique, (4) statistical pattern recognition with soft decision (i.e., probability measure as decisions of classification), and (5) forecast dependent variable(s).

A nice feature of the proposed process is that it extends naturally to forecasting the dependent variable. Since the proposed approach builds models for forecasting during the training phase, each independent variable selected by feature extraction is forecasted via a time series model developed during the training phase. Forecasting for dependent variable is also performed by applying the forecasted independent variables to the trained model. Another novelty of the probabilistic approach is that it provides the possibility to detect advent and cessation well in advance on the forecasting results with confidence intervals for policy relevant analysis [45].

A number of methods are proposed for constructing dynamic models of conflict. Schrodt [58, 59] and Bond et al. [7] discuss a hidden Markov model-based classifier to explain how a country might evolve over time. Subrahmanian et al. [63] develop an agent-based Cognitive Architecture for Reasoning about Adversaries (CARA) using Stochastic Opponent Modeling Agents (SOMA) [36] to build complex models used to forecast a group's behavior. However, the HMM as well as CARA-based modeling methods to social sciences require different models for different organizations/countries. To address this issue, we develop a unified probabilistic model that is applicable to a range of organizations/countries with the attendant computational and modeling efficiencies.

With regard to past work on the MAROB data, the CONVEX algorithm in [47] assesses the similarity between a given situation and information that the group has experienced. Although the algorithm uses distance functions such as k-Nearest Neighbor (kNN) [17] and is computationally efficient, a proper approach to handling numeric data with different scales is not considered to avoid a possible disproportionate impact on the distances obtained. Since the MAROB data has different measurement units or exhibits large differences in variance due to different data sources, an efficient scaling method is employed in this chapter. In addition, Sliva et al. [60] discuss an architecture called CAPE to forecast the conditions under which an organization in the MAROB data will change its behavior over time. However, CAPE is tested on one behavior engaged in by seven organizations, and the architecture still needs to be tested on a much wider range of groups and actions to show its applicability to a range of behaviors. On the other hand, 58 organizations with behavior changes over multiple time periods are tested in this chapter to demonstrate the applicability of our proposed modeling process across organizations and behaviors.

Applying the proposed approach may require further investigation for better forecasting performance because of the following issues. While this is not the case with MAROB, in some cases the data may be too heterogeneous and a single algorithm may not be able to extract all relevant information. A combination of algorithms is a better choice. However, identifying the optimal combination is a challenge. Social science data presents challenges of data aggregation due to various data formats such as numeric sequence, categorical data, free-format text, that are less likely to be present in the engineering arena. Theoretically, a trained model can forecast dependent variables in any period. However, it is a challenging task to optimally determine a forecasting period without losing accuracies. This last challenge is not one that is exclusive to the application of these approaches to the social sciences.

5 Probabilistic Modeling Process Overview

The probabilistic modeling method is an organization-independent statistical classifier model that enables us to predict the probability of a given dependent variable at periodic (e.g., yearly) time intervals. Using training data, the modeling approach identifies organization-independent salient variables for the dependent variable based on the information gain method, and uses these as inputs to a dynamic classifier to estimate the dependent variable. The trained classifier models for the dependent variable are evaluated against validation data using overall accuracy, recall, precision, false positives, and F-measure as performance metrics (discussed later). A block diagram of the approach to analyzing the MAROB data is shown in Fig. 1. The proposed scheme is a five-step process: imputing missing values via time series analysis, factor selection, massaging data such as data transformations, classification/forecast, and validation and performance assessment. These are briefly described with methodologies in the following subsections below although this chapter focuses on forecasting the dependent variable: whether or not an organization used violence in any organizational year.

The original MAROB data has 112 organizations. However, we ended up doing our analysis on 58 organizations and focused on the years between 1995 and 2004. The intent of selecting the period for the analysis in this chapter is as follows: (1) seek to analyze the most complete independent variables (i.e., less number of missing values) and the dependent variables during 1980–2004; (2) the most recent data we have available. In addition, we removed organizations that did not appear during this period of time (removed 5) as well as organizations that either disappeared or appeared during this period of time (removed 25). We also removed all organizations that were missing data on the dependent variable (removed 15).

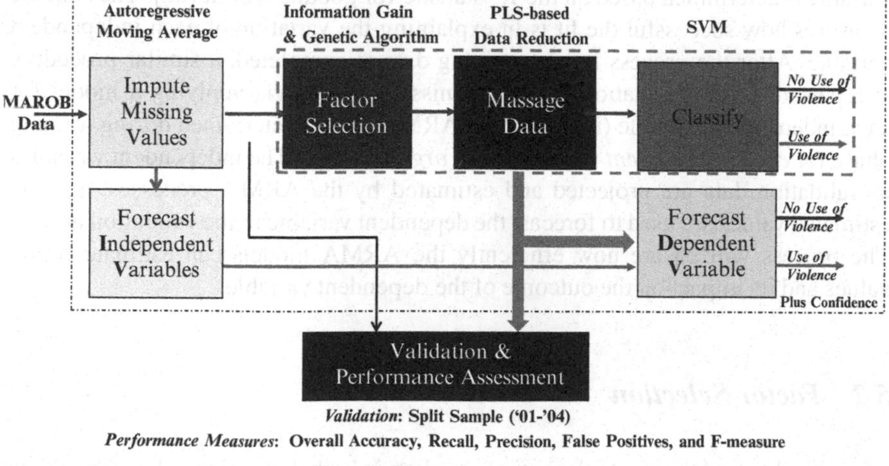

Fig. 1 Process overview

Finally, we removed variables that had no values at all for several of our independent variables (removed 9). However, we plan to experiment not only with early years (i.e., 1980–1994) but also different regions (e.g., North Africa, Post-Communist States, Latin America, etc.) for the probabilistic modeling process to verify the efficacy of our proposed approach. All of the organizations that are in our analysis can be seen in Table A.

5.1 Imputation of Missing Values

Even after we remove these organizations, not all of the MAROB data we consider in this chapter is complete (i.e., independent variables). After obtaining the input data, the imputation operation is performed to fill in missing values for organizations with partial missing data of the independent variables via a times series analysis, Auto Regressive Moving Average (ARMA) [45]. The trend patterns of a time series model can be evident when the series is smoothed using a moving average. In the moving average method, the forecasts for the period $(t + 1)$ are given by averaging the data of the past periods. However, as the series progresses over time, the moving average forecasts are often not accurate due to the effects of random variation.

The random variations can be minimized by adjusting the target by the amount of the past period's mistake. This is the process of auto regression. The ARMA operation is the combination of the Auto Regression (AR) with Moving Average (MA). For training data, the ARMA method is applied to estimate missing values that lie between known sample points. This process is repeated until all the missing values in the training data are filled. If there are missing values after a certain period (e.g., no values after 1998), rolling forecast based on the past values (i.e., use values up to at t and estimate the next value at $t + 1$) is performed. During model development in the training phase, the optimal model for each independent variable is determined based on the R^2 statistic for goodness of fit [62]. The statistic measures how successful the fit is in explaining the variation of each independent variable. After the process for the training data is completed, a similar procedure is repeated on the validation data with missing values via applying a model for each independent variable (i.e., different ARMA model determined during training phase for *each independent variable in an organization*). The independent variables in validation data are projected and estimated by the ARMA processes, and the estimated values are used to forecast the dependent variable in the validation period. The process will ensure how efficiently the ARMA models can estimate future values and its impact on the outcome of the dependent variable.

5.2 Factor Selection

It is critical to find the most relevant information in high dimensional data for a given dependent variable. This process involves identifying salient independent variables

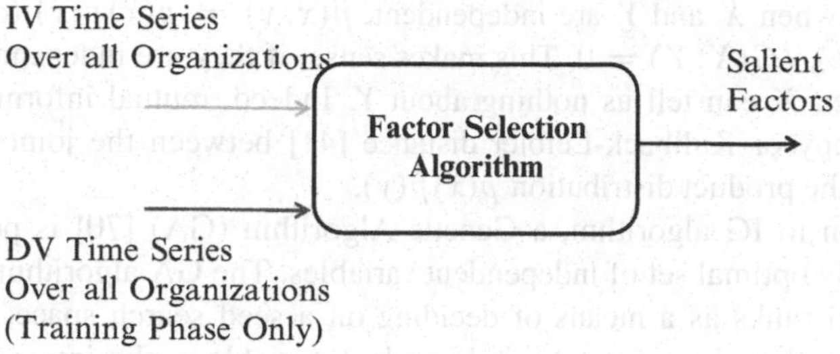

Fig. 2 Factor selection (training phase only)

for the dependent variable *using the entire training and validation data.* This is done once for the dependent variable as shown in Fig. 2. During the validation and forecasting phases, the salient factors identified during *the training period* are used as inputs to the model. Note that the list of salient variables is *not organization-dependent.*

To obtain salient independent variables, we employ the Information Gain (IG) algorithm that is related to the concept of mutual information: how much information does one random variable reveal about another random variable? Mutual information provides a quantitative measure to rank order variables in a decreasing order of mutual information (i.e., information gain). To formalize this concept, we need to understand the information–theoretic concept of entropy.

Entropy is a measure of how "disorganized" a set of data related to a random variable is; it is a measure of uncertainty. Let $y \in Y$ represent a dependent (response, output) variable, where Y is a discrete random variable (binary or multi-valued). Formally, entropy $H(Y)$ is given by

$$H(Y) = -\sum_{y \in Y} p(y) \log_2 p(y). \tag{1}$$

Independent variables (factors) provide information on a dependent variable, thereby reducing uncertainty in our knowledge of the dependent variable. Let $x \in X$ denote an independent variable. Then, entropy of Y after observing X is

$$H(Y|X) = -\sum_{x \in X} \sum_{y \in Y} p(x, y) \log_2 p(y|x). \tag{2}$$

The mutual information (or IG) is related to $H(Y)$ and $H(Y|X)$ via

$$\begin{aligned} IG(X;Y) &= H(Y) - H(Y|X) = H(X) - H(X|Y) \\ &= H(X) + H(Y) - H(X,Y) \\ &= \sum_{x \in X} \sum_{y \in Y} p(x, y) \log_2 \left[\frac{p(x, y)}{p(x)\, p(y)} \right]. \end{aligned} \tag{3}$$

Note that when X and Y are independent, $p(x, y) = p(x)p(y)$ (definition of independence), $IG(X; Y) = 0$. This makes sense: if they are independent random variables, then X can tell us nothing about Y. Indeed, mutual information is the relative entropy or Kullback-Leibler distance [41] between the joint distribution $p(x, y)$ and the product distribution $p(x)p(y)$.

In addition to IG algorithm, a Genetic Algorithm (GA) [70] is performed to find a globally optimal set of independent variables. The GA algorithm also makes use of the IG ranks as a means of deciding on a seed search space, but beyond this, the algorithm does not take independent variable ranks into consideration when searching for an optimal solution. The GA randomly generates sub-groups of variables from the seed population and compares them based on a fitness criterion such as F-measure. The algorithm eventually converges on a combination of factors that produces the best fitness value. The models generated by these optimal variable sets are then used to generate classification and forecasting results.

5.3 *Massage Data*

There are two data massaging operations that we have found to help with classification performance. First, we apply a preprocessing technique – autoscaling – to remove any inadvertent weighting of the independent variables that would otherwise occur. The technique does not affect the relative distribution of the data points in the high dimensional space.

The autoscaling (also known as variances scaling or standardization) method transforms each column's data (i.e., independent variable) into zero mean unity variance data. It is performed as

$$d_{ij}^{as} = \frac{d_{ij} - \mu(D_j)}{\sigma(D_j)}; \quad i = 1, 2, \ldots, I \quad \text{and} \quad j = 1, 2, \ldots, J, \tag{4}$$

where d_{ij}^{as} refers to the auto scaled data element in ith row and jth column, d_{ij} refers to the corresponding original data element, $\mu(D_j)$ refers to the mean of jth column, and $\sigma(D_j)$ refers to the standard deviation of jth column. The autoscaling is recommended when variables have different measurement units or exhibit large difference in variance.

Second, we employ Partial Least Squares (PLS)-based data reduction technique to reduce collinearity in the data [8]. The advantage of PLS as a data reduction technique is that it provides good classification accuracy on high-dimensional datasets, and is also computationally efficient. The goal of the PLS algorithm is to reduce the dimensionality of the input and output spaces to find latent variables which are most highly correlated, i.e., those that not only explain the variation in the $I \times J$ independent training data matrix X, but their variations which are most predictive of the $I \times C$ dependent matrix Y ($C = 2$ classes in our case). While forming the data sets, in each row of Y the dependent variable corresponding to the pattern (sample) in X is represented by 1 in the corresponding column Y.

The basic idea of PLS is to select the weights of the linear combination to be proportional to the covariance between the independent and dependent variables. Once the latent variables are extracted, a least squares regression is performed to estimate the dependent variable. Both matrices X and Y are decomposed into number of components which is known as the model reduction order plus residuals. Each component captures certain amount of variation in the data. The decompositions are given by Bro [8]

$$X = TP^T + E; \qquad T \in R^{I \times L}, P \in R^{J \times L}, E \in R^{I \times J}$$
$$Y = UQ^T + F; \qquad U \in R^{I \times L}, Q \in R^{C \times L}, F \in R^{I \times C}. \tag{5}$$

Here, T and U are score matrices (latent vectors), P and Q are loading matrices, E and F are residuals, and L is the model reduction order which can be determined by cross-validation [32]. PLS can be viewed as a two-phase optimization problem. For simplicity, assume that the response matrix Y is a single column, $\underline{y} \in R^I$. In the first phase, for a given latent vector $\underline{t} \in R^I$, PLS seeks to find a rank 1-matrix $\underline{t}\underline{p}^T$ that is closest to X in Forbenius norm, i.e., $\min_{\underline{p}} \left\| X - \underline{t}\underline{p}^T \right\|_F$. The solution is given by: $\underline{p} = X^T \underline{t}/\underline{t}^T \underline{t}$. In the second phase, PLS seeks to find $\underline{t}$ and a weight vector $\underline{w}$ such that the covariance between $\underline{t}$ and $\underline{y}$ is maximized:

$$\max_{\underline{t},\underline{w}} \underline{t}^T \underline{y}$$
$$\text{subject to:} \quad \underline{t} = X\underline{w} \quad \text{and} \quad \underline{w}^T \underline{w} = 1. \tag{6}$$

The result is:

$$\underline{w} = \frac{X^T \underline{y}}{\left\| X^T \underline{y} \right\|_2} \propto \underline{p} \quad \text{and} \quad \underline{t} = \frac{X X^T \underline{y}}{\left\| X^T \underline{y} \right\|_2}. \tag{7}$$

Since $\underline{t}$ and $\underline{w}$ (or $\underline{p}$) depend on each other, iteration is required. Thus, these processes determine the loading and score vectors which are correlated with Y while describing a large amount of variation in X. PLS regresses the estimated Y score vector $\underline{u}$ to the X score vector $\underline{t}$ by $\underline{u} = r\underline{t}$ where r is the regression coefficient. The score and loading vectors are determined using Nonlinear Iterative Partial Least Squares (NIPALS) algorithm [71], and the reduced space (score matrix) is applied to the classifier to be described in the following section.

5.4 Classification

Support Vector Machine (SVM) [1,6,31,57,68], as a statistical learning theory, has gained popularity in recent years. In fact, the SVM method is different from neural networks but they are closely related. The essential idea of SVM is to transform the data to a higher dimensional feature space, and find an optimal hyperplane that

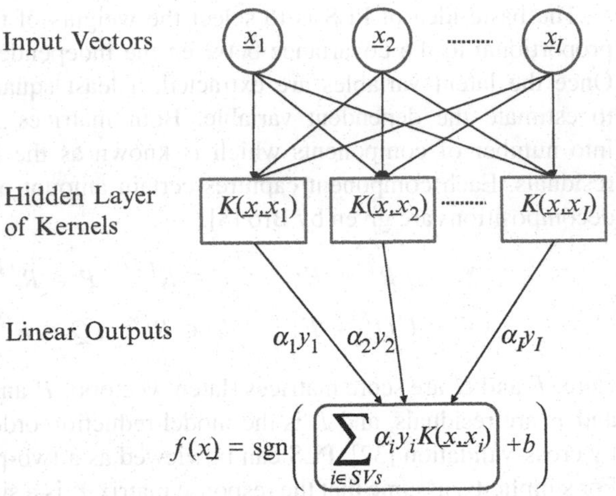

Fig. 3 Schematic of a SVM classifier

maximizes the margin between the classes (whether or not an organization used violence in any organizational year in our case) by solving a quadratic programming problem with linear constraints while neural networks training is performed via solving a non-convex, unconstrained minimization problem [6]. There are two distinct advantages of using the SVM for classification. One is that it is often associated with the physical meaning of data, so that it is easy to interpret. The second advantage is that it requires only a small amount of training data [19, 61]. Recently, this classifier has been widely employed in the social science arena [22, 30, 38, 39]. The SVM classifier for two classes including a kernel function is illustrated in Fig. 3.

We implement a generalized form of SVM for overlapped and nonlinearly separable data. For simplicity, consider a training data matrix, X, with I patterns (as rows) and J variables (as columns) from two classes. Again, samples (patterns) refer to the independent variables over time. Each column, denoting a pattern, belongs to one of two classes: whether or not an organization used violence in any organizational year. The training data for the two classes is arranged as

$$F = [(\underline{x}_1, y_1), (\underline{x}_2, y_2), \cdots, (\underline{x}_I, y_I)]; \quad \underline{x}_i \in R^{I \times J} \quad \text{and} \quad y_i \in \{-1, 1\}, \quad (8)$$

where $y_i = \begin{cases} 1 & \text{if } \underline{x}_i \text{ belongs to the target class} \\ -1 & \text{otherwise.} \end{cases}$

Using a nonlinear mapping Φ, $\underline{x}_i$ is mapped into a feature space and a linear estimate is obtained as

$$y_i = f(\underline{\omega}, \underline{x}_i) = \underline{\omega}^T \Phi(\underline{x}_i) + b, \quad (9)$$

where $\underline{\omega}$ is a vector of the training set, and b is a bias determined by Karush-Kuhn-Tucker (KKT) conditions [35,40]. To determine $\underline{\omega}$ and b for non-separable case, the following regularized risk function is minimized.

$$\Phi\left(\underline{\omega}, \xi\right) = \frac{1}{2}\left\|\underline{\omega}\right\|^2 + G\sum_{i=1}^{I}\xi_i, \tag{10}$$

and a separating hyperplane must satisfy the following constraints

$$y_i[\langle\underline{\omega}\cdot\underline{x}_i\rangle + b] \geq 1 - \xi_i; \quad i = 1, 2, \ldots, I, \tag{11}$$

where $\langle\ \cdot\ \rangle$ denotes the dot (inner) product. In Eq. (10), G is a user-defined regularization parameter to control the trade-off between the two terms. We should note that the larger the G, the more the error is penalized. Thus, G should be chosen with care to keep away from overfitting. ξ is a slack variable that measures degree of misclassification of input patterns, $\underline{x}_i$. The first term in Eq. (10) represents the model complexity, and the second term represents the model accuracy [19]. The solution of Eq. (10) is given by the following dual optimization problem [19]:

$$\max_{\underline{\alpha}} W\left(\underline{\alpha}\right) = \sum_{i=1}^{I}\alpha_i - \frac{1}{2}\sum_{i=1}^{I}\sum_{j=1}^{I}\alpha_i\alpha_j y_i y_j\left\langle\underline{x}_i\cdot\underline{x}_j\right\rangle,$$

$$\text{subject to:} \quad \sum_{i=1}^{I}\alpha_i y_i = 0; \quad \alpha_i \in [0, G] \quad \text{and} \quad i = 1, 2, \ldots, I, \tag{12}$$

where α_i are Lagrange multipliers.

If a nonlinear mapping (or kernel function), $K(\underline{x}_i, \underline{x}_j)$, is chosen a priori, the optimization problem of Eq. (12) becomes

$$\max_{\underline{\alpha}} W\left(\underline{\alpha}\right) = \sum_{i=1}^{I}\alpha_i - \frac{1}{2}\sum_{i=1}^{I}\sum_{j=1}^{I}\alpha_i\alpha_j y_i y_j K\left(\underline{x}_i, \underline{x}_j\right), \tag{13}$$

subject to the same set of constraints as Eq. (12). The nonlinear mapping K is used to transform the original input $\underline{x}$ to a higher dimensional feature space Ω via $K(\underline{x}_i, \underline{x}_j) = \left\langle\varphi(\underline{x}_i)\cdot\varphi(\underline{x}_j)\right\rangle$. The decision function becomes

$$f\left(\underline{x}\right) = sgn\left(\sum_{i\in SVs}\alpha_i y_i K\left(\underline{x}, \underline{x}_i\right) + b\right), \tag{14}$$

where SVs is an index set, which contains the index of the support vectors, and sgn is the negative or positive sign of the function. A kernel function, a Radial Basis Function (RBF), is used for fitting non-linear models by transforming the data into a higher dimensional space (via "Kernels") before finding the optimal hyperplane to separate the classes. The RBF kernel function is a reasonable choice because of two major reasons: (1) the kernel can handle when the relation between independent and dependent variables are nonlinear, and (2) it also has less numerical difficulties. The RBF kernel function is

$$K\left(\underline{x}_i, \underline{x}_j\right) = \exp\left(-\gamma \left\|\underline{x}_i - \underline{x}_j\right\|^2\right), \tag{15}$$

where γ is a nonnegative monotonic parameter that controls the widths of the RBF function, and needs to be tuned with care to avoid overfitting or underfitting issues. G and γ are the design parameters, and these are determined through a simple grid-search using cross-validation.

5.5 Validation and Performance Assessment

As in King and Zeng [37], we employ a split-sample validation procedure. First, the data for the 6-year period from 1995 to 2000 are used as the training set. We train the SVM to learn presence and absence on the training set, and then validate the SVM model on 4-year period validation patterns from 2001 to 2004. The validation results are compared with the dependent variable and examined for 58 organizations. To verify the fitness of the model, we compute five metrics to assess the performance of our proposed approach:

$$\text{Overall Accuracy} = \frac{\text{Number of correctly classified organization-year observations}}{\text{Number of organization-year observations}}$$

$$\text{Recall (R)} = \frac{\text{Number of times dependent variable presence was correctly predicted}}{\text{Number of times dependent variable was present}}$$

$$\text{Precision (P)} = \frac{\text{Number of times dependent variable presence was correctly predicted}}{\text{Number of times dependent variable was predicted}}$$

$$\text{False Positive Rate} = \frac{\text{Number of times dependent variable presence was falsely predicted}}{\text{Number of times no dependent variable was present}}$$

$$\text{F-measure} = \frac{2 \times \text{Precision} \times \text{Recall}}{\text{Precision} + \text{Recall}}$$

We classify each organization-year observation in the out-of-sample periods by the extent of use of political violence (if any) it is likely to experience. Overall accuracy measures the ability of the proposed approach to correctly distinguish between the organizations that do and those that do not use violence as a policy. The recall score pertains to the ability of the approach to correctly forecast or classify on the element of interest in this case. The precision measure refers to the ability of the algorithm to classify without producing too many false positives (e.g., violence that is forecasted to be used but is not used as a policy). F-measure is the weighted harmonic mean of the precision and the recall that is often used as an aggregated performance measure. Our interest is how well the models can accurately forecast presence of violence as a policy without producing too many false positives. Recall, precision, and F-measure are also important in our performance assessment.

6 Sensitivity Analysis

We implemented and experimented with the proposed approach on the MAROB data discussed earlier. The approach – imputing missing values, factor selection, massaging data, classification/forecasting, and validation and performance assessment – was performed on a split-sample validation. Before validation of our model, we performed sensitivity analysis to find how changes in each variable value impact changes in dependent variable probabilities. This information provides insights into the relationship between input variables and forecasting output probabilities. For a notional organization and a given dependent variable, we simulated continuous changes in each variable from its minimum to its maximum value over the training and validation period (1995–2004) while holding all other variables constant at their median value. Table 1 presents the probabilities of violence for each variable selected by factor selection process.

The probabilities and significance of the variables generally support the theoretical argument. Not surprisingly, when the government targets the organization with either violence or oppression the likelihood that the organization will use violence is greatly increased. Interestingly, both kinds of external support increase the likelihood of violence. Religious ideology, an ideology of gender exclusion, cultural grievance, smuggling, and violent rhetoric all have an important and positive impact

Table 1 Probabilities of violence when variable is at either minimum or maximum value – all others held at mean or mode

Independent variable	Minimum (%)	Maximum (%)
Violence by state	10	55
Repression by state	23	60
Violent rhetoric	10	35
Diaspora support	10	45
Humanitarian Support from International Nongovernment Organization	10	40
Religious Organization	12	85
Dominant cultural grievance of the organization	10	38
Democratic	55	28
Gender exclusiveness	10	45
Popularity of organization	32	15
Electoral politics	70	18
Smuggling of consumer goods	10	45

on the choice of violence. Democratic ideology and participating in elections has a strong negative impact on the likelihood of using violence. Interestingly, and perhaps surprisingly, so does the popularity of the organization.

7 Classification and Forecasting Results

Our next effort focused on an out-of-sample analysis (classification). We used data from 1995 to 2000 for training and data from 2001 to 2004 for validation. Table 2 shows the performance in terms of overall accuracy, recall, precision, false positives, and F-measure on training (1995–2000) and validation (2001–2004) patterns.

Various cutoff probability points were experimented with to make predictions of presence/absence of the dependent variable, and we found that a 0.5 cutoff probability provided balanced predictions through the analysis performed in this chapter (the cutoff point was obtained by the ROC curve, discussed later). We believe these measures support the use of this approach and are generally useful if they can be replicated when the analysis is projected out into the future and so that new validation data can be used.

Ward et al. [69] argue that showing an individual variable to be statistically significant in a given model does not always imply that the variable plays an important role to improve the model's predictive power. As an alternative method over statistical tests, Relative (or Receiver) Operating Characteristic (ROC) curve [54] was suggested to achieve a more robust measure of a model's predictive power in social science research [37] for probabilistic forecasting. The ROC curve is a graphical presentation of the relationship between false positives (1-specificity) and true positives (sensitivity or recall) over the range of various cutoff probabilities (between 0 and 1) that make prediction corresponding above which the political violence occurs, and below which the political violence doesn't occur.

As shown in Fig. 4, the lower left corner (point 0,0) and the upper right corner (point 1,1) represent deterministic (and uninteresting) decisions of always deciding absence and presence of events, respectively. The diagonal from the upper right corner to the lower left corner is the line of pure guessing. If a model provides an ROC curve that lies above the diagonal, the model is said to have some expertise level. The upper left corner of the ROC plot represents a perfect model where there are no false positives and only true positives. The area under the curve (AUC) can be used to generate a statistic that indicates the model's predictive power.

Table 2 Classification performance

	Training %	Validation %
Overall accuracy	92.82	92.61
Recall	72.84	77.55
Precision	95.16	86.36
False positives	1.12	3.31
F-measure	82.52	81.72

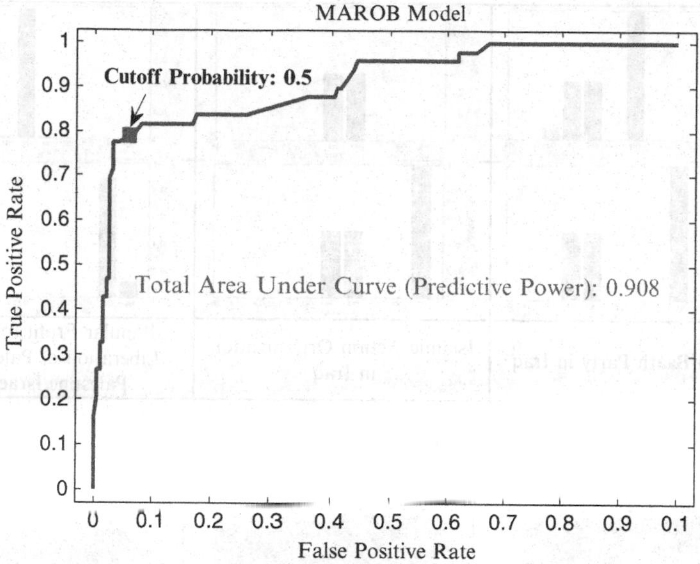

Fig. 4 ROC curve for MAROB model

Figure 4 illustrates the ROC curve for the MAROB model, and the model can be said to have predictive power of 0.908 by calculating the AUC. Since the predictive power of the model is quite close to a perfect model (predictive power of 1 when at point 0, 1), it is evident that the model is a reliable predictor and can be beneficial for further analyses.

The forecasting results from the MAROB model can be used as an indicator of early warning for organizations. For our MAROB model to play a role as an early warning indicator, we derived a change detection based method from Page's test [53] and applied it to detect advent and cessation on presence/absence predictions which enabled us to detect advent/cessation time from the presence/absence forecasted probabilities. Figure 5 illustrates predicted advent and cessation time of each organization in probability. For example, the MAROB model for Popular Front for the Liberation of Palestine in Israel predicted high probability for cessation in 1996 and for advent in 2000 which are 1 year before the true events occurred (early warning). The proposed approach can build a model for both annual occurrences of the dependent variable and transition probabilities to the dependent variable.

Figure 6 illustrates examples of the forecasting that we did in both the in-sample and out-of-sample analysis for selected organizations. Our results are satisfactory although there are clearly some cases we got very wrong (see, for example, the results in Fig. 6 for Hamas). For the most part, the overall accuracy for the validation period is quite high.

The interesting challenge though is to predict into the future. Our next analysis used data from 1995 to 2000 for training and data from 2001 to 2009 for forecasting.

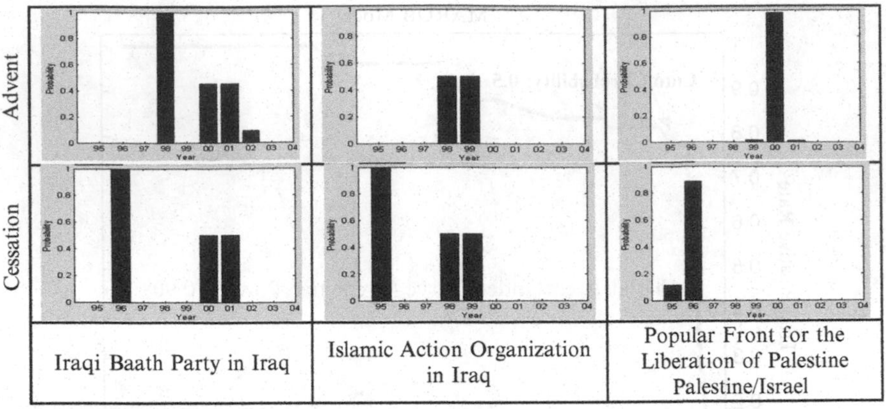

Fig. 5 Advent and cessation probability of the dependent variable over time for selected organizations

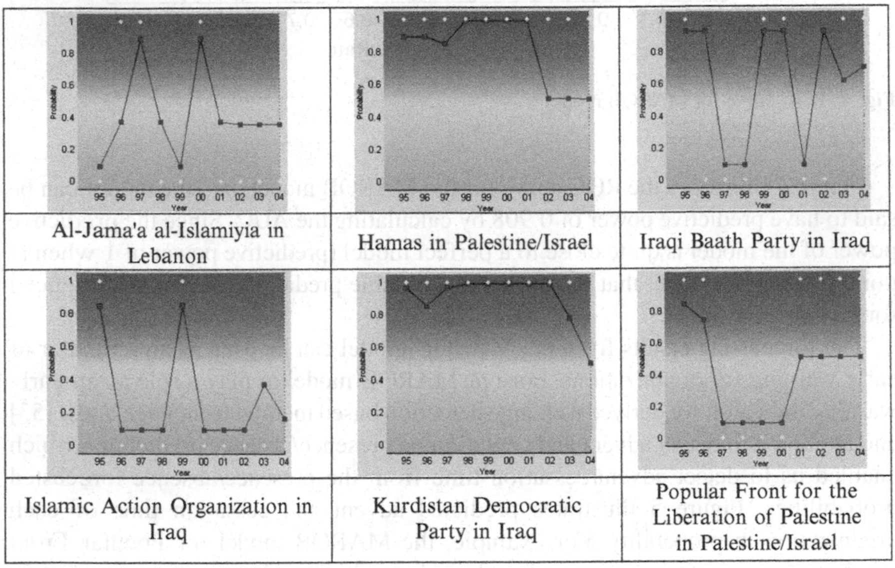

Fig. 6 True dependent variables (*dot*) vs. forecasting results (*solid*) for selected organizations

Since there are no independent values available after 2004, values between 2005 and 2009 were projected and estimated by ARMA processes for the forecasting analysis as discussed in the earlier section. Table A (see Appendix) presents the names of organizations we expect to use violence politically in 2009. Note that six organizations were predicted as 'Uncertain' due to the fact that the forecasting results were at cutoff probability, 0.5.

8 Conclusion

In this chapter, we have proposed a probabilistic modeling framework for the classification and forecasting of violent behavior and experimented with the MAROB data to assess the likelihood that organizations will choose to use violence. We not only achieved high classification accuracy with low false positive rates on both training (1995–2000) and validation (2001–2004) samples but also showed the efficacy of the proposed process by obtaining high predictive power of the model using the ROC curve. Our model holds out the possibility, if more timely data becomes available, of constructing an early warning indicator to identify those organizations representing minorities at risk which are likely to actually use violence.

We found in our analysis that explaining violence by a political organization is not a mono-causal story but that there are key factors that push organizations towards or away from violence. Ideology (specifically religious and patriarchal ideology) and repression and violence directed by a state against a minority group [51] have the largest impact on an organization's choice of violence. External support from a variety of sources (including smuggling) also clearly enables them to engage in violence. We should note that rhetoric also matters – what organizations say they are planning to do does not determine what they will do but it does have an important impact on their behavior. At the same time there are factors that are pushing organizations away from violence. Popular organizations and those that engage in elections are much less likely to engage in violence while democratic ideology also makes them less likely to use violence.

From a policy perspective our analysis has several implications. Analytically and in terms of future planning, our research suggests that the gathering of social science data on a regular basis can have important utility as input in a forecasting environment. Second our research suggests that the behavior of the states (which diplomatic and other efforts can attempt to change) can have an impact on outbreaks of violence. It also suggests that engaging groups in the political process can have a significant impact on the likelihood that their organizations will or will not embrace violence as a policy alternative.

As noted in our introduction, this chapter is part of a new effort to employ forecasting in the social sciences. Most of the other efforts have focused on country level analysis and a wider geographic scope (for example see Goldstone et al. [23] and Rost et al. [56]). There is a serious challenge in comparing across levels of analysis – especially since our unit of analysis is organizational violence which is not likely to reach the level of civil war, the focus of much of the state level analysis. This difference in geographic focus and level as well as temporal periods of analysis preclude some comparisons. Goldstone et al. [23] and Rost et al. [56] peg Africa (specifically sub-Saharan Africa) as the most likely hot spot for future violence. Our analysis of ethnic minorities in the Middle East, when pushed to the country level of analysis, identifies Israel/Palestine, Lebanon and Iraq as problematic countries – but we should note that even here our analysis is limited to only a select number of countries. Clearly it is important to expand the geographic scope of the analysis for comparison across approaches.

We believe that our model has done a good job of explaining why certain organizations are likely to use political violence in any particular year. The main innovation in this study has been our effort at collaboration across the sciences in the service of forecasting. We feel that for our test case of 2001–2004 we have done well and await the results of 2009. Further experiments will be performed by including early years of MAROB data (e.g., 1980–1994) and different regions (e.g., North Africa, Post-Communist States, Latin America, etc.) to verify the efficacy of our proposed approach. An effort using data from 1 year to the next will have to await more near real time coding.

Acknowledgements This material is based upon work supported by the Science and Technology directorate of the U.S. Department of Homeland Security under Grant Award Numbers N00140510629 and 2008–ST–061–ST0004, made to the National Consortium for the Study of Terrorism and Responses to Terrorism (START, www.start.umd.edu). Dr. Choi's and Prof. Pattipati's research was supported by the U.S. Office of Naval Research under contract numbers N00014–09–1–0062 and N00014–12–1–0238. The views and conclusions contained in this document are those of the authors and should not be interpreted as necessarily representing the official policies, either expressed or implied, of the U.S. Department of Homeland Security or START or the Office of Naval Research.

Appendix

Table A General predictions for violence for 2009

Organization	Prediction
Amal in Lebanon	Yes
Fatah/Palestinian Liberation Organization in Lebanon	Yes
Hamas in Israel	Yes
Hezbollah in Lebanon	Yes
Palestinian Islamic Jihad in Israel	Yes
Al-Jama'a al-Islamiyia in Lebanon	Uncertain
Democratic Front for the Liberation of Palestine in Israel	Uncertain
Iraqi Baath Party in Iraq	Uncertain
Islamic Movement in Iraqi Kurdistan in Iraq	Uncertain
Islamic Unity Movement in Lebanon	Uncertain
Supreme Council for the Islamic Revolution in Iraq	Uncertain
Arab Democratic Party in Israel	No
Baath in Syria	No
Bahrain Freedom Movement in Bahrain	No
Conservative Party in Iraq	No
Democratic Front for the Liberation of Palestine in Lebanon	No
Democratic Party in Cyprus	No
Fatah the Uprising in Lebanon	No

(continued)

Table A (continued)

Organization	Prediction
Fatah/Palestinian Liberation Organization in Jordan	No
Front des Forces Socialistes in Algeria	No
Hadash in Israel	No
Hamas in Jordan	No
Hamas in Lebanon	No
Hizb al-Da'wa al-Islamiyya in Iraq	No
Iraqi Homeland Party in Iraq	No
Iraqi National Accord in Iraq	No
Islamic Accord Movement in Iraq	No
Islamic Action Organization in Iraq	No
Islamic Movement in Israel	No
Jordanian People's Democratic Party in Jordan	No
Kurdistan Democratic Party in Iraq	No
Kurdistan Islamic Union in Iraq	No
Kurdistan Socialist Democratic Party in Iraq	No
Kurdistan Toilers" Party in Iraq	No
Muslim Brotherhood/Islamic Action Front in Jordan	No
National Liberation Party in Lebanon	No
National Movement for Change in Israel	No
National Popular Movement in Morocco	No
National Unity Party in Cyprus	No
New Birth Party or New Dawn Party in Cyprus	No
Organization of Revolutionary Toilers of Iranian Kurdistan in Iran	No
Palestine Democratic Union in Israel	No
Palestine Liberation Front in Lebanon	No
Palestinian People's Party in Israel	No
Palestinian Popular Struggle Front in Israel	No
Palestinian Popular Struggle Front in Lebanon	No
Patriotic Union Party in Cyprus	No
Patriotic Union of Kurdistan in Iraq	No
Phalange in Lebanon	No
Polisario in Morocco	No
Popular Front for the Liberation of Palestine – General Command in Lebanon	No
Popular Front for the Liberation of Palestine in Palestine/Israel	No
Popular Movement in Morocco	No
Progressive Socialist Party in Lebanon	No
Rally for Culture and Democracy in Algeria	No
Sons of the Village in Israel	No
Toplumcu Kurtulus Partisi, Cyprus	No
Turkish Republican Party in Cyprus	No

References

1. Alpaydin E (2004) Introduction to machine learning. MIT, Cambridge
2. Andriole SJ, Hopple GW (1984) The rise and fall of event data: from basic research to applied use in the US Department of Defense. Int Interact 10(3):293–309
3. Bankes S, Lempert R, Popper S (2002) Making computational social science effective: epistemology, methodology, and technology. Soc Sci Comput Rev 20(4):377–388
4. Beck N, King G, Zeng L (2004) Theory and evidence in international conflict: a response to de Marchi, Gelpi, and Grynaviski. Am Pol Sci Rev 98(2):379–389
5. Bennett SD, Stam AC (2006) Predicting the length of the 2003 US-Iraq war. Foreign Policy Anal 2(2):101–116
6. Bishop CM (2006) Pattern recognition and machine learning. Springer, New York
7. Bond J, Petrov V, Bond D, O'Brien S (2004) Forecasting turmoil in Indonesia: an application of hidden Markov models. A paper prepared for presentation at the International Studies Association Convention, Montreal, Quebec
8. Bro R (1996) Multiway calibration multilinear PLS. J Chemom 10(1):47–61
9. Bueno de Mesquita E (2005a) Conciliation, counterterrorism, and patterns of terrorist violence. Int Org 59(1):145–176
10. Bueno de Mesquita E (2005b) The quality of terror. Am J Pol Sci 49(3):515–530
11. Bueno de Mesquita E (2005c) The terrorist endgame: a model with moral hazard and learning. J Confl Resolut 49(2):237–258
12. Callaway R, Harrelson-Stephens J (2006) Toward a theory of terrorism: human security as a determinant of terrorism. Stud Confl Terror 29(7):679–702
13. Caprioli M (2000) Gendered conflict. J Peace Res 37(1):51–68
14. Caprioli M (2005) Primed for violence: the role of gender inequality in predicting internal conflict. Int Stud Q 49(2):161–178
15. Chiang LH, Braatz RD (2001) Fault detection and diagnosis in industrial systems. Springer, London
16. Choi K (2011) Intelligent data-driven classification and forecasting processes for complex engineering and social systems. PhD Dissertation, University of Connecticut
17. Duda RO, Hart PE, Stork DG (2001) Pattern classification. Wiley, New York
18. Fearon JD, Laitin DD (2003) Ethnicity, insurgency, and civil war. Am Pol Sci Rev 97(1):75–90
19. Ge M, Du R, Zhang G, Xu Y (2004) Fault diagnosis using support vector machine with an application in sheet metal stamping operations. Mech Syst Signal Process 18(1):143–159
20. Gilpin R (2005) War is too important to be left to ideological amateurs. Int Relat 19(1):5–18
21. Gleditsch KS (2007) Transnational dimensions of civil war. J Peace Res, 44(3):293–309
22. Goldsmith BE, Chalup SK, Quinlan MJ (2008) Regime type and international conflict: towards a general model. J Peace Res 45(6):743–763
23. Goldstone JA, Bates RH, Gurr TR, Lustik M, Marshall MG, Ulfelder J, Woodward M (2010) A global forecasting model for political instability. Am J Pol Sci 54(1):190–208
24. Green KC, Armstrong JS (2007) The ombudsman: value of expertise for forecasting decisions in conflicts. Interfaces 37(3):287–299
25. Gressang DS IV (2001) Audience and message: assessing terrorist WMD potential. Terror Pol Violence 13(3):83–106
26. Gurr TR (1988) Empirical research on political terrorism: the state of the art and how it might be improved. In Slater R, Stohl M (eds) Current perspectives on international terrorism. Palgrave Macmillan, London
27. Gurr TR (2000) People vs. states. United States Institute of Peace, Washington, DC
28. Gurr TR (2004) The minorities at risk (MAR) project, 8 July 2004. Available via DIALOG, http://www.cidcm.umd.edu/mar/. Cited 28 Jan 2012
29. Hollis M, Smith S (1990) Explaining and understanding international relations. Oxford University Press, New York

30. Hopkins DJ, King G (2010) A method of automated nonparametric content analysis for social science. Am J Pol Sci 54(1):229–247
31. Hsu C-W, Lin C-J (2002) A comparison of methods for multiclass support vector machines. IEEE Trans Neural Netw 13(2):415–425
32. Jackson JE (1991) A user's guide to principal components. Wiley, New York
33. Jensen NM, Young DJ (2008) A violent future? Political risk insurance markets and violence forecasts. J Confl Resolut 52(4):527–547
34. Juergensmeyer M (2003) Terror in the mind of God: the global rise of religious violence, 3rd edn. University of California Press, Berkley
35. Karush W (1939) Minima of functions of several variables with inequalities as side constraints. Masters Thesis, University of Chicago
36. Khuller S, Martinez V, Nau D, Simari G, Sliva A, Subrahmanian VS (2007) Computing most probable worlds of action probabilistic logic programs: scalable estimation for $10^{30,000}$ worlds. Ann Math Artif Intell 51(2–4):295–331
37. King G, Zeng L (2001) Improving forecasts of state failure. World Polit 53(4):623–658
38. King G, Zeng L (2007) When can history be our guide? The pitfalls of counterfactual inference. Int Stud Q 51(1):183–201
39. King G, Li Y (2008) Verbal autopsy methods with multiple causes of death. Stat Sci 23(1): 78–91
40. Kuhn HW, Tucker AW (1951) Nonlinear programming. In Neyman J (eds) Proc Second Berkeley Symp on Math Statist and Prob. University of California Press, Berkeley, pp 481-492
41. Kullback S (1987) The Kullback-Leibler distance. Am Stat 41(4):340–341
42. Laitin DD (2002) Comparative politics: the state of the subdiscipline. In Katznelson I, Milner HV (eds) Political science: state of the discipline. Norton, New York
43. Lichbach MI (2003) Is rational choice theory all of social science? The University of Michigan Press, Ann Arbor
44. Lichbach MI, Zuckerman AS (1997) Comparative politics: rationality, culture, and structure. Cambridge University Press, New York
45. Ljung L (1999) System identification: theory for the user. Prentice-Hall, New Jersey
46. Luo J, Pattipati KR, Qiao L, Chigusa S (2008) Model-based prognostic techniques applied to a suspension system. IEEE Trans Syst Man Cybern A 38(5):1156–1168
47. Martinez V, Simari G, Sliva A, Subrahmanian VS (2008) CONVEX: similarity-based algorithms for forecasting group behavior. IEEE Intell Syst 23(4):51–57
48. Mazarr MJ (2007) The Iraq war and agenda setting. Foreign Policy Anal 3(1):1–23
49. McAdam D, Tarrow S, Tilly C (2001) Dynamics of contention. Cambridge University Press, Cambridge
50. McCammon HJ, Granberg EM, Campbell KE, Mowery C (2001) How movements win: gendered opportunity structures and U.S. women's suffrage movements, 1866 to 1919. Am Sociol Rev 66(1):49–70
51. Minorities at Risk Project (2008) Minorities at risk organizational behavior dataset. Center for International Development and Conflict Management, College Park. Available via DIALOG http://www.cidcm.umd.edu/mar/data.asp. Cited 14 Feb 2012
52. O'Brien SP (2002) Anticipating the good, the bad, and the ugly: an early warning approach to conflict and instability analysis. J Confl Resolut 46(6):791–811
53. Page EB (1963) Ordered hypotheses for multiple treatments: a significance test for linear ranks. J Am Stat Assoc 58(301):216–230
54. Peterson WW, Birdsall TG, Fox WC (1954) The theory of signal detectability. Trans IRE Prof Group Inf Theory 4(4):171–212
55. Regan PM, Norton D (2005) Greed, grievance, and mobilization in civil wars. J Confl Resolut 49(3):319–336
56. Rost N, Schneider G, Kleibl J (2009) A global risk assessment model for civil wars. Soc Sci Res 38(4):921–933

57. Schölkopf B, Smola AJ (2002) Learning with kernels: support vector machines, regularization, optimization, and beyond. MIT, Cambridge
58. Schrodt PA (2000) Pattern recognition of international crises using hidden Markov models. In Richards D (eds) Political complexity: nonlinear models of politics. University of Michigan Press, Ann Arbor, pp 296–328
59. Schrodt PA (2006) Forecasting conflict in the Balkans using hidden Markov models. In: Trappl R (ed) Programming for peace: computer-aided methods for international conflict resolution and prevention. Springer, Dordrecht, pp 161–184
60. Sliva A, Subrahmanian VS, Martinez V, Simari G (2009) CAPE: automatically predicting 693 changes in terror group behavior. In: Memon N et al (eds) Mathematical methods in 694 counterterrorism. Springer, New York, pp 253–269
61. Smola AJ, Bartlett PL, Schlkopf B, Schuurmans D (2000) Advances in large margin classifiers. MIT, Cambridge
62. Steel RGD, Torrie JH (1960) Principles and procedures of statistics. McGraw-Hill, New York
63. Subrahmanian VS, Albanese M, Martinez V, Nau D, Reforgiato D, Simari G, Sliva A Wilkenfeld J (2007) CARA: a cultural adversarial reasoning architecture. IEEE Intell Syst 22(2):12–16
64. Taleb NN (2008) The black swan: the impact of the highly improbable. Random House, Inc, New York
65. Tarrow S (1993) Power in movements. Cambridge University Press, Cambridge
66. Tetlock PE (1999) Theory-driven reasoning about plausible pasts and probable futures in world politics: are we prisoners of our preconceptions? Am J Pol Sci 43(2):335–366
67. Tickner JA (2001) Gendering world politics: issues and approaches in the post-cold war era. Columbia University Press, New York
68. Vapnik VN (2000) The nature of statistical learning theory. Springer, New York
69. Ward MD, Greenhill BD, Bakke KM (2010) The perils of policy by p-value: predicting civil 711 conflicts. J Peace Res 47(4):1–13
70. Whitley D (1994) A genetic algorithm tutorial. Stat Comput 4(2):65–85
71. Wold S, Esbensen K, Geladi P (1987) Principal component analysis. Chemom Intell Lab Syst 2(1–3):37–52
72. Zimmermann E (1987) Political violence and other strategies of opposition movements: a look at some recent evidence. J Int Aff 40(2):325–351

Forecasting Changes in Terror Group Behavior

Maria Vanina Martinez, Amy Sliva, Gerardo I. Simari, and V.S. Subrahmanian

1 Introduction

The ability to model, forecast, and analyze the behaviors of other agents has applications in many diverse contexts. For example, behavioral models can be used in multi-player games to forecast an opponent's next move, in economics to forecast a merger decision by a CEO, or in international politics to predict the behavior of a rival state or group. Such models can facilitate formulation of effective mitigating responses and provide a foundation for decision-support technologies.

Behavioral modeling is a computationally challenging problem, since the behavior of an agent is a continuously evolving, dynamic phenomenon. In previous works, several methods were introduced—SOMA Stochastic Opponent Modeling Agents [6], *CONVEX* [7], and *SitCAST* [11]—that can be used to forecast an agent's behavior or contextual situation based on the historical likelihood of particular behavioral patterns in a time series data set. Indeed, most past work on modeling agent behaviors [1, 7, 9, 12, 13] focuses on learning a model of the typical behavior of the agent, and then using that model to predict what the agent might do in the future. However, agent behavior can be adaptive, and under some circumstances can deviate from the statistically "normal" past behavior. Though such

M.V. Martinez (✉) • G.I. Simari
Department of Computer Science, University of Oxford, Oxford OX1 3QD, UK
e-mail: vanina.martinez@cs.ox.ac.uk; gerardo.simari@cs.ox.ac.uk

A. Sliva
College of Computer and Information Science, Northeastern University,
Boston, MA 02115, USA
e-mail: asliva@ccs.neu.edu

V.S. Subrahmanian
Department of Computer Science, University of Maryland College Park,
College Park, MD 20742, USA
e-mail: vs@cs.umd.edu

V.S. Subrahmanian (ed.), *Handbook of Computational Approaches to Counterterrorism*,
DOI 10.1007/978-1-4614-5311-6_11,

occurrences may be sporadic or rare, it is often crucial to understand when an agent's behavior will diverge from the past in order to take mitigating actions or modified policy responses. For example, it is well-known that organization terrorist groups are constantly evolving and adapting their behavior. When a group establishes a standard operating procedure over an extended period of time, the problem of predicting what that group will do in a given situation (real or hypothetical) is easier than the problem of determining when, if, and how the group will exhibit a significant change in its strategic actions.

In this chapter, an architecture called the Change Analysis Predictive Engine (*CAPE*) is proposed as a comprehensive system to effectively predict behavior, including when and how an agent will *change* its behaviors. The *CAPE* algorithms have been tested on about 10 years of real-world data from the MAROB [3, 15] data set for five terrorist organizations in two countries and—in those cases at least—have proven to be highly accurate.

The rest of this chapter details how this forecasting has been accomplished with the *CAPE* algorithms. The general architecture of the *CAPE* system is described in Sect. 2. Intuitively, *CAPE* tries to learn conditions under which an agent has changed its behaviors and use these rules to predict the onset of future behavioral changes. Unlike the focus of systems such as SOMA [5, 6, 10, 12, 13] and *CONVEX* [7], the major object of study is *change in behavior*, and not the *behavior* itself. Section 2.3 focuses then on the problem of forecasting changes in agent behavior. The concept of a *change table* is presented for representing changes, and *change analysis algorithms* are proposed to study this table, generating rules that determine the conditions under which agents changed their behaviors. This section concludes with a formal definition of the *CAPE* algorithm, which builds on top the *SitCAST* and *CONVEX* method. Intuitively, when a behavioral change is predicted, *CAPE* reports this change, otherwise, it will forecast the typical behavior found by the combination of *SitCAST* and *CONVEX* methods.

In Sect. 3, results of experimental evaluations are presented for the *CAPE* system, applied to behavioral time series data for several terror groups. Depending upon the granularity with which the actions are modeled, these algorithms are either 80 % accurate, or about 69.32 % accurate. This result suggests that the granularity at which predictions are made is a key issue to be explored, especially when trying to understand the full picture of agent behavior and the potential onset of strategic changes.

2 *CAPE* Architecture

To forecast agent behaviors directly from data, a formal representation of this problem and the underlying time series is required. Such behavioral time series data sets consist of a relational database describing the behavior of an agent g. Assume the existence of some arbitrary universe $\mathscr{A}$ whose elements are called *attribute names* (attributes for short). Each attribute V_i has an associated domain $dom(V_i)$.

Year	LEAD	ORGPOP	DEMORG	FORSTFINSUP	ARMATTACK	HOSTAGE
1993	4	2	1	0	1	0
1994	3	2	1	0	1	0
1995	3	2	1	0	0	0
1996	3	2	1	0	1	0
1997	3	2	1	0	0	0

Fig. 1 Small subset of actual data for a group in the MAROB data set. LEAD, ORGPOP, DEMORG, and FORSTFINSUP are context attributes, while ARMATTACK and HOSTAGE are action attributes

Behavioral time series data sets contain attributes $\mathscr{A}$ which can be divided into a context schema $CS(g) = \{C_1, \ldots, C_n\}$ of *environmental (independent) variables* describing the context in which an agent g functioned during a given time frame, and an action schema $AS(g) = \{A_1, \ldots, A_m\}$ of *action (dependent) variables* describing actions taken by the agent. Each attribute $V \in \mathscr{A}$ has an associated domain $dom(V)$, and $dom(CS(g))$ and $dom(AS(g))$ denote the domains of the environmental and action variables, respectively. Assume that each environmental variable can be represented as an independent time series $C_i[y]_{y=y_1}^{y_k}$ where $y_1 < y_2 < \cdots < y_k$ are time periods in the data set, which has k rows corresponding to *equally spaced* time periods. $V[y]$ denotes the value of variable $V \in \mathscr{A}$ at time period y. The full behavioral time series for a particular agent or group g is denoted by T_g.

Example 1. The Minorities at Risk Organizational Behavior (MAROB) database [3, 15], which tracks information on violent ethnopolitical organizations, is an example of behavioral time series that meets the above criteria and that is used throughout this chapter. Figure 1 shows a small example of the behavioral time series data T_g for a terrorist organization g from the MAROB data. The table shown represents actual data for a subset of the approximately 175 MAROB attributes. Here, $CS(g) = \{LEAD, ORGPOP, DEMORG, FORSTFINSUP\}$, where the context in which the group is operating is described by categorical attributes denoting the type of group leadership, the amount of popular support, whether the group is internally democratic, and whether they receive financial support from a foreign state, respectively. The action schema $AS(g) = \{ARMATTACK, HOSTAGE\}$ contains two action attributes indicating whether the group uses armed attacks or takes hostages as a strategy at a particular time point. Other examples of environmental variables from MAROB include such things as the level of diaspora support, the level of foreign state financial support, the level of repression experienced by a group, and so forth. Action variables represent the strategic or tactical behaviors of a group, such as whether they used suicide bombings as a strategy, whether they used attacks against domestic security organizations, whether they mounted transnational attacks on foreign groups, etc. The domain of each variable consistent of a subset of the natural numbers, where each value encodes a different situation. For instance, the domain of variable $LEAD$ is the set $\{1, 2, 3, 4\}$ where a value of 1

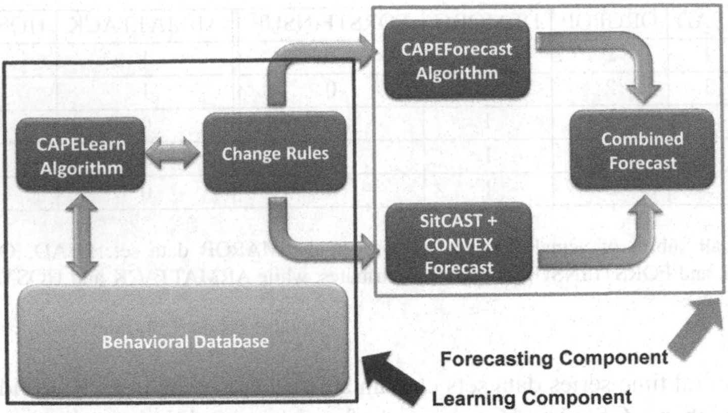

Fig. 2 Architecture of the *CAPE* framework

means that there are *factionalized/competing leaders*, a value of 2 means that the organization has a *weak or decentralized leadership*, value 3 means there is a *strong ruling council*, and a value of 4 states that there is a *strong single leader*. In this data we can see that $LEAD(1993) = 4$, which indicates that in the year 1993 the organization to which this behavioral time series belong had a strong single leader. However, that changed from 1994 onwards, showing a value of 3 instead.

In CONVEX [7], behavioral time series are analyzed to understand what an agent is most likely to do in the future in the average case. In this chapter, we exploit the information contained in the time series to predict when and how an agent is likely to change its behavior. Figure 2 shows the general architecture of the Change Analysis Predictive Engine (*CAPE*), a system for modeling behavioral change. The *CAPE* method consists of two major components—one focusing on *learning* the conditions under which an agent changed its behavior in the past, and another focusing on *forecasting* what the agent will do at some time in the future, including whether or not it will exhibit a change in strategy.

The learning architecture underlying *CAPE* processes a relational behavioral time series T_g to learn a model of behavioral changes for agent g. From the input data, the *CAPE-Learn* algorithm in Fig. 2 constructs a *change table*, which captures that part of a behavioral time series table that changed. A *change analyzer* then analyzes the change table automatically and learns environmental conditions that specify when the behavior of an agent changed. The output of *CAPE-Learn* is a set of *change rules* that indicate conditions under which an agent is likely to alter its behavior.

The forecasting component of *CAPE* in Fig. 2 builds on the *CONVEX* [7] and *SitCAST* [11] forecasters. We briefly recall *SitCAST* below in Sect. 2.1. The *CONVEX* algorithms forecast what an agent might do in a given situation, while *SitCAST* produces a set of possible situations that might be true at time y_{k+s} in the future and a probability distribution over that set of situations. When these possible situations are used as query context vectors for *CONVEX*, *SitCAST*+ *CONVEX* jointly forecasts what actions an agent will take at some time point y_{k+s} in the future, together with a probability. The *CAPE-Forecast* algorithm applies the change rules generated by the learning component of *CAPE* to a given situation to determine the onset of behavioral changes. If the change rules indicate a likely change in behavior, then this is the result predicted by *CAPE*. Otherwise, *CAPE-Forecast* returns the prediction from the combined *SitCAST*+ *CONVEX* method.

2.1 *SitCAST Situation Forecaster*

In this section we recall the basic notions from *SitCAST* [11] and show how it is used by means of an example; we also recall the combination of *SitCAST*+ *CONVEX* methods developed in [11].

Suppose we have a behavioral time series T_g for some group g. Let $CS(g) = \{C_1, \ldots, C_m\}$ and $AS(g) = \{A_1, \ldots, A_n\}$ be context schema and the action schema, respectively. Let us suppose that we have k rows $y_1, \ldots, y_k$ in table T_g corresponding to *equally spaced* time periods $y_1 < y_2 < \cdots < y_k$. Using this data, we may want to predict what the environmental situation might be like for group g during time periods y_{k+s} for $s \geq 1$.

To achieve this goal, we start by building upon any time series prediction algorithm (e.g., a linear regression, quadratic regression, or logistic regression program) ts [2]. Let $C_i[y_{k+s}]$ denote the value of C_i predicted by ts for the time period y_{k+s}.

Definition 1. A *possible situation* at time y_{k+s} according to prediction algorithm ts is a mapping ps from $\{C_1, \ldots, C_m\}$ to the domain of each C_i such that for all $1 \leq i \leq m$, $\lfloor C_i[y_{k+s}] \rfloor \leq ps(C_i) \leq \lceil C_i[y_{k+s}] \rceil$.

In other words, the value of y_{k+s} must be set to either $\lfloor C_i[y_{k+s}] \rfloor$ or $\lceil C_i[y_{k+s}] \rceil$. Let $\mathcal{PS}(ts, s)$ denote the set of all possible situations at time y_{k+s} using time series predictor ts.

Example 2. Consider a simple example shown below for a behavioral time series T_g with four environmental variables.

Time	C_1	C_2	C_3	C_4
1	2	4	1	1
2	2	3	1	2
3	1	3	1	2
4	1	3	1	2
5	1	2	1	2

Suppose we wish to predict what the situation will be like at time 6 and suppose the time series prediction function, ts, used is simple linear regression. In this case, $C_1[y_6] = 0.5$, $C_2[y_6] = 1.8$, $C_3[y_6] = 1$ and $C_4[y_6] = 2.4$.

As all our variables (action and environment) have integer domains, it follows that $C_1[y_6]$ must be either 0 or 1, $C_2[y_6]$ must be either 1 or 2, $C_3[y_6] = 1$ and $C_4[y_6]$ is either 2 or 3. This leads to *eight* possible situations listed below.

Situation	C_1	C_2	C_3	C_4
S_1	0	1	1	2
S_2	0	1	1	3
S_3	0	2	1	2
S_4	0	2	1	3
S_5	1	1	1	2
S_6	1	1	1	3
S_7	1	2	1	2
S_8	1	2	1	3

SitCAST can induce a probability distribution on the set of situations as follows. Let us consider variable C_2 from Example 2 whose predicted value at time 6 is 1.8. One can view this as a probability statement saying that the value of C_2 is 2 with 80 % probability and the value of C_2 is 1 with 20 % probability. If we were to assume that the C_i's values are independent of one another, then we can associate the following probabilities with each situation listed in Example 2 above.

Situation	Prob. calculation	Probability
S_1	$0.5 \times 0.2 \times 1 \times 0.6$	0.06
S_2	$0.5 \times 0.2 \times 1 \times 0.4$	0.04
S_3	$0.5 \times 0.8 \times 1 \times 0.6$	0.24
S_4	$0.5 \times 0.8 \times 1 \times 0.4$	0.16
S_5	$0.5 \times 0.2 \times 1 \times 0.6$	0.06
S_6	$0.5 \times 0.2 \times 1 \times 0.4$	0.04
S_7	$0.5 \times 0.8 \times 1 \times 0.6$	0.24
S_8	$0.5 \times 0.8 \times 1 \times 0.4$	0.16

Thus, we see that the most probable situations at time 6 are S_3 and S_7, each with 24 % probability of occurring.

2.2 *SitCAST* and *CONVEX*

The CONVEX system [7] contains methods to automatically predict what actions a given group will take in a given situation (real or hypothetical) by examining the similarity between the hypothesized situation and past situations in which the group has been. Once *SitCAST* has been used to forecast what a situation might be in the future (e.g. at time 6 in the running example given above), we can use CONVEX to predict what actions the group will take in each of the possible situations.

Suppose we consider a single action A_i whose intensity value we want to predict. That is, we want to predict the value of $A_i[y_6]$ in the example above. In this case, we can apply CONVEX to each possible situation S_i predicted to occur at a future time t. The probability that A_i will have the value v at time y_{k+s} will then be given by the following formula.

$$\mathbf{P}(A_i[y_{k+s}] = v) = \Sigma_{S \in \mathcal{PS}(ts,s) \wedge CONVEX(S,A_i)=v}\, prob(S).$$

In other words, to compute the probability of $A_i[y_{k+s}]$ having the value v, we perform the following steps:

1. Find all possible situations S that can arise at time $k+s$ such that CONVEX predicts that those situations will cause action variable A_i to have value v.
2. The probability of $A_i[y_{k+s}] = v$ is equal to the sum of the probability of all such situations.

Let us apply this to the running example mentioned above.

Example 3. Consider the same eight possible situations given by *SitCAST* in Example 2. Suppose we also have an action A with three possible values: 0,1,2. Let us suppose that CONVEX predicts that:

1. $A = 0$ in possible situations S_1, S_4, S_6.
2. $A = 1$ in possible situations S_2, S_7.
3. $A = 2$ in possible situations S_3, S_5, S_8.

The probability that $A = 0$ above is the sum of the probabilities that either situation S_1, S_4 or S_6 will occur. These probabilities are 0.06, 0.16 and 0.04 respectively, yielding a 26 % probability that $A = 0$ at time 6. This table summarizes the probabilities involved.

Action value	Prob. summation	Total prob.
$A = 0$	$0.06 + 0.16 + 0.04$	0.26
$A = 1$	$0.04 + 0.24$	0.28
$A = 2$	$0.24 + 0.06 + 0.16$	0.46

This result tells us that the *most likely* outcome based on the possible situations at time 6 is that the group in question will engage in the action denoted by $A = 2$ with 46 % probability.

Note that although CONVEX itself runs in polynomial time, the combination of CONVEX and *SitCAST* has an exponential worst case running time, since the number of possible situations to consider may be exponential in the number of environmental variables. One way around this problem is to introduce a window size, w for generating the possible situations. Let w be a real number between 0 and 0.5 inclusive. Whenever an environmental variable is predicted to be a real number r between $\lfloor r \rfloor$ and $\lceil r \rceil$, we do the following.

- Check if $\lfloor r \rfloor + 0.5 - w \leq r \leq \lfloor r \rfloor + 0.5 + w$. If so, then generate two possible situations.
- Otherwise, if $r \leq \lfloor r \rfloor + 0.5 - w$, then reset r's value to $\lfloor r \rfloor$.
- Otherwise, if $r \geq \lfloor r \rfloor + 0.5 + w$, then reset r's value to $\lceil r \rceil$.

To see how this works, suppose $w = 0.1$. This means that if the predicted value $C_i[y_{k+s}]$ is in the range $j.4$ to $j.6$ for an integer j, then two possible situations are generated for $C_i[y_{k+s}]$. Otherwise, if the predicted value is between j and $j.4$, then the value is modified to j. If the predicted value is between $j.6$ and $(j+1)$, then the value is modified to $j+1$.

Example 4. Let us return to our running example where the predicted values are as follows. Let $w = 0.1$.

1. $C_1[y_6] = 0.5$. In this case, no change is made to $C_1[y_6.]$ because $C_1[y_6.]$ has value 0.5 which is between 0.4 and 0.6.
2. $C_2[y_6.] = 1.8$. In this case, $C_2[y_6.]$ is reset to 2 because $C_2[y_6.]$ has a value greater than 1.6.
3. $C_3[y_6.] = 1$. In this case, $C_2[y_6.]$ stays 1 because $C_3[y_6.]$ has a value less than 1.4.
4. $C_4[y_6.] = 2.4$. In this case, no change is made to $C_4[y_6.]$ because $C_4[y_6.]$ has value 2.4 which is in the interval [2.4, 2.6].

It can be seen that by using this method, we will only generate the four possible situations, shown as situation S'_1, S'_2, S'_3, S'_4 in the first table that follows.

Situation	C_1	C_2	C_3	C_4
S'_1	0	2	1	2
S'_2	0	2	1	3
S'_3	1	2	1	2
S'_4	1	2	1	3

This result shows, in this example, a 50 % reduction in the number of generated possible situations. If, on the other hand, we had set $w = 0.31$, then we would only have two situations to consider—see situations $S^\star_1, S^\star_2$ in the second table below.

Situation	C_1	C_2	C_3	C_4
$S^\star_1$	0	2	1	2
$S^\star_2$	1	2	1	2

2.3 The *CAPE* Algorithms

SitCAST+ *CONVEX* does fairly well in predicting what actions an agent will take at a given time [11]; however, both the *CONVEX* algorithms and *SitCAST* focus on how an agent will behave based on *typical past behaviors*. Agents occasionally change their behaviors, and predicting such *changes* is often much more important than predicting that the behavior will conform to what is most normal. In this section, the *CAPE* algorithms are described for learning models of behavioral change and forecasting future behaviors. That is, we want to learn, for any action attribute $A_j \in AS(g)$ for an agent g, the environmental conditions under which the value for A_j will change.

2.3.1 The Change Table

When learning the conditions under which an agent changes its behavior, the first step is to construct a *change table*, a specialized relational data structure for storing changes that occurred in a behavioral time series.

Definition 2 (Change Table). Let T_g be a behavioral time series for agent g with context schema $CS(g)$ and action schema $AS(g)$. The *change table* $CH(g, A_j)$ for agent g w.r.t. $A_j \in AS(g)$ is a table derived from T_g as follows:

1. The set of rows in $CH(g, A_j)$ is given by $\{y_t \mid A_j[y_t] \neq A_j[y_{t-1}]\}$.
2. The set of columns in $CH(g, A_j)$ consists of the column associated with A_j and the set of all $C_i \in CS(g)$ such that $C_i[y_{t-1}] \neq C_i[y_{t-2}]$.
3. If $C_i[y_{t-1}]$ is not eliminated in the previous two steps, then its value is set to the pair $(C_i[y_{t-2}], C_i[y_{t-1}])$.
4. If $A_j[y_t]$ is not eliminated in the first two steps above, then its value is set to the pair $(A_j[y_{t-1}], A_j[y_t])$.

In other words, the change table $CH(g, A_j)$ for agent g w.r.t. action A_j eliminates all rows in the original table T_g where the action A_j did not exhibit a change in value from the previous time period y_{t-1}. In addition, it eliminates all columns except for those environmental variables which changed from $C_i[y_{t-2}]$ to $C_i[y_{t-1}]$. The change table *documents the changes that have occurred in the original table for agent* g*, based on the assumption that changes in an environmental variable from time period* y_{t-2} *to* y_{t-1} *are potentially responsible for a change in the action variable one time period later, i.e., from time period* y_{t-1} *to* y_t. It is easy to account for other time lags in a similar manner.

Example 5. Consider the simple behavioral time series T_g shown below for agent g. The context schema consists of only four environmental variables, i.e., $CS(g) = \{C_1, C_2, C_3, C_4\}$, and the action schema is $AS(g) = \{A_1\}$.

To construct a change table from T_g, the changes in the value of action A_1 are first identified. A_1 changed twice in this data—once in time period 3 and once in time

Time	C_1	C_2	C_3	C_4	A_1
1	2	4	1	1	1
2	2	3	1	2	1
3	1	3	1	2	2
4	1	3	1	2	2
5	1	2	1	2	1

period 5. The resulting change table will have only two rows after the execution of step 1 from Definition 2.

Time	C_1	C_2	C_3	C_4	A_1
3	1	3	1	2	2
5	1	2	1	2	1

Next, all context attributes that did not change in the time interval preceding the action changes get eliminated. Thus, the columns corresponding to attributes C_1 and C_3 are removed because they did not change their values in either of the time periods from 1 to 2 or from 3 to 4. After step 2, we now have the following table.

Time	C_2	C_4	A_1
3	3	2	2
5	2	2	1

Time	C_2	C_4	A_1
3	(4,3)	(1,2)	2
5	(3,3)	(2,2)	1

In step 3, the values of the environmental variables in the above table are replaced by the pairs of values associated with the changes that occurred.

The first row of this table, for example, indicates that the value of C_2 changed from 4 in time 1 to 3 in time 2 and that the value of C_4 changed from 1 in time 1 to 2 in time 2. Finally, this same replacement is made for the action A_1, resulting in the final change table.

Time	C_2	C_4	A_1
3	(4,3)	(1,2)	(1,2)
5	(3,3)	(2,2)	(2,1)

The row associated with time 3 says that there was a change in the action variable A_1 from value 1 in time 2 to value 2 in time 3. The fact that C_2 changed from time 1 (value 4) to time 2 (value 3) and C_4 changed from time 1 (value 1) to time 2 (value 2) mean that these changes in the context are potential causes or indicators of the change in attribute A_1 one time period later.

The final change table, as shown above, encapsulates all past changes of the action A_j. More importantly, the final change table dramatically reduces the number of possible environmental attributes under consideration when modeling the conditions under which these changes occur.

Note that in the form presented above, changes in the environmental variables from time y_{t-2} to y_{t-1} are used to learn changes in the action variables from time period y_{t-1} to y_t. In other words, these change rules can be used to make predictions "one time period ahead." It is easy to learn rules that consider a larger time lag between environmental and action changes via the following simple changes to the change table construction. In the following definition, a time lag of h time periods is considered.

Definition 3 (**h-Change Table**). Let T_g be a behavioral time series for agent g with context schema $CS(g)$ and action schema $AS(g)$. For integer $h \geq 1$ the *h-change table* $CH^h(g, A_j)$ for agent g w.r.t. $A_j \in AS(g)$ is a table derived from T_g as follows:

1. The set of rows in $CH^h(g, A_j)$ is given by $\{y_t \mid A_j[y_t] \neq A_j[y_{t-1}]\}$.
2. The set of columns in $CH^h(g, A_j)$ consists of the column associated with A_j and the set of all $C_i \in CS(g)$ such that $C_i[y_{t-h}] \neq C_i[y_{t-h+1}]$.
3. If $C_i[y_{t-h+1}]$ is not eliminated in the previous two steps, then its value is set to the pair $(C_i[y_{t-h}], C_i[y_{t-h+1}])$.
4. If $A_j[y_t]$ is not eliminated in the first two steps above, then its value is set to the pair $(A_j[y_{t-1}], A_j[y_t])$.

This definition merely changes the parameters associated with the potential environmental indicators C_i to look h time periods prior to an action change in A_j, rather than just one time period back.

2.4 *Learning Change Indicators from the Change Table*

By isolating the changes that occurred in the action and context attributes in a behavioral time series, the change table provides a set of potential correlations, that is, changes in environmental attributes that can be indicators of behavioral changes. When learning conditions that forecast a change in an action variable, only *conjunctive conditions* on context attributes are considered. If C_i is an environmental variable in $CS(g)$ for an agent g, and $c_f, c_t \in dom(C_i)$ are values in its domain s.t. (c_f, c_t) is a pair of values in the change table, then $C_i(c_f, c_t)$ is an *environmental change atom*. If $C_{i_1}(c_{f_1}, c_{t_1}), \ldots, C_{i_n}(c_{f_n}, c_{t_n})$ are environmental change atoms, then $EC = (C_{i_1}(c_{f_1}, c_{t_1}) \wedge \ldots \wedge C_{i_n}(c_{f_n}, c_{t_n}))$ is an *environmental change condition of size n*. There is no loss of generality in the assumption that only conjunctions are considered because *sets* of conditions are being learned and, hence, disjuncts can be easily accounted for. Two metrics, $precision(EC)$ and $recall(EC)$ can be defined to indicate the strength of the correlation between an environmental change condition EC and a particular behavioral change. These definitions use the usual concept of satisfaction of a condition by a tuple [14] w.r.t. the semantics of the value pairs stored in the change table. For example, in a change table, $y \models EC$ for a tuple at time period y and an environmental change condition $EC = (C_{i_1}(c_{f_1}, c_{t_1}) \wedge \ldots \wedge C_{i_n}(c_{f_n}, c_{t_n}))$ means that $C_{i_j}[y] = (c_{f_j}, c_{t_j})$ for all C_{i_j} in EC. On the other hand, in a behavioral time series, $y \models EC$ means that $C_{i_j}[y] = c_{t_j}$ and $C_{i_j}[y-1] = c_{f_j}$ for all C_{i_j} in EC.

Definition 4 (Precision). Let $CH(g, A_j)$ be a change table for agent g w.r.t. an action attribute A_j and EC be an environmental change condition. The *precision* of condition EC w.r.t. a change in A_j from a_f to a_t is defined as

$$precision(EC) = \frac{card(\{y \mid y \in CH(g, A_j) \wedge y \models EC \wedge A_j[y] = (a_f, a_t)\})}{card(\{y \mid A_j[y] = (a_f, a_t)\})}$$

Intuitively, the numerator of the term $precision(EC)$ is the number of time periods y in the change table which satisfy EC and where action variable A_j changed value from a_f to a_t. The denominator represents the number of time periods y in the change table where action variable A_j changed value from a_f to a_t. Thus, $precision(EC)$ computes the conditional probability of EC being true, given that the action variable A_j changed value from a_f to a_t according to the change table. Clearly, this conditional probability should be high for an EC to be considered a good predictor of change in action A_j. Note that the precision is computed without ever looking at the original behavioral table—only the change table is used.

In contrast, the *recall* of an environmental condition EC connects the change table with the original time series.

Definition 5 (Recall). Let T_g be a behavioral time series for agent g and EC be an environmental change condition. The *recall* of condition EC w.r.t. a change in A_j from a_f to a_t with time lag $h \geq 1$ is defined as:

$$recall(EC) = \frac{card(\{y \mid y \in T_g \wedge y \models EC \wedge A_j[y+h-2] = a_f \wedge A_j[y+h-1] = a_t\})}{card(\{y \mid y \in T_g \wedge y \models EC\})}$$

In the above definition, $recall(EC)$ is the ratio of the number of time periods y in the original table T_g which satisfy EC and where action variable A_j changed value from a_f to a_t, over the number of time periods y in T_g that satisfy EC. Intuitively, this is the conditional probability of action attribute A_j changing its value from a_f to a_t, given that environmental condition EC is true according to the *original behavioral table* (not the change table). Obviously, environmental change conditions that are strong predictors will also have a high recall.

The *CAPE-Learn* algorithm given in Fig. 3 learns environmental change conditions from the change table for an agent g that are indicators of a particular

Algorithm *CAPE-Learn*$(T_g, CH(g,A_j), a_f, a_t, sz, pt, rt, STAT\text{-}TESTS)$
1. $Rules = \emptyset$
2. $\mathscr{C} = \{EC \mid EC$ is an environmental change condition of size $\leq sz$ and each atom in $EC \in CH(g,A_j)\}$
3. for each $EC \in \mathscr{C}$ do
4. $precision(EC) = \frac{card(\{y \mid y \in CH(g,A_j) \wedge y \models EC \wedge A_j[y]=(a_f,a_t)\})}{card(\{y \mid A_j[y]=(a_f,a_t)\})}$
5. $recall(EC) = \frac{card(\{y \mid y \in T_g \wedge y \models EC \wedge A_j[y+h-2]=a_f \wedge A_j[y+h-1]=a_t\})}{card(\{y \mid y \in T_g \wedge y \models EC\})}$
6. **if** $precision(EC) \geq pt \wedge recall(EC) \geq rt \wedge STAT\text{-}TESTS(EC)$ **then**
7. add $A_j(a_f, a_t) : [recall(EC)] \leftarrow EC$ to $Rules$
8. **return** $Rules$

Fig. 3 The *CAPE-Learn* algorithm for learning behavioral change rules from an agent's change table

change in action attribute A_j from value a_f to a_t. *CAPE-Learn* also incorporates a special boolean function called $STAT - TESTS$, which takes an environmental change condition EC and tests whether it satisfies certain statistical conditions that a user may wish to impose. There are no restrictions whatsoever on how $STAT - TESTS$ may be implemented—it could compute p-values and ensure that they fall within a given bound, or it might involve a t-test, or confidence intervals, or no additional tests at all, depending on the needs of the user.

In Line 3, *CAPE-Learn* begins to cycle through the set of all environmental change conditions of size sz or less composed of environmental change atoms in the change table. If a condition EC has a sufficiently high precision (exceeding a given threshold pt), a sufficiently high recall (exceeding a threshold rt), and satisfies the designated statistical tests in $STAT - TESTS$, then on Line 7 the *change rule* $A_j(a_f, a_t) : [recall(EC)] \leftarrow EC$ is added to the set $Rules$. The change rules generated by *CAPE-Learn* indicate that, for an agent g, the given change in action attribute A_j from a_f to a_t will occur with a probability of $recall(EC)$ if EC is true.

Proposition 1. *Suppose $CH(g, A_j)$ is a change table for agent g w.r.t. action attribute A_j and $\mathscr{C}$ is the set of all environmental change conditions with maximum size sz composed of the atoms in $CH(g, A_j)$. The complexity of the CAPE-Learn algorithm is*

$$\mathbf{O}(card(\mathscr{C}) \times card(CH(g, A_j))).$$

If $CH(g, A_j)$ contains c columns and each column has at most b values in it, then this complexity boils down to

$$\mathbf{O}(b^c \times card(CH(g, A_j))).$$

Fig. 4 Size of the change table relative to the original behavioral table for 5 MAROB groups

Group	T_g Size	$CH(g,A_j)$ Size	$\frac{CH(g,A_j)\ Size}{T_g\ Size}$
1	1224	40	3.27 %
2	1224	37	3.02 %
3	1176	33	2.81 %
4	1176	51	4.34 %
5	1176	35	2.98 %

Algorithm $\textit{CAPE-Forecast}(T_g, CH^h(g,A_j), dom(A_i), sz, pt, rt, STAT\text{-}TESTS)$

$\quad forecast = null$
$\quad probability = 0$
$\quad$ for each $a_j \in dom(A_j)$ s.t. $A_j[y_k] \neq a_j$ do
$\qquad Rules = \textit{CAPE-Learn}(T_g, CH^h(g,A_j), A_j[y_k], a_j, sz, pt, rt,$
$\qquad\qquad STAT\text{-}TESTS)$
$\qquad$ for each $A_j(a_f, a_t) : [recall(EC)] \leftarrow EC \in Rules$ do
$\qquad\quad$ if $y_{k-h}, y_{k-h+1} \in T_g \wedge y_{k-h+1} \models EC \wedge recall(EC) >$
$\qquad\qquad probability$ then
$\qquad\qquad probability = recall(EC), forecast = a_j$
$\quad$ if $forecast = null$ then
$\qquad forecast = \textit{SitCAST} + \textit{CONVEX}\ forecast$
$\qquad probability = \textit{SitCAST} + \textit{CONVEX}\ probability$
$\quad$ return $(forecast, probability)$

Fig. 5 The *CAPE-Forecast* algorithm for forecasting agent behavior, including possible behavioral changes

Though there is an exponential factor in this computation, c has been quite small in practice. The complexity of the *CAPE-Learn* algorithm is dominated by the computation of the set $\mathscr{C}$ of environmental change conditions. $\mathscr{C}$ is constructed using only environmental change atoms that occur in the change table, which is usually substantially smaller than the original time series. For instance, the size of the change table, as a proportion of the size of the original table for experiments on five terrorist groups in the MAROB data set is given in Fig. 4. In other words, the size of the change table is *very small* compared to the size of the original behavioral time series—typically around 3–4 % of the size.

2.5 The *CAPE-Forecast* Algorithm

The purpose of the change rules generated by the *CAPE-Learn* algorithm is to provide a mechanism for forecasting behavioral change at some time y_{k+s} in the future. The *CAPE-Forecast* algorithm in Fig. 5 uses these change rules to make such a temporal behavioral forecast. Given a behavioral time series T_g containing data for agent g from time periods $y_1, \ldots, y_k$, and a change table $CH^h(g, A_j)$ with time lag h, this algorithm will forecast the value of action variable A_j from its domain $dom(A_j)$ during the next future time period, i.e., y_{k+1}.

CAPE-Forecast first uses *CAPE-Learn* to generate all possible change rules for action attribute A_j going from its current value $A_j[y_k]$ to any other possible value in its domain for time y_{k+1}. Each of the environmental change conditions in these rules is compared to the historical context at y_{k-h}; if an applicable rule is found then it is used to forecast the corresponding change in A_j with the probability $recall(EC)$. In the event of a conflict in what the rules predict, a rule with the highest recall is chosen. If a conflict still exists, the first forecasted value is chosen. If this procedure does not forecast a change in the value of A_j from time y_k to y_{k+1}, then *CAPE-Forecast* will return the value and probability forecast by *SitCAST+CONVEX*. *CAPE-Forecast*, then, is able to produce a comprehensive forecast for each action attribute at a given time point, combining knowledge about behavioral changes with the standard operating procedures of agent g.

3 Implementation and Experiments

The *CAPE* architecture was implemented and applied to study the behavior of seven terror groups in the Asia Pacific region using a specialized subset of the MAROB data [3, 15]. In this data, the contexts and behaviors of these organizations were tracked semi-annually from 1995 to 2006 (i.e., there are 24 time periods in T_g). For these groups, one action was considered: the propensity to engage in rebellion against the state. The goal was to predict whether the group would engage in rebellion at all during a particular time period, as well as the intensity of that rebellious behavior. A time lag of one time period was used when creating the change table and generating the change rules in *CAPE-Learn*.

Two experimental evaluations were conducted. In the first experiment, the data about rebellion by these seven groups indicated only if they were engaged in rebellion or not. In this case, *CAPE-Forecast* accurately predicted the rebellion status of these groups with 80 % accuracy. Looking specifically at the predictions of a change in rebellion status (i.e., instances of going from no rebellion to rebellion or vice versa), 83.3 % of the change forecasts were correct.

In the second experiment, the intensity of rebellion was measured for each time period in T_g on a scale from 0 (no rebellion) to 7 (full-fledged civil war). Sample change rules generated by *CAPE-Learn* for a group in the Philippines are given in Fig. 6. For example, rules 2 and 3 indicate a relationship between changes in

1. $ORGREB(1,0) : [1.0] \leftarrow stateRepression(3,4)$.
2. $ORGREB(1,0) : [1.0] \leftarrow representingInterests(0,1) \wedge terrorAttacks(1,0)$.
3. $ORGREB(0,1) : [1.0] \leftarrow representingInterests(1,0) \wedge terrorAttacks(0,1)$.

Fig. 6 Sample rules indicating conditions when a group will change its rebellious behavior

Table 1 Results from experiments with the *CAPE-Forecast* algorithm. Here the % accuracy is reported for the overall forecast and change forecast for predicting the occurrence of rebellion, the intensity of rebellion, and the intensity ±1

Experiment	Overall forecast accuracy	Change forecast accuracy
Occurrence	80	83.3
Intensity	69.32	61.5
Intensity ±1		69.2

rebellion and changes in the strategy of the group. Specifically, changes in whether they favor representing their interests to officials or terror attacks can be leading indicators of changes in the level of rebellion. When representing their interests becomes a minor strategy, and terrorism is no longer a strategy, the incidence of rebellion declines in the following time period from political banditry (1) to no rebellion (0), and vice versa.

CAPE-Forecast accurately predicted the exact intensity of rebellion 69.32 % of the time; this includes both the "normal" behavioral predictions of the group (i.e., cases where no change rules were applicable and *SitCAST+ CONVEX* was used) as well as the change predictions. Looking only at the cases when a change was predicted, 61.5 % of these forecasts correctly predicted the incidence of behavioral changes. Allowing for a margin of error of ±1 regarding the predicted intensity, the change forecast accuracy increases to 69.2 %. The results from both of these experiments are shown in Table 1.

What these two experiments suggest is that *CAPE-Forecast* does a very good job of making predictions, and that these predictions are significantly more accurate than an arbitrary guess would be (simple guessing would give 50 % accuracy in the first experiment above, and 12.5 % accuracy in the second experiment). However, this also suggests that *CAPE*'s accuracy decreases when highly fine-grained forecasts are required. Further experiments are needed in order to assess *CAPE*'s accuracy for various forecast granularities. Moreover, the above experiments only apply to situations one time period ahead—how *CAPE*'s accuracy changes with longer look-aheads also needs to be further studied.

4 Related Work

There has been intense interest in the last few years in the problem of learning models of the behaviors of groups and applying them to forecast what the group might do in the future [1, 9, 12, 13]. Such works focus on what the group would *normally* do, not how it would change its behaviors. In contrast, *CAPE* focuses on learning conditions under which a group will change its behavior—to date, we are aware of no work on forecasting *changes* in behavior of terror groups.

Time series forecasting is a very diverse field that has been applied in a wide range of contexts, from politics to biology to economics. With a large array of possible statistical models, there have been several attempts to better understand the relationship between these different forecasting procedures. In [4], the author proposes a theoretical framework for unifying a diverse set of forecasting methods for univariate time series. Many of these distinct methods can be integrated into a common general mathematical framework using a structural time series representation and the Kalman filter as a means of approximating the parameters and making forecasts. When estimating a model based on historical data, it is often difficult to determine what the "best" model is, both in terms of fit and predictive ability. The work of [16] presents the AFTER algorithm that attempts to produce better forecasts by assigning weights to several different candidate models and combining the results to achieve more accurate forecasts than through a single model. Raftery et al. [8] address the same problem, using two different Bayesian model averaging techniques to handle uncertainty in selecting predictors in a linear regression model. Bayesian model averaging can reduce the set of candidate models and combine the remaining possible models for purposes of prediction. In terms of accurately fitting the data and in making forecasts, the averaged models outperform other standard model selection techniques that choose a single model.

Schrodt [9] and Bond et al. [1] have come up with methods to build Hidden Markov Models to describe how a country might evolve over time. However, these HMMs are painstakingly constructed over time using data gathered about the country. This is very time-consuming and needs to be repeated for each country or each group, and has a degree of subjectivity to it. Nevertheless, HMMs and their variants (e.g. stochastic automata, stochastic Petri nets, etc.) can form a very valuable modeling tool and mechanism and deserves continued study. Their main advantage is that they are not necessarily based on past data.

Another method to forecast how a group might behave in the future is the Stochastic Opponent Modeling Agents (SOMA) system developed in [5, 6, 10, 12, 13]. SOMA looks at the data collected about a given group and then *automatically* extracts stochastic behavioral rules. This has an advantage over [1, 9] in that our method to build a behavioral model is fully automated and fast, as shown in [10]. These behavioral rules are then used for predictive purposes using a linear programming based algorithm. No detailed analysis of the accuracy of this forecast algorithm is available (the SOMA Forecast Engine is still under development) at this time and the compute time for generating forecasts can be expensive [5, 6].

In contrast to all these methods, the CONVEX [7] system does *not* build a model of a group. It uses data about a group in order to assess the similarity between a given situation (expressed as the query vector) and information about the group's behavior in the past when confronted with similar situations. Similar situations that the group has encountered in the past are then used directly for predictions—the "model building phase" present in related works is therefore completely skipped. Hence, CONVEX predictions can be made solely by examining the data and directly predicting outcomes without a country or group specific model that takes time to build and is often built using subjective methods. However, as in the case of the other work mentioned above, CONVEX, by virtue of examining similarities between a current situation and the past, predicts what is expected behavior of a group, and not how the group changes behaviors. Methods such as *CONVEX* [7] can produce highly accurate forecasts of what a given agent will do in a particular situation of interest based on its past behaviors in similar contexts. These "what if" forecasts can be generalized to temporal predictions of agent behavior by applying the *CONVEX* algorithms jointly with the *SitCAST* (Situation Forecaster) [11], producing a probabilistic forecast of an agent's behavior at some time in the future. However, the ability of such systems to predict when an agent will change its behavior has yet to be proven. Much of the forecasting error incurred by *SitCAST+ CONVEX* [11] arises when the agent in question changes its behavior. Thus, incorporating the ability to predict changes into the behavioral forecasting mechanism will improve the overall accuracy. In this chapter, we build upon CONVEX and show how it is one component of a larger system to predict behavioral changes. The *CAPE* architecture allows to model different aspects of behavior to determine when a behavioral change should be predicted and when forecasts should be made that are consistent with normal behavior.

5 Conclusions

There are numerous applications where it is necessary to predict not only the ordinary behavior of an agent, but also when that agent will change its behavior. For example, financial organizations are interested in tracking the behavior of major investors, political parties are interested in tracking the actions of various special interest groups, and governments are interested in the behaviors of foreign political and military entities.

This chapter has discussed two important contributions. First, the *CAPE-Learn* algorithm was developed to learn conditions under which an agent changes a certain behavior of interest. Second, the *CAPE-Forecast* algorithm was developed to use the rules and conditions generated by *CAPE-Learn* to forecast what a group will do in future time periods, including when they will exhibit a behavioral shift. The *CAPE* architecture was experimentally evaluated on one action (the propensity to engage in rebellion) for seven terrorist groups in the Asia Pacific region, and results indicate that *CAPE* can provide high predictive power, at least in these limited experiments.

The *CAPE* system can be extended in several ways that might make forecasting behavioral changes more accurate. For example, it may be possible to profile clusters of several agents that exhibit similar patterns of behavioral change and pool their change rules to have a potentially more complete model.

The ability to accurately forecast when an agent will change its behavior is a crucial component of developing effective policies and managing contingencies, and is a requirement for any decision-support system. Technologies that might leverage the knowledge provided by *CAPE* for policy analysis.

Acknowledgements Some of the authors of this paper were funded in part by AFOSR grant FA95500610405, ARO grant W911NF0910206 and ONR grant N000140910685.

References

1. Bond J, Petroff V, O'Brien S, Bond D (2004) Forecasting turmoil in indonesia: an application of hidden markov models. In: International studies association convention, Montreal. International Studies Association, USA, pp 17–21
2. Bowerman B, O'Connell R, Koehler A (2004) Forecasting, time series and regression, 4th edn. Southwestern College Publishers, Cincinnati, OH, USA
3. Center for International Development and Conflict Management (2008) Minorities at risk organizational behavior dataset, minorities at risk project. Retrieved from http://www.cidcm.umd.edu/mar
4. Harvey AC (1984) A unified view of statistical forecasting procedures. Int J Forecast 3(3): 245–275
5. Khuller S, Martinez MV, Nau D, Simari GI, Sliva A, Subrahmanian VS (2007) Computing most probable worlds of action probabilistic logic programs: scalable estimation for $10^{30,000}$ worlds. Ann Math Artif Intell, 51(2–4):295–331
6. Khuller S, Martinez MV, Nau D, Simari GI, Sliva A, Subrahmanian VS (2007) Finding most probable worlds of probabilistic logic programs. In: Proceedings of the 2007 international conference on scalable uncertainty management. Lecture notes in computer science, vol 4772. Springer, Berlin/New York, pp 45–59
7. Martinez MV, Simari GI, Sliva A, Subrahmanian V (2008) CONVEX: context vectors as a similarity-based paradigm for forecasting group behaviors. IEEE Intell Syst 23(4):51–57
8. Raftery AE, Madigan D, Hoeting JA (1998) Bayesian model averaging for linear regression models. J Am Stat Assoc 92:179–191
9. Schrodt P (2000) Forecasting conflict in the balkans using hidden markov models. In: Proceedings of the American political science association meetings. American Political Science Association, USA
10. Sliva A, Subrahmanian VS, Martinez MV, Simari GI (2008) The SOMA terror organization portal (STOP): social network and analytic tools for the real-time analysis of terror groups. In: Proceedings of the 2008 first international workshop on social computing, behavioral modeling and prediction. Lecture notes in computer science. Springer, New York/Berlin, pp 9–18
11. Sliva A, Subrahmanian VS, Martinez MV, Simari GI (2009) CAPE: automatically predicting changes in group behavior, In: Memon N, Farley JD, Hicks DL, Rosenorn T (eds) Mathematical methods in counterterrorism. Springer, Wien, pp 247–263
12. Subrahmanian VS (2007) Cultural modeling in real-time. Science 317(5844):1509–1510
13. Subrahmanian VS, Albanese M, Martinez MV, Nau D, Reforgiato D, Simari GI, Sliva A, Wilkenfeld J (2007) Cara: a cultural adversarial reasoning architecture. IEEE Intell Syst 22(2):12–16
14. Ullman J (1989) Principles of database and knowledge base systems, vol 2. Computer Science Press, Rockville
15. Wilkenfeld J, Asal V, Johnson C, Pate A, Michael M (2007) The use of violence by ethnopolitical organizations in the middle east. Technical report, National Consortium for the Study of Terrorism and Responses to Terrorism
16. Zou H, Yang Y (2004) Combining time series models for forecasting. Int J Forecast 20:69–84

Using Temporal Probabilistic Rules to Learn Group Behavior

John P. Dickerson, Gerardo I. Simari, and V.S. Subrahmanian

1 Introduction

The ability to reason about the past, present, or future state of the world is widely applicable to many fields. Additionally, considering uncertainty over the precise time at which events occurred or will occur increases realism, but also increases theoretical and computational intractability. This sort of probabilistic temporal reasoning is important in domains like those listed below.[1]

1. The advent of the Internet has clearly affected traders' reasoning about past and future movements in the **stock market**. For example, Fujiwara et al. [7] and De Choudhury et al. [5] discuss how stories in newspapers, blogs, and miscellaneous websites move prices in the stock market. A variety of data mining and machine learning techniques are used by investment banks and hedge fund managers to predict future stock movements based on past patterns in the values of various indicators. Formally, an investor could learn rules like, "the probability that the

[1]These examples and others are discussed in depth in work by Dekhtyar et al. [6] and Shakarian et al. [18, 19]. We omit some discussion due to space; for more information, see these articles.

J.P. Dickerson (✉)
Gates-Hillman Center, School of Computer Science, Carnegie Mellon University, Pittsburgh, PA 15213, USA
e-mail: dickerson@cs.cmu.edu

G.I. Simari
Department of Computer Science, University of Oxford, Oxford OX1 3QD, UK
e-mail: gerardo.simari@cs.ox.ac.uk

V.S. Subrahmanian
Department of Computer Science, University of Maryland College Park, College Park, MD 20742, USA
e-mail: vs@cs.umd.edu

V.S. Subrahmanian (ed.), *Handbook of Computational Approaches to Counterterrorism*,
DOI 10.1007/978-1-4614-5311-6_12,

stock of IBM will rise by at least 3 % at time $(t + \Delta)$ is 90 % given that, at time t, early coverage from blogs is positive about an upcoming earnings call and, also at time t, IBM does not announce fresh layoffs." The processes required to collect this data are outside the scope of this chapter, but a large financial institution has the resources to scrape web and print sources from which such rules could be learned.

2. Advances in both electronic record keeping and large-scale data analysis have introduced the "big data" mentality into **medicine**. For example, the Dartmouth Atlas of Health Care [22] aggregates US health care data across multiple dimensions (time, location, socioeconomic status, gender, severity, etc.); however, like other such large projects, it is largely sourced from billing data. Billing data alone is both incomplete and inaccurate, so reasoning over such uncertain temporal data is difficult. For instance, medical practitioners or policy analysts may wish to write rules of the sort, "the probability that a patient will return to the hospital before time $(t + \Delta)$ is 10 % if the patient was not in the ICU at time t and the patient's visit lasted less than 1 hour."
3. Large-scale data collection regarding **environmental phenomena** has resulted in a deluge of noisy, temporal data available to the public. For example, a government warning agency may wish to announce that, "if a forest fire occurs at time t and the amount of rain at time t is less than 0.1cm, then the probability that the fire will continue at time $(t + \Delta)$ is at least 85 %."
4. The Minorities at Risk research project [23] monitors the conflicts and activities of minority ethnicities, religious sects, and **terrorist groups** around the world. Our group at the University of Maryland has worked extensively with this data, and published analyses of some of these groups' behaviors (e.g., Hezbollah [10] and Hamas [11]). We built the SOMA Terror Organization Portal [14], which has registered users from over 12 US government agencies and contains thousands of (automatically) extracted rules about various groups' behaviors. Analysts engaged in counter-terrorism efforts need to be able to reason with such rules and make appropriate forecasts; in separate work, we have also done extensive work on making such forecasts [13, 15]. In this chapter, we formulate a running example in the context of the terrorist group Lashkar-e-Taiba.

In this chapter, we discuss two related types of logic programs that allow for logical reasoning in situations that involve temporal uncertainty. In Sect. 2, we first discuss *temporal probabilistic logic programs* (TPLPs), originally formulated by Dekhtyar et al. [6] as an extension to the generalized annotated programs (GAPs) of Kifer and Subrahmanian [9]. TPLPs allow for reasoning about point probabilities over time intervals using temporal probabilistic rules (tp-rules). In Sect. 3, we present an algorithm for automatically learning tp-rules from data, as detailed in [20]. We are also present a method for making policy recommendations by employing standard integer programming techniques to the automatically learned rules. Then, in Sect. 4, we describe a large-scale system we recently built to analyze terror groups using tp-rules. Using this system, we automatically learn rules about the south Asian terrorist group Lashkar-e-Taiba. In Sect. 5, we conclude with a discussion of future research directions, including an adaptation of our architecture to the recently introduced *annotated temporal probabilistic* (APT) logic programs, an extension to TPLPs that does not make independence assumptions about the underlying features and allow for reasoning over probability intervals over time periods, rather than just point probabilities.

2 Modeling Group Behavior with Temporal Probabilistic Logic Programs

Temporal probabilistic logic programs (TPLPs) were first introduced by Dekhtyar et al. in [6]. The system provides a framework within which a logic programmer can express *tp-rules* of the form "If some condition is true, then some atom is also true at some time/time interval with some probability distribution over the points in the time interval." Dekhtyar et al. [6] also provided a syntax and semantics for temporal probabilistic logic programs, as well as initial complexity results. In this section, we overview TPLPs and tp-rules in the context of modeling group behavior.

2.1 Database Schema for a Group's Past Behavior

Before defining the general temporal probabilistic logic, we introduce a running example that focuses on Lashkar-e-Taiba (LeT), a well-known, active South Asian terrorist group. The example uses real data collected by the *Computational Modeling of Terrorism* (CMOT) codebook [17], a research project that records past and current activities of multiple terrorist groups including LeT.

We view the data as a single relation consisting of tuples with two types of attributes: *environmental* and *action*. Environmental attributes correspond to aspects of the environment in which the group operated, while action attributes correspond to the various types of actions taken by a group, along with their intensities. Each tuple corresponds to the set of these attributes' values for a given a month. Example 1 gives a very small subset of the raw data collected on LeT.

Example 1. The table below shows four attributes of CMOT data collected for Lashkar-e-Taiba across 12 months in 2004.

The first column is a date labeling each tuple. The next column corresponds to the action attribute `attackCiv`, a binary variable that is activated if LeT both attacked civilians during a given month and that attack resulted in casualties.[2]

[2]The CMOT codebook tracks fine-grained aspects of violent group behavior. Other civilian attack-related attributes include attacks on civilian transportation, attacks on civilians without civilian casualties, and attacks specifically targeting civilian minorities.

Date	AttackCiv	Religious	Raided	PersonnelKilledJK
Jan 2004	1	1	0	13
Feb 2004	0	1	1	23
Mar 2004	0	1	0	10
Apr 2004	0	1	0	8
May 2004	0	1	0	15
Jun 2004	0	1	0	7
Jul 2004	0	1	0	14
Aug 2004	0	1	0	13
Sep 2004	0	1	0	11
Oct 2004	0	1	0	25
Nov 2004	0	1	0	16
Dec 2004	0	1	1	9

The next three columns correspond to environmental attributes. The attribute `religious` is set to 1 if LeT operated as a religious organization during a specific month. We see that LeT operated as a religious organization during every month of 2004. The attribute `raided` is a binary variable that is set to 1 if the government of a host country raided LeT during a specific month. The data shows that this occurred in February and December of 2004. Finally, the last column, `personnelKilledJK`, is an integral variable that takes nonnegative values corresponding to how many members of Lashkar-e-Taiba were killed in the northernmost Indian state of Jammu and Kashmir.

Example 1 considers a subset of the database where each attribute has a value for each time period. This need not be the case; attribute values can be left unset if they are unknown. For instance, the CMOT database considers group behavior over many decades; data on some attributes may no longer be available (e.g., pertaining to the number of kidnappings that occurred, or whether or not LeT actively lobbied the government of Pakistan).

We will now define the formal syntax for temporal probabilistic logic, through which we will be able to learn tp-rules with which we can reason about a group's past and future behavior.

2.2 Syntax

We assume the existence of a first order logical language with finite set $\mathcal{L}_{cons}$ of constant symbols, finite set $\mathcal{L}_{pred}$ of predicate symbols, and infinite set $\mathcal{L}_{var}$ of variable symbols. Each predicate symbol $p \in \mathcal{L}_{pred}$ has an *arity* (denoted *arity(p)*). A (ground) *term* is any member of $\mathcal{L}_{cons} \cup \mathcal{L}_{var}$ (resp. $\mathcal{L}_{cons}$); if $t_1, \ldots, t_n$ are (ground) terms, and $p \in \mathcal{L}_{pred}$, then $p(t_1, \ldots, t_n)$ is a (ground) atom.

In the context of the behavioral data discussed in Sect. 2.1, every attribute corresponds to a predicate symbol. In fact, each attribute in the example (and, in fact, the entire CMOT codebook) represents a *unary* predicate symbol. Although our formalization is easily generalized, we will thus concentrate only on predicates p such that $arity(p) = 1$. Let p be the predicate corresponding to an attribute, and t a term in the domain of p. Then $p(t)$ is an *action atom* when p corresponds to an action attribute, and an *environmental atom* when p corresponds to an environmental attribute. Finally, if $X \in \mathscr{L}_{var}$ and $Y \in \mathscr{L}_{cons}$, then $X = Y$, $X < Y$, $X > Y$, $X \leq Y$, and $X \geq Y$ are called *comparison atoms*.

Example 2. In this example, we use the table of data shown in Example 1. In this table, `attackCiv` is an action attribute with domain 0 and 1. In January 2004, Lashkar-e-Taiba's attacks on civilians resulted in civilian casualties; we represent this using the ground atom `attackCiv`(1). Similarly, `personnelKilledJK` is an environmental attribute whose domain is the non-negative integers. If $X \in \mathscr{L}_{var}$ ranges over the non-negative integers, then `personnelKilledJK`(X) can be instantiated to represent any number of LeT personnel killed in Jammu and Kashmir. For example, in January 2004, we instantiate $X = 13$ to return the ground atom `personnelKilledJK`(13).

We now formally introduce the concept of time. Let $T = \{1, \ldots, \tau_{max}\}$ denote the entire set of time points in which we are interested. We require a fixed time window size ranging over T, but allow τ_{max} to be arbitrarily large. The user may choose both the granularity of T and τ_{max} in an application-specific way. For instance, in the stock market example given in Sect. 1, a user may be interested in reasoning about 15-min segments (when the market is open) over the course of 10 years, and would set τ_{max} to around 78,000 to represent 30 periods per day over roughly 2,600 trading days. On the other hand, our terrorism application does not require such a fine-grained temporal resolution. The CMOT codebook records data on the order of months, so we use a t_{max} of approximately 240 to reflect an interest in events over the past 20 years.

Given time period $\tau \in T$ and probability $\rho \in [0, 1]$, we call $[\tau, \rho]$ a *temporal-probabilistic annotation* (or *tp-annotation*). Intuitively, a tp-annotation $[\tau, \rho]$ refers to some unspecified event occuring exactly τ time periods after a given time, with a probability of ρ.

We now syntactically connect time to our fledgling logic. Given a tp-annotation $[\tau, \rho]$ and an action (environmental) atom $p(t)$, we call $p(t) : [\tau, \rho]$ an action (environmental) *tp-annotated atom*. If $p(t)$ is ground then $p(t) : [\tau, \rho]$ is called *ground* as well. Intuitively, $p(t) : [\tau, \rho]$ says that $p(t)$ will occur with probability ρ exactly τ time intervals after some fixed time. Example 3 gives sample tp-annotated atoms in the context of our running example.

Example 3. The action tp-annotated atom `attackCiv`(1) : [3, 0.9] states that there is a 90 % chance of Lashkar-e-Taiba carrying out deadly attacks against civilians in 3 time units after some fixed time. The environmental tp-annotated atom `personnelKilledJK`(4) : [1, 0.5] states that there is a 50 % chance that personnel belonging to Lashkar-e-Taiba will be killed in Jammu and Kashmir in 1 time unit after some fixed time.

We are now ready to introduce the main basic reasoning tool used in our analysis.

Definition 1 (Temporal probabilistic rule). If $p(t) : [\tau, \rho]$ is a tp-annotated atom and $A_1, A_2, \ldots, A_n$ are atoms (or comparison atoms), then

$$p(t) : [\tau, \rho] \leftarrow A_1 \wedge A_2 \wedge \ldots \wedge A_n$$

is a *temporal-probabilistic rule* (tp-rule). The *head* of the rule is $p(t)$, and the *body* of the rule is $A_1 \wedge A_2 \wedge \ldots \wedge A_n$.

Intuitively, such a tp-rule r states that if each atom in *body*(r) is true at a fixed time, then the *head*(r) atom will be true with probability ρ at a time τ units afterward.

Definition 2 (Temporal probabilistic logic program). A *temporal probabilistic logic program* (TPLP) is a finite set of tp-rules.

Example 4 presents a small TPLP consisting of a subset of tp-rules learned about LeT from real data.

Example 4. The following tp-rules, $\{r_1, \ldots, r_9\}$, form a small TPLP that focuses on the attack patterns of LeT. These rules were learned from the full set of CMOT data available for LeT (of which Example 1 displays a small subset).

r_1.`attackCiv(1)` : [1, 1.0] ← `religious(1)` ∧ `leadersDied(X)` ∧ $X \leq 2$.
r_2.`attackHin(1)` : [3, 0.909] ← `terrClaims(0)` ∧ `leadersDied(X)` ∧ $X \leq 2$.
r_3.`attackCiv(1)` : [1, 1.0] ← `religious(1)` ∧ `raided(X)` ∧ $X \leq 12$.
r_4.`attackSym(0)` : [3, 0.909] ← `locIndia(1)` ∧ `leadersDied(X)` ∧ $X \leq 5$.
r_5.`attackSym(0)` : [2, 0.976] ← `locIndia(1)` ∧ `leadersDied(X)` ∧ $X \leq 4$.
r_6.`attackSym(0)` : [3, 0.909] ← `locIndia(1)` ∧ `personnelRel(X)` ∧ $X \leq 9$.
r_7.`attackHol(1)` : [2, 0.917] ← `remInfluenceJK(1)` ∧ `personnelKilled(X)` ∧ $X \leq 8$.
r_8.`attackHol(1)` : [2, 0.909] ← `personnelArrested(X)` ∧ $X \leq 8$.
r_9.`attackHol(1)` : [2, 0.917] ← `advChangeLife(1)` ∧ `personnelKilled(X)` ∧ $X \leq 8$.

Temporal probabilistic rule r_1 states that, at time t, if Lashkar-e-Taiba is operating as a religious group and the number of group leaders who died during this time interval is at most 2, then with 100 % probability LeT will perform deadly attacks against civilians at time $t+1$. The environmental atom `religious(1)` ensures that, when *body*(r_1) is true, LeT is operating as a religious group. Similarly, the environmental atom `leadersDied(X)` and the comparison atom $X \leq 2$ combine to ensure that, when *body*(r_1) is true, at most two leaders of LeT died during this time interval. Finally, *head*(r_1) is an action atom stating that LeT performs deadly attacks against civilians when *body*(r_1) is true.

As another example, rule r_6 states that with 90.9 % probability, LeT will *not* attack symbolic sites at time $t+3$ if at time t LeT has active locations across the border of India and at most 10 LeT personnel were released by the government during the time interval t. Unlike rule r_1, rule r_6 states that LeT will *not* perform an attack (with high probability); this is specified in $head(r_6)$, where the ground term serving as an argument for predicate `attackSym` has value 0 instead of 1.

While the examples above focus primarily on describing the attack patterns of a terrorist group, we emphasize that this temporal probabilistic logic can easily be used in other domains. Regardless of the domain, it is clear that *manually* determining tp-rules and TPLPs from historical data or expert opinions would quickly grow intractable. In the next section, we present a method to learn tp-rules automatically from historical data, as well as a general method for extracting *policy recommendations* (e.g., "reduce funding to LeT" or "sell stock in APPL but buy stock in GOOG") from these learned tp-rules.

3 Automatically Learning Rules from Historical Data

In Sect. 2, we formally introduced temporal probabilistic rules (tp-rules) and temporal probabilistic logic programs (TPLPs). In this section, we present a general method for automatically learning tp-rules from historical data. We then describe an integer programming-based method to derive "good" policy recommendations based on these learned tp-rules.

3.1 Automatic Extraction of TP-Rules

Temporal probabilistic reasoning is important in many domains (see Sect. 1), and tp-rules are one natural way for analysts and reasoning agents to formally write down their expert knowledge. However, *manually* constructing tp-rules from historical data is tedious in the small, infeasible in the large, and subject to human error and bias. For these reasons, it is necessary to remove the human from the tp-rule creation process in favor of automatically learning tp-rules from historical data.

3.1.1 SOMA Rules

Our method for learning tp-rules from historical data is heavily based on one by Subrahmanian and Ernst [20]. This algorithm was originally motivated by the need to mathematically model the behavior of terrorist groups, and operates on the first (to our knowledge) model used toward this end. The algorithm uses *Stochastic Opponent Modeling Agent rules* (SOMA-rules), which provide probabilistic but not temporal reasoning about a group. In fact, SOMA-rules are syntactically very similar (although they do not consider time) to the tp-rules discussed in this chapter, making statements of the form, "When conditions C are true in the environment in which a terror group G operates, there is a probability of between $l\,\%$ and $u\,\%$ that G will take actions A at some intensity level L." We formalize this notion in Definition 3.

Definition 3 (SOMA-rule). If $A_1, A_2, \ldots, A_n$ are environmental or comparison atoms, $p(t)$ is an action atom, and $l, u \in [0, 1]$, then

$$p(t) : [l, u] \leftarrow A_1 \wedge A_2 \wedge \ldots \wedge A_n$$

is a *SOMA-rule*. As with tp-rules, the *head* of the rule is $p(t)$, and the *body* of the rule is $A_1 \wedge A_2 \wedge \ldots \wedge A_n$.

Recently, SOMA-rules have been used to formally present the behaviors of many terrorist groups. In the past 5 years, work by Mannes, Subrahmanian, and others has automatically learned expressed SOMA-rules about Hezbollah [10], Hamas [11], and Lashkar-e-Taiba [12]. These projects accessed historical data about their respective terrorist groups through the CMOT codebook, and have shown confirmed predictive power. For example, the work by Mannes et al. [10] covering Hezbollah made predictions about the group's behavior in early 2009 before the Lebanese elections. Hezbollah then made public comments in the Beirut Daily Star expressing skepticism about the predictions; however, the group proceeded to operate exactly as predicted in early 2009.

Formally, SOMA-rules use a constrained version of the syntax of probabilistic logic programs [16]. However, for the purposes of this section, we can think of SOMA-rules as tp-rules with no temporal offset and a point probability; that is, the tp-annotation $[\tau, \rho]$ will always have $\tau = 0$, and the corresponding SOMA-annotation $[l, u]$ will always have $l = u$. Intuitively, the non-trivial temporal offsets of tp-rules can be thought of as adding a notion of causality to SOMA-rules. This is accomplished by clearly separating the time interval during which the body of a tp-rule takes place (i.e., interval t) and the time interval during which the head of a tp-rule files (i.e., interval $t + \tau$).

We are now ready to present the algorithm by Subrahmanian and Ernst [20], as well as our straightforward augmentation to allow the algorithm to work with temporally-aware TPLPs.

3.1.2 Subrahmanian-Ernst Algorithm: Preliminaries

We now describe a method for automatically extracting SOMA-rules from a database, first proposed by Subrahmanian and Ernst [20]. We call this the Subrahmanian-Ernst (SE) algorithm. Afterward, we describe the small tweak required to adapt the method to extract tp-rules.

Definition 4 (Bi-conjunct). If p is a predicate, $X \in \mathscr{L}_{var}$, and $l, u \in \mathscr{L}_{cons}$, then

$$p(X) \wedge l \leq X \leq u$$

is a *bi-conjunct*.

The SE algorithm generates a specific type of SOMA-rules whose bodies consist of bi-conjuncts. We formally define these *bi-SOMA-rules* in Definition 5.

Definition 5 (Bi-SOMA-rule). If $B_1, B_2, \ldots, B_n$ are bi-conjuncts, $p(t)$ is an action atom, and $l, u \in [0, 1]$, then

$$p(t) : [l, u] \leftarrow B_1 \wedge B_2 \wedge \ldots \wedge B_n$$

is a *bi-SOMA-rule*. As with standard SOMA-rules, the *head* of the rule is $p(t)$. The *bi-body* of the rule is $B_1 \wedge B_2 \wedge \ldots \wedge B_n$. The *dimension* of a bi-body is the number of bi-conjuncts in it.

Clearly, the set of all bi-SOMA-rules is a subset of the set of all SOMA-rules, as the definition is identical to that of the SOMA rule with the added constraint of a specific combination of environmental and comparison atoms in the body of the rule. We now induce equivalence classes on the set of all bi-bodies of bi-SOMA-rules.

Definition 6 (Equivalence of bi-bodies). If r_1 and r_2 are bi-SOMA-rules with bi-bodies b_1 and b_2, then b_1 and b_2 are *equivalent* if and only if:

- The bi-conjuncts in bi-bodies b_1 and b_2 always co-occurred (i.e., the set of time intervals in which b_1 is true is identical to the corresponding set of time intervals for b_2); and
- The *environmental* atoms in both bi-bodies are identical (but not necessarily their respective comparison atoms).

The SE algorithm requires a *tight canonical member* from each equivalence class. Informally, if $B^* = \{B_1, B_2, \ldots, B_n\}$ is an equivalence class such that B_i contains some bi-conjunct $p(X) \wedge l_i \leq X \leq u_i$, then the tight canonical member chosen must contain the bi-conjunct:

$$p(X) \wedge \min_{i=1,\ldots,n} (l_i) \leq X \leq \max_{i=1,\ldots,n} (u_i)$$

The tight canonical member must contain similar "tight" (with respect to the equivalence class B^*) bi-conjuncts for each unique environmental atom in the bi-bodies $B_1, \ldots, B_n$.

Example 5. The bi-bodies B_1, B_2, and B_3 each have two bi-conjuncts (and thus $dimension(B_i) = 2$). Each bi-body references two environmental attributes, the binary-valued `religious` and nonnegative integral-valued `leadersDied`.

B_1. [religious(X_1) $\wedge\, 0 \leq X_1 \leq 1$] $\wedge$ [leadersDied(X_2) $\wedge\, 0 \leq X_2 \leq 2$]
B_2. [religious(X_1) $\wedge\, 1 \leq X_1 \leq 1$] $\wedge$ [leadersDied(X_2) $\wedge\, 0 \leq X_2 \leq 6$]
B_3. [religious(X_1) $\wedge\, 0 \leq X_1 \leq 1$] $\wedge$ [leadersDied(X_2) $\wedge\, 1 \leq X_2 \leq 12$]

Assume the bi-conjuncts in each bi-body always co-occurred. Since each bi-body contains the same environmental atoms (although their respective comparison atoms are different), they are also in the same equivalence class B. Then a tight canonical member of B^* is B_t, as shown below.

B_t. [religious$(X_1) \wedge 0 \leq X_1 \leq 1$] $\wedge$ [leadersDied$(X_2) \wedge 0 \leq X_2 \leq 12$]

To aid in reasoning over tight canonical members of equivalence classes, the SE algorithm also induces an ordering on bi-bodies, formalized in Definition 7 below.

Definition 7 (Simpler than). If B_1 and B_2 are bi-bodies and $p(t)$ is an action atom, then B_1 is *simpler than* B_2 (denoted $B_1 \gg B_2$) if:

- $dimension(B_1) \geq dimension(B_2)$,
- $conf(B_1) \geq conf(B_2)$; and
- $sup(B_1) \geq sup(B_2)$.

The confidence of bi-body B_i, $conf(B_i)$, with respect to the action atom of interest $p(t)$ is defined as follows:

$$conf(B_i) = \frac{\#\text{intervals when } B_i \text{ was true and } p(t) \text{ was true}}{\#\text{intervals when } B_i \text{ was true}}$$

The support, $sup(B_i)$, is just the numerator of the $conf(B_i)$ fraction.

We now define the structure computed as the end goal of the SE algorithm.

Definition 8 (Up-set). If B is a bi-body and d is a positive integer, then the *up-set* of B (denoted $up(B)$) is:

$$up(B) = \{B' \mid B' \text{isatightbi} - \text{body} \wedge dimension(B') \leq d \ \wedge \ B' \gg B\}$$

Intuitively, given some bi-body B and a maximum dimension d, the up-set of B is the set of all bi-bodies of dimension at most d that are also simpler than B. The SE algorithm computes layers of sets of bi-bodies based on these up-sets as follows:

Definition 9 ($Tp \uparrow k$). If d is a positive integer, then $\forall k \in \mathbb{Z}^+$ we define $Tp \uparrow k$ iteratively as follows:

$$Tp \uparrow 1 \quad = \{B \mid B \text{isatightbi} - \text{body} \wedge dimension(B) \leq d \ \wedge \ up(B) = \emptyset\}$$
$$Tp \uparrow (k+1) = \{B \mid B \text{isatightbi} - \text{body} \wedge dimension(B) \leq d \ \wedge \ up(B) \subseteq Tp \uparrow k\}$$

The set $Tp \uparrow 1$ is then the set of all bi-bodies B with d or fewer bi-conjuncts in the body, such that no other bi-body B' with d or fewer bi-conjuncts in the body is strictly simpler than B. The subsequent $Tp \uparrow i$ for $i > 1$ are "looser" versions of each parent set. The computation of these sets is the main purpose of the SE algorithm; however, naïvely computing all such sets (up to some constant integer k) would be intractable. To this end, we define the workhorse of the SE algorithm, the *condition graph* (COG).

Definition 10 (Condition graph). A condition graph (COG) is a graph $G = (V, E)$ such that $\forall v \in V$:

- $v.bibody$ is a label referencing a single, tight bi-body
- $v.level$ is a label that is set to 0 if there is no vertex v' such that $\exists(v', v) \in E$; otherwise, it is defined as $\max_{v' in V, v \neq v} level(v') + 1 \mid (v', v) \in E$.

Let $K \in \mathbb{Z}^+$ represent the maximum desired level of a COG. Then, for each bi-body $B \in TP \uparrow K$, there is exactly one vertex $v \in V$ such that $v.bibody = B$. This completely defines the set V.

The set E is defined as follows:

$$\begin{aligned} E = \{(v, v') \mid & v, v' \in V \wedge \\ & (v.bibody \gg v'.bibody) \wedge \\ & \nexists w \in V \text{s.t.} (v.bibody \gg w.bibody \gg v'.bibody)\} \end{aligned}$$

Building the complete COG is a computationally difficult problem. To alleviate some of the computational complexity, the SE algorithm takes as a parameter a user-defined outcome (in our terror group example, an action atom) of interest, and computes only the portion of the COG relevant to that outcome. This is done by determining if a given bi-body references the outcome and, if it does not, ignoring it. Once this outcome-specific version of the COG is fully constructed, we need only extract the vertices from the COG that fall within the desired (user-specified) confidence and support intervals. We describe this process formally in the next section.

3.1.3 The Subrahmanian-Ernst Algorithm and an Adaptation to TPLPs

In this section, we formally describe the Subrahmanian-Ernst (SE) algorithm. We also adapt the algorithm to the temporal probabilistic logic presented in Sect. 2. This section builds on the formalizations of Sect. 3.1.2.

Algorithm 1 formally presents the Subrahmanian-Ernst algorithm. The algorithm takes as input:

- A database (DB) whose schema mirrors that discussed in Sect. 2.1. In the case of our running LeT example, this is a database whose rows correspond to months and columns correspond to action and environmental attributes.
- A list of environmental attributes (ENV). In the case of the LeT example, this is just the indices of the columns corresponding to environmental attributes.
- A positive integer d, the maximum dimension of a bi-body. For example, if $d = 3$, then all bi-bodies computed by the algorithm will have dimension at most 3.
- A positive integer k, determining the maximum level a vertex in the COG can attain.

Algorithm 1: Subrahmanian-Ernst algorithm

Data: Database DB, environmental attributes ENV, action atom Outcome, maximum dimension $d \in \mathbb{Z}^+$, maximum level $k \in \mathbb{Z}^+$
Result: Set of bi-bodies relevant to Outcome that satisfy pre-defined support and confidence levels

```
begin
    Set COG = (V, E) with V = E = ∅
    foreach combination φ of d or fewer attributes in ENV do
        SatTuples = BuildDataStructure(DB, ENV, φ, Outcome)
        NotSatTuples = BuildDataStructure(DB, ENV, φ, ¬Outcome)
        TightBibodies = GenerateTightBibodies(φ, SatTuples)
        foreach vertex v ∈ TightBibodies do
            numNotSat = CountQuery(v.bibody, NotSatTuples)
            v.confidence = v.support/(v.support + numNotSat)
            COG = InsertCOG(v, COG, k)
        end
    end
    return ExtractBibody(COG)
end
```

Algorithm 1 references five undefined procedures. We describe them here.

BuildDataStructure. Informally, this procedure splits the DB into two subsets of rows: those satisfying an outcome and those not satisfying an outcome. In the algorithm, after calling *BuildDataStructure* with "Outcome" as a parameter, the "SatTuples" variable contains the projection of DB on attributes in the combination ϕ for specific tuples that satisfy the user-defined outcome atom. The "NotSatTuples" variable then contains the projection that do not satisfy the outcome atom, since it is the product of calling *BuildDataStructure* with "¬Outcome".

GenerateTightBibodies. This procedure generates the support of all tight bi-bodies associated with the combination ϕ. A set of vertices corresponding to these tight bi-bodies is returned, such that for each vertex the *confidence*, *support*, and *bibody* fields are set properly.

CountQuery. This procedure counts the total number of tuples that satisfy the bi-body of a specific vertex, but do not satisfy the user-specified outcome atom.

InsertCOG. This procedure is called once per vertex returned by the *GenerateTightBibodies* procedure. The procedure first checks the level of the vertex and, if the level is at most k, inserts the vertex into the COG. The procedure also propagates the *level* value to neighbors of the vertex. If this cascade of updates forces any vertex's level to exceed k, the vertex is removed from the COG.

ExtractBibody. This procedure checks every vertex in the COG and, if the vertex satisfies some user-defined confidence and support criteria (e.g., "only report bi-bodies with support above 10 and confidence above 90 %"), reports the corresponding bi-body. The set of all such bi-bodies is then returned by the algorithm.

Algorithm 2: OffsetDB algorithm

```
Data: Database DB, temporal offset τ ∈ ℤ
Result: Temporally-augmented database DB′
begin
    Set DB′ = DB
    foreach row r_i in DB′ do
        if i < τ then
            Delete r_i from DB′
        else
            foreach environmental attribute E do
                Let r′_{i−τ} be row i − τ in the original database DB
                Replace r_i(E) with r′_{i−τ}(E)
            end
        end
    end
    return DB′
end
```

As presented, Algorithm 1 does not take time offsets into account. In other words, it will always return tp-rules that have a trivial temporal component. In Algorithm 2 (the OffsetDB algorithm), we provide a simple way to augment our database DB such that the SE algorithm returns general tp-rules.

Informally, Algorithm 2 takes as input the raw database of historical data DB, and outputs an augmented database DB′ such that each environmental attribute in DB′ has been "pushed up" τ rows. In this way, the temporal offset is built into the database DB′. The SE algorithm is then called with DB′ as the data source, and proceeds normally.

For a specific time offset τ, by invoking the OffsetDB algorithm followed by the SE algorithm once for every outcome of interest (e.g., for every action attribute corresponding to LeT performing violent attacks), an analyst can derive all possible tp-rules that satisfy specified support, confidence, and dimension levels for an offset of τ. Then, for all time offsets of interest (e.g., between 0 and 5 months), an analyst can derive all possible tp-rules for any time offset.

3.2 Toward Converting TP-Rules into Policy Recommendations

The SE algorithm presented in Sect. 3.1.3 automatically learns expressed tp-rules from historical data. These learned tp-rules can be analyzed manually by area experts and used to determine policies of actions; however, as with the creation of the tp-rules themselves, this is both intractable in the large and subject to both human bias and mental capacity constraints. For instance, in our running example focusing on attacks by Lashkar-e-Taiba, an immediately obvious policy for reducing attacks in one dimension might have unforeseen repercussions at different points of time or with different types of attacks. In this section, we present a method for automatically extracting desirable policies from a database of tp-rules. The method makes a few assumptions that should be relaxed in future work; we discuss these as well.

3.2.1 Computational Policies

Informally, a *policy* is a specific setting of a (subset of the) world that, when present, triggers desirable properties elsewhere in the world. For example, in the context of effecting change in a terrorist group's behavior, a governing body or advisory committee may be interested in understanding what changes it could make to the environment in which a group operates (e.g., cutting down on foreign aid or increasing raids) so that the group behaves differently (e.g., no longer attacks civilians).

Before formally defining a policy in the language of our temporal probabilistic logic, we discuss a fairly strong assumption: that the tp-rules over which we are reasoning can be represented in *propositional logic*. That is, terms in the body of each rule are all ground. The assumption that each body term is ground lets us view the body of each rule, consisting of atoms $A_1 \wedge A_2 \wedge \ldots \wedge A_n$, as a conjunction of *literals*. We can then reason about these literals and their negations in the standard way. For example, `religious`, which has domain $\{0, 1\}$, can be viewed as two complementary literals `religious(1)` and `religious(0)`. In our experience learning real tp-rules from data, this assumption is not too confining (in fact, as we will discuss in Sect. 4, our recent study focusing on preventing attacks by Lashkar-e-Taiba used only rules of this type). Future research will relax this requirement.

For the rest of this section, we will assume the existence of a set of tp-rules *RDB* (called a *rule database*). This set of tp-rules could have been learned automatically using techniques like those presented in Sect. 3.1 or constructed manually. Let *body*(*RDB*) be the set of all literals appearing in the body of any tp-rule in the rule database *RDB*. Furthermore, let ¬*body*(*RDB*) be the set of all literals $\{\neg\ell \mid \ell \in body(RDB)\}$. We now formally define a policy.

Definition 11 (Policy). Given a set of tp-rules *RDB* (called a *rule database*) and a set of action atoms A, a *policy that potentially eliminates* A is a consistent subset of ¬*body*(*RDB*) that satisfies the following:

1. $\forall r \in RDB$ such that $head(r) \in A$, $\exists \ell \in P$ such that $\neg\ell \in body(r)$
2. $\nexists P' \subset P$ such that P' satisfies the preceding condition

Intuitively, given a database of tp-rules *RDB* and a set of action atoms that we would like to prevent, a policy is a way to set environmental variables such that no tp-rules pertaining to the specific set of action atoms fire. Furthermore, it is the "simplest" such set in that no strict subset of the policy would result in none of the desired tp-rules firing. Since, by definition, the policy is a consistent subset of

$\neg body(RDB)$, it cannot contain both literals ℓ and $\neg\ell$; if this were not the case, it would be impossible to implement the policy.

Example 6. The following set of tp-rules, $\{r_1, r_2, r_3\}$, forms a small rule database *RDB* that focuses on the attack patterns of LeT toward civilians.

r_1.`attackCiv(1)` : [1, 0.99] $\leftarrow$ `terrClaims(0)` $\wedge$ `religious(1)`
r_2.`attackCiv(1)` : [3, 0.909] $\leftarrow$ `terrClaims(0)`
r_3.`attackCiv(1)` : [1, 0.916] $\leftarrow$ `remInfluenceJK(1)` $\wedge$ `advChangeLife(1)`

Let $A = \{$`attackCiv(1)`$\}$, representing a desire to prevent LeT from attacking civilians. There are two possible policies that potentially eliminate A:

- $P_1 = \{$`terrClaims(1)`, `remInfluenceJK(0)`$\}$
- $P_2 = \{$`terrClaims(1)`, `advChangeLife(0)`$\}$.

Clearly, any policy must include `terrClaims(1)`, since this is the only way to prevent rule r_2 from firing. This also prevents rule r_1 from firing. Finally, we can choose to negate either of the components in $body(r_3)$. Thus, both P_1 and P_2 prevent all rules $r \in RDB$ pertaining to the set of action atoms A from firing; furthermore, no strict subset of either P_1 or P_2 satisfies this statement, and both P_1 and P_2 are consistent, so both P_1 and P_2 are policies that potentially eliminate A.

3.2.2 Iteratively Computing All Policies

We now describe the computational method used to automatically generate policies from a set of tp-rules. The algorithm we will describe builds upon integer linear programming techniques for computing the set of all minimal models of logic programs, originally discussed in Bell et al. [1]. We now explain its straightforward adaptation to the case of temporal probabilistic logic.

First, we define a set of linear constraints (*LC*) that enforce the formal rules of a policy, as defined above. Assume we have a tp-rule database RDB' and a set of action atoms A; for convenience, denote $RDB = \{r \in RDB' \mid head(r) \in A\}$. For each literal $\ell \in body(RDB)$, let X_ℓ be a binary variable representing whether or not literal ℓ is included in a policy. Similarly, define binary variable X_a for each $a \in A$. Then we define the set of linear constraints *LC* as follows:

1. For each rule $a \leftarrow \ell_1 \wedge \ell_2 \wedge \ldots \wedge \ell_n$, add a constraint

$$X_a + \sum_{i=1}^{n}(1 - X_{\ell_i}) \geq 1$$

   Intuitively, this constraint forces either the head of the rule (represented by X_a) to be true, or at least one of the literals in the body to be false.
2. For each pair of complementary literals ℓ and $\neg\ell$, add a constraint

---

Algorithm 3: Policy computation algorithm

Data: Database of tp-rules *RDB*, set of action atoms *A*
Result: Set of policies *P*
begin
    $P = \emptyset$
    $(RDB', LC) = MakeConstraints(RDB, A)$
    while *true* **do**
        $S = CalculateHS(RDB', LC)$
        if *S exists* **then**
            $P = P \cup \{\neg\ell \mid \ell \in S\}$
            $LC = LC \cup \{\sum_{\ell \in S} X_\ell \leq card(S)\}$
        else
            return *P*;
        end
    end
end

$$X_\ell + X_{\neg\ell} \leq 1$$

This ensures consistency; that is, at most one of the complementary literals is included in a policy.

3. For each rule r and $a \in A$, if $a \in head(r)$, add the constraint

$$X_a = 0$$

This ensures that no rule (of interest) fires.

4. Ensure that each X_a and X_ℓ variable is binary by adding a constraint

$$X_{\{a,\ell\}} \in \{0, 1\}$$

The savvy reader will notice that we can combine the constraints in item 1 with those in item 3, removing the need for the X_a variables ranging over the action atoms in the heads of tp-rules entirely. We choose to present *LC* in a more general way. In the event that *LC* is over-constrained (that is, there is no policy *P* such that *no* tp-rule in the rule database fires), a policy analyst could relax the constraint in item 3 and then try to minimize the number of tp-rules that fire (instead of requiring that none fire at all).

Second, using this initial set of linear constraints *LC*, we iteratively solve a series of integer programs (minimizing the number of activated X_ℓ variables), adding constraints to *LC* until the program becomes infeasible. The solution to each intermediary integer program represents a legal policy that potentially eliminates *A*, given some set of action atoms *A*. Algorithm 3 formalizes this process.

Algorithm 3 makes use of two previously undefined functions:

MakeConstraints. Given a rule database and a set of action atoms, this returns the initial set of linear constraints *LC* as defined earlier in the section, as well as a filtered rule database RDB' containing only tp-rules pertinent to the set of action atoms.

CalculateHS. This function calculates a minimum hitting set for the bodies of the pertinent tp-rules in the filtered rule database RDB', subject to the constraint defined by LC. The minimum hitting set can be calculated using the linear integer program:

$$\min \sum_{\ell \in body(RDB')} X_\ell$$
$$\text{s.t. } LC$$

Intuitively, Algorithm 3 iteratively produces minimum hitting sets consisting of a literals in rule bodies such that, were those literals to be negated, no tp-rules (in the filtered rule database RDB') would fire. After each successful solve of the integer program, a new constraint is added to LC preventing any strict superset of the most recently determined policy from being found in the future. In this way, we ensure that only legal policies are found. Finally, once all policies are found, the integer program becomes infeasible and the algorithm returns the set of all policies that potentially eliminate the user-specified set of action atoms.

In the next section, we provide an extensive application of Algorithms 1–3 to a large, real-world database representing the actions and operating environment of Lashkar-e-Taiba, an active terror group in southern Asia.

4 Policy Recommendations and Lashkar-e-Taiba

In this section, we apply the techniques discussed in Sect. 3 to study environments that provoke attacks by Lashkar-e-Taiba (LeT), a terror group in South Asia. Over the last two decades, LeT has been responsible for many terrorist attacks in India, Kashmir, Pakistan, and Afghanistan. In 2006, LeT operative Faheem Lodhi was arrested and convicted of planning sophisticated attacks on Australia's power grid [4], demonstrating the potential global threat of this organization.

We learn a set of tp-rules from real-world data collected by the *Computational Modeling of Terrorism* (CMOT) codebook [17], a research project that tracks past and current activities (recording data at a granularity level of months) of multiple terrorist groups including LeT. We then determine a set of policies that could help prevent further attacks by LeT. A far more in-depth discussion of these results in can be found in [21].

4.1 Experimental Methodology and Learned Rules

The CMOT codebook tracks hundreds of environmental and action variables for Lashkar-e-Taiba, recording intensity levels on a month-by-month basis. A few examples of environmental variables include those relating to:

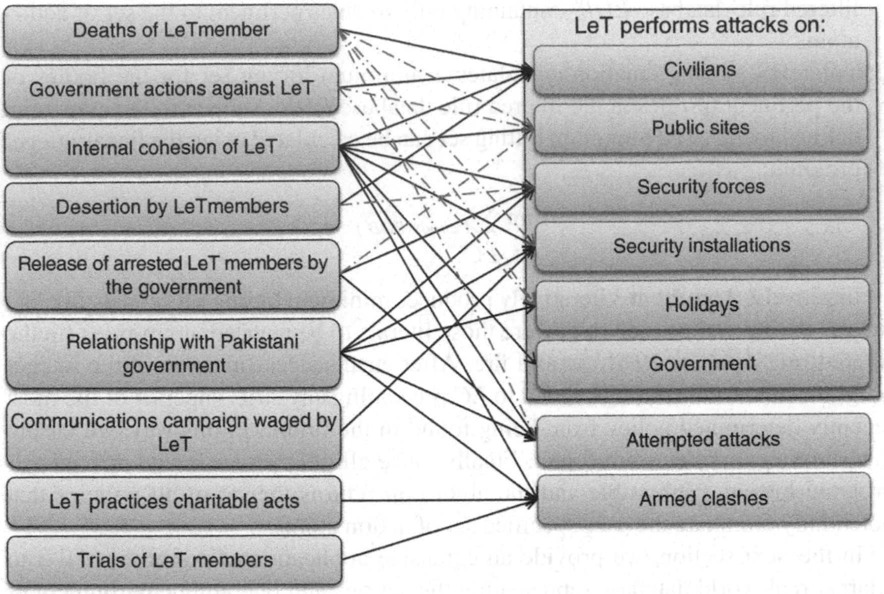

Fig. 1 A graphical summary of high support, high probability tp-rules learned about Lashkar-e-Taiba. *Solid black lines* from an environmental attribute to an action attribute represent a positive correlation, while *dashed blue lines* represent negative correlation

- The internal politics and activities of LeT (e.g., "What level of intra-organizational conflict exists in LeT?");
- The level of local and international monetary, military, and political support for LeT (e.g., "At what level is Pakistan's military supporting LeT?"); and
- Information about the group's operating facilities and staffing.

Examples of action variables tracked by the CMOT codebook include those relating to:

- Armed and suicide attacks against military forces, security forces, or civilians;
- Hijackings and abductions/kidnappings; and
- Attacks on military targets, government facilities, tourist sites, or symbolic sites.

We learned tp-rules using all of the action and environmental variables tracked by the CMOT codebook. These rules were learned automatically using Algorithms 1 and 2. We then filtered these rules to include only those with high *support* in the data and *probability* of occurring. Figure 1 shows a summary of the learned rules.

For example, Fig. 1 states that increases in the environmental variable tracking the deaths of LeT members is positively correlated with increased attacks on civilians, while increases in the same environmental attribute is negatively correlated with increased attacks on public sites. A much more in-depth discussion of the data, experimental methodology, and set of learned tp-rules can be found in upcoming work by Subrahmanian et al. [21]. These results clearly show the expressive power of tp-rules and the promise of the methods for automatically learning them from real-world data.

4.2 Policies That Potentially Eliminate or Reduce Violent Attacks by Lashkar-e-Taiba

Using the filtered tp-rules described above and the policy computation methodology described in Algorithm 3, we computed a set of policies that potentially eliminate or reduce violent attacks by Lashkar-e-Taiba. Critically, we are *not* claiming that instituting these policies in reality will stop all attacks by LeT; rather, they may be effective at changing LeT's behavior. These policies are based on tp-rules forming a behavioral model of LeT based on past behavior; in reality, terrorist groups frequently change their behavior in response to counter-terror strategies [8]. In this light, it is imperative that policies adapt to the changing actions and strategies of groups like LeT.

The set of tp-rules produced eight policies. The policies were overall quite similar to one another, varying individually in a subtle ways. We very briefly describe them now. Overall, each of the policies suggest:

- Targeting LeT's internal cohesion;
- Targeting the Pakistani military's support of LeT;
- Targeting LeT's training facilities;
- Targeting any communication campaigns launched or run by LeT;
- Pushing for the resignation of senior LeT leaders;
- Keeping LeT prisoners (i.e., preventing the release of LeT prisoners by the governments that hold them);
- Reconsidering targeted efforts and long-term campaigns to kill or arrest LeT[3]; and
- Not explicitly encouraging low-level personnel to defect from LeT.

Individual variability amongst the policies was low. Individually, the policies suggested taking one or some of the following actions (in addition to those listed above):

- Targeting social and medical services run on the local level by LeT;
- reducing media coverage and publicity of trials of LeT members (especially in Australia);
- Maintaining or pushing for a government ban by Pakistan on LeT; and

[3]This is an interesting point. We emphasize that this is not discouraging governments or groups from working to arrest active LeT members. Rather, explicit campaigns to arrest members can lead to mixed and sometimes dangerous responses.

- Disrupting or targeting relationships between LeT and other Islamic organizations.

Clearly, no one policy offers a simple and deterministic route to preventing violent attacks by LeT. Furthermore, these policies would need to adjust to the constantly adapting strategies and actions of the active terror group. As statisticians George Box and Norman Draper wrote, "essentially, all models are wrong, but some are useful [3]." It is our hope that the policies presented here will be useful.

5 Conclusions and Directions for Future Research

Many applications require logical reasoning about situations that involve temporal uncertainty, including predicting movements in the stock market, assessing the potential future damage of environmental disasters, and reasoning about the behavior of terror groups. In this chapter, we overviewed temporal probabilistic logic programs (TPLPs), through which logic programmers can formally express rules that have both temporal and probabilistic aspects. We provided a general method to derive TP rules from databases of categorical and numerical variables based on work by Subrahmanian and Ernst [20]. We also presented a general method to provide "good" policy recommendations based on these automatically learned rules. Finally, we presented recent work that led to a successful, large-scale application of these techniques to model Lashkar-e-Taiba, an active militant terrorist group.

The framework we described in this chapter automatically finds expressed causal rules within historical data and presents the end user with a set of suggestions (e.g., policies in the case of terror groups) based on the rules found in the data. This framework could easily be adapted to handle different types of temporal reasoning systems. For instance, a recent extension to temporal probabilistic logic called *annotated probabilistic temporal* (APT) logic increases the expressiveness of tp-rules [18, 19]. Like TP logic, APT logic does not make independence assumptions; however, it provides bounds on probabilities as opposed to using only point probabilities. This generality could provide, for example, a more expressive system for policy recommendations. To our knowledge, systems based on APT logic have not yet been implemented in the large.

The integer programming-based method for finding desirable policies given a set of tp-rules can, as we found while doing experiments on the real-world LeT data, become overconstrained. This is due in part to the fact that *real-world groups are not rational*, leading to seemingly contradictory actions which leads to an infeasible hitting set problem. Expert knowledge could be used to cut out contradictory rules from a TPLP; however, manual interaction with large sets of tp-rules can be difficult, and this would be prone to human error and bias. Instead, a policy analyst could relax the objective function from preventing *all* rules from firing to discovering the *largest* subset of rules that could be prevented from firing. One technique to do this, again using integer programming, is suggested by Bell et al. [1, 2]. This problem is equivalent to the maximum Boolean satisfiability problem (MAX-SAT), is NP-complete, and could still be solved using an industry-standard integer programming solver. We suspect a method like this will likely be necessary when dealing with large sets of tp-rules learned about imperfectly rational groups.

Acknowledgements Some of the authors were funded in part by AFOSR grant FA95500610405, ARO grant W911NF0910206 and ONR grant N000140910685.

References

1. Bell C, Nerode A, Ng R, Subrahmanian V (1994) Mixed integer programming methods for computing nonmonotonic deductive databases. J ACM 41(6):1178–1215
2. Bell C, Nerode A, Ng R, Subrahmanian V (1996) Implementing deductive databases by mixed integer programming. ACM Trans Database Syst 21(2):238–269
3. Box G, Draper N (1987) Empirical model-building and response surfaces. Wiley, New York
4. Brenner J, Frazzetto M (2011) America the vulnerable
5. De Choudhury M, Sundaram H, John A, Seligmann DD (2008) Can blog communication dynamics be correlated with stock market activity? In: Proceedings of the 19th ACM conference on hypertext and hypermedia (HC-08). ACM, New York, pp 55–60
6. Dekhtyar A, Dekhtyar MI, Subrahmanian VS (1999) Temporal probabilistic logic programs. In: ICLP 1999. MIT, Cambridge, MA, pp 109–123
7. Fujiwara I, Hirose Y, Shintani M (2008) Can news be a major source of fluctuation: a Bayesian DGSE approach, vol. Discussion Paper Nr. 2008-E-16. Institute for Monetary and Economic Studies, Bank of Japan
8. Ganor B (2005) The counter-terrorism puzzle: a guide for decision makers. Transaction Publishers, New Brunswick
9. Kifer M, Subrahmanian V (1992) Theory of generalized annotated logic programming and its applications. J Log Program 12:335–367
10. Mannes A, Michaell M, Pate A, Sliva A, Subrahmanian V, Wilkenfeld J (2008) Stochastic opponent modelling agents: a case study with hezbollah. In: Proceedings of the 2008 first international workshop on social computing, behavioral modeling and prediction. Springer, Berlin/New York
11. Mannes A, Sliva A, Subrahmanian V, Wilkenfeld J (2008) Stochastic opponent modeling agents: a case study with hamas. In: Proceedings of the 2008 international conference on computational cultural dynamics. AAAI, Menlo Park, pp 49–54
12. Mannes A, Shakarian J, Sliva A, Subrahmanian V (2011) A computationally-enabled analysis of Lashkar-e-Taiba attacks in Jammu & Kashmir. In: Proceedings of European intelligence and security informatics conference (EISIC-2011)
13. Martinez V, Simari G, Sliva A, Subrahmanian V (2008) CONVEX: similarity-based algorithms for forecasting group behavior. IEEE Intell Syst 23(4):51–57
14. Martinez V, Simari G, Sliva A, Subrahmanian VS (2008) The SOMA terror organization portal (STOP): social network and analytic tools for the real-time analysis of terror groups. In: Liu H, Salerno J (eds) Proceedings of the first international workshop on social computing, behavioral modeling and prediction. Springer, New York/Berlin
15. Martinez V, Simari G, Sliva A, Subrahmanian V (2009) CAPE: automatically predicting changes in terror group behavior. In: Memon N (ed) Mathematical methods in counterterrorism. Springer, Wien
16. Ng RT, Subrahmanian VS (1992) Probabilistic logic programming. Inf Comput 101(2): 150–201

17. Shakarian J, The CMOT codebook (2012). Available from the laboratory for computational cultural dynamics (LCCD), University of Maryland Institute for Advanced Computer Studies, University of Maryland, College Park
18. Shakarian P, Parker A, Simari GI, Subramanian V (2011) Annotated probabilistic temporal logic. ACM Trans Comput Log 12:14:1–14:44
19. Shakarian P, Simari GI, Subramanian V (2012) Annotated probabilistic temporal logic: approximate fixpoint implementation. ACM Trans Comput Log 13
20. Subrahmanian V, Ernst J (2009) Method and system for optimal data diagnosis
21. Subrahmanian V, Mannes A, Sliva A, Shakarian J, Dickerson JP (2012) computational analysis of terrorist groups: Lashkar-e-Taiba. Springer, New York
22. Wennberg J, Cooper M, Fisher E, Goodman D, Skinner J, Bronner K (1996) The Dartmouth atlas of health care. Dartmouth Institute for Health Policy, Hanover
23. Wilkenfeld J, Asal V, Johnson C, Pate A, Michael M (2007) The use of violence by ethnopolitical organizations in the middle east. Technical report, National Consortium for the Study of Terrorism and Responses to Terrorism

Part III
Terrorist Network Analysis

Leaderless Covert Networks: A Quantitative Approach

Bart Husslage, Roy Lindelauf, and Herbert Hamers

1 Introduction

It is hypothesized that many of the current covert organizations organize according to leaderless principles, see [15]. According to [4] international law enforcement pressure is forcing criminal and terrorist organizations to decentralize their organizational structures, e.g., Mexican law enforcement efforts are causing drug cartels in Mexico to break into smaller units. It is also known that terror organizations exist that are a mix of hierarchical and decentralized structures, i.e., think of Hezbollah and Peru former's Shining Path (cf. [4]). Clearly this is done to frustrate intelligence agencies that try to disrupt such organizations by taking out key leaders. If the networks are flat rather than hierarchical it becomes very difficult to determine who the leaders are based on network principles alone. The study of covert networks has received high levels of attention from the modeling community in the last decade. Among others, covert networks have been formally characterized by Tsvetovat and Carley [18], McAllister [13] and McCormick and Owen [14], and their optimal network structures have been analyzed and approximated by Lindelauf et al. [9]

B. Husslage (✉)
Department of Mathematics, Fontys University of Applied Sciences,
P.O. Box 90900, 5000 GA Tilburg, The Netherlands
e-mail: b.husslage@fontys.nl

R. Lindelauf
Military Operational Art and Science, Netherlands Defense Academy,
P.O. Box 90002, 4800 PA Breda, The Netherlands
e-mail: rha.lindelauf.01@nlda.nl

H. Hamers
Center and Department of Econometrics and Operations Research,
Tilburg University, P.O. Box 90153, 5000, LE Tilburg, The Netherlands
e-mail: h.j.m.hamers@tilburguniversity.edu

V.S. Subrahmanian (ed.), *Handbook of Computational Approaches to Counterterrorism*, DOI 10.1007/978-1-4614-5311-6_13,

and Enders and Su [5]. Other approaches concern covert network destabilization strategies, see [6] and [3], and tools to identify the most important members of the corresponding organizations, see [8, 12] and [17].

It is interesting to note that many of the models that focus on covert network structures deal with global network properties, such as their optimality with regard to the secrecy versus information trade-off. The methodology that is being developed to aid in the attack of covert networks, however, focuses on the identification of key players, i.e., is mostly concerned with methods of centrality and thus of a local nature. In this chapter both this global and local approach to covert network modeling will be combined. Recent research points to certain network structures that are more 'optimal' in the sense of balancing the need for secrecy and the ability to exchange information, see [9]. The question becomes whether such networks still contain 'important' individuals. Clearly, the intuitive answer is that such covert networks do not contain any individual that is much more important than any other individual in the network. The current evolution of global terrorist networks reinforces this statement, i.e., it is well known that many of such networks are flat and do not contain any leaders, see [15]. The entities making up terrorist networks can be large organizations that work together without any common hierarchy or central commanding authority between them. Whatever the components of the network, what makes it a network is the absence of this central authority or control, see [7]. The question of importance of an individual has been dealt with extensively in the social network domain (cf. [19]) and has also been applied to the covert network domain, see [11]. In this chapter we investigate the relationship between the variance in game theoretic centrality values of the individuals in the network and the optimality of the respective network. This is interesting because a low variance implies that it becomes hard to identify key players and, henceforth, options for law enforcement to engage the network using kingpin strategies might not be useful. On the other hand, it provides a new measure for the optimality of a covert network.

Essentially we combine two approaches to the analysis of covert networks, i.e., we combine the models that describe the overall covert network topology with the analysis of individual members of those networks. We do this by analyzing centrality in approximate optimal covert network structures. That is, we analyze the centrality values of individuals in both homogeneous and heterogenous approximate optimal covert networks. The outline of this chapter is as follows. In Sect. 2 we introduce the rationale behind covert network models and the various centrality measures that are used to investigate the importance of the individuals in the network. We focus on game theoretic centrality measures, since such measures are able to incorporate additional information about the network and about the individuals involved in the network. We analyze approximate optimal covert networks in Sect. 3 and the corresponding variance of the game theoretic centrality values of the individuals in the network. In Sect. 4 we repeat this analysis but now for heterogeneous networks and in Sect. 5 we apply our method to Jemaah Islamiyah's operational network responsible for the 2002 Bali bombing. We end with a conclusion in Sect. 6.

2 Covert Network Models and Centrality

In this section we describe the ideas behind a model of covert networks and introduce game theoretic centrality measures. Covert organizations have to make a trade-off between efficient coordination and control on the one hand and maintaining secrecy on the other (cf. [2]). This is intuitively clear: if everybody in the covert organization knows everybody else, then the security risk to the organization is very high because the exposure of a single individual potentially exposes the entire organization. On the other hand, a very sparsely connected organizational network topology is difficult to coordinate and control, simply because efficient communication between individuals in such an organization is hard. This is quantified by use of an information measure, a secrecy measure, and a balanced trade-off measure, see [9]. The information measure reflects the fact that the ability to transfer information between individuals in a network is inversely proportional to the number of edges in the shortest path between those individuals. On the other hand, the secrecy measure reflects the fraction of individuals in the network that is expected to remain unexposed upon capture of a single individual according to a realistically chosen probability distribution. Finally, the total performance of a covert organization in dealing with the information versus secrecy trade-off dilemma is reflected by a multi-objective optimization based function. The higher the value of this total performance measure a network attains, the better the network does in balancing secrecy and information. A detailed description of these measures can be found in Sect. 7.

Two settings can be considered in constructing the information and secrecy measure. The first setting, the homogeneous case, only considers the communication structure and does not take the nature of interaction into account. The second setting, the heterogeneous case, explicitly takes the nature of interaction into account, since, for example, the frequency or duration of interaction may differ between individuals in a network. This difference is modeled by assigning 'weights' to the links between individuals, representing the risk of that interaction. In Sect. 7 details of these quantitative measures can be found.

Game theoretic centrality measures can be used to determine the key player in a terrorist network, see [11]. Using such centrality measures we let the value for each possible coalition be defined by the network structure of the coalition. The game theoretic centrality values are obtained by computing the Shapley values (see [16]) for the connectivity game v^{conn} (see Sect. 7) of the network. This leads to a ranking of the members of the covert organization, with the key players ranked on top. The connectivity game considers coalitions of individuals in the network and their respective lines of communication. If the members of the coalition are able to communicate using only the links present within the coalition we say that the corresponding subnetwork is connected and assign a value of 1 to the coalition. If not all members in the coalition are able to communicate then we say that the subnetwork is not connected and we therefore assign a value of 0 to this coalition. A coalition consisting of a single individual obtains a value of 0 by definition. The strength of game theoretic centrality measures lies in the fact to incorporate additional information that is available for the coalitions into the modeling. To do so, we introduce a class of weighted connectivity games v^{wconn}. The weight of a coalition now not only depends on the internal communication structure (as presented by the subnetwork formed by the individuals in the coalition) but also on the additional data that is available on the individuals and their relationships present in the respective coalition.

3 Homogeneous Networks

In a homogeneous network only the structure of the network is taken into account, i.e., who communicates with whom. Each network of n persons has a balanced trade-off performance measure μ associated with it, see Sect. 7. In [9] optimal homogeneous networks were derived with respect to this μ-measure for networks up to seven persons and approximate optimal networks for up to ten persons. These networks are depicted in Fig. 1 and the corresponding μ-values can be found in Table 1.

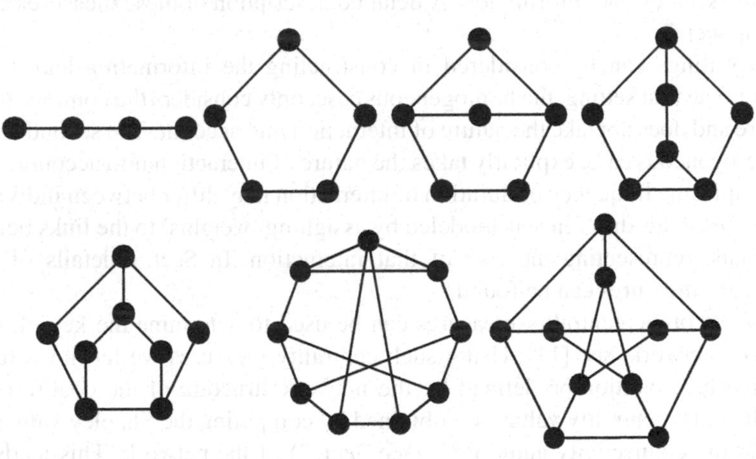

Fig. 1 The (approximate) optimal homogeneous networks for $n = 4, \ldots, 10$ persons

Table 1 Value of μ versus σ^2 for (approximate) optimal homogeneous networks

n	4	5	6	7	8	9	10
μ	0.2100	0.2667	0.2826	0.3055	0.3229	0.3355	0.3600
σ^2	0.0625	0	0.0069	0.0049	0.0069	0.0009	0

In [9] it was shown that covert organizations favor network structures that attain high values of μ. Obtaining game theoretic centrality values for the members of such a network structure, we can construct rankings of the persons involved. The more the centrality values of individuals differ the easier it becomes to differentiate between persons in the ranking. A natural way to quantify differences in individual values by a single number is by computing the variance (denoted by σ^2) of the centrality values. A larger variance then corresponds to a more differentiated ranking. Qualitatively it can be seen that recent covert networks adopt leaderless structures which can be translated to networks that have low variance (i.e., low σ^2) in the centrality values of the respective members of the network. Here we investigate whether covert networks that have high trade-off performance values (μ) attain low values for the variance (σ^2) in centrality of its members. We expect covert organizations to adopt a network structure with a large value for μ and a small value for σ^2.

A first approach in this chapter is to use simulation techniques to investigate the trade-off between μ and σ^2 for networks up to ten persons. First we randomly select $10,000$ connected networks with n nodes ($n = 4, \dots, 10$). Then for each network we compute the trade-off performance measure μ and the variance σ^2 of the centrality values. Figure 2 shows the resulting $10,000$ pairs (μ, σ^2) for $n = 4, \dots, 10$. In each chart, the square (□) depicts the pair (μ, σ^2) corresponding to the (approximate) optimal network, as given in Fig. 1. The corresponding values of μ and σ^2 for the (approximate) optimal networks are presented in Table 1. Note that $\sigma^2 = 0$ for a regular network, i.e., a network in which each node has the same number of links. This holds, for example, for the regular networks with five and ten nodes, see also Fig. 1. A variance of 0, however, does not guarantee an optimal value of μ, see, for example, the optimal networks with four and six nodes.

Looking at Table 1 it can be seen that networks that attain high values for μ indeed have low variance in the centrality of its members. Thus this first simulation approach confirms the conjecture on high μ and low σ^2.

In Fig. 3 we show some networks with nine persons corresponding to different (μ, σ^2) pairs. In network Fig. 3a all nodes have approximately the same number of links. Hence, the value of σ^2 is small and it is hard to differentiate between persons in the ranking. For a covert network this is a desirable property. However, the large number of links results in a small value of μ, i.e., upon capture of a single individual most of the remaining organization is exposed. Hence it can be concluded that only a low value for σ^2 in the centrality of its members is not necessarily advantageous for a covert organization. If we reduce many of the links we arrive at network Fig. 3b resulting in a significant improvement of the value of μ. However, now the variance σ^2 in the centrality of the individuals is high, making it easier to identify key members of the organization. A slightly more uniform distribution of the links in network Fig. 3c takes care of this problem. The number and the placement of links in the approximate optimal network Fig. 3d not only maximizes the value of μ, but at the same time reduces the variance even further.

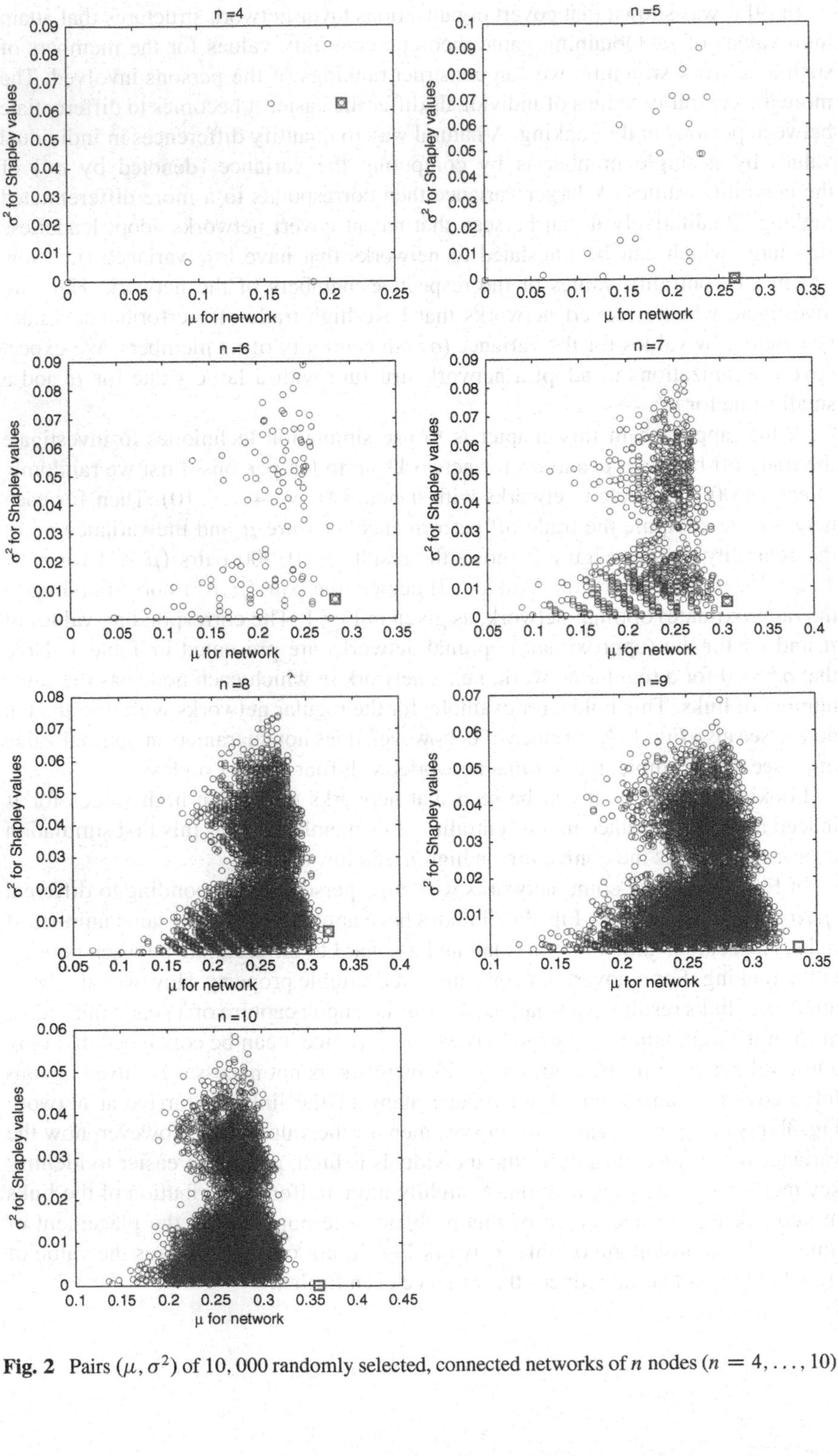

Fig. 2 Pairs (μ, σ^2) of 10,000 randomly selected, connected networks of n nodes ($n = 4, \ldots, 10$)

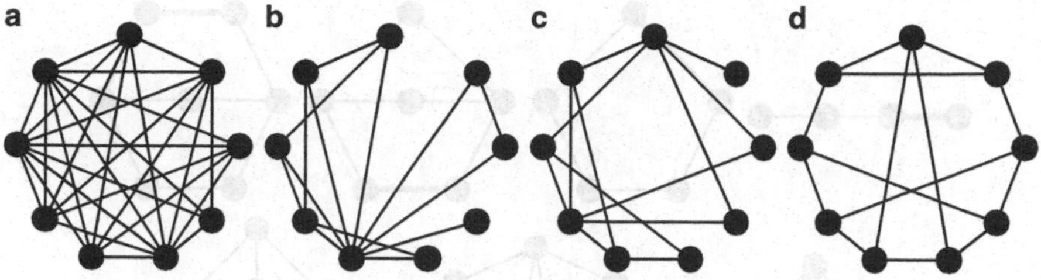

Fig. 3 Networks for the (μ, σ^2) pairs $(0.1222, 0.0021)$ (**a**), $(0.2759, 0.0683)$ (**b**), $(0.2929, 0.0346)$ (**c**) and $(0.3355, 0.0009)$ (**d**)

4 Heterogeneous Networks

In the previous section we analyzed covert organizations by looking at their network structure but ignored the activities that such individuals undertake. In this section we repeat the analysis as presented in the previous section but now we introduce the fact that members of a covert organization interact with one another in different ways. For example, some members may communicate more frequently than other members or may use different (insecure) communication channels. The nature of these interactions may result in a higher risk of exposing (part of) the organization. A heterogeneous network assigns weights to links in the network in order to reflect these risks. Lindelauf et al. [10] consider networks with a single high risk interaction pair, i.e., heterogeneous networks where exactly one link has a higher weight than the other links. They show that in this case the high risk interaction pair should have the least connection to the remainder of the network in order to obtain the largest value of the trade-off performance measure μ. Furthermore, they obtain approximate optimal networks for up to ten nodes using a greedy algorithm and starting from an initial network structure. In these networks the high risk interaction pair is assigned a weight of 2, whereas all other links are assigned a weight of 1. We improve upon these networks by considering the (approximate) optimal homogeneous networks of Fig. 1 and selecting the high risk interaction (with a weight of $w = 2$) in each network as stated above. Furthermore, using the simulation techniques of Sect. 3 no better networks were found, except for the case where the number of nodes is 7 or 8. Figure 4 depicts the resulting networks. In these networks the link with weight 2 is highlighted. In line with the homogeneous case, different (μ, σ^2) pairs will lead to different network structures.

As in the homogeneous case we investigate the trade-off between the values of μ and σ^2 for networks up to ten persons. The game theoretic centrality values are obtained by computing the Shapley values for the weighted connectivity game v^{wconn}. This game is a generalization of the connectivity game. If the members of a coalition are able to communicate using only the links present within the coalition we now assign a value equal to the maximal weight of the links present to the coalition. In our case this implies that either a value of 0, 1 or 2 is assigned to

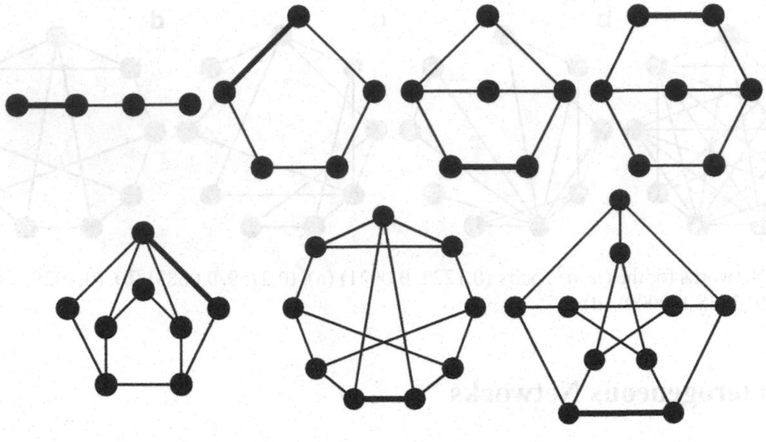

Fig. 4 The (approximate) optimal heterogeneous networks (with $w = 2$) for $n = 4, \ldots, 10$ persons

Table 2 Value of μ versus σ^2 for (approximate) optimal heterogeneous networks (with $w = 2$)

n	4	5	6	7	8	9	10
μ	0.2125	0.2667	0.2866	0.3067	0.3233	0.3360	0.3600
σ^2	0.1458	0.0483	0.0391	0.0298	0.0418	0.0329	0.0288

each coalition. The variance σ^2 is computed for the centrality values resulting from the Shapley value of the weighted connectivity game. Graphical representations of the pairs (μ, σ^2), resulting from 10, 000 randomly selected, connected networks for $n = 4, \ldots, 10$, closely resemble the ones depicted in Fig. 2, and, henceforth, are not repeated here. The values of μ and σ^2 for the (approximate) optimal networks are presented in Table 2.

To investigate the effect of links crucial to the covert network, we consider the above stated heterogeneous case with the difference that the high risk interaction pair is assigned a weight of 10, instead of only 2. Simulation techniques yield the (approximate) optimal networks depicted in Fig. 5 and the results in Table 3. Note that the increase of the weight leads to different heterogeneous networks only in the cases of 7, 8 or 9 nodes. As is to be expected, the variance (σ^2) increases significantly, enabling identification of the important persons in the network. It thus may be concluded that optimal homogeneous network structures are not robust with respect to the addition of high risk interaction pairs. For a covert organization this implies that it should not adopt a static operational structure, but should instead dynamically let its operational structure reflect the current status of its high risk interaction pairs. On the other hand, intelligence agencies should use additional information available on (suspected) members of covert organizations and the relationships present between these members to identify key members of such organizations.

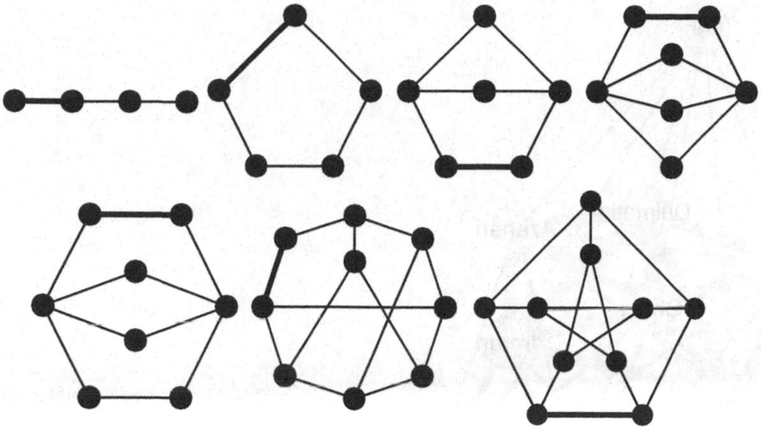

Fig. 5 The (approximate) optimal heterogeneous networks (with $w = 10$) for $n = 4, \ldots, 10$ persons

Table 3 Value of μ versus σ^2 for (approximate) optimal heterogeneous networks (with $w = 10$)

n	4	5	6	7	8	9	10
μ	0.2196	0.2667	0.3032	0.3214	0.3329	0.3427	0.3600
σ^2	3.8125	3.9150	3.2702	2.9110	2.1520	2.3213	2.3311

5 Case: Jemaah Islamiyah's Bali Bombing

In the year 2002 the extremist group Jemaah Islamiyah perpetrated one of the deadliest attacks in Indonesia's history. This attack took place on the island of Bali and resulted in the death of 202 people. The operational network conducting the attack consisted of 17 individuals, divided into three teams (cf. [8]). Figure 6 depicts the heterogeneous operational network and the three teams: a team of bomb builders (gray), a support team (lightgray) and a team responsible for coordinating the attack (white). Koschade [8] uses recordings of interaction between members of the network prior to the attack, and in particular the *transactional content* and *frequency and duration* of these interactions, to assign weights between 0 and 5 to each link. In Fig. 6 these weights are visualized by the thickness of the lines connecting the individuals in the operational network; i.e., the thicker the line the higher the weight assigned to the corresponding interaction.

Using the weighted connectivity game of Sect. 4 (but now assigning each coalition a value from 0 up to 5), we find that $\mu = 0.2740$ and $\sigma^2 = 0.8661$. A closer inspection of Fig. 6 reveals several high risk interaction pairs, i.e., links with a significant higher weight than the other links. Assume that all lines of communication present in the network were necessary to successfully plan and conduct the attack, i.e., we assume that the network structure is fixed. Using simulation techniques we try to find a distribution of the weights over the links

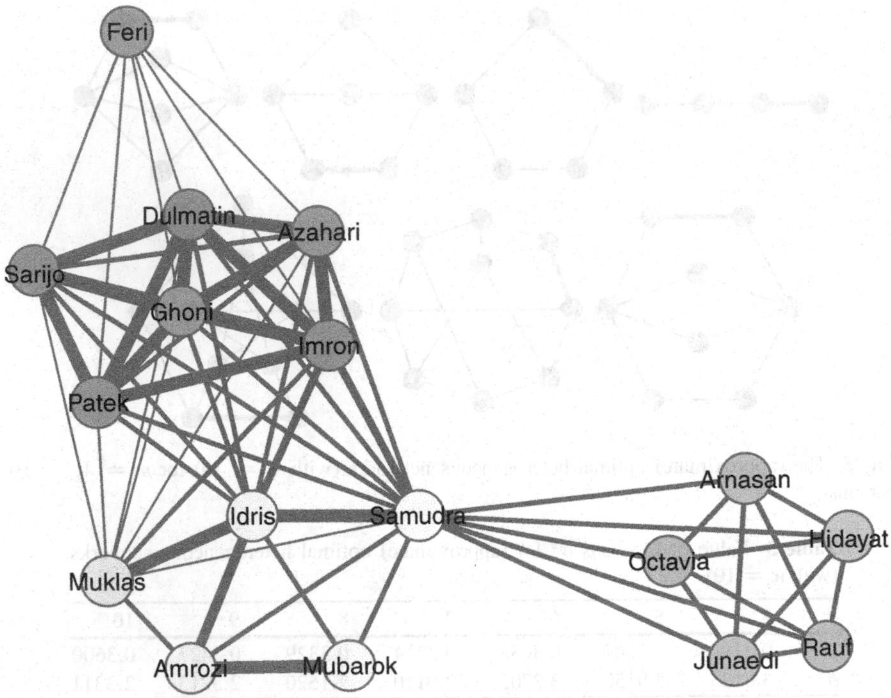

Fig. 6 Operational network of Jemaah Islamiyah's Bali attack. Coordination team (*white*), support team (*lightgray*) and bomb building team (*gray*)

such that the value of μ is maximal. Constructing operational networks with a random distribution of the weights we find a network with $\mu = 0.2996$ and $\sigma^2 = 2.0120$. Hence, the value of μ increases slightly, whereas the variances increases significantly with respect to the real-world case. On the other hand, searching for a distribution of the weights that minimizes the variance does not result in an improvement (i.e., no network with $\sigma^2 < 0.8661$ is found). From these observations it can be concluded that, given the network structure, the placement of high risk interaction pairs was such that it not only facilitated both the secrecy and efficiency of the operational network but also ensured that the variance among the centrality of the members in the network was small, i.e., it maximized success at avoiding identification of the key members.

6 Conclusion

In this chapter we investigated the conjecture that (approximate) optimal covert networks have low variance in the centrality of their respective members. Qualitative theory hypothesizes that current terrorist networks organize according to leaderless principles. We show that indeed this is optimal, i.e., in Sect. 3 we show that homogeneous covert networks that are (close to) optimal have low variance in the centrality of their members. However, if we investigate heterogenous networks in which some members have much higher interaction with each other than others then the variance in the centrality increases drastically. Thus for covert networks this implies that they not only should adopt certain network structures, i.e., organize according to leaderless principles, but additionally that they should adopt a communication policy that is flat. Finally, we investigated whether covert organizations actually adopt such a communication policy. We analyzed Jemaah Islamiyah's interaction structure and concluded that the communication policy utilized did indeed closely resemble the optimal flat policy of the given network.

7 Methods Summary

A covert network is modelled by a graph $g = (N, E)$, where N represents the set of members of the organization and E represents the links (or relationships) present among these members. We set $|N| = n$ and $|E| = m$. The set of all such networks is indicated by $\mathbb{G}(n, m)$.

7.1 Information Measure I

The information measure of a graph $g \in \mathbb{G}(n, m)$ is defined by the normalized reciprocal of the total distance in g, i.e.,

$$I(g) = \frac{n(n-1)}{T(g)}.$$

Here $T(g)$ equals the total geodesic distance, i.e., $T(g) = \sum_{(i,j)\in N^2} l_{ij}(g)$ with $l_{ij}(g)$ the geodesic (or shortest) distance between vertex i and vertex j. It follows that $0 \le I(g) \le 1$.

7.2 Homogeneous Secrecy Measure S_{hom}

The homogeneous secrecy measure of a graph $g \in \mathbb{G}(n, m)$ is defined by

$$S_{\text{hom}}(g) = \frac{2m(n-2) + n(n-1) - \sum_{i\in N} d_i^2(g)}{(2m+n)n}.$$

Here $d_i(g)$ equals the degree of vertex i in graph g. It follows that $0 \le S(g) \le 1$.

7.3 Heterogeneous Secrecy Measure S_{het}

Define the weighting function $w : E \mapsto [1, \infty)$ such that $w_{ij} > w_{kl}$, $ij, kl \in E$, is interpreted as interaction between individuals i and j presenting a higher risk to the organization than interaction between individuals k and l. We denote the set of all such weighting functions by $\mathbb{W}$. Let $w_i = \sum_{j \in \Gamma_i(g)} w_{ij}$ where $\Gamma_i(g) = \{j \in V | ij \in E\}$ and define

$$W = \sum_{i \in V} w_i = 2 \sum_{ij \in E} w_{ij}.$$

The heterogeneous secrecy measure of a graph $g \in \mathbb{G}(n, m)$ is defined by

$$S_{\text{het}}(g) = \frac{n^2 - 2m - n + W(n-1) - \sum_{i \in V} d_i w_i}{n(W + n)}.$$

7.4 Balanced Trade-Off Performance Measure μ

For $g \in \mathbb{G}(n, m)$ it holds that

$$\mu(g) = I(g)S(g).$$

7.5 Game Theoretic Centrality

A cooperative game is a pair (N, v), where N denotes the set of players. These players can cooperate and form different coalitions. A map v assigns a value $v(S)$ to each possible coalition $S \subseteq N$, which reflects the potential 'power' coalition S represents. By definition $v(\emptyset) = 0$. Let the subgraph S_g consist of the players in coalition S and the lines of communication between these players. If the players in coalition S are able to communicate using only the relationships present within coalition S we say that subgraph S_g is connected and assign a value of 1 to coalition S. If not all players in coalition S are able to communicate then we say that subgraph S_g is not connected and we therefore assign a value of 0 to this coalition. A coalition consisting of a single player obtains a value of 0 by definition. Henceforth, the connectivity game v^{conn} (cf. [1]) is defined as

$$v^{\text{conn}}(S) = \begin{cases} 1 & \text{if } S_g \text{ is connected,} \\ 0 & \text{otherwise.} \end{cases}$$

Fig. 7 Example of a network of four persons

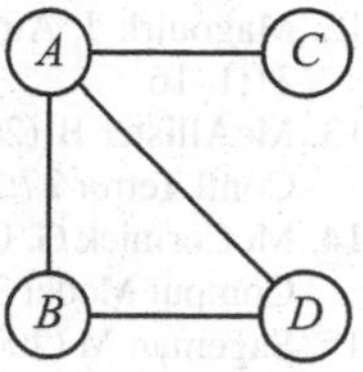

The game theoretic centrality of player i is found by computing the Shapley value $\varphi_i(v)$ (see [16]):

$$\varphi_i(v) = \sum_{S \subseteq N, i \notin S} \frac{|S|!(|N| - 1 - |S|)!}{|N|!} \cdot [v(S \cup \{i\}) - v(S)],$$

where $|S|$ is the number of players in coalition S.

Consider the network depicted in Fig. 7. The subgraph corresponding to, for example, coalition $\{A, B, C\}$ is connected, whereas the subgraph corresponding to coalition $\{B, C, D\}$ is not. Hence, $v^{\text{conn}}(\{A, B, C\}) = 1$ and $v^{\text{conn}}(\{B, C, D\}) = 0$. Computing the Shapley values of the 4 players in the connectivity game v^{conn} results in $\varphi_{\text{A}}(v) = 0.6667$, $\varphi_{\text{B}}(v) = 0.1667$, $\varphi_{\text{C}}(v) = 0$ and $\varphi_{\text{D}}(v) = 0.1667$, which leads to the ranking: A, B and D, C.

References

1. Amer R, Giminez J (2004) A connectivity game for graphs. Math Methods Oper Res 60: 453–470
2. Baker W, Faulkner R (1993) The social organization of conspiracy: illegal networks in the heavy electrical equipment industry. Am Sociol Rev 58:837–860
3. Carley K, Reminga J, Kamneva N (2003) Destabilizing terrorist networks. In: NAACSOS conference proceedings, Pittsburgh
4. Dishman C (2005) The leaderless nexus: when crime and terror converge. Stud Confl Terror 28:237–252
5. Enders W, Su X (2007) Rational terrorists and optimal network structure. J Confl Resolut 51:33–57
6. Farley J (2003) Breaking al qaeda cells: a mathematical analysis of counterterrorism operations. Stud Confl Terror 26:399–411
7. Fukuyama F, Shulsky A (1997) The "virtual corporation" and army organization. RAND corporation, Santa Monica
8. Koschade S (2006) A social network analysis of jemaah islamiyah: the applications to counterterrorism and intelligence. Stud Confl Terror 29:559–575
9. Lindelauf R, Borm P, Hamers H (2009) The influence of secrecy on the communication structure of covert networks. Soc Netw 31:126–137
10. Lindelauf R, Borm P, Hamers H (2009) On heterogeneous covert networks. Springer, Vienna
11. Lindelauf R, Hamers H, Husslage B (2011) Game theoretic centrality analysis of terrorist networks: the cases of jemaah islamiyah and al qaeda. CentER Discussion Paper No. 2011-107, Tilburg

12. Magouirk J, Atran S, Sageman M (2008) Connecting terrorist networks. Stud Confl Terror 31:1–16
13. McAllister B (2004) Al qaeda and the innovative firm: demythologizing the network. Stud Confl Terror 27:297–319
14. McCormick G, Owen G (2000) Security and coordination in a clandestine organization. Math Comput Model 31:175–192
15. Sageman M (2008) Leaderless Jihad: terror networks in the twentyfirst century. University of Pennsylvania Press, Philadelphia
16. Shapley L (1953) A value for n-person games. Ann Math Stud 28:307–317
17. Sparrow M (1991) The application of network analysis to criminal intelligence: an assessment of the prospects. Soc Netw 13:251–274
18. Tsvetovat M, Carley K (2005) Structural knowledge and success of anti-terrorist activity: the downside of structural equivalence. J Soc Struct 6, Vol. 6, No. 2
19. Wasserman S, Faust K (1994) Social network analysis, methods and applications. Cambridge University Press, Cambridge

Link Prediction in Highly Fractional Data Sets

Michael Fire, Rami Puzis, and Yuval Elovici

1 Introduction

In recent years, online social networks have grown in scale and variability and offer individuals with similar interests the possibility of exchanging ideas and networking. On the one hand, social networks create new opportunities to develop friendships, share ideas, and conduct business. On the other hand, they are also an effective media tool for plotting crime and organizing extremists groups around the world. Online social networks, such as Facebook, Google+, and Twitter are hard to track due to their massive scale and increased awareness of privacy. Criminals and terrorists strive to hide their relationships, especially those that can associate them with a executed terror act.

A large portion of recent research in social network analysis has been targeted at identifying these hidden relationships in social networks. These research endeavours are usually referred to as *link prediction* methods. These methods are able to detect existing social ties that have not been established in a particular social network [8, 10, 17, 23, 29, 33]. In the security and counter-terrorism domains link prediction can assist in identifying hidden groups of terrorists or criminals [17]. However, link prediction is also useful for civil applications, such as friend-suggestion mechanisms embedded in online social networks. For example, in bioinformatics, link prediction can be used to find interactions between proteins [3], and in e-commerce, it can help build recommendation systems [19].

M. Fire (✉) · Y. Elovici
Department of Information Systems Engineering, Telekom Innovation Laboratories, Ben-Gurion University of the Negev, Beersheva, Israel,
e-mail: mickyfi@bgu.ac.il; elovici@bgu.ac.il

R. Puzis
University of Maryland Institute for Advanced Computer Studies (UMIACS), College Park, MD, USA
e-mail: puzis@umd.edu

V.S. Subrahmanian (ed.), *Handbook of Computational Approaches to Counterterrorism*, DOI 10.1007/978-1-4614-5311-6_14,

Several different methods for solving the link prediction problem have been proposed in recent years. These days, the majority of solutions are based on supervised machine learning methods, such as Bayesian probabilistic models, relational Bayesian networks, and linear algebraic methods. Further details on these approaches can be found in a thorough survey written by Hasan and Zaki [18]. We briefly describe a selected few in Sect. 2.2.

We focus on link perdition methods based on machine learning classifiers trained on a set of topological features. We use a set of well-known features, such as connectivity features, intersection and union of the friends groups, *Jaccard's-Coefficient* [34], *Preferential-Attachment* score [8, 17], and the *Friends-Measure*, introduced in [14]. The latter is a variation of the Katz measure [20] and estimates how well the friends of two users know each other.

Many link prediction methods are applied to large social networks where most of the links are assumed to be known. This is a reasonable assumption when the main objective is to predict ties that are not yet established in the social network. However, terrorist social networks mined from open sources are typically small and very partial due to the efforts of their subjects to obfuscate their activity.

In this study, we investigate the effects of dataset partiality on the effectiveness of link prediction by gradually reducing the number of visible links in the test networks. The effectiveness of classifiers was evaluated using six different types of data sets: a group of Facebook users sharing the same employer, a researchers community from Academia.edu [12], Friends and Family SMS messages social network [4], Students Cooperation Network [13], AnyBeat social network,[1] and the Profiles in Terror (PIT) dataset [31]. Information on these social networks is presented in Sect. 3. Evaluation results presented in Sect. 5 demonstrate that classification quality (in terms of Area Under the ROC Curve (AUC)) degrades with the number of visible links. Nevertheless, even a small fraction of visible links (5–20 %) helps in solving the link prediction problem with chances significantly higher than random.

The remainder of the paper is organized as follows. In Sect. 2, we review previous studies on terrorist's social networks and link prediction. In Sect. 3, we describe the datasets used in this study. Experimental setup and the features extracted from the structure of social networks are described in Sect. 4.1. We present the experimental results in Sect. 5, and conclusions in Sect. 6.

[1] http://www.anybeat.com

2 Background

2.1 Social Networks of Terrorists

Over the last two decades, social networks have been studied fairly extensively in the general context of analyzing interactions between people and determining the important structural patterns of such interactions [1]. In the previous decade, even before September 11, 2001, social network analysis was recognized as a tool for fighting the war against criminal organizations in an age where there is no well-defined enemy with a formal hierarchical organization [5]. Moreover, after the September 11, 2001 events, social network analysis became a well-known mainstream tool to help the fight against terror [26].

Several studies have analyzed terror organization social networks based on graph structural features. In the winter of 2002, Krebs [21] studied Al-Qaeda's network structural properties by collecting publicly available data on the Al-Qaeda hijackers. Rothenberg [28] conjectured on the structure of the al Qaeda network based on public media sources. After the Madrid bombing, in March 11, 2004, Rodriguez [27] used public sources to construct and study the terrorists' network. He showed that the terror organization network included mainly weak ties that are hard to detect. In 2004, Sageman [30] used various public sources, mostly records of trials, to collect and analyze 400 terrorist biographies. He discovered that 88 % of the terrorists had friendship or family bonds to the Jihad. In 2005, Basu [7] studied terrorists' organization in India. He used social network analysis, such as the betweenness measure, to identify major groups of terrorists and key players. In 2010, Wiil et al. [35] studied a recent Denmark terror plan. By using data mining techniques, they were able to construct, from public sources, the social network of David Coleman Headley, one of the terror plan conspires.

Attempts to reconstruct the social networks of terrorists requires a significant effort spent on mining the Web for publicly available information and free text analysis. This typically results in the ability to obtain small networks only with a high likelihood of missing information. In this study, we attempt to predict links inside social networks where a substantial amount of the network's links data were missing. A similar idea was studied by Dombroski et al. [9]. Their study examined the possibilities of using the inherent structures observed in social networks to make predictions of networks using limited and missing information.

2.2 Link Prediction

In this study, we focus on link prediction methods based on supervised machine learning algorithms. These methods were first introduced by Liben-Nowell and Kleinberg in 2003 [23], who used graph topological features in a study on five co-authorship networks, each containing several thousands of vertices. In 2006, Hasan et al. [17] increased the scale of analyzed networks to hundreds of thousands of nodes by analyzing the DBLP and BIOBASE co-authorship networks. Supervised learning was also applied by many other researchers to solve the link prediction problems, for example in [10,22,29]. Initially, the proposed link prediction solutions were tested on bibliographic or on co-authorship data sets [10, 17, 23, 29].

In 2009, Song et al. used matrix factorization to estimate the structural similarity between profiles in online social network services, such as Facebook and MySpace [33]. In 2010, Leskovec et al. [22] studied a similar problem of predicting links' signs. Recently, Zaki and Hasan [18] published through survey on link prediction in social networks.

In 2011, the IJCNN social network challenge [24] inspired several publications on link prediction using topological network analysis. These publications proposed and evaluated different methods for predicting links in social networks. Narayanan et al. won the challenge by using a method that combined machine-learning algorithms with de-anonymization [25]. Cukierski et al. [8] took second place by extracting 94 distinct features for each one of the of several thousands of vertex pairs in the training data and analyzing it with the Random Forest algorithm. Recently, Fire et al. presented a method for predicting links inside communities; their methods used supervised learning ensemble classifiers constructed by using small training sets only which consisted of several hundreds of examples [12].

When no information on the social network besides its structure is available, it is crucial to define and calculate the features that are as informative as possible. On the one hand, large networks containing millions of vertices and links pose a scalability challenge and require the use of easy to compute topological features extracted from the neighbourhoods of the tested vertices. For example, Facebook has more than 901 million registered users and each month many new users are added [11]. On the other hand, small networks, such as the terrorist datasets available today (e.g., [21, 36]), pose a different challenge. They contain too few links and vertices to construct a large training set. Moreover, these networks are much more prone to noise than their huge counterparts because the existence or absence of every link may significantly change the values of the extracted features. In this study, we take the latter challenge to the extreme and evaluate structural link prediction methods while gradually removing random links from organizational, group, and terrorist affiliation networks.

3 Social Network Datasets

In this study, we apply link prediction classifiers to six labeled social network datasets (see Table 1), namely, Profile in Terror (PIT) [36], AnyBeat network,[2] a group of Facebook[3] users, Academia.edu [14, 15], Friends and Family study [2], and Students' Cooperation Social Network [13].

---

[2] http://www.anybeat.com

[3] http://www.facebook.com

Table 1 Datasets

Network	Directed	Vertices	Links	Positive examples	Visible networks	Features
PIT	No	244	840	420	200	8
AnyBeat	Yes	12,645	67,053	25,000	40	12
Co-Workers	No	165	722	361	200	8
Researchers	Yes	207	702	351	200	12
F&F	Yes	103	281	140	200	12
Students	No	185	311	155	200	8

Profile in Terror (PIT) [31] is a data set that captures intelligence information extracted from publicly available sources collected by the MIND Lab at UMD.[4] This data set contains 851 labeled relationships among terrorists and was previously used in multi-label link classification in [36]. Each relationship can have one or more labels from: colleague, congregate, contact, or family. In this study, we disregard the relationship labels and treat this data set as a flat network which contains 840 links and 244 vertices. Figure 1 depicts the graph topology.[5]

AnyBeat. "AnyBeat is an online community; a public gathering place where you can interact with people from around your neighborhood or across the world". AnyBeat is a relatively new social network in which members can log in without using their real name and members can follow any other member in the network. In this study, we evaluated our algorithm on a major part of the network's topology, which was obtained using a dedicated web crawler. The topology contained 12,645 vertices and 67,053 links (see Fig. 2). Among the networks studied in this study, AnyBeat is outstanding in its size. We included a network of more than 10,000 vertices in order to shade the results of the study and focused mainly on small partially visible social networks. As presented in Sect. 5, link prediction appears to be significantly easier for this network.

A Facebook group of co-workers (Co-Workers). Facebook is a website and social networking service that was launched in February 2004. As of March 2012, Facebook has more than 901 million registered users [11]. Facebook users may create a personal profile, add other users as friends, and interact with other members. Friendship ties in Facebook are reciprocal, therefore, we refer to the underlying group's social network as undirected. We extracted a community of co-workers who, according to their Facebook profile, worked for the same well-known high-tech company. The graph representing the co-workers' community network contains 165 vertices and 726 links and was obtained using a web crawler in the beginning of February 2012 (see Fig. 3).

[4] http://www.mindswap.org/

[5] All the social networks figures in this paper were created by Cytoscape software [32].

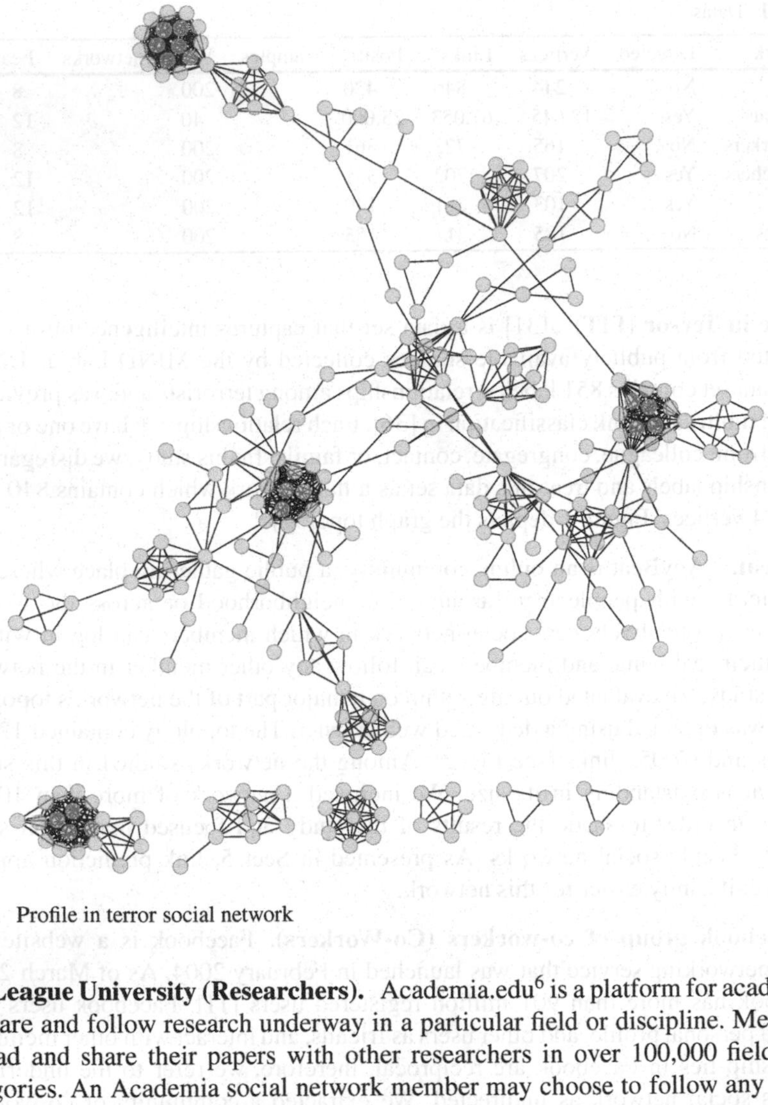

Fig. 1 Profile in terror social network

Ivy League University (Researchers). Academia.edu[6] is a platform for academics to share and follow research underway in a particular field or discipline. Members upload and share their papers with other researchers in over 100,000 fields and categories. An Academia social network member may choose to follow any of the network's members; hence, the directed nature of the links within this network. We evaluated our classifiers for a small community of researchers who, according to their Academia.edu profiles, belonged to the same Ivy League University. The researchers' community network graph contained 207 nodes and 702 links (see Fig. 4) and was obtained using a web crawler.

Friends and Family (F&F). The Friends and Family dataset contains rich data signals gathered from the smart-phones of 140 adult members of a young-family

[6]http://wwww.academia.edu

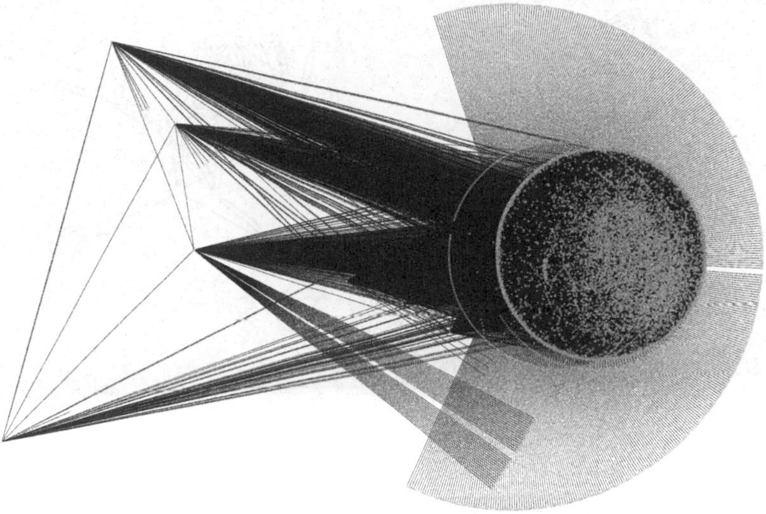

Fig. 2 AnyBeat social network

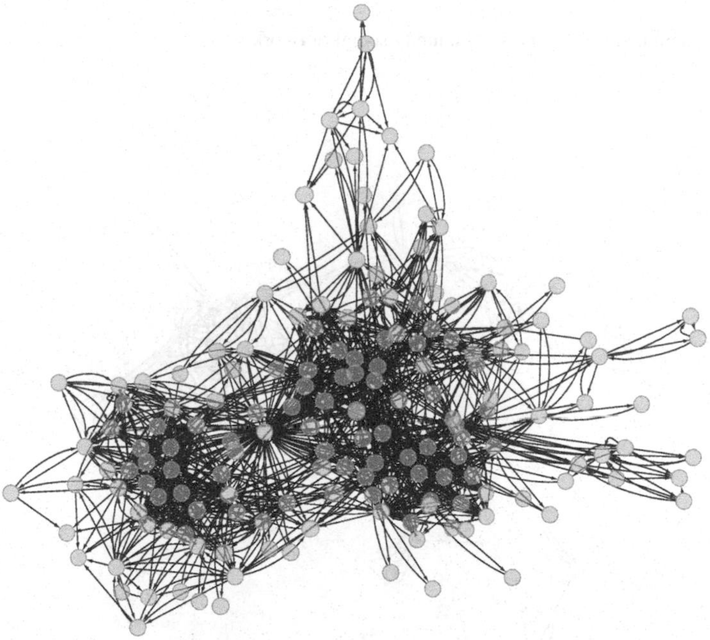

Fig. 3 Facebook coworker community social network

residential community. The data were collected over the course of 1 year [2]. We evaluated our classifiers on a social network that was constructed based on SMS messages sent and received by the members. The SMS messages social network directed graph contained 103 nodes and 281 links (see Fig. 5).

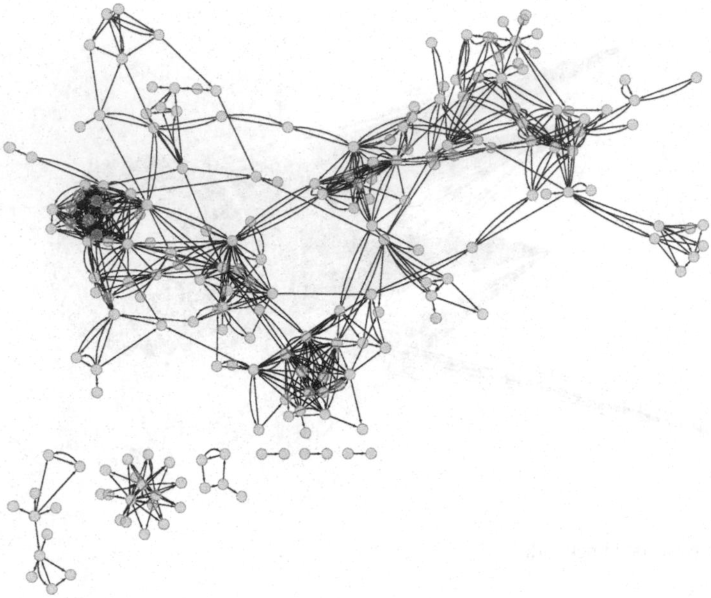

Fig. 4 Academia.edu researchers community social network

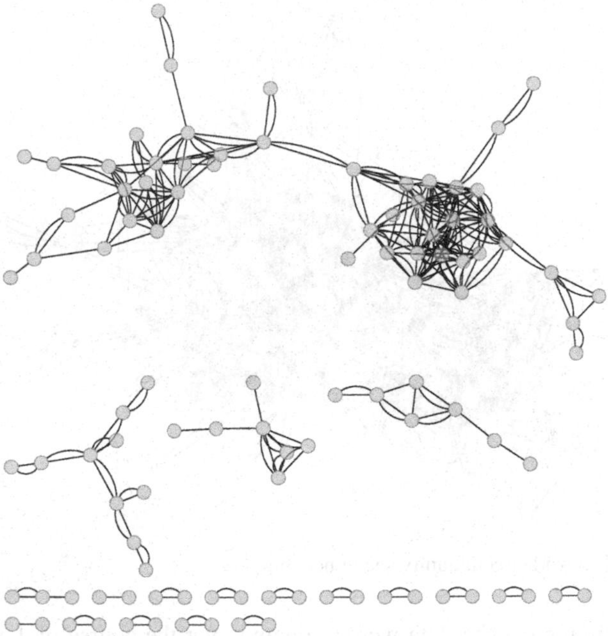

Fig. 5 Friends and family SMS messages social network

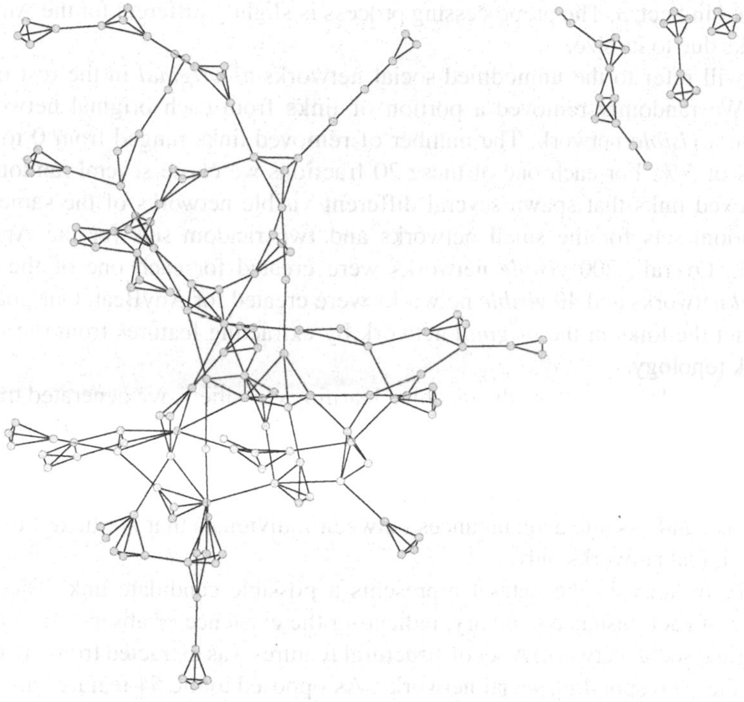

Fig. 6 Students' cooperation social network

Students' Cooperation Social Network (Students). The students' cooperation social network was constructed from data collected during a "Computer and Network Security" course; a mandatory course taught by two of this paper's authors at Ben-Gurion University [13]. The social network contains data collected from 185 participating students from two different departments. The course's social network was created by analyzing the implicit and explicit cooperation among the students while doing their homework assignments. The Students' Cooperation graph contained 185 nodes and 311 links (see Fig. 6).

4 Methods and Experiments

4.1 Experimental Setup

The goal of structural link prediction is to identify a set of hidden links within a social network's structure by analyzing the topology of the known (visible) network. As a first step in our experiments, we preprocessed each one of the social networks described in Sect. 3. The preprocessing process is slightly different for the AnyBeat networks due to its size.

We will refer to the unmodified social networks as *original* in the rest of this paper. We randomly removed a portion of links from each original network to generate a *visible* network. The number of removed links ranged from 0 to 95 % in steps of 5 %. For each one of these 20 fractions we chose several random sets of removed links that spawn several different visible networks of the same size; ten random sets for the small networks and two random sets for the AnyBeat network. Overall, 200 *visible* networks were created for each one of the small *original* networks and 40 *visible* networks were created for AnyBeat. Our goal was to predict the links in the *original* network by extracting features from the *visible* network topology.

Next, in order to evaluate the machine learning classifiers, we generated training and testing data sets from each visible network with the original network treated as the ground truth. Note that due to privacy profiles and imperfect data acquirement methods, the original networks may have missing links as well. We will disregard these links and assume acquaintances between individuals that are linked together in the original networks only.

Every instance in the dataset represents a possible candidate link. The target attribute of each instance is binary, indicating the existence or absence of a link in the original social network. A set of structural features was extracted from the *visible* part of the corresponding social networks. As opposed to the 54 features discussed in [15], we have only 7 and 11 of the most efficient features for undirected and directed networks, respectively in this study. These features are briefly described in Sect. 4.2.

In order to create a balanced data set for training the classifiers, we included the same number of positive and negative examples. Positive examples are vertex pairs that are connected in the *original* network. Negative examples include the same number of vertex pairs, but this time, random pairs of vertices not connected in the *original* network. The method for selecting the negative examples in this study corresponds to the *easy* dataset, as described in [15]. Note that the features describing each positive or negative example were extracted from the structure of the *visible* networks. Table 1 summarizes the datasets used in this study.

WEKA [16], a popular suite of machine learning software written in Java and developed at the University of Waikato, New Zealand, was used as the machine learning platform for this study. We used a WEKA's J48 classifier with ten minimum objects that showed effective performance in past experiments. We evaluated our results using a ten-folds cross validation method.

Next, we describe the set of features extracted from the visible social network graphs. The features used in this study are a subset of a more extensive set of features investigated in [15].

4.2 Feature Extraction

This section describes the features extracted from the social network structure in order to build our link prediction classifiers. The extracted features are based primarily on the Friends-features subset, as suggested by Fire et al. [15].

Let $G = < V, E >$ be the graph representing the structure of a social network. Links in the graph are denoted by $e = (u, v) \in E$ where $u, v \in V$ are vertices in the graph. Our goal is to construct classifiers capable of computing the likelihood of $(u, v) \in E$ or $(u, v) \notin E$ for every two vertices $u, v \in V$. To achieve this goal, we extracted the following features for each pair, (u, v), in our datasets.

1. **Vertex degree:** Let $v \in V$ be some vertex, we can define the neighborhood of v by:

$$\Gamma(v) := \{u|(u, v) \in E \text{ or } (v, u) \in E\}$$

If G is a directed graph, we can also define the following neighborhoods:

$$\Gamma_{in}(v) := \{u|(u, v) \in E\}$$
$$\Gamma_{out}(v) := \{u|(v, u) \in E\}$$

We define the degree features for directed and undirected networks as the sizes of the respective neighborhoods:

$$degree(v) := |\Gamma(v)| \quad (1)$$

$$degree_{in}(v) := |\Gamma_{in}(v)| \quad (2)$$

$$degree_{out}(v) := |\Gamma_{out}(v)| \quad (3)$$

The degree features measure the number of friends v has. If we look at a directed graph, such as Academia.edu, the meaning of the degree feature is how many other members of the community v follows (out-degree), and how many members of the community follow v (in-degree).

2. **Common-Friends:** Let $u, v \in V$ be a pair of vertices in the network; we define the common friends of u and v to be all the vertices in the network that are friends of both u and v. Formally, common friends is the size of the intersection of the respective neighborhoods:

$$Common\text{-}Friends(u, v) := |\Gamma(v) \cap \Gamma(u)| \quad (4)$$

The Common-Friends feature was widely used in previous works for predicting links in different datasets [8, 14, 19, 23, 29, 33].

3. **Total-Friends:** Let $u, v \in V$ be a pair of vertices; we can define the number of distinct friends of u and v as the size of the union of the respective neighborhoods:

$$Total\text{-}Friends(u, v) := |\Gamma(u) \cup \Gamma(v)| \quad (5)$$

4. **Jaccard's-Coefficient:** Jaccard's-Coefficient is a well-known feature for link prediction [8, 14, 19, 23, 29, 33]. This feature, which measures the similarity among sets of nodes, is defined as the size of the intersection divided by the size of the union of the sample sets:

$$Jaccard's\text{-}Coefficient(u,v) := \frac{|\Gamma(u) \cap \Gamma(v)|}{|\Gamma(u) \cup \Gamma(v)|} \tag{6}$$

In our approach, this measure indicates whether two community members have a significant number of common friends regardless of their total number of friends. A higher value of the Jaccard's-Coefficient typically indicates a stronger connection between two nodes in the network.

5. **Preferential-Attachment-Score:** The Preferential-Attachment-Score indicates the likelihood of a new link to be formed between the vertices u and v, according to the preferential attachment model [6]. It is defined as the multiplication of the number of friends of u and v.

$$Preferential\text{-}Attachment\text{-}Score(u,v) := |\Gamma(u)| \cdot |\Gamma(v)| \tag{7}$$

The Preferential-Attachment score was used many times in past research, for examples see [8, 14, 17].

6. **Friends-Measure:** Let $u, v \in V$ be two vertices; the Friends-Measure of u and v is the extent to which their friends are interconnected. The higher the number of connections between u and v's friends, the greater the chance that u and v know each other.

$$Friends\text{-}Measure(u,v) := \sum_{x \in \Gamma(u)} \sum_{y \in \Gamma(v)} \delta(x,y) \tag{8}$$

where $\delta(x, y)$ is defined as:

$$\delta(x,y) := \begin{cases} 1 & \text{if } x = y \text{ or } (x,y) \in E \text{ or } (y,x) \in E \\ 0 & \text{otherwise} \end{cases}$$

The Friends-Measure was first presented by Fire et al. [14].

5 Results

For each undirected and directed visible network, we extracted 7 and 11 features, respectively, (see Sect. 4.2) and evaluated the specified machine-learning algorithms (see Sect. 4.1) using a ten-fold cross-validation approach. The AUC results of the J48 algorithm on all the networks where no edges were removed from the networks are presented in Fig. 7.

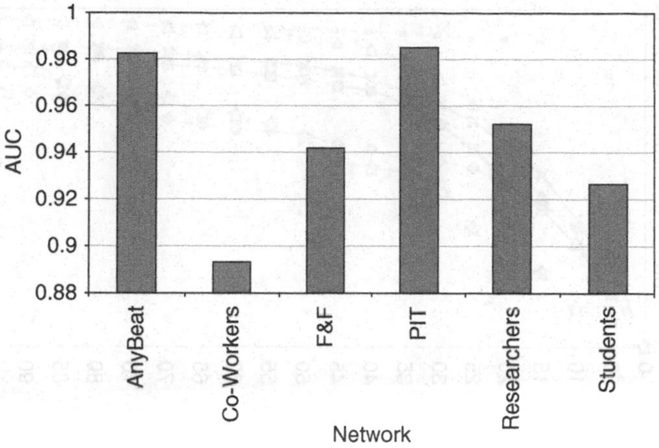

Fig. 7 AUC Results – J48 with no removed links

Table 2 Effectiveness of link prediction in fractional networks

Network	Percent of visible links				
	100 %	80 %	60 %	40 %	20 %
PIT	0.985	0.980	0.966	0.940	0.789
Co-Workers	0.893	0.858	0.820	0.791	0.738
Researchers	0.952	0.942	0.920	0.895	0.799
F&F	0.941	0.928	0.908	0.877	0.744
Students	0.926	0.877	0.841	0.762	0.629
AnyBeat	0.982	0.979	0.972	0.966	0.954

Table 2 and Fig. 8 present the AUC results of the J48 algorithm for each social network with various numbers of removed links. As expected, we noticed a degradation in the AUC values as more edges are removed from the network. The slope of the AUC as a function of the visible network size increases with the number of removed vertices. In the terrorists affiliation network (PIT), the AUC remains above 0.9, even when features in the dataset are computed according to 40 % of the original links.

Another notable observation is that the classification results (with the exception of the Students' Cooperation network) are significantly above random, even when only 5–10 % of the *original* links are *visible* with respect to the computation of structural features. The Students' Cooperation network was based on links between students in a single class taught for a short period of time. Therefore, the network did not contain as many cliques as in the other social networks. We believe that the existence of large cliques, (as in PIT network) or densely connected parts in the network, makes it easier to predict ties randomly removed from the network.

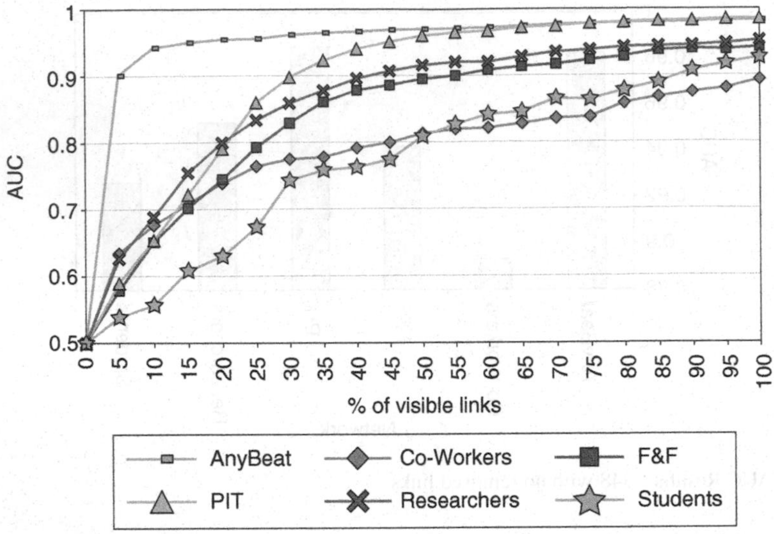

Fig. 8 AUC as a function of the fraction of visible links for various social networks

A final interesting observation is that the best AUC values were obtained for the AnyBeat network. Link prediction in this network is much easier than in the smaller networks when significant portions of the network are removed. The AUC remains above 0.9, even when there are only 5 % of visible links. Note that 5 % of visible links in AnyBeat is 3,352, which is a few orders of magnitude higher than in the other networks.

To obtain an indication of the usefulness of the various features, we analyzed their importance using Weka's Information Gain attribute selection algorithm. The the InfoGain results for different datasets are presented in Tables 3 and 4. In order to check the effect of link removal on the usefulness of various features, we computed the InfoGain values for the original networks and for networks with only 5 % of visible links. The results indicate that InfoGain of all features drops and the set of useful features changes when random links are removed from the dataset. Only the *Preferential-Attachment-Score* is useful for both collections of datasets. Nonetheless, the InfoGain values of connectivity degree features drop in Table 4 and they become more useful relative to the other features. This effect is even more apparent for the *Total-friends* feature that has the highest InfoGain value in Table 4. Yet, this value is more than two times lower than the respective value in Table 3. In contrast, we can clearly see that *Common-friends*, *Friends-Measure*, and *Jaccard's Coefficient* have high InfoGain values for networks where all links are visible and the lowest InfoGain values for networks with only 5 % of visible links.

Table 3 InfoGain of the extracted features for various social networks

	|Inner-subgraph(u,v)|	|nh-subgraph(u,v)|	|nh-subgraph+(u,v)|	Common-Friends(u,v)	d(u)	Density(nh-subgraph(u,v))	dout(u)	**Friends-measure(u,v)**	Jaccard's-coefficient(u,v)	Opposite-direction-friends(u,v)	Preferential-attachment-score(u,v)	scc(Inner-subgraph(u,v))	scc(nh-subgraph(u,v))	scc(nh-subgraph+(u,v))	Shortest-path(u,v)	Total-Friends(u,v)	Transitive-Friends(u,v)
Academia	0.6	0.2	0.3	0.5	0.3	0.2	0.3	**0.7**	0.5	0.3	0.4	0.2	0.2	0.2	0.4	0.3	0.4
Facebook	0.7	0.2	0.2	0.7	0.3	0.2		**0.7**	0.7		0.4	0.0	0.0	0.5	0.7	0.3	
Flickr	0.3	0.2	0.3	0.1	0.3	0.2	0.2	**0.4**	0.1	0.3	0.7	0.7	0.7	0.7	0.0	0.7	0.0
TheMarker	0.5	0.5	0.5	0.5	0.5	0.6	0.3	**0.6**	0.8	0.3	0.8	0.6	0.7	0.7	0.0	0.6	0.2
YouTube	0.5	0.4	0.5	0.3	0.5	0.4	0.5	**0.6**	0.3	0.6	0.7	0.5	0.5	0.4	0.2	0.6	0.3
Average	0.51	0.32	0.38	0.42	0.38	0.32	0.36	**0.61**	0.47	0.39	0.60	0.40	0.43	0.50	0.26	0.48	0.22

Table 4 InfoGain of the extracted features for fractional social networks with 5 % of visible links

	|Inner-subgraph(u,v)|	|nh-subgraph(u,v)|	|nh-subgraph+(u,v)|	Common-Friends(u,v)	d(u)	Density(nh-subgraph(u,v))	dout(u)	**Friends-measure(u,v)**	Jaccard's-coefficient(u,v)	Opposite-direction-friends(u,v)	Preferential-attachment-score(u,v)	scc(Inner-subgraph(u,v))	scc(nh-subgraph(u,v))	scc(nh-subgraph+(u,v))	Shortest-path(u,v)	Total-Friends(u,v)	Transitive-Friends(u,v)
Academia	0.1	0.2	0.2	0.3	0.2	0.1	0.3	**0.3**	0.3	0.3	sri	0.1		0.1	0.2	0.1	0.4
Facebook	0.2	0.2	0.2	0.4	0.3	0.2		**0.3**	0.3		0.2	0.0	0.0	0.0	0.0	0.1	
Flickr	0.5	0.5	0.5	0.5	0.5	0.6	0.3	**0.6**	0.8	0.3	0.8	0.6	0.7	0.7	0.0	0.6	0.2
TheMarker	0.4	0.4	0.4	0.3	0.4	0.4		**0.4**	0.2		0.3	0.0	0.0	0.0	0.0	0.2	
YouTube	0.2	0.3	0.3	0.5	0.3	0.2	0.3	**0.6**	0.5	0.6	0.3	0.1		0.2	0.2	0.2	0.5
Average	0.28	0.32	0.35	0.41	0.35	0.31	0.30	**0.42**	0.38	0.38	0.35	0.17	0.25	0.20	0.09	0.23	0.36

6 Conclusion

Today's terror acts are conducted by groups of individuals who link with facilitators and other groups. Together, such a network has the resources, the means, and the insights to execute these attacks [35]. It is important to establish connections between the participating parties in order to understand their group dynamics and effectively mitigate their activity [30].

In this chapter, we presented the results of a study on link prediction in small scale, highly fractional networks, such as terrorists networks obtained from an analysis of publicly available media. We employed machine learning classifiers trained on a set of features extracted from the network structure. The results indicate that, in contrast to large networks where effective link prediction can be performed even when the vast majority of links are hidden, small scale networks require at least 30–50 % visible links to get AUC values above 0.8. We also notice that networks containing large cliques, or densely connected parts, are less vulnerable to random removal of links.

In conclusion, as expected, our results show that Information Gain values of the structural features drop when most of the links are removed from the network. However, it should be noted that the decrease in the Information Gain is not uniform. Features that were the most useful for predicting links in the original network become the least useful when only 5 % of the links are visible. These features are the number of common friends, the Friends-Measure [14], and Jaccard's Coefficient [34]. All these measures refer to the connections and similarities between vertices in both neighborhoods of the two ends of the link being tested. When there are enough connections, these measures make sense, however, when only few links are visible, the most useful information is the number of acquaintances each vertex has.

Availability

An anonymous version of the Co-Workers, Researchers, Students, and the AnyBeat social network topologies crawled as a part of this study are available on our research group website http://proj.ise.bgu.ac.il/sns/.

References

1. Aggarwal C (2011) Social network data analytics. Springer, New York
2. Aharony N, Pan W, Ip C, Khayal I, Pentland A (2011) Social fMRI: investigating and shaping social mechanisms in the real world. Pervasive and Mobile Computing 7(6):643–659
3. Airoldi E, Blei D, Fienberg S, Xing E, Jaakkola T (2006) Mixed membership stochastic block models for relational data with application to protein-protein interactions. In: Proceedings of the international biometrics society annual meeting. ENAR, Tampa, FL, USA

4. Altshuler Y, Fire M, Elovici Y, Pentland A (2012) How many makes a crowd? On the evolution of learning as a factor of community coverage, Social Computing, Behavioral – Cultural Modeling and Prediction, Lecture Notes in Computer Science, (7227), Springer, Berlin/Heidelberg, pp 43–52
5. Arquilla J, Ronfeldt D (2001) Networks and netwars: the future of terror, crime, and militancy. 1382. Rand Corp
6. Barabasi AL, Albert R (1999) Emergence of scaling in random networks. Science 286:509–512
7. Basu A (2005) Social network analysis of terrorist organizations in india. In: North American Association for Computational Social and Organizational Science (NAACSOS) conference. CASOS, Notre Dame, Indiana, USA, pp 26–28
8. Cukierski WJ, Hamner B, Yang B (2011) Graph-based features for supervised link prediction. International joint conference on neural networks. IEEE, San Jose, California
9. Dombroski M, Fischbeck P, Carley K (2003) Estimating the shape of covert networks. In: Proceedings of the 8th international command and control research and technology symposium
10. Doppa JR, Yu J, Tadepalli P, Getoor L (2009) Chance-constrained programs for link prediction. In Proceedings of workshop on analyzing networks and learning with graphs at NIPS conference
11. Facebook-Newsroom. http://www.facebook.com
12. Fire M, Katz G, Rokach L, Elovici Y (2012) Links reconstruction attack using link prediction, Security and Privacy in Social Networks , pp 181–196, Springer, Berlin/Heidelberg
13. Fire M, Katz G, Elovici Y, Shapria B, Rokach L (2012) Predicting student exam's scores by analyzing social network data, AMT 2012, LNCS 7669, pp. 584–595. Springer, Heidelberg
14. Fire M, Tenenboim L, Lesser O, Puzis R, Rokach L, Elovici Y (2011) Link prediction in social networks using computationally efficient topological features. In: Privacy, security, risk and trust (PASSAT), 2011 IEEE third international conference on and 2011 IEEE third international confernece on social computing (SocialCom). IEEE, Washington, DC, USA pp 73–80
15. Fire M, Tenenboim L, Puzis R, Lesser O, Rokach L, Elovici Y. Computationally efficient link prediction in variety of social networks, (working paper)
16. Hall M, Frank E, Holmes G, Pfahringer B, Reutemann P, Witten IH (2009) The weka data mining software: an update. SIGKDD Explor Newsl 11:10–18. doi:http://doi.acm.org/10.1145/1656274.1656278
17. Hasan MA, Chaoji V, Salem S, Zaki M (2006) Link prediction using supervised learning. SDM workshop of link analysis, counterterrorism and security. SIAM, Lake Buena Vista, Florida
18. Hasan MA, Zaki MJ (2011) Social network data analytics. Springer, New York
19. Huang Z, Li X, Chen H (2005) Link prediction approach to collaborative filtering. Proceedings of the 5th ACM/IEEE-CS joint conference on digital libraries. ACM, Denver, CO, USA
20. Katz L (1953) A new status index derived from sociometric analysis. Psychometrika 18(1):39–43. http://ideas.repec.org/a/spr/psycho/v18y1953i1p39-43.html
21. Krebs V (2001) Mapping networks of terrorist cells. Connections 24(3):43–52
22. Leskovec J, Huttenlocher D, Kleinberg J (2010) Predicting positive and negative links in online social networks. In: Proceedings of the 19th international conference on World wide web. ACM, New York, NY, USA pp 641–650
23. Liben-Nowell D, Kleinberg J (2007) The link-prediction problem for social networks. J Am soc Inf Sci Technol 58(7):1019–1031
24. Nachbar D (2010) IJCNN social network challenge. http://www.kaggle.com/c/socialNetwork/Data
25. Narayanan A, Shi E, Rubinstein B (2011) Link prediction by de-anonymization: How we won the kaggle social network challenge. In: The 2011 international joint conference on neural networks (IJCNN). IEEE, Washington, DC, USA pp 1825–1834
26. Ressler S (2006) Social network analysis as an approach to combat terrorism: Past, present, and future research. Homel Secur Aff 2(2):1–10
27. Rodriquez J (2005) The march 11th terrorist network: in its weakness lies its strength, VIII Congreso Espaol de Sociologa, Alicante, Spain

28. Rothenberg R (2001) From whole cloth: making up the terrorist network. Connections 24(3):36–42
29. Sa HR, Prudencio RBC (2010) Supervised learning for link prediction in weighted networks. III international workshop on web and text intelligence. São Bernardo do Campo, Brazil
30. Sageman M (2004) Understanding terror networks. University of Pennsylvania Pr
31. Sen P, Namata GM, Bilgic M, Getoor L, Gallagher B, Eliassi-Rad T (2008) Collective classification in network data. AI Mag 29(3):93–106
32. Shannon P, Markiel A, Ozier O, Baliga N, Wang J, Ramage D, Amin, N, Schwikowski B, Ideker T (2003) Cytoscape: a software environment for integrated models of biomolecular interaction networks. Genome Res 13(11):2498–2504
33. Song HH, Cho TW, Dave V, Zhang Y, Qiu L (2009) Scalable proximity estimation and link prediction in online social networks. In: Proceedings of the 9th ACM SIGCOMM conference on Internet measurement conference, IMC '09. ACM, New York, pp 322–335. doi:http://doi.acm.org/10.1145/1644893.1644932
34. Tan PN, Steinbach M, Kumar V (2005) Introduction to Data Mining. Addison Wesley, Boston, Massachusetts, USA
35. Wiil U, Memon N, Karampelas P (2010) Detecting new trends in terrorist networks. In: 2010 international conference on advances in social networks analysis and mining (ASONAM). IEEE, Washington, DC, USA pp 435–440
36. Zhao B, Sen P, Getoor L (2006) Event classification and relationship labeling in affiliation networks. In: Proceedings of the workshop on statistical network analysis (SNA) at the 23rd international conference on machine learning (ICML). Pittsburgh, Pennsylvania

Data Analysis Based Construction and Evolution of Terrorist and Criminal Networks

Khaled Dawoud, Tamer N. Jarada, Wadhah Almansoori, Alan Chen, Shang Gao, Reda Alhajj, and Jon Rokne

1 Introduction

The wide-spread usage of network and graph based approaches in modeling data has been approved to be effective for various applications. The network based framework becomes more powerful when it is expanded to benefit from the widely available techniques for data mining and machine learning which allow for effective knowledge discovery from the investigated domain. The underlying reason for the substantial efficacy in studying graphs, either directly (i.e., data is given in graph format, for example, the "phone-call" network in studying social evolutions) or indirectly (network is inferred from data by predefined method or scheme, such as co-occurrence network for studying genetic behaviors), is the fact that graph structures emphasize the intrinsic relationship between entities, i.e., nodes (or vertices) in the network (in this chapter, the terms network and graph are used interchangeably). For the indirect case information extraction techniques may be adapted to investigate open sources of data in order to derive the required network structure as reflected in the current available data. This is a tedious process but effective and could lead to more realistic and up-to-date information reflected in the network. The latter network will lead to better and close to real-time knowledge discovery in case online information extraction is affordable and provided. Estimating network structure has attracted the attention of other researchers involved in terrorist network analysis, e.g. [9].

K. Dawoud (✉) • T.N. Jarada • W. Almansoori • A. Chen • S. Gao • J. Rokne
Department of Computer Science, University of Calgary, Calgary, AB, Canada
e-mail: krdawoud@ucalgary.ca; tjarada@gmail.com; wadmansoori@gmail.com; aclchen@ucalgary.ca; shangao@ucalgary.ca; rokne@ucalgary.ca

R. Alhajj
Department of Computer Science, University of Calgary, Calgary, AB, Canada

Department of Computer Science, Global University, Beirut, Lebanon
e-mail: alhajj@ucalgary.ca

V.S. Subrahmanian (ed.), *Handbook of Computational Approaches to Counterterrorism*, DOI 10.1007/978-1-4614-5311-6_15,

By studying relationships within a network we hope to better understand the patterns and dynamics of entities in scope. From the network perspective, patterns refer to well observed phenomena such as modularity (clustered nodes), hubs (nodes with high degree) and small-world effect (nodes are separated by small distances), whereas dynamics roughly refer to the mechanism that drives the evolution of the network. For instance, using the topological information of the network we wish to predict the appearance/disappearance of links in other time slots for better vision about the future evolution of the network, e.g., [18, 24, 27]. This requires employing a sophisticated link prediction technique that works with high accuracy.

Networks in general and social networks in particular have been recently heavily investigated in the scientific community [19]. Their importance as essential source for knowledge discovery has been well realized by practitioners and researchers, including law enforcement agencies and intelligence services. The social network model is powerful enough to be applied to various domains, including software engineering, performance analysis, study of customer behavior, project management, terror and criminal networks, etc. To build the social network model, it is necessary to decide on the actors first and then the links are added based on the particular domain to be studied. For instance, in software engineering actors could be modules and a directed link from module A to module B may reflect the number of time A calls B; in another setting links may be undirected and may reflect the number of projects in which the two modules coexist [7, 22, 25].

Social Network Analysis (SNA) is an approach commonly used to analyze relationships between people or objects structured as a network formed by nodes and links [26]. The aim is to ameliorate the understanding of social relationships, and this goal is achieved by summarizing patterns and modeling dynamics as aforementioned. The analysis may lead to valuable information which otherwise could not be acquired by traditional means. The development in technology allowed for better accuracy and scalability by using computer based tools.

In this chapter, we study terror and criminal networks, which is an old problem with renewed interest. This problem has received considerable attention for long time and the interest in developing more effective approaches has considerably increased over the past two decades. The wide availability of raw data which could be analyzed to derive entities and relationships encouraged the development of more sophisticated techniques and tools. Tools should be capable of extracting from some given sources of raw data (including social media and traditional Web sites, an even manually collected and maintained documents) all entities and relationships that could lead to the social network to be analyzed. Researchers and developers are working closely to produce more effective solutions with the ultimate goal defined as to develop an early warning system that could identify terrorists and criminals with the capability of alerting for upcoming activities. This will lead to avoiding and preventing potential terror and criminal activities instead of post acting after the incident is over with increased number of victims as well as social and economic destruction. Actually, the study of terror and criminal activities is of central interest in modern society and the practical implications are immense. Intelligence services, law enforcement agencies and security consulting firms have great interest in crossing the criminal patterns and predicting the behavior of terrorists. However, the challenges are also difficult to overcome. Questions like what patterns can be thought of as significant or how to validate the predicted results have been posed in both empirical and theoretical studies [8, 17, 25].

To contribute to the ongoing activities in the field, we introduce a novel framework which integrates network methods and data mining techniques to model and analyze terrorism networks. We first describe a tool which is capable of computationally constructing the terror/criminal network from publicly available data, including online social media and Web-based documents. Our approach depends on searching the co-occurrences of some keywords which are provided by a domain expert. Alternatively, the tool includes the option of extracting the keywords from the given sources by analyzing a sample text from the given sources. We partition the constructed network into subnetworks by considering different link types. then we study the subnetworks in details in order to find hidden links. We also study the dynamics of the terror/criminal network by link prediction. We describe a method for link prediction by classifying links using node and interaction features particular to the network under investigation. Overall, through the study we demonstrate an automated systemic way to handle the work flow of analyzing terror and criminal networks. Our approach can be applied to other types of social networks in different settings.

The rest of this chapter is organized as follows. Section 2 describes the procedure and the tool which have been used to construct the terrorist/criminal network using text extraction and mining techniques. In Sect. 3, we present a network partitioning method to study the subnetworks which could be extracted from the original network by considering different link types and this will allow us to find hidden patterns. The employed link prediction method is introduced in Sect. 4; we show how the prediction model is constructed using existing network features. We conclude the investigation in Sect. 5 with future research directions.

2 Network Construction

In general, large volumes of valuable data available for analysis and research is in raw text format. Such data is useless unless it is transformed into a format that can be processed by sophisticated computing techniques for knowledge discovery. We assume manual processing is unaffordable as the size of the data grows large. Finding the target entities within the data and deriving the interactions from the data required great effort. Thanks to data mining and machine learning techniques that facilitate the process smoothly. Our framework described in this chapter contains a basic component for raw data analysis in order to derive the required entities and relationships.

At an early stage of the analysis, relevant content from the raw data need to be identified and mapped into another format which makes the analysis easier and more applicable for automated approaches. Once analyzed effectively, sources of raw data related to terrorists and criminals could lead to valuable information and hence will make it possible to derive a social network that summarizes all the entities and relationships available within the data and related to the particular problem under investigation.

Network based representation of the data opens the door for various data mining techniques to be applied on the produced network for effective and useful knowledge discovery which is otherwise hard to achieve. A social network presentation of information/data contains a set of nodes and set of edges, edges represent specific relationships to connect nodes together. Nodes of a network can be any set of entities. For example in a social network that represents data related to terrorism nodes could represent individuals or even activities, while edges could reflect various types of communication. Once a network is obtained from the raw source of data, more automated data mining techniques can be applied. Also when the required information is in a raw source of data, a text mining technique is needed to extract the desired nodes and relationships [26].

Data mining is generally described as the process of discovering patterns implicitly available in the data. Thus, data mining could be seen as the process of extracting implicit, unknown, and potentially useful information from data. The problem is the unavailability of this hidden information for automatic consumption leading to effective decision making. Text is just as opaque as raw data when it comes to extracting information. The extracted information should comprehensively cover the analyzed domain and this depends heavily on the employed text mining technique and the postprocessing of the outcome.

While it is possible to find raw data related to almost every application domain. In this work we concentrate on the data that could be used to derive criminal/terror networks. Criminal/terrorist related data analysis is becoming widely necessary in the research community in order to support the ongoing efforts to dissolve these networks and hence eliminate their threat. The structure of communication among criminals/terrorists is known to be complicated and deceiving. Criminals may minimize their interactions to avoid attracting attention and they try to hide patterns in their activities. Incorrectness and incompleteness of the information related to criminal activities is another problem that researchers and professionals suffer from. Incorrectness may be due to several reasons. For instance, it can be due to the fact that criminals tend to hide their identities and hence they use false identities; then realizing all possible identities of the same person would be the challenge in order to eliminate the incorrectness. It can also be due to an error from the investigators themselves who are trying to collect information about criminal activities. The former may be resolved by employing some sophisticated automated techniques that base their analysis on a number of features including location and time.

Conducting an analysis on criminal/terror networks requires these networks to be in specific format in which criminals/terrorists are nodes and associations between them are represented as links. However, these association relationships are not quite clear in a raw data format. Therefore special effort is required to extract information from raw data sources in an efficient way. Analyzing the network becomes easier once the data is analyzed the right way and the target network is derived such that it reflects well the current state as present in the analyzed data. The more the network represents the data the more concrete results will be produced from the analysis. Even if the required network has been extracted successfully and efficiently, researchers are still facing another obstacle which is due to the dynamic structure of criminal/terror networks. Networks are always subject to change in a fast pace. Thus, a successful data transformation and data analysis technique should consider the dynamics of criminal/terror networks [24, 25, 28]. These problems are tackled in this chapter.

The rest of this section describes the first stage of the proposed framework. We extract the required entities and reconstruct a network that includes relationships between terrorists, related events and terms collected from a raw text source of information. We mainly used a web crawler to collect relevant sources from the Web and then utilized the content for further analysis. To achieve the target of network construction from raw data using text mining, we have built a tool that discovers and associates a set of keywords with a set of nodes to build a social network representation of the information extracted from the given text. We named the tool "Net-Builder". There is also some existing tools as well such as [2] that allows users to extract instances associated with a user-specified schema, independently of the domain about which users wish to extract data.

Net-Builder is a tool developed by us, to build networks from raw text based on given parameters and constraints.

- **Design:** Net-Builder consists of five options tabs, the work flow starts from specifying the nodes to be included (searched for) within the text (documents, PDFs, HTML). Nodes are expected to be received in CSV format (comma separated). The next step in the work flow is entering/importing a collection of keywords which will be the base of the co-occurrence relationships between nodes. Keywords occurrences constraint can be tweaked depending on the type of the searched documents; Net-Builder gives the ability to limit keyword occurrences to the same paragraph or to the entire document. A relationship (connection) between two nodes is determined to exist if one or more of the keywords within the list of keywords provided exist along with the two nodes, within the same paragraph or within the same document as per specifications and settings.
- **Accuracy:** Net-Builder detects relationships (connections) in a base of words-words occurrences. The degree of confidence of a relationship existence is relying on the assumption that stating the name of a node aside with a special word within the same paragraph indicates some sort of a relationship. The relationship is determined from the keyword list, and thus the accuracy of detecting true connections between nodes is partially dependent on the accuracy of keywords choice, non-conflicting choice of keywords, and relevancy to the desired connections. Another factor affecting accuracy of connections detection is the domain in which we are detecting connections for and the source of data, the more the text is written in a standardized way the more reliable the detection

becomes; as an example: text mining scientific publications will be more reliable in extracting regulatory relationships in biomedical research as it would be in other sources or in other non-scientific sources or sources that do not following standardized formats to some level.

- **Confidence Score:** A confidence score reflects the repetition of a connection existence between two nodes and a keyword; the more frequent a connection exists the higher the score goes up. Confidence score is presented as a sequential number equal to the number of repetitions. Networks built from the constructed connections project the confidence score in the thickness of the connection edges.

2.1 Network Re-construction

Net-Builder is designed to adapt to changes by allowing the user to set various input parameters. Input parameters in Net-Builder play important role from the stage of extracting information from raw data until the formation of the required association network. Many features in Net-Builder serve to reduce some of the criminal/terror network structure complexity. Some of the features and how they play a role in criminal/terror network analysis are illustrated next.

- **Nodes:** Nodes in Net-Builder either can be entered manually by the user or can be imported from a file. Then association relationships are built to connect the nodes. Criminal/terror networks have fuzzy boundaries due to their volatility and dynamism. It is very likely to have ambiguous boundaries in such networks. This makes it harder for the analyst to decide on which nodes should be included and which nodes should be excluded at a certain stage. Using Net-Builder, different networks for different nodes (criminals/terrorists) can be built with the least amount of human involvement. This allows researchers and law enforcement agencies to study networks of group of criminals/terrorists individually. The flexibility and the ease of creating different networks provide the chance to overcome the problem of ambiguity in their boundaries. This may also allow researchers and practitioners to study how the networks could evolve over time.
- **Keywords:** Keywords in Net-Builder form the basic source depending on which the binding links between criminals/terrorists will be built. Keywords provided to Net-Builder differ depending on the objective of building the network and what the researchers/analysts are looking for. For instance, if the association network is intended to represent the regulatory relationship between genes, i.e., gene-gene relationship and interaction [3], then keywords such as regulate and interact should be provided. Keywords/rules will indicate which binding link should be associated with genes/nodes. However, nothing useful will be achieved in case the same gene regulatory network related keywords are supplied to Net-Builder with the raw data changed to include criminal/terror network

related stuff. Therefore, it is necessary to depend on a domain expert to provide the relevant keywords. Alternatively, Net-Builder gives the opportunity to extract the keywords from the sources by preprocessing some sample data.

Full-sentence parsing algorithms are used here to relate keywords to nodes within the raw text documents. We allow the user/researcher to add individual weight for each keyword for either being found in the same document or the same paragraph. The outcome will be reflected on the thickness of the binding link. Keywords/rules can be either set manually using Net-Builder GUI or imported from a text file. The need to provide the ability of importing the keywords along with their weights enforces us to create the following specific format to reprcsent keywords and weights:

<Keyword1>[Document:<DocumentWeight>,Paragraph:<ParagraphWeight>] ||/&&
<Ke-word2>[Document:<DocumentWeight>,Paragraph:<ParagraphWeight>]
...
<Keyword(n)>[Document:<DocumentWeight>,Paragraph:<ParagraphWeight>]

One of the main obstacles in building criminal/terror networks is defining the relationship between criminals/terrorists. Relationships are dependent on the investigated case, the investigation process, previous knowledge about some actors in the network, and finally what investigators already know about communication techniques between criminals/terrorists. This information is mostly incomplete and may become available incrementally. The use of keywords in Net-Builder to extract information from raw data with the ability to associate different strength to different keywords allows the researchers/analysts to analyze the structure of networks that have been built from different sets of input data. For example, a criminal/terrorist group may have a communication code to facilitate smooth and undiscovered interaction between the members. This code related fact may be found at some point during the investigation process. Communication codes become critical to the process of building relationships once they have been discovered. Further, a tool to extract information from raw data must consider the dynamic changes in the information acquired by investigators. Studying how the network changes is crucial to keep track of its members and how they change role and communicate.

- **Dates:** The ability to build networks in Net-Builder by considering the input data collected during different time slots, allows criminal/terror network analysts to compare activities and communications between a set of groups or suspects before and after the crime date, or even before and after other special dates related to crime events that have been discovered during the investigation. This also allows analysts and law enforcement agencies to watch the evolution of the network over time. Further, knowing how the trend of the criminal/terrorist network has evolved before and after the crime date would guide researchers, investigators and analysts in their effort to develop early warning techniques.

We used the Net-Builder tool (as shown in Fig. 1) to extract terrorism related information from terrorism related documents, mostly Web-based sources and

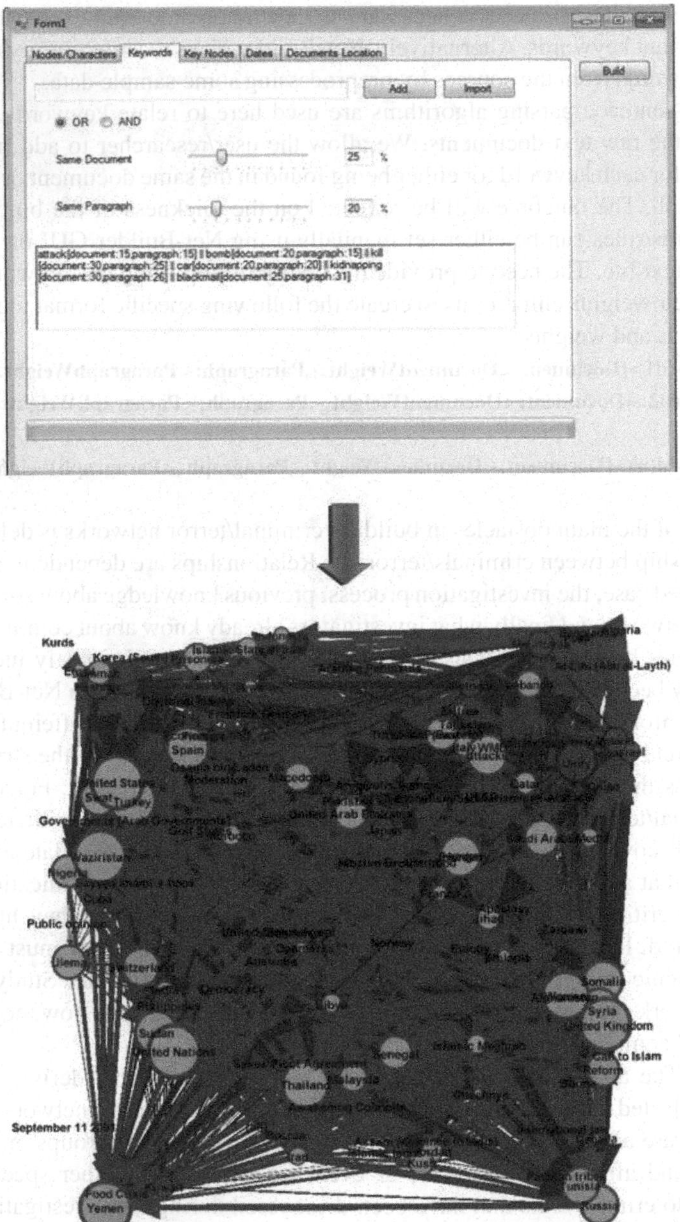

Fig. 1 Screen shot of the network builder feature of Net-Builder (keywords settings)

social media sources. To construct the terror network, we used the Worldwide Incidents Tracking System (https://wits.nctc.gov) and extracted the entire incident data separated by month and by year, ranging from January 2004 to September 2011.

The dataset totally contains 84,922 incidents worldwide. Details of each incident include subject, summary, date, incident type, weapon type and victim list, etc. The keywords that we use in our framework have been extracted from Al-Qaeda Statements Index (http://people.haverford.edu/bmendels/keyword.html) to associate with the incident dataset for the network construction.

The target is to construct a network that reflects the relationships between criminals/terrorists in a way to represent the occurrence of a set of keywords (individuals, groups and terms) within the same context. A relationship (link) is constructed between two keywords if they appear within the same context. Frequency of occurrences of the same pair of keywords (link) has been also tracked and associated as a weight assigned to the link. To distinguish between different types of links, we kept track of each document and the date it has been released. Each link is also associated with a link type that refers to the source and date of release. The constructed network contains 2,403 links with 21 link types (therefore highly clustered) as shown in Fig. 1. The link types are essential for discovering hidden links in the network as described in the next section.

3 Network Partitioning

Social network analysis provides an effective and useful way for analysts and law enforcement agencies to investigate criminal/terrorist activities in order to identify key persons, groups, agencies, countries etc. [11, 19] who are backing and playing various roles in order to keep the momentum of the terror and horror. However, here the investigators face a major problem which is missing and hidden links in the network [18]. This may lead to partial information and hence not very precise estimation. This problem is known as incompleteness, which Sparrow identified along with two other main problems for using social network analysis for criminal activities [25].

Incompleteness in social network analysis for criminal/terror activities is defined as having missing nodes and/or links in the network. This is very common and natural problem that cannot be avoided because criminal and terror networks are volatile and dynamic such that it is hard to keep a full coverage of them at any time. This is due to intelligence services and the law enforcement agencies' inability to uncover every relevant node and link. Further, the problem can arise because criminals may attempt to hide their ties to other criminals or events in order to minimize their connection with criminal activities [17]. For example, one simple method for criminals to conceal their connection with another person or event is to have a middleman who will handle all the communication in a well articulated way that will never lead to the two parties to know each other. In this case, there would be a link between the criminals and the middleman, as well as a link between the middleman and the other person or event. However, there will be no direct link between the criminals and the other person or event. Furthermore, criminals can have several middlemen in between in order to reduce the weights of the links by spreading their connection with all the middlemen, thus making it less obvious that there is any connection between the criminals and the other person or event. This makes it harder to discover persons in a criminal or terror network. In addition, criminals can place several middlemen in between themselves and the other person or event, thus further distancing themselves, and making it more difficult for intelligence services and law enforcement agencies to discover the connection. These are simple methods which can be employed by the criminals, but they result in effectively eliminating or hiding the link in the social network and hence prevent effective analysis by investigators and law enforcement agencies.

The problem of finding the missing or hidden links requires some knowledge about the status and various known components of the existing network. Some historical information will also help. The problem can be considered similar to the problem of predicting links for the future state of a social network [19]. The main difference is the context of the uncovered link, whether or not the link exists in the current time-frame. Although there are methods that uncover links, which represent either links likely to exist in a later time-frame or hidden links in the current time-frame such as Backstrom and Leskovec's work [5]; this research work focuses only on uncovering hidden links within the time-frame of the network data. Leskovec also has done similar work in predicting positive and negative links [14], and although signed networks are not explored here, the method can still be applied to uncover positive or negative links. This can be accomplish by splitting the network into two with the positive and negative links occurring only in one network, thus when the method is applied to each network, it will uncover hidden positive links in the positive network and hidden negative links in the negative network. Once hidden links are estimated or found, they can be integrated in the network which then can be used for more comprehensive analysis [12, 16, 21]. The solution to this problem will guide analysts and law enforcement agencies to concentrate on the unconnected entities of a social network, and find the number of different indirect links in the different social networks between the unconnected entities. A higher number of indirect links between two unconnected entities than the average would indicate a potential hidden link, and thus should be further investigated by the analysts or the law enforcement agencies. This however can be a daunting task when the social network grows large, as there can be too many factors to manually keep track of the whole process. In this case, an automated solution becomes a more appealing method. An automated solution will lead to more concrete and robust results. Trivial errors that may be committed by humans performing manual analysis will be eliminated when an automated system is developed and utilized.

The general idea for the automated solution is to use the various sources of data to create social networks based on the type of relationships between the entities. This differs from some other methods that focus only on one type of relationship, e.g., [6, 20, 24]. Starting from these social networks, we can generate more networks that represent all the possible hidden links. In these networks, the weights between the nodes indicate the number of indirect links between the nodes in the generated social

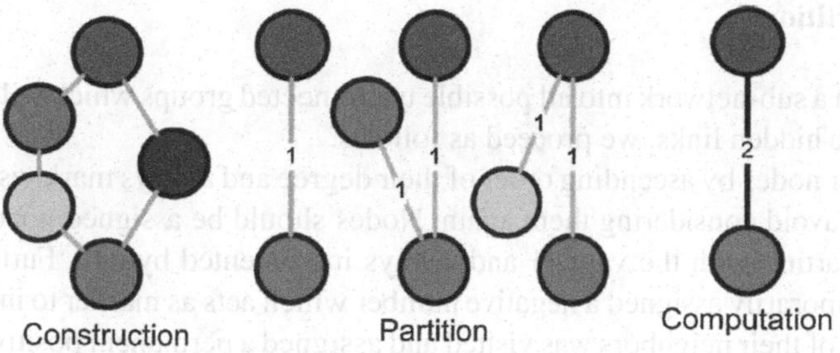

Fig. 2 Network partitioning process

networks and based on which the type of relationships is revealed. Afterwards, we look for links with the highest weights. These links would indicate that the entities do not have a direct link, but do have many indirect links across the different social networks generated based on relationship types available in the raw data. These are the links that are significant, and thus they are in need of more investigation.

3.1 Method

The overall process employed in the proposed framework is depicted in Fig. 2. We detail the method in the sequel.

3.1.1 Construction

The hidden link prediction method uses data that is structured to connect two entities with a specific relationship type. It is a general method that can work for any type of social network. The idea is to start by creating a main social network of all the entities with a link for specifying the different relationship types connecting the two entities. The next step is to split the main social network up into several smaller sub-networks based on relationship types such that each relationship type has its own social network. Links in the latter sub-network represent the existence of the corresponding relationship type in the main network. Afterwards we need to partition each of the sub-networks into groups of nodes that could lead to potential hidden links. The partitioning method presented below more closely resembles graph coloring techniques rather than graph partitioning approaches such as the multi-level partitioning method used in the METIS graph partitioning framework [15]. The main difference is that the partitioning method creates sets of nodes with no connections in the sets, whereas the multi-level method used in METIS generates sets of nodes with minimal connections between the sets [15].

3.1.2 Partition

To partition a sub-network into all possible unconnected groups which will represent the possible hidden links, we proceed as follows.

Consider nodes by ascending order of their degree and always mark visited nodes in order to avoid considering them again. Nodes should be assigned nondecreasing numbers starting with the value 1 and always incremented by one. Further, nodes may be temporarily assigned a negative number which acts as marker to indicate that at least one of their neighbors was visited and assigned a permanent positive number. At the end of the process, each assigned positive value, say v_a will represent a set which will include all the nodes which were assigned the value v_a. In other words, the number of sets will be the same as the number of assigned values. To find the sets of nodes, the following steps are applied.

1. Start at the next unvisited node with the lowest degree, let this node be n_u.
2. Assign to n_u as its number the next unassigned value, say v_u. This means node n_u belongs to set v_u.
3. Locate all direct neighbors of n_u.
4. Assign the value $-v_u$ to each direct neighbor of n_u. The number assigned to each neighbor of n_u is the negative counterpart of the positive number assigned to n_u. This negative number assignment is required to indicate that the neighbor shouldn't be considered as part of the same set with n_u since they are directly connected.
5. Locate all non-neighbors of n_u.
6. Assign the value v_u to each located non-neighbor of v_u. This means v_u and all its non-neighbors got the same value and hence are in the same set v_u.
7. Repeat steps 1–6 for every neighbor of n_u and to neighbors of already considered neighbors in either a breath first or depth first manner. It is worth noting that in case a node has already been assigned a number, excluding negative numbers, there is no need to assign to it a new number because it is already going to be considered in a set.

After the above steps are applied to every node in a component of the sub-network, we can start building the sets of nodes, which represent groups of nodes that have potential hidden links with each other. We build a set for each number assigned to the nodes. Nodes holding the same number are included in the set dedicated for their number such that there is no negative assignment of the same number. We ignore nodes that are assigned numbers, but in addition, are also assigned the negative of a number because this case indicates the need to further break the network down before assigning the mentioned nodes to a set.

The next step is to create a subset for each of the sets that correspond to numbers which have nodes assigned the positive and negative of the number. These subsets are filled by applying the same steps enumerated above on the sub-graph of nodes that were assigned both the positive and negative values of the same set number. The sets that are created from the sub-graph will be considered as the subsets of the original set number, and this recursion will continue until the sub-graph is empty.

After the above steps are completed for each component of the sub-network, we will get a hierarchy of sets with nodes, and from this we can use a depth first method to create the hidden groups. All the hidden groups represent the maximal of all possible groupings of nodes which are not directly connected in the sub-network. These grouping of nodes will be referred to as the hidden sets.

This partitioning method should also be applied to the main network, where all the different relationship types are considered for links. Each hidden group generated by this method can be treated as a complete graph such that all the nodes in the hidden group are connected to all the other nodes in the hidden group with a link weight of 1. These new networks will be referred to as the hidden networks.

3.1.3 Computation

The final phase of the hidden link prediction process is to update the weights of the links in the hidden networks. This will reflect the number of indirect links between two nodes in a hidden network across the different social networks generated based on the type of relationship. This is achieved by comparing every hidden set generated from each sub-network to every hidden network. If there is a link in the hidden network and the two linked nodes are both in the hidden set, then increase the weight of the link by some value. This is done as long as the weight has not already been increased for the run through that sub-network hidden set comparison. The value to be used for incrementing the weight is set to 1 in the conducted experiment and the results are reported in the next section.

When the increment is set to 1 for all situations, then this simply counts the number of times the nodes are indirectly linked in the social network of relationship types. Although the increment can take different values for different relationship types, one relationship type should be given more weight if there is an indirect connection with that relationship type. The increment can even become a decrement for relationship types that lower the likelihood of a hidden link. When the computation of the weights for the links of the hidden network is completed, the last step is to locate the links with the highest weights for investigation.

3.2 *Results*

The automated hidden link prediction method described above was applied to the social network mined in Sect. 2 and shown in Fig. 1. Originally the method produced inconclusive results as all the possible hidden links in the hidden groups generated had a weight of 1. Looking at the original mined social network, this result can be attributed to the fact that too many insignificant links and nodes were included in the main network. The range of the frequency of occurrences or the weight of the links was between 112 and 3,012, and this means that many irrelevant links were being used to create a concrete link. The consequence of this property is creating a dense network, and this essentially tries to consider all missing links with even small possibilities of being hidden as being actual links in the network. This is not a bad strategy when having many false positive links in the network is tolerable to analysts and law enforcement agencies, and missing an actual hidden link is not an option. In this case, the automated method is not necessary as all the possible hidden links are already being considered as actual links in the network. Therefore there's no need to find potential hidden links. In other words, running the method that highlights the potential hidden links will confirm the strong connectivity of the network in hand. It will find any potential hidden links in case they were missed by the network generation process. Therefore, it can be considered as a postprocessing step that guarantees the completeness of the network construction step based on the current data which was made available.

To better show the effectiveness of the automated hidden links prediction method, the original network was filtered to only include significant links, and disregard all other links and any isolated entities. Significant links were defined as links with weights above the average of the weights which was computed as 920. Data with less than a 920 frequency of occurrence was disregarded and considered insignificant because these frequencies were too low compared to the rest of the data.

After applying the automated hidden links prediction method described in this section to the filtered network, the results showed that the majority of the weights for the hidden groups generated were 1 or 2. Potential hidden links with a weight of 1 accounted for 38.9 % of the total, and potential hidden links with a weight of 2 accounted for 51.2 % of all the possible hidden links. In total, the majority of the weights of 1 and 2 were 90.1 %. This is because these two low weights make up such a large portion of the potential hidden links that they are all considered insignificant. On the other hand, there were also potential hidden links with a weight of 3 which made up 6.4 % of the total, and potential hidden links with a weight of 4 which made up 3.5 % of all the possible hidden links. As these two weights were of higher value than the majority of the weights for the other links, they can be considered as outliers, and thus significant. They are significant because in the filtered social network the unconnected nodes are normally only indirectly linked in one or two other social networks when broken down based on the type of relationships. For potential hidden links with a weight of 3 and 4, this means that the nodes are indirectly linked in three or four different social networks when broken down based on the type of relationships; the latter cases should be considered as outliers as they have more indirect links than what is expected.

Potential hidden links of weight 4 were indicated by the method for links between Canada and Japan, Canada and Kush, Canada and China, Italy and Kush, as well as between Japan and Kush. In the original unfiltered network, all the links were present, but of course with a frequency below 920, which justifies why they did not show up in the filtered network. This means that the occurrences of these keywords in the terrorism documents could potentially be under-represented, and there may be some dependencies related to terrorist activities between these regions. In particular, each of the uncovered hidden links with a weight of 4 involves either Canada or Kush, which means that these two regions may be an important base of operations related to the documented terrorism events. The automated hidden links prediction method indicates that these links should be further investigated, as they might be significant for the filtered social network. The potential hidden links of weight 3 may also be further investigated, but the rest of the potential hidden links with weights of 1 and 2 should be ignored.

The automated hidden links prediction method creates hidden networks that give weights to all the possible hidden links. This provides valuable information for investigators and law enforcement agencies as they can use the weights to help determine which hidden links are more likely to be real for further investigation, and therefore reduce the number of hidden links in their network. The method also provides flexibility to analysts and law enforcement agencies in that it allows for modifications for different scenarios, as the increment weight can be adjusted for different relationship types. Accordingly, this justifies why the method is an excellent tool for law enforcement agencies to reduce the number of hidden links in criminal networks; this will improve the results of the analysis by addressing one of the major issues for these types of networks.

4 Link Prediction

Link prediction, introduced by Liben-Nowell and Kleinberg [19], refers to a problem underlying a given social network in order to predict a link (relation) between two nodes for a different snapshot of time in the future. Given a snapshot of a social network at time t and a future time t', the problem is to predict links that reflect new relationships which are likely to appear in the network within the time interval $[t, t']$.

The problem of link prediction has attracted considerable attention in the research community, e.g., [7, 10, 13, 27]. Once achieved with high accuracy it gives insight about the evolution of the social network and hence allows for better planning. However, researchers focused mostly on predicting how a social network may change by adding new relationships or links between nodes. In other words, most of the previous research efforts related to link prediction have concentrated on predicting relations that will be added to nodes in the future given that these relations do not exist between the two nodes in the current state of the network. Moreover, researchers have limited the implementation of the link prediction model in specific domains such as co-authorship and limited the prediction to the addition of new links to the network. To the best of our knowledge, there is no work that has been done before to predict links which are expected to be missed or dropped from the network in the future. This a very common case that should be considered in order to produce a more comprehensive coverage of the investigated domain.

In this section, we describe a link prediction model called positive link prediction (PLP). We have already demonstrated the applicability and effectiveness of this model for link prediction in health informatics and systems biology [4]. In this section, we apply the PLP model to criminal/terrorist networks in order to accurately predict potential links that may appear between nodes in the future. To achieve this, we consider features that are based on the link structure of the network. In addition, we introduce a negative link prediction model (NLP) to accurately predict which links may be dropped from the network in the future. When combined together, both PLP and NLP will lead to a more realistic projection of the network in the future.

Studying the network model derived from terrorist/criminal activities has recently attracted the attention of researchers who try to produce robust techniques to the satisfaction of intelligence services and law enforcement agencies. Researchers mostly try to track the terrorists' activities by studying network measures such as betweenness, centrality, closeness, density, etc. The work described in [8] introduced a measure to study the danger of a terror network structure based on social network analysis measures to discover terrorist leaders based on the location of the nodes in the network which could be well tracked to avoid any future possible terrorist/criminal activities by their groups.

Problem Formulation: Given two nodes of terrorist organizations that appeared in the same terror related article, the process described in Sect. 2 constructed a social network as a one-mode network graph $G = (V_1; V_2; E)$, where V_1 is an organization, V_2 is another organization and E is the connection between them reflecting the appearance of the two organizations in the same terror related articles, i.e., the existence of a link between organization A and organization B means that both organizations have been mentioned in the same terror related articles at least once. Then, we seek to predict hidden relationships between two organizations that did not appear in any article in the past. We try to predict how the two organizations are expected to be linked in the same social network constructed at a different time in the future. In addition, we seek to predict if a relationship connecting two nodes will be removed from the network given that the two nodes have a relationship in the current state.

4.1 Link Prediction Method

The link prediction method employed in the framework described in this chapter consists of two parts, PLP and NLP as described next.

Positive Link Prediction (PLP)

In the positive link prediction task, our goal is to accurately predict future hidden relations between terrorists given that the relationship does not exist in the current network structure. We tested this model by using terror network data. This task is achieved by applying the following steps:

- **Data Preprocessing:** We extracted terrorist organizations that appear in the same investigated articles and form an adjacency matrix where rows columns represent organizations that appeared in the considered articles. Entries in the matrix are either 0 or 1 to represent whether the two organizations appear in the same article or not. We then construct the social network where nodes represent organizations and links indicate that the two organizations appeared in the same terror related article.
- **Feature Extraction:** To apply the link prediction model, a number of features have to be extracted in order to be used in building the classification model which will accomplish the prediction task. Choosing the right features forms the key towards a successful solution for the link prediction problem. We have built our classification data set based on the following network structure features [1, 23]: authority, betweenness, clique count, closeness, density clustering, cognitive distinctiveness, cognitive expertise, centrality column degree, eigenvector, and degree. As the obtained results were satisfactory, we decided to avoid overloading the model with more features.
- **Classification:** After we extracted the features, we present the features in a matrix form so that rows mark a pair of nodes and columns represent feature values. The last column of the data is the class label which is considered as either 1 or 0 to indicate where there is a current link between the two nodes or not, respectively. Then the matrix is applied to the selected classifiers in order to perform training and testing.

Negative Link Prediction (NLP)

The aim of this model is to accurately predict the links that are more likely to be dropped from the network at some point in the future. To form the matrix we used the same data preprocessing steps described above for the PLP. Then we applied the following steps:

1. Replace 0 by 1 and vice versa in the original matrix.
2. Extract the features of the inverted matrix. We use the same set of features we extracted for the PLP model.
3. Predict the new links based on the inverted matrix.
4. Invert the new matrix that resulted from the previous step.
5. Discover which links are reported to be missing by comparing the matrix from step 4 to the original matrix.

We tested both the PLP and NLP models using the network shown in Fig. 1. Next, we describe the testing results of both models. Some of the discovered links are shown in Tables 1 and 2 for the positive and negative links, respectively, where 0 indicates that the link is missing between the nodes and 1 indicates that the link exists between the two nodes.

Table 1 Positive link prediction example

Nodes		Original link status	Predicted link status
Org1	Org11	0	1
Org2	Org12	0	1
Org3	Org13	0	1
Org4	Org14	0	1
Org5	Org15	0	1
Org6	Org16	0	1
Org7	Org17	0	1
Org8	Org3	0	1
Org9	Org18	0	1
Org1	Org16	0	1
Org1	Org12	1	1
Org3	Org19	1	1
Org2	Org20	1	1
Org4	Org12	1	1
Org10	Org16	0	0
Org6	Org21	1	1
Org7	Org22	1	1
Org8	Org1	0	0
Org9	Org23	1	0

Table 2 Negative link prediction example

Nodes		Original link status	Predicted link status
OrgA	OrgT	1	0
OrgB	OrgU	1	0
OrgC	OrgV	1	0
OrgD	OrgW	1	0
OrgE	OrgX	1	0
OrgF	OrgY	1	0
OrgG	OrgZ	1	1
OrgH	OrgAA	1	1
OrgI	OrgY	1	1
OrgJ	OrgBB	1	1
OrgK	OrgCC	1	0
OrgL	OrgS	1	1
OrgM	OrgW	1	0
OrgN	OrgDD	1	1
OrgO	OrgEE	1	1
OrgP	OrgDD	1	1
OrgQ	OrgEE	1	0
OrgR	OrgW	1	0
OrgS	OrgQ	1	0

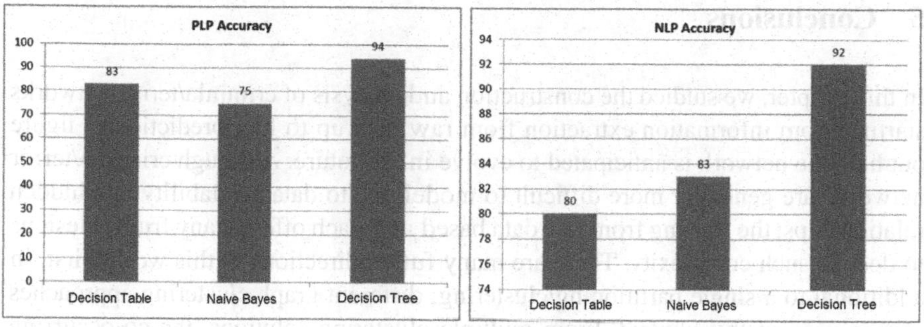

Fig. 3 PLP accuracy and NLP accuracy

4.2 Results and Discussions

In the testing we used three classifiers, namely decision table, decision tree and naive Bayes model. The results are reported in this section.

4.2.1 Success Criteria

The success criteria are expressed in terms of percentages of accuracy for the PLP and in terms of graph statistics for the NLP. Accuracy is computed as the ratio of the sum of all data points classified correctly over the number of all data points.

$$Accuracy = (TP + TN)/AllPoints$$

Figure 3 shows the classification accuracy as a result of PLP and NLP, respectively. The accuracy is encouraging as reflected by the results reported by the naive Bayes model: 94 % for the PLP model and 92 % for the NLP model. The excellent success reported by the classifiers demonstrates the importance of the selected features based on the graph statistics and the location of the terrorist related organizations in the network. Well understanding the structure of the criminla/terror network is not enough to know the hidden terrorist relationship that may be established in the future since terror network structure is volatile and dynamic. Predicting the hidden information is essential. For example, predicting that there is a terrorist relationship or activity between two organizations will be very helpful to stop the threat and prevent future terror attack. On the other hand, knowing when a terrorist relationship is going to end is very crucial because it gives an indication of a strategy change in the plan of criminals/terrorists.

5 Conclusions

In this chapter, we studied the construction and analysis of criminla/terror networks starting from information extraction from raw data up to link prediction to figure out how the network is anticipated to evolve in the future. Although criminla/terror networks are generally more difficult to model due to data availability and hidden relationships, the starting from raw data based approach offers many fruitful lesions to decode such complexity. There are many future directions in this work. First, in additional to a single partitioning/clustering, different graph clustering approaches can be used in this context. From multiple clustering solutions, the co-occurring relationship can be statistically quantified, i.e., how many entities occur in k number of clustering solutions. This way, we will be able to further deepen the clustering model to incorporate multi-object optimization techniques, and build a Pareto front to find equivalences of optimal solutions for analyzing terrorist relationships. Second, in view of link prediction, it will be advantageous to consider the multi-mode nature of the network besides simple co-occurrences. A multi-mode network is a graph that involves a large number of relationship types. Although we have considered the relationship types in the partitioning process, link prediction does not take the multi-modes into consideration. The purpose of this multiplexity is to further account for more complex relationships in criminal/terror networks. However, to include the factor of multiplexity in predicting links is a much harder problem, which leaves an alluring interesting problem for us to investigate.

References

1. http://www.fmsasg.com/socialnetworkanalysis/
2. Albanese M, Subrahmanian VS (2007) T-REX: A System for Automated Cultural Information Extraction. In Proceedings of the First International Conference on Computational Cultural Dynamics (ICCCD 2007), pp 2–8, AAAI Press, College Park, Maryland, USA, August 27–28
3. Albert I, Albert R (2004) Conserved network motifs allow protein-protein interaction prediction. Bioinformatics 20(18):3346–3352
4. Almansoori W, Gao S, Jarada TN, Elsheikh AM, Murshed AN, Jida J, Alhajj R, Rokne J (2012) Link prediction and classification in social networks and its application in healthcare and systems biology. Netw Model Anal Health Inf Bioinf 1(1–2):27–36
5. Backstrom L, Leskovec J (2011) Supervised random walks: predicting and recommending links in social networks. In Proceedings of the fourth ACM international conference on Web search and data mining. ACM, New York, pp 635–644
6. Chen ACL, Gao S, Karampelas P, Alhajj R, Rokne JG (2011) Finding hidden links in terrorist networks by checking indirect links of different sub-networks. In: Wiil UK (ed) Counterterrorism and open source intelligence. Springer, Wien/New York, pp 143–158
7. Clauset A, Moore C, Newman MEJ (2008) Hierarchical structure and the prediction of missing links in networks. Nature 453(7191):98–101
8. Dawoud K, Alhajj R, Rokne JG (2010) A global measure for estimating the degree of organization of terrorist networks. In: Proceedings of advances in social networks analysis and mining. IEEE Computer Society Washington, DC, pp 421–427

9. Dombroski MJ, Carley KM (2002) Netest: estimating a terrorist network's structure. Comput Math Organ Theory 8(3):235–241
10. Dunlavy DM, Kolda TG, Acar E (2011) Temporal link prediction using matrix and tensor factorizations. ACM Trans Knowl Discov Data 5(2):1–27
11. Farely D (2003) Breaking al qaeda cells: A mathematical analysis of counterterrorism operations. Stud Confl Terror 26(6):399–411
12. Garreau J (2001) Disconnect the dots: maybe we can't cut off terror's head but we can take out its nodes. In: Washington Post Online. http://edge.org/response-detail/2328/what-now-
13. Getoor L, Diehl CP (2005) Link mining: a survey. SIGKDD Explor Newsl 7(2):3–12
14. Jure Leskovec DH, Kleinberg J (2010) Predicting positive and negative links in online social networks. In Proceedings of the 19th international conference on World wide web (WWW '10). ACM, New York, pp 641–650
15. Karypis G, Kumar V (1995) Unstructured graph partitioning and sparse matrix ordering system. http://dm.kaist.ac.kr/kse625/resources/metis.pdf
16. Klerks P (2001) The network paradigm applied to criminal organizations. Connections 24(3):53–65
17. Krebs V (2002) Mapping networks of terrorist cells. Connections 24:43–52
18. Latora V, Marchiori M (2004) How the science of complex networks can help developing strategies against terrorism. Chaos Solitons Fractals 20(1):69–75
19. Liben-Nowell D, Kleinberg J (2007) The link-prediction problem for social networks. J Am Soc Inf Sci Technol 58(7):1019–1031
20. Magdon-Ismail M, Goldberg M, Wallace W, Siebecker D (2010) Locating hidden groups in communication networks using hidden markov models. In: Chen H, Miranda R, Zeng D, Demchak C, Schroeder J, Madhusudan T (eds) Intelligence and security informatics. Lecture notes in computer science, vol 2665. Springer, Berlin/Heidelberg, p 958
21. Qin J, Xu J, Hu D, Sageman M, Chen H (2005) Analyzing terrorist networks: a case study of the global salafi jihad network. Lecture Notes in Computer Science, 2005, Vol 3495, pp 287–304
22. Ressler S (2006) Social network analysis as an approach to combat terrorism: past, present, and future research. Homeland Security Affairs 2(2)
23. Santos EE, Pan L, Arendt D, Pittkin M (2006) An effective anytime anywhere parallel approach for centrality measurements in social network analysis. In: IEEE international conference on systems, man and cybernetics, 2006. SMC '06, vol 6. pp 4693–4698, IEEE, Taipei, Taiwan, October 8-11, 2006
24. Shaikh M, Wang J (2006) Discovering hierarchical structure in terrorist networks. In: Proceedings of the international conference on emerging technologies. pp 238–244. IEEE, Peshawar, Pakistan, 13-14 November 2006
25. Sparrow M (1991) The application of network analysis to criminal intelligence: An assessment of the prospects. Soc Netw 13(3):251–274
26. Strogatz SH (2001) Exploring complex networks. Nature 410(6825):268–276 (2001)
27. Tylenda T, Angelova R, Bedathur S (2009) Towards time-aware link prediction in evolving social networks. In: Proceedings of the 3rd workshop on social network mining and analysis, SNA-KDD '09. ACM, New York, pp 1–10
28. Xu JJ, Chen H (2005) CrimeNet explorer: a framework for criminal network knowledge discovery. ACM Trans Inf Syst 23(2):201–226. http://dx.doi.org/10.1145/1059981.1059984

CrimeFighter Investigator: Criminal Network Sense-Making

Rasmus Rosenqvist Petersen and Uffe Kock Wiil

1 Introduction

Criminal network investigation involves a number of complex knowledge management tasks such as collection, processing, and analysis of information. Synthesis and sense-making are core analysis tasks; analysts move pieces of information around, they stop to look for patterns that can help them relate the information pieces, they add new pieces of information and iteration after iteration the information becomes increasingly structured and valuable. Synthesizing emerging and evolving information structures is a creative and cognitive process best performed by humans. Making sense of synthesized information structures (i.e., searching for patterns) is a more logic-based process where computers outperform humans as information volume and complexity increases. CrimeFighter Investigator is a novel tool that supports sense-making tasks through the application of advanced software technologies such as hypertext structure domains, semantic web concepts, known human-computer interaction metaphors, and a tailorable computational model rooted in a conceptual model defining first class entities that enable separation of structural and mathematical models.

We focus on criminal network sense-making and how tailoring can leverage transparency and ownership, increasing trust in information provided by sense-making algorithms. CrimeFighter Investigator is part of the CrimeFighter Toolbox for counterterrorism [70]. Besides the Investigator tool, CrimeFighter consists of the Explorer tool targeted at open source collection and processing and the Assistant tool targeted at advanced structural analysis and visualization. The remainder of this paper is organized as follows: In Sect. 2, we discuss criminal network sense-making.

R.R. Petersen (✉) • U.K. Wiil
The Maersk Mc-Kinney Moeller Institute, University of Southern Denmark, Odense M, Denmark
e-mail: rrp@mmmi.sdu.dk; ukwiil@mmmi.sdu.dk

V.S. Subrahmanian (ed.), *Handbook of Computational Approaches to Counterterrorism*, DOI 10.1007/978-1-4614-5311-6_16,

A generic process model for human-centered, target-centric criminal network investigation is proposed. A list of sense-making tasks that guided the development of the tool is also presented. Section 3 describes our developed models and how selected sense-making tasks are supported by the CrimeFighter Investigator tool. In Sect. 4, we describe a scenario in which a criminal network investigator focus on European terrorist networks and applies some of the sense-making algorithms described in this paper. Section 5 reviews related work and compares it with the CrimeFighter Investigator approach to sense-making. Finally, Sect. 6 concludes the paper and outlines future work.

2 Criminal Network Sense-Making

The nature of modeling something as complex and diverse as crime is an ongoing and potentially open-ended process that demands for an interactive modeling approach [8]. What complicates everything is that the picture constantly changes. With every interaction, people change, group dynamics change, and social dynamics change [6]. If we are to think seriously about this sort of complexity, and reason effectively about it, some sort of simplified map of reality, some theory, concept, model, paradigm, is necessary [25]. The CrimeFighter Investigator approach to synthesis is based on three first class entities, which, combined with hypertext structure domains[1] are used to support a set of synthesis tasks.

Our modeling approach must also embrace frequent customization and extension through robustness and scalability of the underlying mathematical framework. At the beginning of an investigation it is not clear what sense-making approach will be required to understand and reason about a certain criminal network. Sometimes more than one measure has to be calculated for the criminal network or maybe some measures are used as input for an algorithm providing yet another measure. It is impossible to know beforehand what information attributes (meta data) will be the deciding factors for a criminal network investigation. First of all, information attributes are emerging over time, just like the information entities. Second, investigators have to decide if they will try to predict missing information entities in the network based on for example an individual's record of supplying weapons or a measure of each individual's centrality in a criminal network.

Taking a computational approach to criminal network sense-making, claiming that investigators will benefit from the information provided, raises concerns about

[1]Structure domains (or simply structures) play a vital role in the design and development of CrimeFighter Investigator. The Hypertext research community has developed a number of structure domains over the years: **navigational structures** allow arbitrary pieces of information (entities) to be linked (associated); **spatial structures** were designed to deal with emergent and evolving structures of information which is a central task in information analysis; **taxonomic structures** can support various classification tasks. See [42] for further details.

user acceptance of this computed information.[2] Experienced investigators with the skills to manually derive the computed information (given more time) might question how exactly the information has been automatically computed and they might be inclined not to trust this computed information enough to base their decisions on it [49]. For computational sense-making to be effective, decision makers must consider the information provided by such systems to be trustworthy, reliable [35], and credible. The calculations are not the hard part; the challenge is to find a good way to use the data and understand them.[3] Simply by turning to the computer when confronted with a problem, we limit our ability to understand other solutions. The tendency to ignore such limitations undermines the ability of non-experts to trust computing techniques and applications [49] and experienced investigators would be reluctant to adopt them.

We propose sense-making transparency and ownership as key CrimeFighter Investigator concepts necessary to achieve trust in sense-making provided information. Sense-making transparency can be achieved through intelligent dissemination of computed sense-making information, e.g., in terms of updating criminal network investigators who did not participate in the sense-making process.

Sense-making ownership is achieved through interaction, i.e., investigators must be able to interact with sense-making algorithms and their sub-functionality (algorithm steps) before, during, and after their execution. We believe that interactions are more powerful than algorithms: "Smartness in mechanical devices is often realized through interaction that enhances dumb algorithms so they become smart agents" [65].

The structure of criminal network information is emerging and evolving [62] and successful sense-making is dependent on humans (investigators) [42, 70] to select and tailor the sense-making algorithms relevant for the case they are investigating. We have chosen our research approach according to the concepts of tailorability, transparency and ownership: *We aim to support criminal network*

---

[2]We have found three studies evaluating user acceptance of intelligence and security informatics technology (COPLINK [24], COPLINK Mobile [23], and POLNET [76]) all based on the Technology Acceptance Model [14]. However, none of these studies ask the users to what degree they trust the information provided by the systems and how that affects their acceptance of the technology.

[3]Computer security expert Clifford Stoll spent a year studying at a Chinese observatory with Professor Li Fang. Li studied star observations and used a Fourier transform, the standard tool of astronomers everywhere, to hunt for periodic motions. Li, however, did the Fourier transform completely by hand! Stoll decided to show Li how his new Hewlett Packard HP-85 could be used to calculate some 50 coefficients for the polar wandering in under a minute. The task had taken Professor Li 5 months. When presented to the computer's results, Li smiled and said: "When I compare the computer's results to my own, I see that an error has crept in. I suspect it is from the computers assumption that our data is perfectly sampled throughout history. Such is not the case and it may be that we need to analyze the data in a slightly different manner". Stoll realized that Li had not spent 5 months doing rote mechanical calculations. Instead, he had developed a complex method for analyzing the data that took into account the accuracy of different observers and ambiguities in the historical record [56].

sense-making by applying a computational model that facilitates tailoring, and thereby ownership and transparency, of sense-making algorithms, to increase end user trust in the provided information. Two main challenges are associated with our chosen approach:

The first challenge is the conflicting nature of criminal network synthesis and sense-making. Synthesizing emerging and evolving (complex) information structures is a creative and cognitive process best performed by humans while making sense of the resulting structures (i.e., searching for patterns) is a process where computers outperform humans as information volume and complexity rises. Typically, this conflict is resolved by developing measures for one specific structure domain, e.g., centrality measures for (navigational) nodes and links structures. However, our studies of complex criminal networks indicate that more than one structure domain is often required to synthesize the network; and if only one structure domain is required, it is not possible to decide which one at the beginning of a criminal network investigation. Challenge 1: *can we provide helpful sense-making algorithms for criminal networks that are synthesized using multiple structure domains?*

The second challenge is the potential conflicts between sense-making algorithms that alter the structure of criminal networks (transformative) and sense-making algorithms that provide a measure of individual entities of criminal networks in their current state (measuring). Often, criminal network sense-making requires that multiple measures are used in combination. In order to understand a criminal network, maybe the investigators require three measures of centrality displayed simultaneously. Or maybe a prediction of missing information entities is followed by an overview of changes in the importance of links in the criminal network. Challenge 2: *can we support the combination of transformative and measuring sense-making algorithms in a suitable way?*

2.1 Criminal Network Investigation Model

Based on a specific target-centric model for intelligence analysis [11], we propose a generic process model for human-centered, target-centric criminal network investigation [41, 42]. The model is presented in Fig. 1.

The customer requests information about a specific target. The investigators request information from the collectors (that may also be investigators). Information related to the target is acquired in disparate pieces over time. The investigators use the acquired information to build a model of the target (synthesis) and extract useful information from the model (sense-making). The extracted information results in changes to the model (synthesis). The synthesis – sense-making cycle is continued throughout the investigation as new information is acquired and extracted from the model. The investigators both work individually and cooperatively as a team. The results of the investigation are disseminated to the customer at the end of the investigation or at certain intervals (or deadlines).

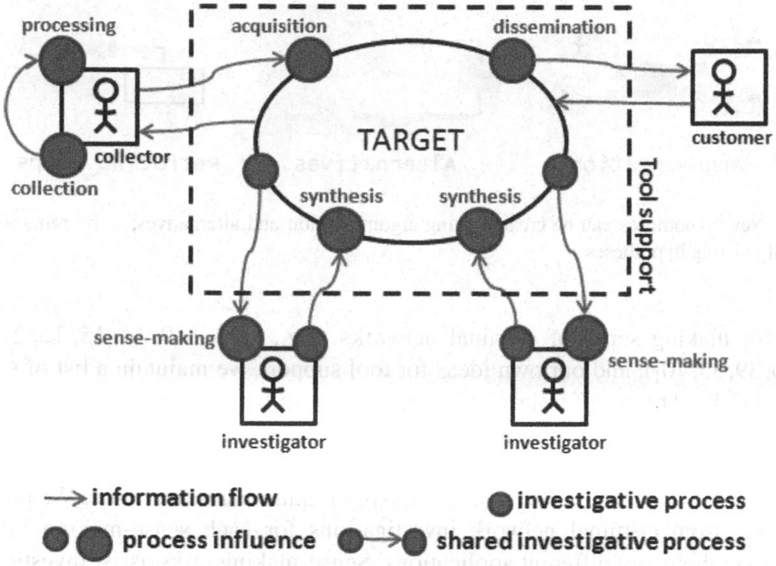

Fig. 1 Human-centered, target-centric criminal network investigation

Criminal network investigation is a human-centered process. Investigators (and collectors) rely heavily on their past experience (tacit knowledge) when conducting investigations. Hence, these processes cannot be fully automated. The philosophy of the CrimeFighter Toolbox is that the humans (in this case the investigators) are in charge of the investigative tasks and the software tools are there to support them [70]. The tools should be controlled by the investigators and should support the complex intellectual work (e.g., synthesis and sense-making) to allow the investigators to reach better results faster. CrimeFighter Investigator focuses on providing human-centered, target-centric support for criminal network investigations (acquisition, synthesis, sense-making, and dissemination). Tool support for collection and processing is beyond the scope of this paper. The CrimeFighter Explorer tool focuses on this type of tool support. Tool support for advanced structural analysis and visualization of the generated target model is also beyond the scope of this paper. The CrimeFighter Assistant tool focuses on this type of tool support.

2.2 Sense-Making Tasks

Based on cases and observations of criminal network sense-making, contact with experienced end-users from various investigation communities (intelligence, police, and journalism), examination of existing process models and existing

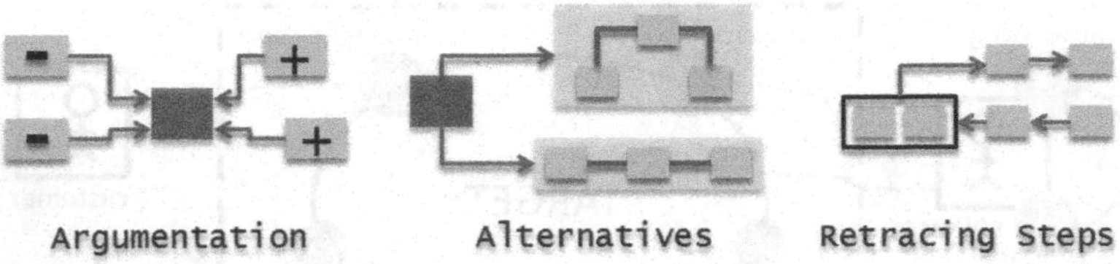

Fig. 2 New hypotheses can be created using argumentation and alternatives, or by retracing the steps of existing hypotheses

tools for making sense of criminal networks (e.g., [1, 3–5, 9, 11, 15, 16, 27, 29, 31, 36, 39, 55, 70]), and our own ideas for tool support, we maintain a list of sense-making tasks. The list of tasks can be seen as a wish list of requirements which the sense-making part of a tool for criminal network investigation should support; the list serves as the basis for our tool development efforts. The list is not exhaustive; we expect to uncover additional sense-making requirements over time. We provide examples from criminal network investigations for each sense-making task to emphasize the many different applications. Sense-making tasks assist investigators in extracting useful information from the synthesized target model.

Creating Hypotheses

Generating hypotheses and competing hypotheses is a core task of sense-making that involves making claims and finding supporting and opposing evidence. Investigators often retrace the steps of their investigation to see what might have been missed to evolve an existing hypothesis or start a new one (see Fig. 2).

Journalist Daniel Pearl was kidnapped in Karachi in early 2002. The criminal network investigators followed the hypothesis that the leader of a radical Islamist group, Shaikh Gilani, masterminded the kidnapping, since Pearl was scheduled to meet him on the day of his disappearance. One day the investigative team receives an email, profiling a shadowy character suspected of having bankrolled the 9/11 attacks, Omar Saeed Sheikh: "Omar has a particular specialty: he kidnaps Westerners". But the team finds nothing linking Omar to Daniel's disappearance (besides this specialty), and the current state of their hypothesis has a lot more supporting arguments pointing towards Gilani [31, 39, 58].

On February 5, 2003, secretary of state (Colin Powell) presented to the United Nations council the US hypothesis on Saddam Hussein's weapons of mass destruction programme. The supporting arguments were primarily based on one human intelligence source, an Iraqi defector who manufactured a story based on open source United Nations reports and his previous work as a chemical engineer [16, 67].

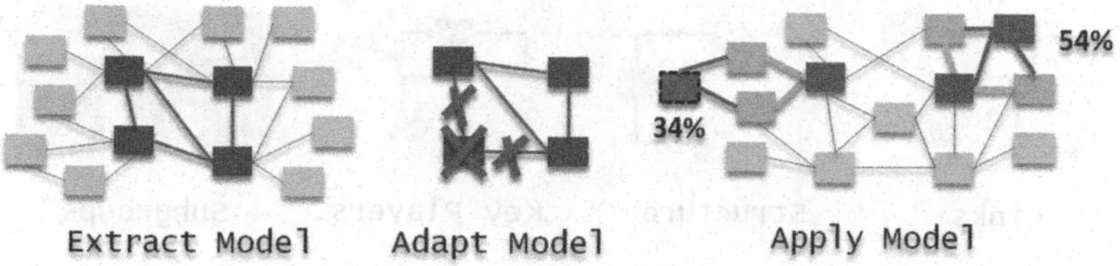

Fig. 3 Extracting a model from a criminal network investigation, adapting the model to a new situation, and then applying the model to the same or another criminal network

Adaptive Modeling

Representing the expected structure of networks for pattern and missing information entity detection is a proactive sense-making task. Adaptive modeling embeds the tacit knowledge of investigators in network models for prediction and analysis (see Fig. 3).

Many authors have described the evolution of terrorist networks related to AQAM and plotting to hit targets in Europe have evolved through four phases. Vidino outlines the evolution of these European networks during the first three phases, and provides a detailed description of the fourth phase including characteristics in terrorism related to AQAM [62] resembling a model.

Sageman's work on structural patterns in "terror networks" (phase 1 and 2) [47], found that people had joined the jihad in small groups, also known as cliques, where every node is connected to every other node. Several individuals lived together for a while and had intense discussions about the jihad. When one of the friends was able to find a bridge to the jihad, they often went as a group to train in Afghanistan. Examples in Sageman's sample includes: the Montreal group, the Hamburg group, the Khamis Mushayt group, and the Lackawanna group.

Nesser outlines a typology of terrorist networks in United Kingdom and Europe based on a survey of a number of al-Qaeda associated or al-Qaeda inspired terrorist cells that planned, prepared, and in three instances managed to launch attacks in European countries in the period 1998 until the present (phase 2 and 3) [37]. Nesser identified a distinct set of profiles amongst those involved that recurred across the cases. A typical cell includes an entrepreneur, his protege, misfits, and drifters which also explains Sageman's "bunch of guys" concept where the entrepreneur is both the bridge to the jihad and the hub for incoming information. The relations among cell profiles as well as meta data characteristics for each profile (e.g., education, marital status, children, and age) are described in great detail by Nesser.

Prediction

The ability to determine the presence or absence of relationships between and groupings of people, places, and other entity types is invaluable when investigating a case. Prediction based on different information entities, i.e., information elements, relations, composites, and their attributes is preferable (see Fig. 4).

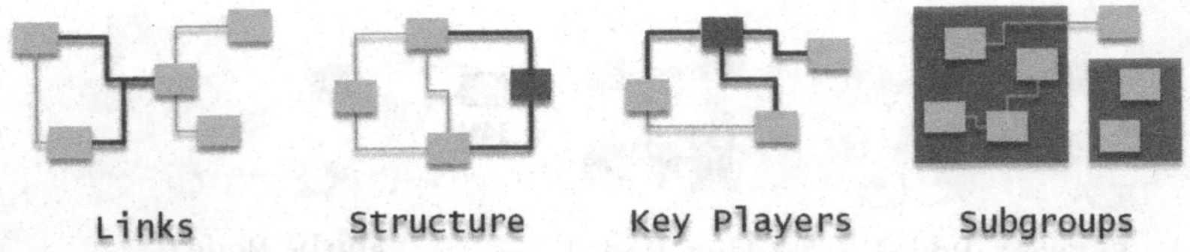

Fig. 4 Predicting missing information entities: links, structures, key players, and subgroups

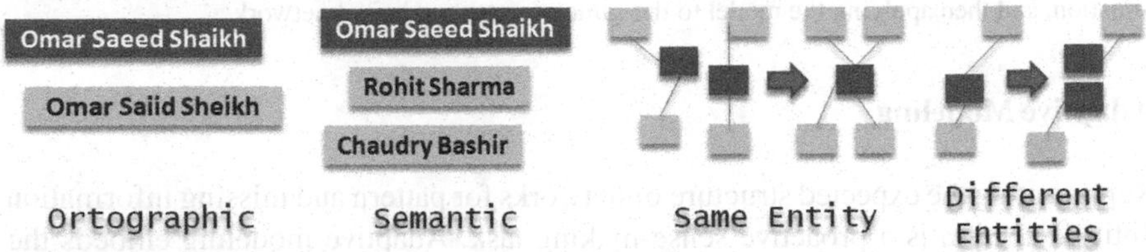

Fig. 5 Detecting semantic and orthographic aliases to analyze if two entities are in fact the same, or if a single entity was in fact two different entities

The value of a prediction lies in the assessment of the information entities that will shape the state of the criminal network and thereby future events [11]. After Operation Crevice [27] a list with 55 suspects linked to the case was created, but MI5 did not have enough resources for surveillance of everybody on the list. They selected (predicted) the 15 individuals they thought were a threat to national security, missing key individuals behind the July 7th bombings [72].

In an 2011 interview, Alex Strick van Linschoten, author of several books on Afghan Taliban (e.g., [32]), suggested the prediction of missing links between Afghan Taliban members based on knowledge about their andiwali[4] system, "where groups tend to gather based on prior connections. Young men from the same village could group together in one cell; madrassas also allow young men to form ties. Some groups may have blood relations that bring them together in a group of andiwali" [34].

Alias Detection

Network structures may contain duplicate or nearly duplicate entities. Alias detection can be used to identify multiple overlapping representations of the same real world object (see Fig. 5).

An extreme example is the mastermind behind the kidnapping of journalist Daniel Pearl, Omar Saeed Shaikh, who used up to 17 aliases [31]: "You run up against the eternal problem of any investigation into Islamist groups or al-Qaeda in particular: the extreme difficulty of identifying, just identifying, these masters of

[4]"Andiwal" is the Pashto (Afghani language) word for "friend".

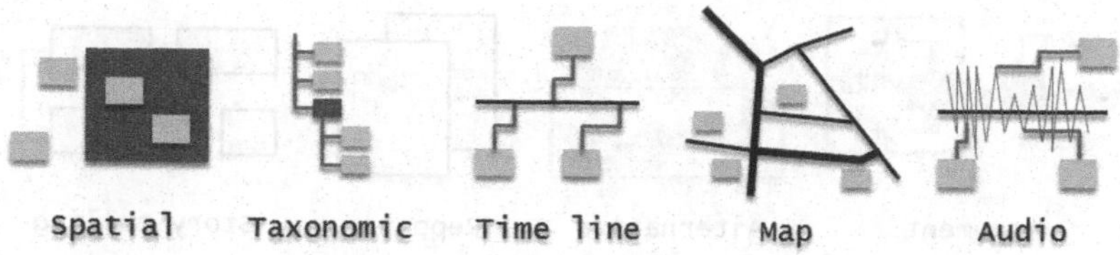

Fig. 6 Alternatives to the often used navigational (link) perspective are the spatial, taxonomic, time line, map, and audio perspectives

disguise, one of whose techniques is to multiply names, false identities, and faces". In the UK investigation of whether or not the July 7th bombings in London (2005) could have been prevented, MI5 had come across different variations of the name "S. KHAN" (the name of the plot ringleader, Mohammed Siddique Khan). They consequently believed the name could have been an alias "due to a combination of both the multiple spellings and lack of traces on databases" [27]. Aliases are inherently also a problem when analyzing online violent radical milieus: "the Internet allows for the virtual construction and projection of personalities that may or may not be accurate reflections of the physical lives controlling those avatars" [7].

Exploring Perspectives

To reduce the cognitive biases associated with a particular mind set, the exploration of different perspectives (views) of the information is a key criminal network investigation task (see Fig. 6).

During the Daniel Pearl investigation a chronology of events (time line) is created simultaneously with the criminal network (link chart) of involved individuals who were potentially linked to the crime [39]. When Colin Powell presented the United States' hypothesis on Saddam Hussein's weapons of mass destruction programme, he used both augmented satellite photos (images/maps) and texted recordings of intercepted phone calls (audio) [63, 77].

Decision-Making

During an investigation, decisions have to be made such as selecting among competing hypotheses. Auto-generated reports and storytelling can also be used for higher-level decision-making (see Fig. 7).

As mentioned, a list with 55 individuals was created after Operation Crevice, and it had to be decided how to focus limited resources [27, 72]. In the case of CIA's investigation into possible weapons of mass destruction in Iraq, the CIA based their decision on uncorroborated evidence (arguments) [16, 67]. The team investigating the kidnapping of Daniel Pearl decides to focus resources on the alleged mastermind Sheikh Gilani, the man who Pearl was scheduled to interview on the day of his disappearance [31, 39, 58].

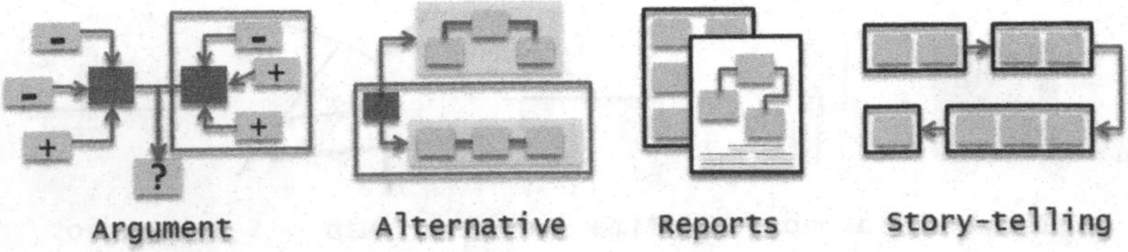

Fig. 7 Decision-making is typically done by selecting arguments and alternatives, or it is based on reports and storytelling

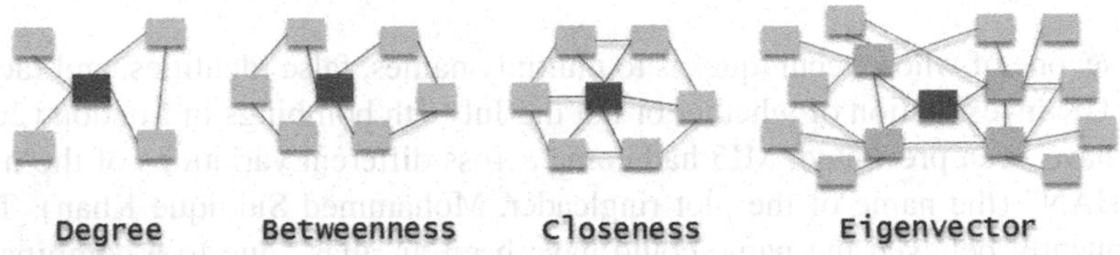

Fig. 8 Degree, betweenness, closeness, and eigenvector measures of centrality

Social Network Analysis

Social network analysis measures such as degree, betweenness, closeness, and eigenvector can provide important criminal network insights (see Fig. 8). These and similar measures are often used as input for other more advanced and specialized sense-making algorithms, either producing new measures or transforming the network.

Slate reporter Chris Wilson has described how the US military used social network analysis to capture Saddam Hussein [71]: "In Tikrit, players were captured, killed, and replaced at a low enough rate that the network was able to cohere. The churn rate is likely much higher in an extremist group like al-Qaeda". In one assessment of destabilization tactics for dynamic covert criminal networks, it is pointed out that in standard social network analysis node changes are the standard approach to network destabilization [9].

> MI5 [...] decided not to continue surveillance of Khan and Tanweer because the quantity of Khan and Tanweer's links to the fertilizer bomb plotters targeted in Operation Crevice were less than 0.1% of the total links. Their argument failed to take into account the betweenness centrality of Khyam. Betweenness centrality refers to relationships where one individual provides the most direct connection between two or more groups. These individuals *bridge* networks, or subnetworks. In the case of Khan and Tanweer, Khyam was likely serving a *liaison* role rather than a broker role, meaning his betweenness was not likely critical to their plot but was *indicative* of Khan and Tanweer's intelligence value [28].

Terrorist Network Analysis

Sense-making measures specifically developed for terrorist networks such as level of secrecy (covertness) and efficiency can provide more focused insights due to their

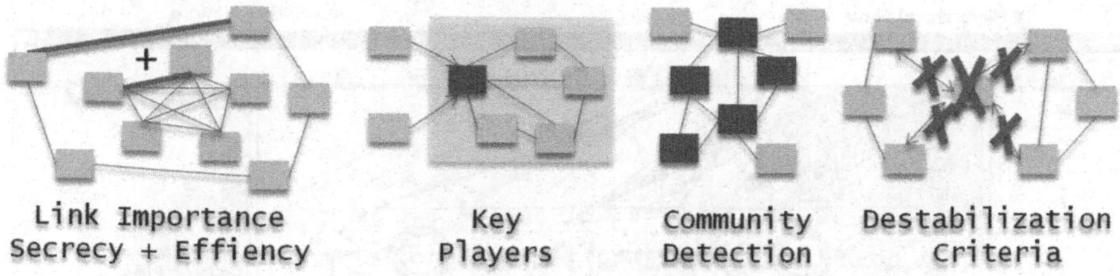

Fig. 9 Terrorist network measures includes secrecy and efficiency for measuring link importance, and detection of key players and communities (subgroups). Terrorist network destabilization criteria are often used to determine the success or failure of such measures

domain focus. Terrorist network measures are used to understand and subsequently destabilize networks (e.g., to reduce the flow of information through the network or to diminish the network's ability to reach consensus as a decision-making body) or to search for specific entities or patterns in the network (e.g., key players or hierarchical organization). Examples are shown in Fig. 9.

The link importance measure has been shown to offer new insights into the 9/11 and Bali bombing terrorist networks by pointing out links that are important to the network [68]. Community (subgroup) detection has been applied to a network of 60 criminals dealing with drugs [75] and prediction of missing key players has been tested on the Greek terrorist network November 17 [44].

3 CrimeFighter Investigator

CrimeFighter Investigator [40–42] is based on a number of concepts (see Fig. 10). At the center is a shared information space. Spatial hypertext research has inspired the features of the shared information space including the support of investigation history [42]. The view concept provides investigators with different perspectives on the information in the space and provides alternative interaction options with information (hierarchical view to the left (top); satellite view to the left (bottom); spatial view at the center; algorithm output view to the right). Finally, a structural parser assists the investigators by relating otherwise unrelated information in different ways, either based on the entities themselves or by applying algorithms to analyze them (see the algorithm output view to the right). In the following, central CrimeFighter Investigator sense-making features are presented. Due to the space limitations not all implemented features can be presented and the presented features cannot be described in full detail.

Finally, we describe our developed conceptual and computational models (see Fig. 11). We have separated structural concerns from the default mathematical

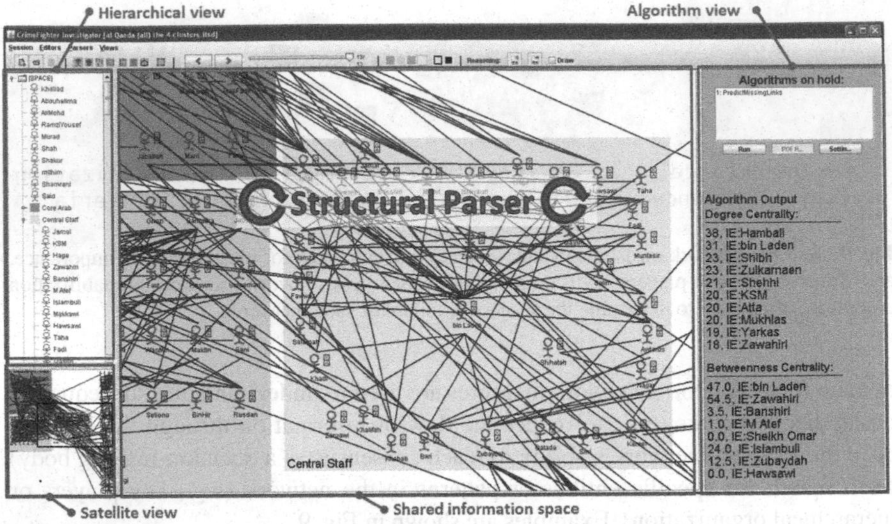

Fig. 10 CrimeFighter Investigator screen shot with overlays

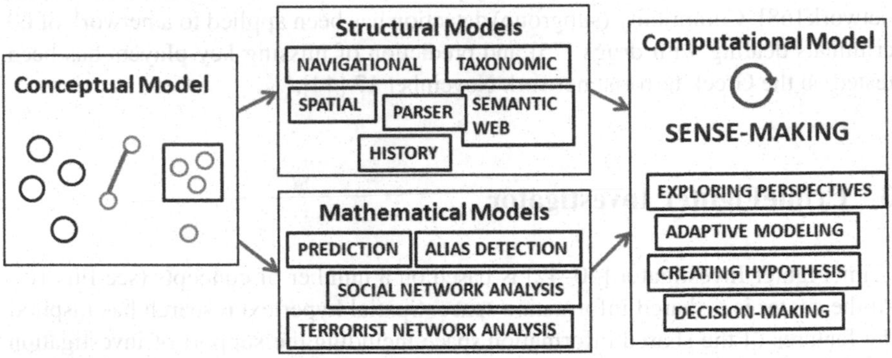

Fig. 11 Conceptual, structural, mathematical and computational models

models,[5] since the mathematical models should be able to process or adapt to any structural model they are faced with; not only the traditional navigational (link) structures.

The *predict missing links* and *betweenness centrality* algorithms are used to explain the sense-making concepts that we find essential for criminal network sense-making. We focus specifically on how to handle the knowledge management perspective; how to customize algorithm input and output, algorithm steps and

[5]Structural models are typically embedded in mathematical models. See for example [8].

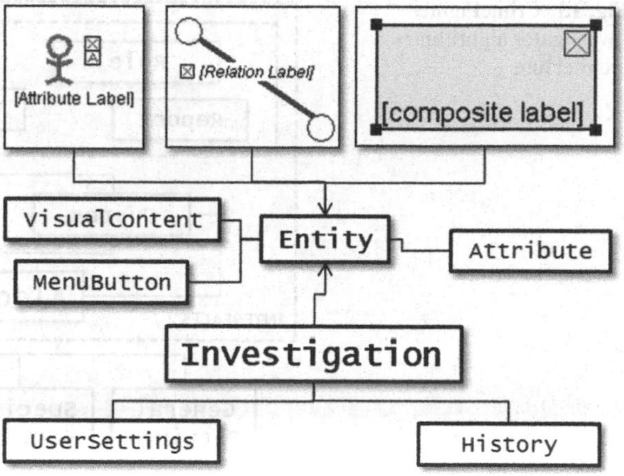

Fig. 12 CrimeFighter Investigator conceptual model

combinations thereof, in a way that matches our chosen approach, i.e., tailorability of the computational model. The mathematical inner workings of algorithms, albeit its inherent relevance for the computed information, is not our focus.

3.1 Conceptual Model

The CrimeFighter Investigator conceptual model consists of three first class information entities: information elements, relations and composites (see Fig. 12). Previous research on criminal network sense-making has to a large degree focused on making sense of nodes. Links are seldom first class objects in the terrorism domain models with the same properties as nodes. This is in contrast to the fact that the links between the nodes provide at least as much relevant information about the network as the nodes themselves [20]. The nodes and links of criminal networks are often laid out at the same level in the information space. Composites are first class entities that add depth to the information space. Navigable structures and entities (including composites) are useful for synthesis tasks such as manipulating, re-structuring, and grouping entities [42].

An information entity comprises several components. Each entity has a set of dynamic *attribute(s)* (meta data). Currently three types of attributes are supported: strings (single line of text), text areas (multiple lines of text), and enumerations (a defined set of allowed values). The visual abstraction of an entity is computed from its *visual content* and *menu button(s)*. The visual content is used to create the default information elements available in CrimeFighter Investigator, which are all composed using geometric shapes (circles, lines, rectangles, and polygons). A number of menu buttons can be added to entities to create a link to a specific functionality. The examples shown in Fig. 12 are the *delete* button (X symbol) and the *attributes* button (A symbol).

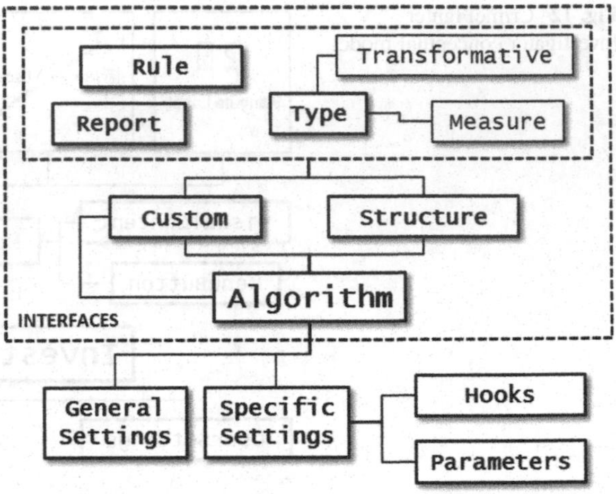

Fig. 13 CrimeFighter Investigator algorithm architecture

3.2 Computational Model

CrimeFighter Investigator supports three structure algorithm types: measures (e.g., entity centrality), transformative algorithms (e.g., entity prediction), and combinations of these (custom algorithms) as shown in Fig. 13. Custom algorithms are templates of specific criminal network investigation workflow, e.g., understanding the secondary effects of entity removal or insertion. All algorithms implement the *report* interface, where an algorithms' report elements and design is defined. *Rules* are used to describe entity-to-entity relations, attribute cross products, etc. Each algorithm has a set of *general settings* and *specific settings*. Specific settings include *algorithm hooks*, i.e., the entity attributes that algorithms base their computations on, and customizable algorithm *parameters*.

Prediction

Prediction techniques include extrapolation, projection, and forecasting based on past and current states of a criminal network. These three predictive techniques follow the approach of assessing forces that act on an entity [11]. The value of prediction lies in the assessment of the forces that will shape future events and the state of the criminal network. An extrapolation assumes that those forces do not change between the present and future states; a projection assumes that they do change; and a forecast assumes that they change and that new forces are added.

Bayesian inference is a (forecasting) prediction technique based on meta data about individuals in criminal networks. A statistical procedure that is based on Bayes' theorem can be used to infer the presence of missing links in networks. The process of inferring is based on a comparison of the evidence gathered by investigators against a known sample of positive (and negative) links in the network,

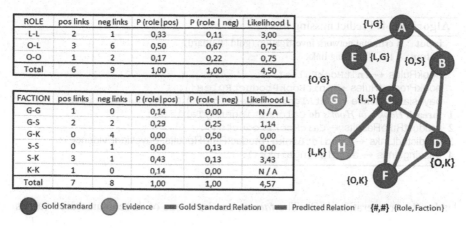

ROLE	pos links	neg links	P (role|pos)	P (role | neg)	Likelihood L
L-L	2	1	0,33	0,11	3,00
O-L	3	6	0,50	0,67	0,75
O-O	1	2	0,17	0,22	0,75
Total	6	9	1,00	1,00	4,50

FACTION	pos links	neg links	P (role|pos)	P (role | neg)	Likelihood L
G-G	1	0	0,14	0,00	N / A
G-S	2	2	0,29	0,25	1,14
G-K	0	4	0,00	0,50	0,00
S-S	0	1	0,00	0,13	0,00
S-K	3	1	0,43	0,13	3,43
K-K	1	0	0,14	0,00	N / A
Total	7	8	1,00	1,00	4,57

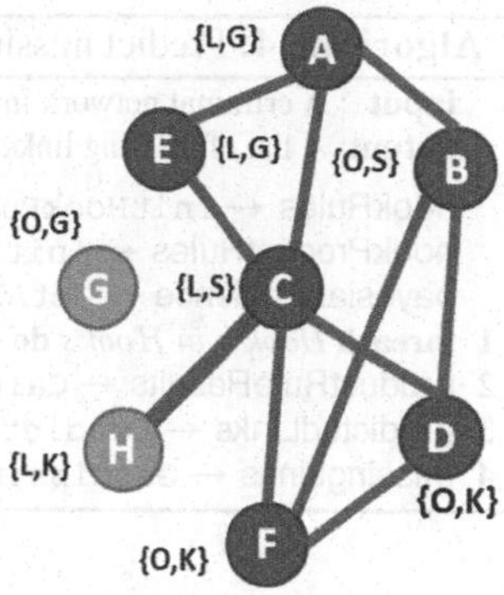

Fig. 14 A predict missing links example

where positive links are those links that connect any two individuals in the network whereas negative links are simply the absence of a link. The objective is often to assess where links may be present that have not been captured in the collected and processed criminal network information.

Prediction of covert network structure [46] is useful when you have a list of individuals suspected to be part of your current criminal network investigation. The algorithm indicates probable covert members on the list and how they are linked to the existing structure. The predict missing links algorithm [45] starts prediction based on the current criminal network structure. The likelihood of a link being present between all node pairs in the network is calculated based on the attribute data of the remaining individuals. Links that have a missing likelihood higher than a pre-determined value (calculated from the product of individual attribute likelihoods) are predicted as new links in the network. Links are predicted in the same way by the covert network structure algorithm, using a Bayesian inference method.

In the following example[6] we describe CrimeFighter Investigator support of the Bayesian inference method described in [45]. The major steps involved in the calculation are shown in Algorithm 4 and the network we predict missing links for, is shown in Fig. 14. The network has six nodes and seven (positive) links.

Part of the customization of this algorithm (see Sect. 3.3) is to select the entity attributes (algorithm hooks) for the prediction algorithm. Only enumerated attributes are accepted as algorithm hooks, i.e., name is not eligible since it can have basically any value.

The first step of the algorithm (line 1), is to calculate the contingency table for each of the selected algorithm hooks. We will explain how to calculate the contingency table for a *role* hook which can have one of two enumerated values:

[6]The network nodes and attributes used in this example are inspired by the Greek criminal network November 17 (see [45] for more details).

Algorithm 4: Predict missing links

```
input : A criminal network investigation (gold standard)
output: A list of missing links
hookRules ← InitHookRules();
hookProductRules ← InitHookProductRules();
bayesianEvidence ← GetAlgorithmSettings().GetBayesianEvidence();
1 foreach Hook h in Hooks do CalcContingencyTable(h);
2 productRuleResults ← CalcHookProducts();
3 predictedLinks ← PredictLinks(productRuleResults, bayesianEvidence);
4 missingLinks ← GetMissingLinks();
```

leader (L) or operational (O). The *faction* can have one of three enumerated values (G, S, or K), each named after an individual within that respective faction. The contingency table records the relation between positive and negative links in the gold standard (purple nodes in Fig. 14).

The second step is to calculate the products of different hook relations if more than one hook is added to the inference[7] (line 2):

- $L-L \times G-S = 3,00 \times 1,14 = 3,42$
- $L-L \times S-K = 3,00 \times 3,43 = 10,29$
- $O-L \times S-K = 0,75 \times 3,43 = 2,57$
- $O-O \times S-K = 0,75 \times 3,43 = 2,57$

The third step is the actual prediction of missing links based on the likelihood products calculated above together with the likelihoods for individual algorithm hooks (line 3). The second input to the prediction of links is the evidence, that is the attributes and their values for all individuals in the network.[8] From the likelihoods we see that $L-L$ and $S-K$ relations are above the cut-off value, together with the products mentioned under the second step above. We see that entities sharing both $L-L$ and $S-K$ relations are especially likely to be connected, hence the thicker red line between C and H in Fig. 14.

The fourth step is a simple clean-up function which will remove those links already in the network prior to the prediction, leaving only new (missing) links (line 4).

The result of a missing links prediction on a sampled version of 20 individuals from the al-Qaeda network is shown in Fig. 15. The investigator can decide to append the predictions to the network or simply discard them.

[7] Only the products above a cut-off value of 2,14 are included. The cut-off value is calculated as the total possible links in the gold standard divided by the existing links.

[8] If we chose to apply the *predict covert network structure* algorithm then the evidence could also be information about individuals not in the network. These individuals would be added if a link (relation) to them is predicted from within the gold standard network.

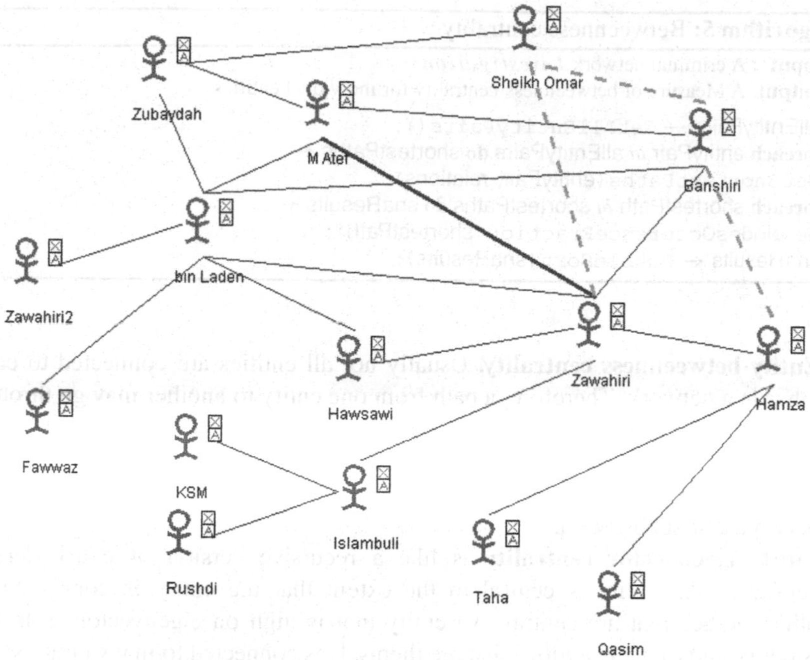

Fig. 15 The result of a missing links prediction on a sampled version of 20 individuals from al-Qaeda central staff [47]. *Blue solid lines* are true positives while *green dashed lines* indicate false positives

Social Network Analysis

Techniques from social network analysis and graph theory can be used to identify key entities in criminal networks [64]. Information about key entities (individuals, places, things, etc.) is helpful for network destabilization purposes [9], or as input for other criminal network sense-making algorithms. Relevant social network analysis measures include [28, 64]:

- **Entity degree centrality.** An entity is central when it has many links (associations) to other entities in the network. This kind of centrality is measured by the degree of the entity, the higher the degree, the more central the entity. Degree centrality can be divided into *in-degree* centrality and *out-degree* centrality, referring to the number of incoming and outgoing links an entity has. A social network with high degrees of both is a highly cohesive network.
- **Entity closeness centrality.** Closeness centrality indicates that an entity is central when it has easy access to other entities in the network. This means that the average distance (calculated as the shortest path) to other entities in the network is small.

Algorithm 5: Betweenness centrality

input : A criminal network *Investigation*
output: A Measure of betweenness centrality for individual entities

```
1 allEntityPairs ← GetAllEntityPairs();
2 foreach entityPair in allEntityPairs do shortestPaths ←
  GetShortestPaths(entityPair, relations);
3 foreach shortestPath in shortestPaths do snaResults ←
  GetNodesOccurenceFraction(shortestPath);
4 snaResults ← BubbleSort(snaResults);
```

- **Entity betweenness centrality.** Usually not all entities are connected to each other in a network. Therefore, a path from one entity to another may go through one or more intermediate entities. Betweenness centrality is measured as the frequency of occurrence of an entity on the geodesic connecting other pairs of entities. A high frequency indicates a central entity. These entities bridge networks, or sub networks.
- **Entity eigenvector centrality** is like a recursive version of entity degree centrality. An entity is central to the extent that the entity is connected to other entities that are central. An entity that is high on eigenvector centrality is connected to many entities that are themselves connected to many entities.

CrimeFighter Investigator supports dangling endpoints during synthesis (empty relation endpoints), and the social network analysis algorithms are therefore extended to include this aspect in calculations if it is found necessary for the investigation. We focus on *betweenness centrality* and describe how this centrality measure is implemented (see Algorithm 5) and how the algorithm is customized to suit different needs.

The *betweenness algorithm* starts by creating a set of all entity pairs in the criminal network (line 1). Then the shortest path between each pair of entities is calculated (line 2). For each entity pair, we determine the fraction of shortest paths that pass through each entity on those paths (line 3). The betweenness of each entity is the sum of all these fractions across the entire network. The results are bubble sorted with for example highest centrality first before it is presented to the user (line 4).

The betweenness centralities of a sampled version of 20 individuals from the al-Qaeda network are shown in Fig. 16. The investigator has decided to append the predicted links shown in Fig. 15 to the network before calculating the centralities.

Combining Algorithms

The CrimeFighter Investigator tool supports a node removal algorithm (described in more detail in [40]) assisting criminal network investigators by combining two default algorithms to make sense of a network post the removal of an individual

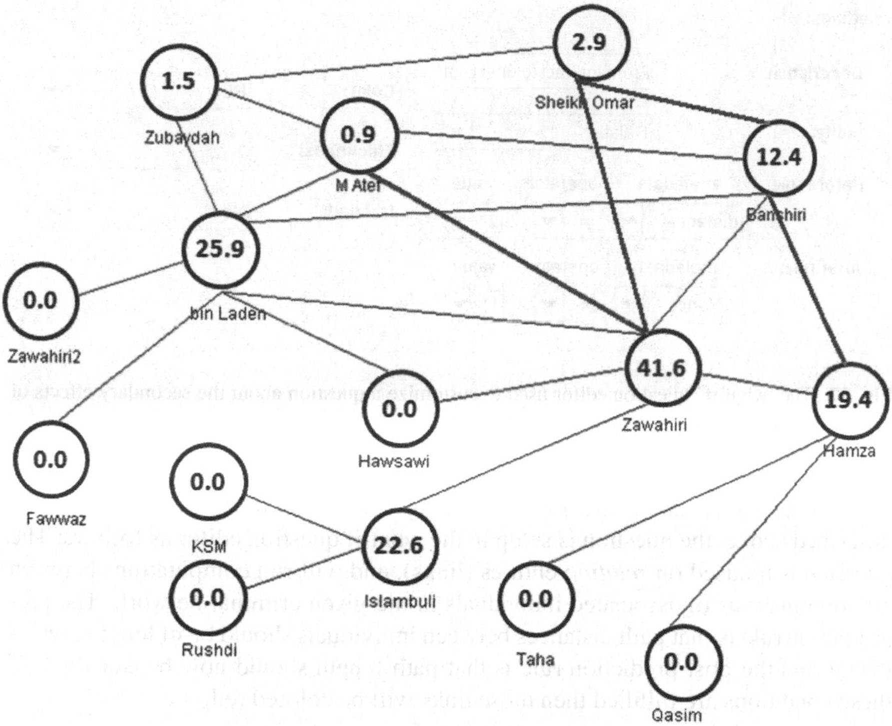

Fig. 16 Betweenness centrality for the individuals in Fig. 15, with four added links (*thick blue*)

with high centrality. This is achieved by combining an inference-based prediction of new probable links (using the *predict missing links* algorithm) and changes in entity *degree centrality*. Many interventions against criminal (and other covert) networks often take place in the context of a multi-agency effects-based operations doctrine [54]. Consequently, it is imperative to assist analysts and investigators in assessing the likely impact and consequences of interventions against proposed targets in complex socio-technical systems.

Node removal is a well-known technique for destabilization of criminal networks [9, 10]. Deciding which node or group of nodes to remove is dependent on available intelligence and the topology of the criminal network (hierarchical, cellular, etc.), complicating the prediction of secondary effects following a node removal. Inference-based prediction and social network analysis provides different perspectives on criminal networks, thereby assisting investigators in their decision making by answering the 'what if' questions they inherently would like to ask. An example of such a 'what if' question could be: "*Are individuals who were not directly associated before the node removal directly associated afterwards?*". Computational 'what if' questions can be setup using a special editor (see Fig. 17).

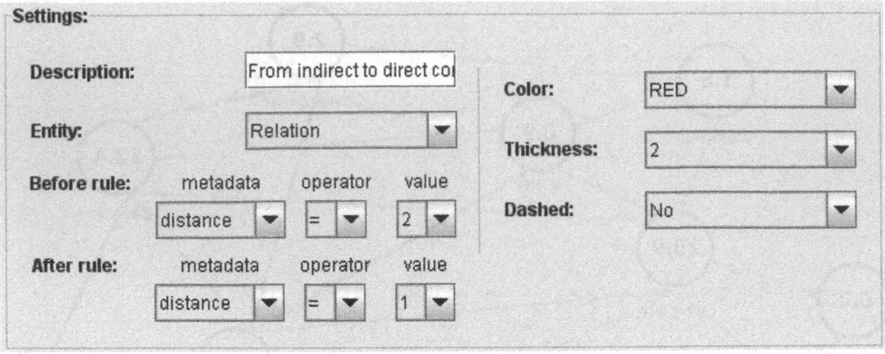

Fig. 17 The 'what if' question editor used to customize a question about the secondary effects of removing a central entity from a criminal network

In order to visualize the links matching the 'what if' question constraints described above, the question is setup in the what-if question editor as follows: The question is focused on *relation* entities (links), and will run computations between all combinations of associated individuals in the given criminal network. The pre-prediction rule is that path distances between individuals should be of length greater than 1 and the post-prediction rule is that path-length should now be exactly 1. If these conditions are fulfilled then those links will be colored red.

3.3 Structural Parser

CrimeFighter Investigator algorithms are managed by a structural parser (Fig. 18), where investigators can select different algorithms to run and control the order in which they are applied, for example either simultaneously or sequentially.

Figure 18 (left, top-frame) shows tabs for different algorithm types. The *SNA* tab covers social network analysis measures such as degree, closeness, and betweenness [28, 64]. The terrorist network analysis measures on the *TNA* tab are part of our future work, supporting integration with the CrimeFighter Assistant [69, 70]. The default prediction algorithms include *predict covert network structure* and *predict missing links* [45, 46]. Figure 18 (left, bottom-frame) shows the algorithms selected by the investigators to run. The structural parser will indicate if there is a potential conflict between the selected algorithms. If a prediction algorithm is selected to run on every network event, it could create a loop (since it is transformative). Similarly, if algorithms are running sequentially the position of an entity centrality measure before or after a transformative algorithm is quite important. In Fig. 10, criminal network investigators have decided to run two different centrality measures simultaneously while displaying the results.

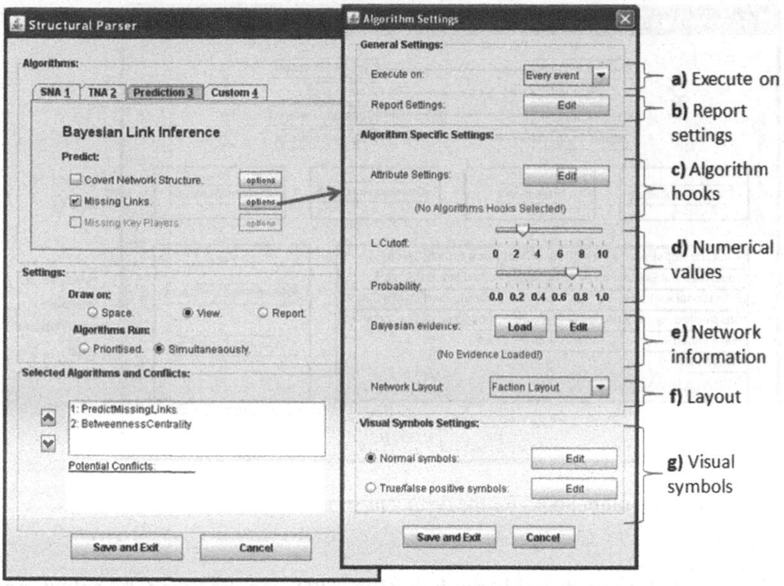

Fig. 18 The structural parser (*left*) and the *predict missing links* algorithm customization window (*right*)

Algorithm settings, both general and specific, are accessed by clicking on the *options* button shown in Fig. 18 (left, top-frame). The *predict missing links* customization window is also shown in Fig. 18 (on the right). Algorithms can run on every system event or when the investigator requests it (Fig. 18, top right).

Tailoring Prediction of Missing Links

The user can customize when and how often a prediction algorithm should compute (Fig. 18a). One option is to automatically run the algorithm every time a change is made to the criminal network. But the *predict missing links* algorithm is a transformative algorithm, and would continue to predict missing links, since each transformation of the network would start the algorithm again. Therefore, an option to run algorithms when clicking a button has been added (see Fig. 10, right side).

Next is the selection of algorithm hooks (Fig. 18c). A special drag and drop view is used for this task (Fig. 19). Both entity attributes and centrality measures can be selected as algorithm hooks.

Numerical algorithm variables are customized using standard input fields such as text fields (any number or text), sliders (bounded numbers), and drop down boxes (enumerated values) as shown in Fig. 18d. Network information (evidence) is what the prediction algorithms base their inferences on (Fig. 18e). For *predict missing links*, it will be all entities currently in the network.

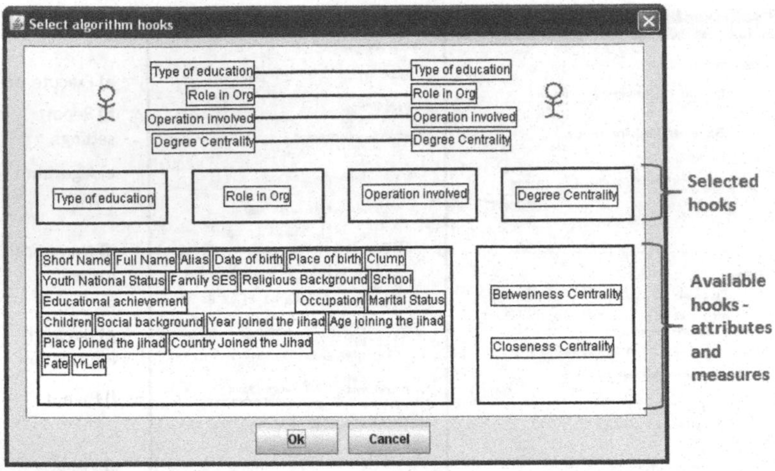

Fig. 19 Selecting algorithm hooks for the *predict missing links* algorithm

The *network layout* drop down box (Fig. 18f) can be used to select one of several default layout algorithms that will be applied after the prediction. Finally, the investigators can customize what visual symbols (color, thickness, etc.) to apply to the predicted links (Fig. 18g).

Tailoring Measure of Betweenness Centrality

The interface for customizing measures of centrality is structured in the same way as the interface for transformative algorithms described above (Fig. 20). There are however a couple of important differences which we would like to emphasize, using *betweenness centrality* as an example:

- **Entities?** The investigator should decide which entities to include for the calculation of betweenness centrality, all or only selected entities (e.g., persons)? If not all entities are included, what should the algorithm do if it encounters a non-included entity when tracing shortest paths? Should it skip the entity and then continue on the other side if the path continues, or simply not count the path?
- **Associations?** The investigator has to decide how to deal with for example empty relation endpoints in terms of calculating betweenness centrality. If a relation endpoint is expected to contain a person-entity, but it is not yet known who, then it might be relevant to include that empty endpoint in the measure of centrality anyway.
- **Results?** Often it is an advantage to normalize the measure of betweenness centrality for all entities for comparison purposes, but not always. Also, in some situations it might be relevant to only list the first 10 or 20 results and in other

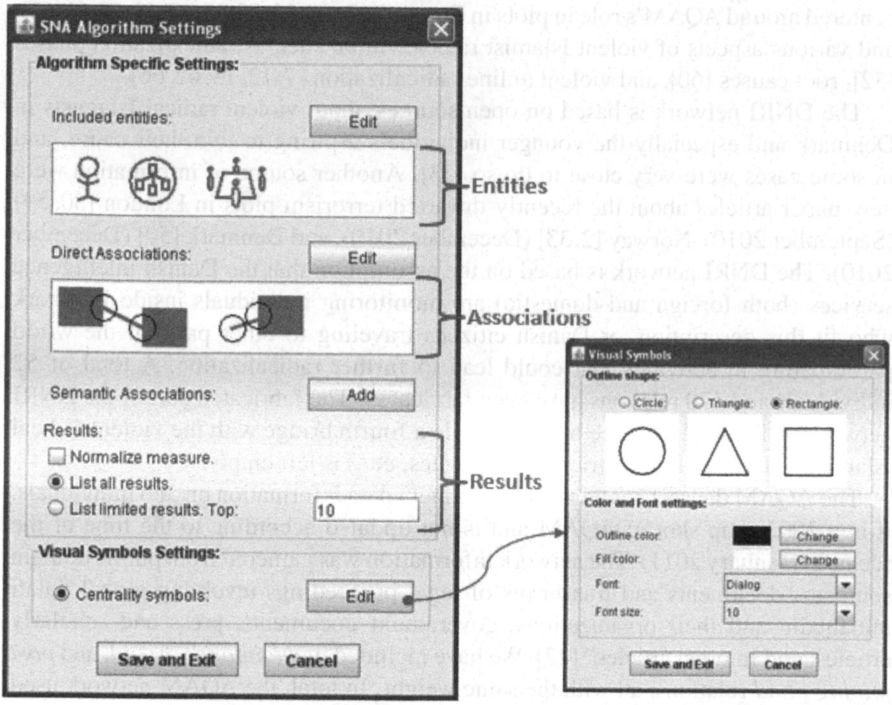

Fig. 20 The user can customize which entities and associations to include, how to display results, and the visual symbols for betweenness centrality

situations all measures are required for further sense-making. Finally, it could be useful to emphasize the entity (or entities) with the highest degree centrality, using color, relative size, or other forms of visual symbols in the information space.

Figure 20 shows the interface for customizing SNA measures of centrality (left) and the sub-interface for setting up visual symbols for visualization of results in the information space (right).

4 Scenario: Investigating Linkage Between DNRI and AQAM

The purpose of this scenario is to demonstrate how the calculations are not the hard part of criminal network sense-making; the challenge is to find a good way to use the data and understand it. The scenario describes a proactive investigation into potential linkage between aspiring extremists in a fabricated Danish network of radical Islamists (DNRI) and al-Qaeda and affiliated movements (AQAM). The scenario is centered around AQAM's role in plots in Europe [17–19,22,28,37,47,48,57,61,62], and various aspects of violent Islamist radicalization such as radicalization phases [52], root causes [60], and violent online radicalization [7, 12, 13, 62, 66].

The DNRI network is based on open sources about violent radical Islamists in Denmark and especially the younger individuals aspiring to join their cause, and in some cases were very close to do so [53]. Another source of information were newspaper articles about the recently thwarted terrorism plots in London [30, 59] (September 2010), Norway [2, 33] (December 2010), and Denmark [50] (December 2010). The DNRI network is based on the assumption that the Danish intelligence services (both foreign and domestic) are monitoring individuals inside Denmark who fit this description, or Danish citizens traveling to other parts of the world participating in activities that could lead to further radicalization. A total of 52 individuals and 170 relations have been fabricated. The fabricated part of the DNRI network is divided into three bridges, while a fourth bridge with the violent radical Islamists relations (family, friends, colleagues, etc.) is left empty.

The AQAM data set contains elaborate meta data information on 366 individuals. It is a 2003 snap shot of AQAM and is not updated according to the time of the scenario (January 2011). The network information was gathered from public domain sources: "documents and transcripts of legal proceedings involving global Salafi mujahedin and their organizations, government documents, press and scholarly articles, and Internet articles" [47]. We have included *acquaintance*, *friend*, and *post joining jihad* relations, all with the same weight. In total, the AQAM network used has 999 links.

It is important to note that the vast majority of EU-wide terrorist attacks in 2010 were carried out by traditional separatist terrorists and not violent radical Islamists as some might expect [13]. More precisely, three Islamist terrorist attacks were carried out within the European Union. However, 249 terrorist attacks in total were reported, and of 611 arrests for terrorism-related offenses, 89 individuals were arrested for the preparation of attacks. Islamist terrorists continue to undertake attack planning against member states, as Europol concludes in their *EU Terrorism Situation and Trend Report 2011* [19].

4.1 The Scenario

It is January 2011, and Mark enters the office as usual. He has been working for the al-Qaeda section of the Danish counterterrorism unit (Danish CTU)[9] since late 2000. The section is daily assessing the risk that al-Qaeda associated or affiliated movements (AQAM) will strike the Danish homeland and they use CrimeFighter Investigator for different work tasks.

[9]The Danish CTU is "invented" for this scenario and is not related to the Danish Security and Intelligence Service's Center for Terror Analysis or other Danish counterterrorism units.

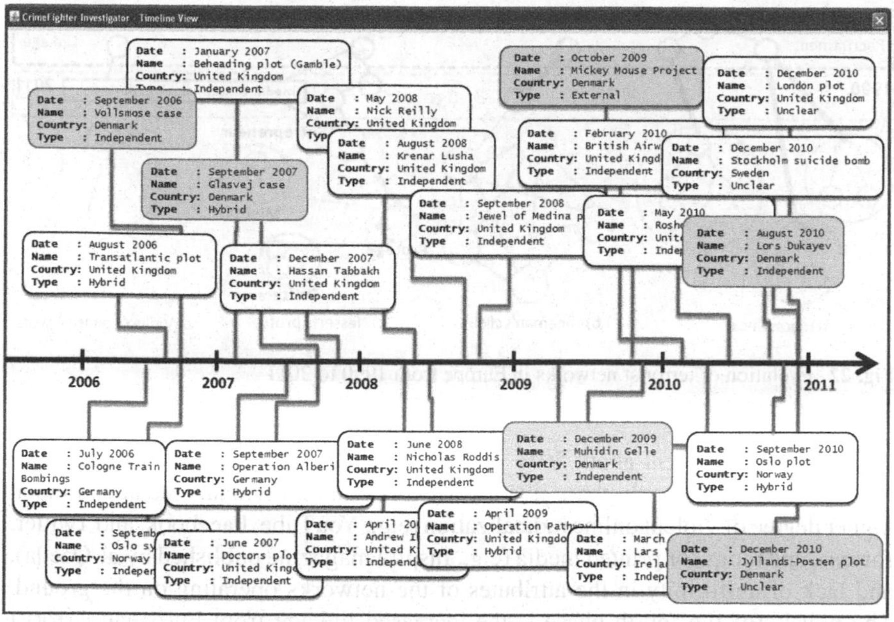

Fig. 21 (mock-up) CrimeFighter Investigator timeline view with all plots against targets inside Denmark, Sweden, Norway, United Kingdom, and Germany from January 1, 2006, to December 31, 2010 [62]

Mark and his fellow investigators have been synthesizing a chronology of AQAM related terrorism plots in selected European countries. The time line provides them with an interactive overview of all the plots (entities), and clicking one entity will open the corresponding CrimeFighter Investigator *information space*, showing the networked information related to the case. They can organize the entities spatially, and *filtering* is applied to only show the desired information (*date*, *name*, *country*, and *type*) and highlight Danish plots with a red color. The *time line* is shown in Fig. 21.

Mark's area of expertise is terrorism information structures and how they evolve over time. He has studied existing literature on AQAM structure and organization primarily in Europe. From Sageman describing the global violent radical Islamism (phase 1–2) [47] and how European terrorist networks are radicalized and associated with AQAM, over Nesser's profile of AQAM terrorist networks in Europe [37] to the most recent fourth phase of plots and attacks [62].

The fourth phase of terrorism plots in Europe (from about 2006 to present) is characterized by a bottom-up approach defined as *linkage*, in which terrorist networks get associated with AQAM in different ways. They are not recruited by AQAM or other transnational networks. However, in the majority of plots (about 75 %) the plotters worked independently (and amateurishly), while about one third were hybrid plots connecting to AQAM. But the hybrid plots pose a higher security

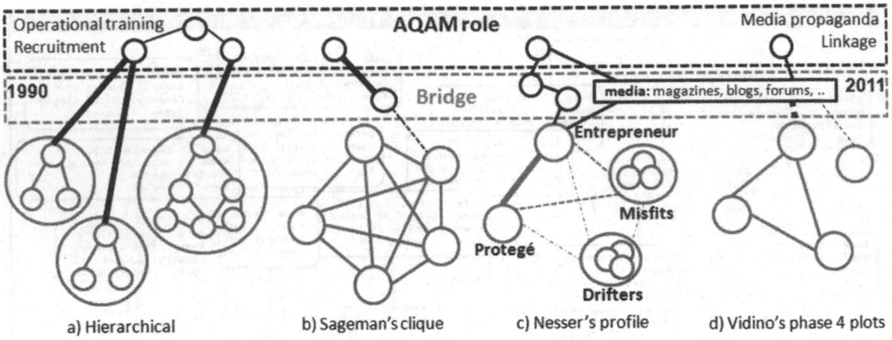

Fig. 22 Evolution of terrorist networks in Europe from 1990 to 2011

risk and they represent about 50 % of the most lethal plots. Other characteristics of the homegrown fourth phase jihadist networks and individuals include: a much higher degree of violent online radicalization (e.g., YouTube, Facebook, and Twitter, forums, and blogs) or printed media (e.g., Inspire magazine published by al-Qaeda), and lack of uniformity in the attributes of the networks operating on the ground. The novelty of the fourth phase is the increased linkage from European terrorist networks to AQAM. Finally, these characteristics of terrorist networks also differ significantly from country to country and, in many cases, within each country from region to region and from city to city. Marks analysis of the evolution of terrorist network cells in Europe is outlined in Fig. 22.

The recent arrests in September and December 2010 just confirmed Mark's analysis of 4th phase plots: the London Stock Exchange plot, the Oslo plot, and the Denmark/Sweden Jyllands-Posten plot. Mark believes strongly in the bridging concept (connecting two network clusters) and the novel observation of bottom-up linkage in European terrorist networks as opposed to top-down recruitment. Mark is certain that if a radicalized individual has a large network of close and likely-minded friends and relatives, other members for a future network cell could come from that group of people. Mark decides to use a measure of betweenness centrality as an extra condition for predicting links between two individuals in adjacent bridges. He thinks that if an individual is peripheral to a network in terms of betweenness centrality, the probability of linkage from this individual to an individual in the bridge above is low.

Mark starts creating his prediction model by first dividing the violent radical part of the DNRI network under surveillance into three bridges. He places the relations (who are not known to be violent radicals) of these individuals in a fourth bridge. Mark thinks there is a potential for top-down recruitment, where violent radical Islamists could radicalize family, friends, or colleagues in Bridge 4 because of their close ties. Mark's classification of individuals in Bridge 1–3 is shown below.

- **Bridge 1** contains individuals that can provide ideological approval of violent radical Islamism and linkage to AQAM. Mark places known radical Islamic scholars in this bridge. Retired violent radicals and other individuals who received operational training could provide linkage to AQAM because of their skills or knowledge about previous operations. Established al-Qaeda media individuals are also placed in Bridge 1.

- **Bridge 2** is the radical violent milieu in Denmark – self-proclaimed imams, online "celebrity shayks" who preach violent radical Islamism, and individuals who sell radical Islamist propaganda like books, magazines, CDs, and DVDs etc. Finally, self-established online recruiters are also made members of this bridge.
- **Bridge 3** is by volume the largest. Individuals aspiring to become violent radical Islamists are placed here. This aspiration may have been externalized through online expression of desire to contribute violently. It could be individuals somehow alienated from society or otherwise non-integrated (e.g., a group of young individuals living together or meeting in an apartment). Bridge 3 individuals are often rather entrepreneurial in their approach. They might be consumers of violent radical online and printed propaganda, or they might be creating such propaganda themselves, pretending to be an established al-Qaeda media organization.

AQAM and the four bridges in the DNRI network constitute four sub-networks each containing two bridges: the 'Bridge 1 → AQAM', 'Bridge 2 → Bridge 1', 'Bridge 3 → Bridge 2', and 'Bridge 3 → Bridge 4' networks. The four networks are encapsulated in collapsed composites. For each of these sub-networks Mark defines a set of attributes he believes could enable linkage from individuals in the lower bridge to individuals in the upper bridge (see Fig. 23):

- **Bridge 1 → AQAM:** Information about *previous operations* is a relevant linkage attribute for this bridge, since Bridge 1 individuals might have participated in the same militant operations in the past. Information about *operational training* may very well overlap with the previous operations, but also covers training camps and similar. A *school* attribute could indicate that the same madrassas, universities, or other schools have been attended at the same time. A *weapons* attribute would cover similar skills in use of weapons; guns, explosives etc. *andiwaal group*, albeit an Afghan concept, it applies to many societies (tribal, western, asian, etc.) that if you were part of a group in your teens, you will have strong relations to those individuals the rest of your life.
- **Bridge 2 → Bridge 1:** Mark decides that *family*, *friend*, and *school* information are linkage attributes from Bridges 2 to 1.
- **Bridge 3 → Bridge 2:** Key linkage attributes from Bridges 3 to 2 are: *Local area* in which random meetings could happen, *online violent radical milieu* that is what forum, chat room or social network site the Bridge 3 individual reads and posts comments to, and who reads it from Bridge 2. *Mosque* and *Sunday school* could be other places for random meetings or radicalizing preachings.
- **Bridge 3 → Bridge 4:** Mark defines key recruitment attributes from Bridges 3 to 4 to be: *school, hobby, workplace, mosque,* and *current residence*. Mark's argument is that the aspiring violent radical Islamists might meet and influence individuals at these places.

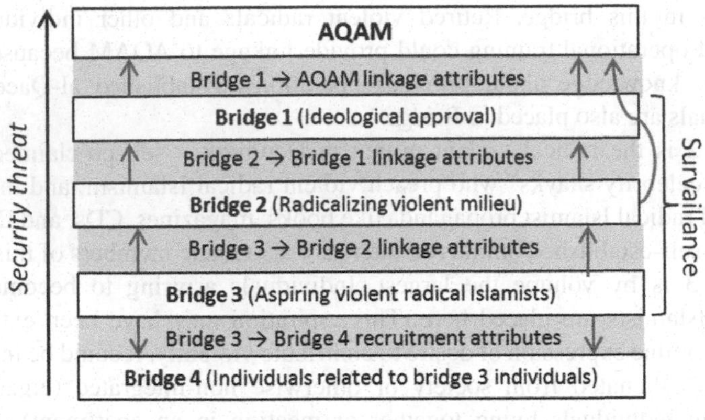

Fig. 23 Mark's prediction model: the DNRI bridges with linkage and recruitment attributes in between adjacent bridges

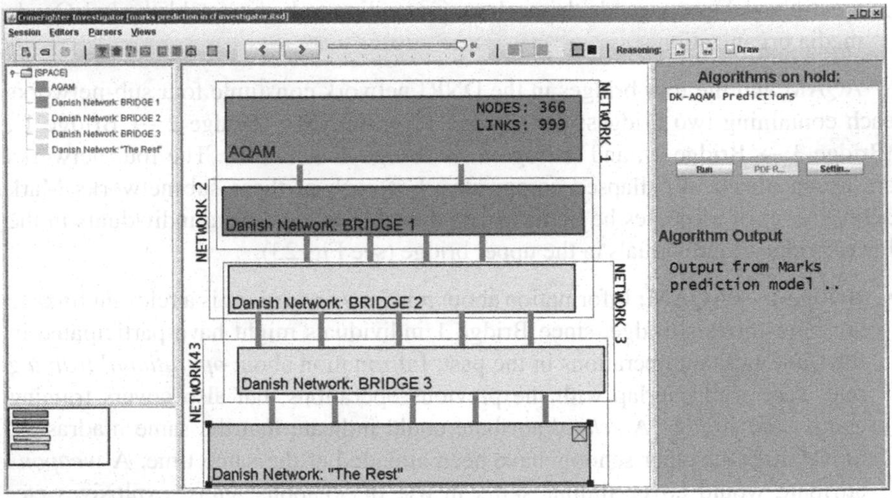

Fig. 24 (semi mock-up) CrimeFighter Investigator showing the AQAM and DNRI bridges and predicted links between them

Mark decides to use the Oslo, London, and Denmark/Sweden networks, whose plots were thwarted in late 2010, as the gold standard for his predictions. After feeding these networks to his prediction model, he predicts missing links for each of the four sub networks, and asks CrimeFighter Investigator to merge individuals with the same names to see if there is probable linkage which forms networks spanning all bridges. A mock-up of predicted links between the four collapsed bridges is shown in Fig. 24.

Mark's prediction model computes four cells (the second cell is shown in Fig. 25) to have linkage potential with AQAM. Before retrieving a pdf report with the information he has requested, he marks the second cell as being of particular

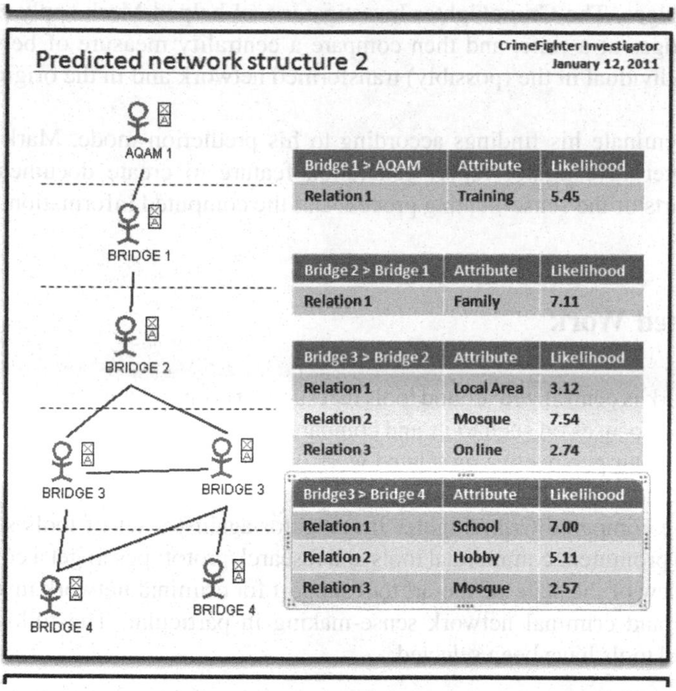

Fig. 25 (mock-up) One of the predicted network structures as shown in the report generated based on the prediction model

interest, since the predicted links here have the highest likelihoods of linkage. Plus, the individuals in the network seem to have skills necessary to carry out a small scale attack. Mark summarizes his findings in an email to his decision-making superiors and attaches the computed pdf report.

4.2 *Summary*

Mark used his knowledge about terrorist networks in Europe to design a prediction model that could solve the specific problem at hand. Later, he tailored existing CrimeFighter Investigator functionality to actual apply his sense-making approach to a network of established and aspiring violent radical Islamists living in Denmark from which future (terrorist) networks could form and pose a threat to Danish society.

Mark's first step towards applying his understanding of these networks was to use CrimeFighter Investigator synthesis functionality to divide the DNRI network and related individuals into four bridges, that he believed were actually functioning as linkage bridges. The CrimeFighter Investigator tool helped Mark apply prediction to two bridges at a time, and then compare a centrality measure of betweenness for each individual in the (possibly) transformed network and in the original DNRI network.

To disseminate his findings according to his prediction mode, Mark used the CrimeFighter Investigator report generation feature to create documentation of relevant parts of the sense-making process and the computed information.

5 Related Work

Related work is centralized around tools that support criminal network investigation. This section focuses on such tools and compares our approach against existing work in that area. Our perspective on related work is narrow and focuses on transparency of process and ownership of information.

We have compared CrimeFighter Investigator against a set of tools selected to cover both prominent commercial tools and research prototypes to get a comprehensive overview of the state-of-the-art tool support for criminal network investigation in general and criminal network sense-making in particular. The following four commercial tools have been selected:

- Analyst's Notebook 8.5 is part of i2 Limited's analysis product line and aims at supporting a rich set of analysis and visualization capabilities to support analysts in quickly turning large sets of disparate information into high-quality and actionable intelligence to prevent crime and terrorism [26].
- Palantir Government 3.0 is a platform for information analysis designed for environments where the fragments of data that an analyst combines to tell the larger story are spread across a vast set of starting material. Palantir is currently used in various domains such as intelligence, defense, and cyber security [38].
- Xanalys Link Explorer 6.0 (previously Watson [1]) allows investigators to apply powerful query and analysis techniques to their data, presenting the answers in a range of visualizations such as link charts, time lines, maps, and reports [74].
- COPLINK is designed for both general policing and specialist use for detectives/crime analysis [15]. The tool consists of three modules: "Connect" database, "Detect" criminal intelligence, and "Collaboration" [21]. With the merger between Knowledge Corporate Computing and i2 in 2009, COPLINK became a separate product line within i2 Limited.

Also, three research prototype tools have been selected for the comparison:

- Aruvi is the prototype implementation of an information visualization framework that supports the analytical reasoning process [51].
- Sandbox is a flexible and expressive thinking environment that supports both ad-hoc and formal analytical tasks [73].

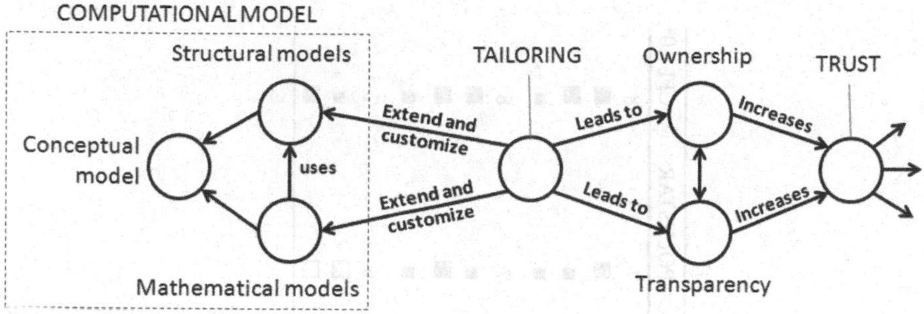

Fig. 26 Proposed computational modeling concepts and their interrelationship

- POLESTAR (POLicy Explanation using STories and ARguments) is an integrated suite of knowledge management and collaboration tools for intelligence analysts [43].

Evaluation of tools for criminal network sense-making from a task perspective requires well-designed usability experiments and reporting of statistical tendencies. It is a very broad perspective to take. Instead, we will assess state-of-the-art according to tailorability of the computational model. We have previously defined ownership of information and transparency of process to be direct results of tailorability, meaning the ability to extend and customize existing functionality for a specific purpose. In our chosen approach, we claimed that the level of tailorability depends on the computational model. We proposed a computational model that separated structural and mathematical models, both utilizing a conceptual model offering three first class entities.

The evaluation and comparison of the selected tools was made based on the concepts developed for our approach to criminal network sense-making. These concepts are summarized in Fig. 26. At the center is tailoring, a concept that facilitates extension and customization of structural and mathematical models. Tailoring leads to transparency of the sense-making process and ownership of sense-making computed information. Transparency and ownership increases trust in the provided information, which will increase the likelihood of that information being used for operational decision-making or similar.

A thorough examination of each tool has been made by the authors based on the available research literature, books, manuals, and other publicly available information. The results can be seen in Table 1.

Each tool is rated against each concept (model) and sub-concept in the list. A judgment has been made whether the tool provides full support (■), partial support (▪), or no support (□) for this task, indicated using the shown icons in the table. Based on the support for individual sub-concepts, each tool has been given a score for each concept (conceptual model, structural models, and mathematical models) based on a judgment of how many of the sub-concepts that they support. This score is between 0 (no support), 1–2 (fragmentary support), 3–7 (partial support), and

Table 1 The authors' assessment of computational modeling concepts

	AN 8.5[a]	PG 3.0[a]	XLE 6.0[a]	COPLINK	Aruvi	Sandbox	POLESTAR	CFI 1.0[a]
Conceptual model	**5**	**7**	**5**	**2**	**5**	**7**	**7**	**8**
First class information elements	■	■	■	▪	■	■	■	■
First class relations	▪	▪	▪	□	▪	▪	▪	■
First class composites	□	▪	□	□	□	▪	▪	▪
Structural models	**4**	**7**	**5**	**2**	**5**	**6**	**7**	**8**
Navigational structure	■	■	■	▪	▪	▪	▪	■
Spatial structure	▪	▪	▪	□	■	■	■	■
Taxonomic structure	□	▪	□	□	□	▪	▪	▪
Mathematical models	**5**	**5**	**5**	**0**	**2**	**0**	**0**	**7**
Transformative[b]	□	□	□	□	□	□	□	▪
Measuring	■	■	■	□	▪	□	□	■

[a] *AN* Analyst's Notebook, *PG* Palantir Government, *XLE* Xanalys Link Explorer, *CFI* CrimeFighter Investigator

[b] Filtering is not included

8–9 (full support). Fragmentary support means that the core concept is in theory supported by the tool through the combination of various features (not the listed sub-tasks), but it is found to be too complicated to be really useful in terms of tailorability.

We observe two tendencies in our assessment of computational modeling concepts in commercial tools and research prototypes for criminal network investigation. Separating commercial tools from research prototypes, we see that the research prototypes are slightly more diverse in their support of first class entities. Tools and research prototypes are equally strong in terms of structure domains supported; the commercial tools are strong on navigational structures, where the research prototypes have better support for spatial structures. Finally, the commercial tools outperform the research prototypes in terms of mathematical models (measures) supported. CrimeFighter Investigator has better support of first class entities (conceptual model), structure domains (structural models), and transformative and measuring algorithms (mathematical models) than the state-of-the-art tools and research prototypes analyzed for this comparison.

In our "invented" scenario, Mark used sense-making tailoring to be able to understand and reason about the network information he was asked to analyze. More specifically he customized a prediction algorithm to base its inferences on different information element attributes for different parts of the network. He also extended the actual prediction of links to be conditioned by the betweenness centrality of the individuals between who links where predicted, prior to that prediction. The tailoring in CrimeFighter Investigator made the process transparent and helped Mark to gain a feeling of ownership towards the information provided. In other words, he trusted the sense-making provided information enough to forward his findings to his decision-making superiors.

6 Conclusion and Future Work

The CrimeFighter Investigator approach to criminal network sense-making has been developed based on three primary types of analysis work:

- **Exploring methods.** We have explored analytical practices, processes, and techniques related to making sense of criminal networks.
- **Studying related work.** We have found inspiration from existing tools supporting criminal network sense-making and from relevant investigations.
- **Evaluation.** We have compared the sense-making approach of existing tools with CrimeFighter Investigator.

Two challenges associated with our chosen research approach were initially listed together with a question for each challenge. We found that:

1. The sense-making algorithms supported by CrimeFighter Investigator are applicable to criminal networks that are synthesized using multiple structure domains.

We developed a computational model that separates structural models from mathematical models and is based on a conceptual model defining first class information entity types.
2. CrimeFighter Investigator supports both transformative and measuring sense-making algorithms. To achieve this, a structural parser was implemented to provide an interface to these algorithms.

The novelty of the CrimeFighter Investigator approach to criminal network sense-making is the underlying tailorable computational model. Tailorability was (partially) achieved with a structural parser that provides the user with an interface to customize and combine sense-making algorithms. The approach introduces transparency of the sense-making process and ownership of the computed information. In our review of and comparison with state-of-the-art commercial tools and research prototypes, we found that CrimeFighter Investigator has better support of first class entities (conceptual model), structure domains (structural models), and transformative and measuring algorithms (mathematical models).

Currently, many of the envisioned sense-making tasks are supported. As part of our near term future work, we will provide support for the remaining sense-making tasks. We then plan to thoroughly test the tool in cooperation with experienced criminal network investigators.

Acknowledgements This work builds upon previous publications by the authors in hypertext and security informatics conference proceedings [40–42].

References

1. Adderly R, Musgrove P (2001) Police crime recording and investigation systems – a user's view. Int J Police Strateg Manag 24(1):100–114
2. Amland BH. (2012) 2 convicted in al-Qaeda terror plot in Norway. Associated Press, New York City, NY
3. Atzenbeck C, Hicks DL, Memon N (2011) Supporting reasoning and communication for intelligence officers. Int J Netw Virtual Organ 8(1/2):15–36
4. Badalamente RV, Greitzer FL (2005) Top ten needs for intelligence analysis tool development. In: Proceedings of the 2005 international conference on intelligence analysis, MITRE, (McLean, Virginia) online proceedings only, (http://cryptome.org/intel-2005.htm)
5. Bier EA, Card SK, Bodnar WJ (2010) Principles and tools for collaborative entity-based intelligence analysis. IEEE Trans Vis Comput Graph 16(2):178–191
6. Brachman JM. (2009) Global jihadism: theory and practice. Routledge, London/New York
7. Brachman JM, Levine A (2011) You too can be awlaki! Fletcher Forum World Aff 35:25–46
8. Brantingham P, Glässer U, Jackson P, Vajihollahi M (2009) Modeling criminal activity in urban landscapes. In: Mathematical methods in counterterrorism. Springer, Wien, pp 9–31
9. Carley KM. (2003) Destabilizing dynamic covert networks. In: Proceedings of the 8th international command and control research and technology symposium. Evidence Based research, Electronic publication. Vienna, VA
10. Carley KM, Lee JS, Krackhardt D (2001) Destabilizing networks. Connections 24:31–34
11. Clark R (2007) Intelligence analysis: a target-centric approach. CQ Press, Washington, DC

12. Conway M (2008) Jihadi video and auto-radicalisation: evidence from an exploratory youtube study. In: Intelligence and security informatics. Lecture notes in computer science (LNCS). Springer, Wien, pp 108–118
13. Conway M (2012) From al-zarqawi to al-awlaki: the emergence of the internet as a new form 861 of violent radical milieu. Working paper, Dublin City University (available from http://www.isodarco.it/courses/andalo12/doc/Zarqawi%20to%20Awlaki_V2.pdf,lastvisited)
14. Davis F (1989) Perceived usefulness, perceived ease of use and user acceptance of information technology. MIS Q 13:319–340
15. Dean G, Gottschalk P (2007) Knowledge management in policing and law enforcement. Oxford University Press, Oxford/New York
16. Drogin B (2008) Curveball. Ebury Press, London
17. Europol (2009) TE-SAT 2009: EU terrorism situation and trend report. Europol, The Hague, Netherlands
18. Europol (2010) TE-SAT 2010: EU terrorism situation and trend report. Europol, The Hague, Netherlands
19. Europol (2011) TE-SAT 2011: EU terrorism situation and trend report. Europol, The Hague, Netherlands
20. Gloor PA, Zhao Y (2006) Analyzing actors and their discussion topics by semantic social network analysis. In: Proceedings of information visualization. IEEE Computer Society. Washington, DC, pp 130–135
21. Hauck RV, Chau M, Chen H (2002) Coplink: arming law enforcement with new knowledge management technologies. In: Advances in digital government: technology, human factors, and policy. Kluwer, Boston, pp 163–179
22. Hoffman B (2006) Inside terrorism. Columbia University Press, New York, NY
23. Hu PJH, Chen H, Hu H, Larson C, Butierez C (2011) Law enforcement officers' acceptance of advanced e-government technology: a survey study of coplink mobile. Electron Commer Res Appl 10:6–16
24. Hu PJH, Lin C, Chen H (2005) User acceptance of intelligence and security informatics technology: a study of coplink. Am Soc Inf Sci Technol 56:235–244
25. Huntington SP (1996) The clash of civilizations and the remaking of world order. Simon & Schuster, New York
26. i2 limited, analyst's notebook (2012) http://www.i2group.com/
27. Intelligence and Security Committee, United Kingdom (2009) Could 7/7 have been prevented? Review of the intelligence on the London terrorist attacks on 7 July 2005
28. Irons LR. (2008) Recent patterns of terrorism prevention in the united kingdom. Homel Secur Aff 28:1–19
29. Kebbell MR, Muller DA, Martin K (2010) Understanding and managing bias. Dealing with uncertainties in policing serious crime. ANU E Press, Canberra, pp 87–97
30. Laville S (2012) Al-Qaida-inspired plotters planned attacks on high-profile London targets. The Guardian (online), London
31. Levy BH. (2003) Who killed Daniel Pearl? Melville, Brooklyn, NY
32. Linschoten AS, Kuehn F (2012) An enemy we created: the myth of the Taliban/Al-Qaeda merger in Afghanistan:1970–2010. Hurst, London
33. MacDougall I (2012) Norway 'bomb plot' highlights al-Qaida problems. Associated Press, New York City, NY
34. MacKensie J (2010) The battle for aghanistan: militancy and conflict in helmand. New America Foundation, Washington, DC
35. McBride M, Morgan S (2010) Trust calibration for automated decision aids, Institute for Homeland Security Solutions, Durham, NC
36. National Commission on Terrorist Attacks upon the United States, United States (2004) The 9/11 Commission report (executive summary). http://www.9-11commission.gov/report/911Report_Exec.pdf

37. Nesser P (2006) Structures of jihadist terrorist cells in the uk and europe. In: Proceedings of the joint FFI/King's college conference on "the changing faces of jihadism", It is actually a presentation, not proceedings authors notes. Kings College, London
38. Palantir (2012) http://www.palantirtech.com/government
39. Pearl M (2004) A mighty heart. Virago, London
40. Petersen RR, Rhodes CJ, Wiil UK (2011) Node removal in criminal networks. In: Proceedings of European intelligence and security informatics conference. IEEE, Washington, DC, pp 360–365
41. Petersen RR, Wiil UK (2011) Crimefighter investigator: a novel tool for criminal network investigation. In: Proceedings of European intelligence and security informatics conference. IEEE, Washington, DC, pp 360–365
42. Petersen RR, Wiil UK (2011) Hypertext structures for investigative teams. In: Proceedings of the 22nd ACM conference on hypertext. ACM, New York, NY, pp 123–132
43. Pioch NJ, Everett JO. (2006) Polestar: collaborative knowledge management and sensemaking tools for intelligence analysts. In: Proceedings of the international conference on information and knowledge management. ACM, pp 513–521
44. Rhodes C (2011) The use of open source intelligence in the construction of covert social networks. In: Wiil UK (ed) Counterterrorism and open source intelligence. Lecture notes in social networks (LNSN 2). Springer, Wien, pp 159–170
45. Rhodes CJ, Jones P (2009) Inferring missing links in partially observed social networks. J Oper Res Soc 60(10):1373–1383
46. Rhodes CJ, Keefe CMJ (2007) Social network topology: a bayesian approach. J Oper Res Soc 58(12):1605–1611
47. Sageman M (2004) Understanding terrorist networks. PENN, Philadelphia, PA
48. Sageman M (2008) Leaderless jihad. PENN, Philadelphia, PA
49. Saunders-Newton D, Scott H (2001) "but the computer said!": credible uses of computational modeling in public sector decision making. Soc Sci Comput Rev 19:47–65
50. Security D, (PET), I.S. (undated) Terror arrests in Copenhagen. http://www.pet.dk/Nyheder/morkhoj-uk.aspx
51. Shrinivasan YB, Wijk JJ (2008) Supporting the analytical reasoning process in information visualization. In: Proceedings of the 26th conference on human factors in computing systems. ACM, New York, NY
52. Silber MD, Bhatt A (2007) Radicalisation in the west: the homegrown threat. NYPD, New York, NY
53. Skjoldager M (2009) Truslen indefra: de danske terrorister. Lindhardt & Ringhof, Copenhagen, Denmark
54. Smith EA. (2006) Complexity, networking, & effects-based approaches to operations. CCRP, Washington, DC
55. Sparrow MK. (1991) The application of network analysis to criminal intelligence: An assessment of the prospects. Soc Netw 13:251–274
56. Stoll C (1995) Silicon snake oil: second thoughts on the information highway, Doubleday. New York, NY
57. Taarnby M (2006) Jihad in Denmark: an overview and analysis of jihadi activity in denmark 1990–2006. Danish institute for international studies, DIIS. Copenhagen, Denmark
58. Todd BF, Nomani A (2011) The truth left behind: inside the kidnapping and murder of Daniel Pearl. The Pearl Project, Georgetown University
59. N/A (2012) London terror bomb plot: the four terrorists. The Telegraph (online), London
60. Veldhuis T, Staun J (2009) Islamist radicalisation: a root cause model. Institute of International Relations, The Hague, Netherlands
61. Vidino L (2005) Al Qaeda in Europe: the new battleground of international jihad. Prometheus Books, Amherst, NY
62. Vidino L (2011) Radicalization, linkage, and diversity: current trends in terrorism in Europe, RAND Corporation. Santa Monica, CA and Arlington, VA and Pittsburgh, PA

63. Vogel KM. (2008) 'iraqi winnebagos^{TM} of death': imagined and realized futures of us bioweapons threat assessment. Sci Public Policy 35:561–573
64. Wasserman S, Faust K (1994) Social network analysis: methods and applications. Cambridge University Press. Cambridge, United Kingdom
65. Wegner P (1997) Why interaction is more powerful than algorithms. Commun ACM 40:80–91
66. Weiman G (2010) Terror on facebook, twitter, and youtube. Brown J World Aff 16:45–54
67. Weiner T (2008) Legacy of ashes: the history of the cIA. Anchor Books, New York, NY
68. Wiil UK, Gniadek J, Memon N (2010) Measuring link importance in terrorist networks. In: proceedings of the international conference on advances in social networks analysis and mining. IEEE, Washington, DC, pp 225–232
69. Wiil UK, Gniadek J, Memon N, Petersen RR. (2011) Knowledge management tools for terrorist network analysis. In: Knowledge discovery, knowledge engineering and knowledge management. Lecture notes in communications in computer and information science (LNCCIS). Springer, Wien
70. Wiil UK, Memon N, Gniadek J (2011) Crimefighter: a toolbox for counterterrorism. Lecture notes in communications in computer and information science (knowledge discovery, knowledge engineering and knowledge management), Springer Berlin Heidelberg, vol 128. pp 337–350
71. Wilson C (updated 2010) Searching for saddam: a five-part series on how the u.s. military used social networking to capture the iraqi dictator. http://www.slate.com/articles/news_and_politics/searching_for_saddam/2010/02/searching_for_saddam_5.single.html
72. Woo G (2009) Intelligence constraints on terrorist network plots. In: Memon N, Farley JD, Hicks DL, Rosenorn T (eds) Mathematical methods in counterterrorism. Springer, Wien, pp 205–214
73. Wright W, Schroh D, Proulx P, Skaburskis A, Cort B (2006) The sandbox for analysis: concepts and methods. In: Proceedings of the conference on human factors in computing systems. ACM, New York, NY, pp 801–810
74. Xanalys (2012) http://www.xanalys.com/
75. Xu J, Chen H (2005) Criminal network analysis and visualization. Commun ACM 48(6): 100–107
76. Yalcinkaya R (2007) Police officers' adoption of information technology: a case study of the turkish polnet system. Ph.D. thesis, University of North Texas
77. Youtube (2012) General colin powell un speech on iraq part 1 of 5. http://www.youtube.com/watch?v=Nt5RZ6ukbNc. Last visited on February 19th 2012

Part IV
Systems, Frameworks, and Case Studies

The NOEM: A Tool for Understanding/ Exploring the Complexities of Today's Operational Environment

John J. Salerno, Jason E. Smith, Warren M. Geiler, Patrick K. McCabe, Aleksey V. Panasyuk, Walter D. Bennette, and Adam Kwiat

1 Introduction

> Counterinsurgency is a proactive approach involving all elements of national power; even down to the tactical level. COIN operations strive to achieve unity of effort amongst many joint, interagency, intergovernmental, and multinational organizations. COIN includes tactical planning; intelligence development and analysis; training; materiel, technical, organizational assistance; advice; infrastructure development; tactical-level operations; and information engagement [1].

In interviews with commanders and reported by [41] in an article in Small Wars Journal "... commanders that have faced this operational environment, all agreed on the complexity of the contemporary battlefield and the challenges in understanding and managing it. There was also wide recognition of lost opportunities due to poor information management and failures to identify the second and third order effects of actions. They recognized that the large volume of data available to them needs to be made manageable. They want to understand possible outcomes, and the ripple effects from the actions of their units and others across the theater. There is a strong need for 'what if analysis' at the small unit level. A good example is the organization of the Company Intelligence Support Teams. This strong need for analysis comes at a time when these commanders feel that staffs are losing analytical skills in basic

J.J. Salerno (✉) • W.M. Geiler • P.K. McCabe • A.V. Panasyuk • W.D. Bennette
Air Force Research Laboratory, Rome Research Site, 525 Brooks Road, Rome, NY, 13441
e-mail: john.salerno@rl.af.mil; warren.geiler@rl.af.mil; patrick.mccabe@rl.a.mil; aleksey.panasyuk@rl.af.mil; walter.bennette@rl.af.mil

J.E. Smith
Air Force Research Laboratory, Rome Research Site, 525 Brooks Road, Rome, NY, USA,
e-mail: jason.e.smith@itt.com

A. Kwiat
CUBRC, 725 Daedalian Drive, Rome, NY, 13441
e-mail: adam.kwiat@cubrc.org

V.S. Subrahmanian (ed.), *Handbook of Computational Approaches to Counterterrorism*,
DOI 10.1007/978-1-4614-5311-6_17,

warfighting functions, which hurts their ability to adapt. While they know that 'every soldier is a sensor,' they recognize that this puts a premium on timely analysis that leads to not just situational awareness, but situational understanding."

Major General Michael T. Flynn, US Army, in his report, *Fixing Intel: A Blueprint for Making Intelligence Relevant in Afghanistan* [37], makes the case that in order to make an actual difference in defeating an insurgency, one must place focus on the local populace, their needs, and the environment in which they live, in addition to seeking and destroying the adversary. "Eight years into the war in Afghanistan, the U.S. intelligence community is only marginally relevant to the overall strategy. Having focused the overwhelming majority of its collection efforts and analytical brainpower on insurgent groups, the vast intelligence apparatus is unable to answer fundamental questions about the environment in which U.S. and allied forces operate and the people they seek to persuade. Ignorant of local economics and landowners, hazy about who the powerbrokers are and how they might be influenced, incurious about the correlations between various development projects and the levels of cooperation among villagers, and disengaged from people in the best position to find answers—whether aid workers or Afghan soldiers—U.S. intelligence officers and analysts can do little but shrug in response to high level decision-makers seeking the knowledge, analysis, and information they need to wage a successful counterinsurgency." (Flynn et al. 2010) Many believe that traditional methods of warfare are being phased out, and being replaced with an emphasis on human security. In traditional warfare the goal has been to destroy the will of the populace in order to elicit surrender. Under the concept of human security the goal is to build up the confidence of the local populace so that they may stand up for themselves and withdraw support for the local insurgents.

The key concept is that in order to effectively rout out an insurgent or terrorist network, one must first win over the hearts and minds of the host populace. If the local populace has the proper level of security in place and their basic needs and expectations are being met, they will be able to focus more clearly on improving their own lives and will be emboldened to reject and resist the insurgents holding their community hostage. For the purpose of providing sufficient security, basic needs, and a sense of hope for the local populace, we require more than a purely military-based solution.

How does a nation-state end up in such dire straits that quite often force a rushed military response? Dr. Thomas Barnett, in his book *The Pentagon's New Map; War and Peace in the Twenty-First Century* [2], suggests that a capitalist economic worldview drives global interdependency forward, subtly forcing states to continually integrate economically, politically, culturally, and on other levels into a core global whole. Connectivity is thus developed between functioning parts of the global community as globalization extends throughout the world. Those areas that do not connect into this global economy (for various reasons) are often found to be the areas where security problems arise (failed states). Barnett, on page 3 [2], defines state failure as:

> A state is 'failing' if it either cannot attract or build itself the connectivity associated with globalization's progressive advance or if it essentially seeks to retard or deny the development of such connectivity out of the desire to maintain strict political control over its population.

Recent events around the world, in particular the Arab Spring and the 2009 Iranian post-election violence, have demonstrated the increasing importance of connectivity, as Barnett describes it. In these instances, various means of social media (Facebook and Twitter), were used by protesters/activists to organize protests or incite violence. Social media has now become another medium of connectivity that authoritarian regimes are now looking to crack down upon.

Terrorist networks seek out failed states for a variety of reasons. Regimes in failed states often do not have full or adequate control over their own territory (e.g., Somalia, Afghanistan), thereby allowing groups to arrange agreements with local leaders and set up operations without fear of a national government response. Failed states often experience political turmoil, which is then exploited by terrorists to foster a supportive and friendly environment (e.g., Sudan and potentially other sub-Saharan African states in the near future). Failed states are often ruled by authoritarian regimes offering support to terrorist networks (e.g., Iran's support of Hezbollah, Liberia's offer of sanctuary in return for cash). On page 4 of [2] it is suggested that an underlying rejectionist ideology amongst the terrorist leadership fighting against the "reality of globalization's progressive advance into traditional Islamic societies" as the driving rationale behind the current geopolitical situation.

Barnett directly states that there is a continuing and pressing need for nation-building as the US seeks to rehabilitate and reintegrate the failed states into the global economy. In fact, in his book, Barnett expounds a controversial vision of a US military force-shaping evolution that is necessary to successfully conduct the potentially numerous upcoming nation-building operations.

Such operations are usually planned by small cells of Diplomatic, Information, Military, Economic (DIME) or Political, Military, Economic, Social, Infrastructure and Information (PMESII) trained officers in a pseudo-ad-hoc fashion. However, the officers, although trained, usually are woefully limited in their understanding of the implications or ramifications that specific operations will have on the overall state of regional stability. Defining stability operations can change in scope or definition depending on whether we are monitoring a nation or region, engaging in full-out Military Operations, or rebuilding a nation-state.

Careful planning is necessary to conduct all operations required in order to achieve stability. As stated earlier, current planning practices today are highly manual processes, usually performed in a pseudo-ad-hoc fashion and in a relatively short timeframe around the actual operational events. Tools that can help decision-makers and planners effectively plan for nation-state stability during any/all phase(s) of the conflict are needed. These tools should possess the ability to take the entire environment into consideration and produce timely analysis of policy sets that can be used to achieve mission objectives while at the same time securing a stable environment quickly and at minimal cost.

1.1 A Step Forward

> America is now threatened less by conquering states than we are by failing ones [3].

> The Army on Monday will unveil an unprecedented doctrine that declares nation-building missions will probably become more important than conventional warfare and defines 'fragile states' that breed crime, terrorism and religious and ethnic strife as the greatest threat to U.S. national security [4].

> The new American way of war, says Defense Secretary, Robert Gates, will be to prevent costly and controversial military interventions. That will be achieved not by being isolationist, but by helping governments in weak nations root out terrorists... The most likely catastrophic threats to our homeland – for example, an American city poisoned or reduced to rubble by a terrorist attack – are likely to emanate from failing states than from aggressor states [5].

It is clear that failing and failed states are a primary concern of the US's national interests. As stated on page 4 of the National Security Strategy, the primary tenet of the United States national security strategy elucidates American determination to act preemptively, if necessary, in exercising the right of self-defense against purveyors of global terrorism. Therefore, denying support and sanctuary to terrorists across the globe where enabling conditions for terrorism prevail is a primary driver for future military conflicts. The potential threat to U.S. national security posed by these failing states, which are inadvertently or actively providing safe harbor to global terrorist organizations, provides a galvanizing motivation to explore the complexities involved in "convincing or compelling states to accept their sovereign responsibilities".

According to a Government Accounting Office (GAO) report: "Military Operations: Actions Needed to Improve DoD's Stability Operations Approach and Enhance Interagency Planning," [45] the DoD has instituted a fundamental shift in its policy by elevating stability operations as a core mission comparable to combat operations and emphasize that military and civilian efforts must be integrated. Why is stability so important? [34], defines "stability operations" as an overarching term encompassing various military missions, tasks, and activities conducted outside the US in coordination with other instruments of national power to maintain or reestablish a safe and secure environment, provide essential governmental services, emergency infrastructure reconstruction, and humanitarian relief. Also stated in the GAO report is that DOD's involvement in stability operations activities has been significant and on the rise.

> Stability operations are conducted to help establish order that advances U.S. interests and values. The immediate goal is often to provide the local populace with security, restore essential services, and meet humanitarian needs. The long-term goal is to help develop indigenous capacity for securing essential services, a viable market economy, rule of law, democratic institutions, and a robust civil society [6].

Joint Publication 3-0, defines six phases during a joint campaign or operation: Phase 0—Shape, Phase I—Deter, Phase II—Seize Initiative, Phase III—Dominate,

Phase IV—Stabilize, and Phase V—Enable Civil Authority. Operations and activities in the "shape" and "deter" phases normally are outlined in Strategic Capability Plans and those in the remaining phases are outlined in Joint Strategic Capability Plan (directed operation plans). **Shaping** is inclusive of normal and routine military activities — and various interagency activities are performed to dissuade or deter potential adversaries and to assure or solidify relationships with friends and allies. They are executed continuously with the intent to enhance international legitimacy and gain multinational cooperation in support of defined military and national strategic objectives. This in many cases is referred to either "Theater Security Operations" or Deterrence.

When Phase 0 activities fail, we move into phases I–III. The intent of the **Deter** stage (Phase I) is to stop undesirable adversary action by demonstrating the capabilities and resolve of the joint force. It differs from deterrence that occurs in the "shape" phase in that it is largely characterized by preparatory actions that specifically support or facilitate the execution of subsequent phases of the operation/campaign. Phase II or **Seize Initiative:** one seeks to seize the opportunity through combat and noncombat situations through the application of appropriate joint force capabilities. In combat operations, this involves executing offensive operations at the earliest possible time, forcing the enemy to offensive culmination and setting the conditions for decisive operations. **Dominate** or Phase III focuses on breaking the enemy's will for organized resistance or, in noncombat situations, control of the operational environment.

Once a secure environment has been created and maintained, we would then move into Phase IV. In Phase IV or **Stabilize,** there is limited or no functioning legitimate civil governing entity present. The military may have to stand in and may be required to perform limited local governance, integrating the efforts of other supporting or contributing multinational, Other Government Agencies (OGAs), Inter-Governmental Organizations (IGO), or Non-Governmental Organizations (NGOs) participants until legitimate local entities are functioning. This includes providing or assisting in the provision of basic services to the population. The last Phase, **Enable Civil Authority** is predominantly characterized by US support to legitimate civil governance. This support will be provided to the civil authority with its agreement at some level, and in some cases especially for operations with the United States, under direction of the civil authority.

1.2 Supporting Stability Operations

As described in Joint Publication 3-07 Stability Operations [35], the environment is a composite of the conditions, circumstances, and influences that affect the employment of capabilities bearing on the decisions of the commander. Characterized as a complex system, achieving situational awareness in such a complex domain requires increasingly sophisticated techniques. Currently, there is a limited ability to forecast stability/instability within the environment in sufficient detail so that

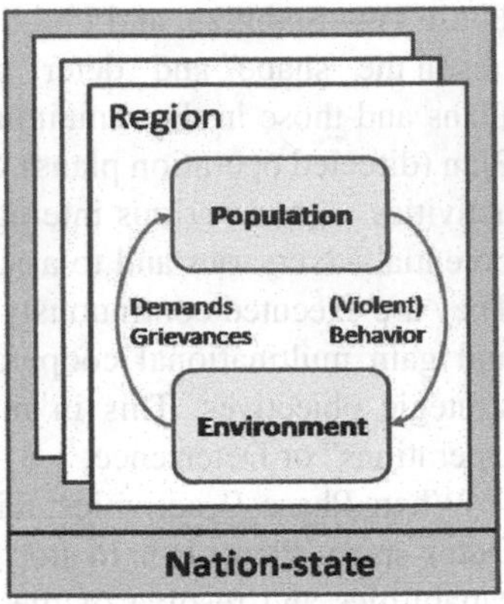

Fig. 1 Population/ Environment model

either "corrective" action(s) can be investigated or the ramifications of such actions can likewise be identified. The inherent complexity of the state space makes this problem difficult. No rigorous process exists to identify the leverage points within the environment that can be manipulated to bring an unstable national system back into a state of stability. Making matters even worse, there is no general consensus for defining stability or what it might entail. Many existing definitions rely on the level of conflict or violence and are used primarily to measure civil unrest. The rise of failing states has directly led to constantly evolving U.S. national (e.g., 9/11) and global security concerns. Failing states tend to be the originators of regional violence and aggression, and often preclude or include international terrorism. The U.S. recognizes the threat of these failing states prior to, during, and after conflict. The current U.S. strategy to combat failing states is to stabilize them through stability operations.

To elaborate further, Stability Operations (SOPs) research has identified the components of the environment that provide services and commodities to the population, as well as explored, perceived, and real grievances. It is the host nation government's responsibility to ensure the current and future well-being of its population in this regard. From the viewpoint of this chapter, we can focus on a population's (violent) behavior as a consequence of the realization of (dis)satisfaction of demands and redress of grievances. This behavior in turn affects the population's own environment and may help or hinder the satisfaction of its demands and grievances. We refer the reader to Fig. 1, which illustrates this relationship.

A population places demands on the system within its region (and on some national level, capabilities, as well). The population seeks security, rule of law, governance, and social and economic well-being. The regions and/or nation-state exist to supply the demanded services and commodities and address grievances. The success or failure of the supporting system in supplying the population to sufficiently meet its demands will impact the population's behavior.

A population's behavior (defined to an extent by their age, occupation, affiliations in interest groups, and the relationships between these groups) is determined in part by the way it perceives external events. The population (from a collective group paradigm) will perceive and act on the fact that its demands are or are not being met.

1.2.1 Modeling and Simulation Support to Stability Ops

Through the course of developing this paper, it has been generally perceived that current and past modeling efforts can be divided into one of two different classes: Early Warning and Large Scale Models & Simulation programs. In the following paragraphs, we provide a number of examples of each. We note here that the examples presented are only exemplary and by no means exhaustive.

The first class attempts to learn which observables are important and then map those observables to some output measure. These are commonly referred to as Strategic Early Warning Capabilities. Their purpose is to direct your attention towards the nation-states that seem to be the most fragile or unstable. These capabilities (which have included input from analysts within several organizations) produce a priority list.

In his paper, Dixon [7] provides an excellent summary for more than 46 published studies looking at civil war and unrest. Dixon categorizes the various causes into: Demography and Diversity; Geographic and Environmental; Greed; Economic; Conflict History and International Insecurity; and Government Type and Regime Change. We will come back to this paper in the next section when we discuss indicators. There are a number of other works that Dixon does not cover. Among them and one that seems to receive much press is the Political Instability Task Force (PITF) tool developed by Dr Jack Goldstone [8]. PITF is based on four indicators: (1) Regime Type: Partial Democracy with Factionalism, Partial Democracy, Partial Autocracy; (2) Poor Socioeconomic Conditions (higher infant mortality rate); (3) State-Led Discrimination; and (4) Bordering States with Armed Conflicts, and outputs the forecasted level for conflict. It uses the previous year's data to project conditions for the following year. For example, in 2008, they use the 2007 data to project the 2009 forecast. Other researchers have proposed similar models. Such examples include: Analyzing Complex Threats for Operations and Readiness (ACTOR), Forecast and Analysis of Complex Threats (FACT), and DARPA's Integrated Crisis and Early Warning System (ICEWS). The Analyzing Complex Threats for Operations and Readiness (ACTOR) modeling effort was designed to assist policy makers and analysts with determining which countries around the world were at risk for political instability, perhaps requiring some form of intervention. ACTOR can be likened most to the Political Instability Task Force (PITF) model; however, rather than producing a binary output—instability or stability within a time-frame of 2 years—ACTOR sets out to predict the degree of instability that a nation may encounter within a 5- to 15-year time span. Thus, not only does ACTOR boast the same level of overall accuracy in forecasts (about

80%), but it may categorize the severity of instability across four levels: none, low, moderate, and high.

FACT classified patterns of nation state instability in an attempt to identify common instability factors. Standard pattern classification algorithms were used to identify instability patterns, and features characterizing state instability were identified. The study included over 100 countries and forecasted future levels of instability and potential conflict and conflict potential characterized ten plus years into the future. ICEWS is an on-going DARPA program begun in 2007, incorporating a variety of modeling approaches. The goal is to more accurately and precisely forecast the effects of national or regional events on U.S. national security interests. Each modeling approach relied heavily on a specific social-theoretic construct, and the hypothesis was that the approaches would complement each other, resulting in a more actionable product [40]. The general concern with this class of models is in their inability to provide insight into what needs to be done to correct the potential instability. They typically cannot explain why problem(s) exist, nor allow one to investigate the ramifications of any potential actions on the environment. In fairness, these types of models are only intended to provide limited insight as to whether the possibility of an unstable situation may occur.

The second class of models attempts to provide detailed models that simulate portions of a nation-state or region. The majority of these capabilities can be described in terms of PMESII. In this section we can categorize the models based on their focus: (1) Physical components (critical infrastructure); (2) Social Systems and (3) Behavior Components. In the following sections we provide examples of a number of modeling environments. We note here that we are not endorsing any single model but providing a small subset of examples. There are many more models out there and we encourage the reader, if interested in any specific model or area to investigate these ideas further.

Models of Infrastructure

The Department of Homeland Security's (DHS) National Infrastructure Simulation and Analysis Center (NISAC) program attempts to model how attacks on critical infrastructure disrupt the US government and businesses. Cascading effects can be seen that reach far beyond the targeted sector and physical location of the incident. Eighteen critical infrastructure sectors are modeled: Food and Agriculture; Banking and Finance; Chemical; Commercial Facilities; Communications; Critical Manufacturing; Dams; Defense Industrial Base; Emergency Services; Energy; Government Facilities; Healthcare and Public Health; Information Technology; National Monuments and Icons; Nuclear Reactors, Materials and Waste; Postal and Shipping; Transportation Systems; and Water. Some of the planning analysis examples include: Pandemic influenza; Chemical supply chain; High-voltage transformers; New Madrid seismic zone; Hurricane scenario analyses for Atlantic and Gulf Coasts; Food defense; Global finance and payment systems; California water systems; Power grid; Southern California earthquakes; and Infrastructure asset prioritization.

TRANSIMS is an integrated set of tools to conduct regional transportation system analyses based on a cellular automata microsimulator. It uses a new paradigm of modeling individual travelers and their multi-modal transportation based on synthetic populations and their activities. Compared to other transportation aggregate models, TRANSIMS represents time consistently and continuously, as well as detailed persons and households. Its time-dependent routing and person-based microsimulator also differ from other aggregate models. This uses cellular data and then, for example, helps route individuals based on previous data and time of day. It is not designed to work for third world countries. However, there could be a great deal to learn as the simulation is built using real, continuous data and hence likely generates a very realistic model. It is open source and is actively used by Google Earth, for instance. A primary use is to analyze detailed traffic patterns and the need for new roads.

The Water Infrastructure Simulation Environment (WISE) is an integrated software framework for modeling, simulating, and analyzing interdependent water infrastructures. WISE allows quick and accurate management and analysis of large infrastructure networks across the dams and the water sector. WISE models not only the distribution network of water pipes, but also attempts to model flooding, rainfall runoff, dams, etc. Thus, it is capable of modeling flood risks from a range of events. Flood risks are used to assess the effects to critical infrastructure and homes. This is developed by NISAC and fits into the DHS-IP program. EpiSims models the spread of disease in urban areas, allowing for the assessment of prevention, intervention, and response strategies by simulating the daily movements of synthetic individuals within an urban region. EpiSims allows the user to specify the effects in detail of a pathogen on a specific person, and to assign different effects to various people based on demographic characteristics. In conjunction with population mobility models it can represent behavioral reactions to an outbreak, including official interventions. It has been developed using the activity generator from TRANSIMS. The activity generator is used to simulate household activities and preferences and has been constructed using surveys. The authors claim "The emergent person-to-person social network and mixing patterns are consistent with published survey studies [47]" so that some validation work has been performed.

Models of Social Systems

Senturion uses simulations to analyze political dynamics within and among local, regional, and international contexts to predict policy position evolution over time. It is agent based and utilizes approaches from game theory, decision theory, spatial bargaining, and microeconomics. Senturion was used in three case studies; Operation *Iraqi Freedom* and Aftermath, Iraqi Elections, January 2005, and Palestinian Leadership Transition after Yasser Arafat's Death [9]. Soar Technology's Advanced Global Intelligence and Leadership Experiment (AGILE) [42] allows for the building of executable models and conducting regional analysis. AGILE is a simulation framework that enables the specification of simulation parameters, models, and agents, and simulations, as well as provides a capability to carry out post-run analysis of the results. Analysts are given tools to model the dynamics of a geopolitical crisis, and then play out the sort of role-playing and wargaming that is traditionally carried out manually. CAESAR was developed as a framework to assist analysts to develop and evaluate courses of action in the context of a set of end state effects. Key components include an influence net modeler and an executable model generator and associated simulator [46].

Models of Behaviors

Modeling behavior of people has received increased interest over the past few years and is being addressed from many sides. We have seen some models focus on insurgencies' vs coalitions' interactions (Combat and Non-Combat), the creation and maintenance of insurgent groups, general populace actions/reactions to the flow of information between groups/individuals. While there has been much research in this area, there is room for significantly more. Our understanding and ability to provide robust behavior models has a long way to go. In this section, we provide just a few examples of such research.

Anderson [10] proposed a system dynamics model to gain insight into the dynamic behavior of insurgencies. His model is also based on FM 3-24. Similar to other models, Anderson starts with what he calls "potential" insurgents, or the pool of people who might become insurgents if the conditions are right. Based on FM 3-24, only a small fraction of potential insurgents ever pick up arms and fight. Anderson models this fraction as a decreasing function based on the popular support for the government. Indicated insurgents join the insurgency over time and become active insurgents. Insurgents can move to an inactive state based on age (retirement), killed in action or detained. This reduction is determined by the number of COIN combat patrols, the combat efficacy of the COIN troops and the effect of insurgent density. The insurgent recruitment portion of the model is just that, only a small portion of the overall model. Anderson uses a series of casual loops to describe the remaining portions. The number of insurgents is affected by the "suppression" loop. As the number of insurgents increase, so will the number of incidents (or attacks), which in turn increases the pressure to reduce them. To reduce the number requires an increase in the number of patrols (fraction combat versus intelligence patrols).

Choucri et al. [33] developed a state stability model exploiting system dynamics. The goal of the effort was to forecast stability/instability of a nation-state. They divided the population into three: general population, dissidents, and insurgents; and provided two direct ways to curb these transitions: appeasement and removal. First, dissidents might become appeased by actions performed by the state or government and move from becoming a dissident back to the general populace; secondly, dissidents who become insurgents can be removed by the state through arrests, detentions or violent actions. The core of their model is the idea of anti-regime messages and the flow of communications, and this flow is considered within the model. In their paper, the authors investigated two scenarios: (1) a reduction in the regime resilience and (2) removing insurgents versus reducing anti-regime messages.

Epstein et al. [11] developed an agent-based computational model to investigate civil violence. In [11], the authors present two models. In the first model there are two types of agents—general population and cops. All agents are wandering randomly on a landscape grid. Each individual in the general population has a grievance level. If its level of grievance against the central authority is high enough, and its perception of the risks of being involved in open rebellion is sufficiently low, it will openly rebel. The cops, acting on behalf of the central authority, seek to suppress the rebellion by arresting openly rebelled persons and throwing them in jail. The grievance level is calculated by the agent's perceived hardship multiplied by the government's illegitimacy.

There have also been a number of intelligence operations (IO) and military (force on force) modeling tools created. The Synthetic Environment for Analysis and Simulation [38], developed by Purdue University/Simulex for JFCOM, is a simulated environment that models several aspects of the economy, including the government, competition, foreign and public policy, and other international economies. SEAS contains geography and physical details of the space, such as road networks, traffic patterns, structures, and pedestrian dispersion. Within this simulated environment, agents are able to communicate, sense, and interact at various levels of abstraction. Unlike most modeling frameworks, SEAS allows for human input during simulations, where users can participate as an actor in the virtual environment.

Integrated Approaches

Pierson, Barge and Crane [12] provide an excellent overview as to the objectives of and functionality being offered in the COunter INsurgency (COIN) model. The COIN model was developed by the Warfighting Analysis Division of the Joint Staff (J8) to portray the details of the FM 3-24 to support a COIN campaign. Similar to others, the modelers started with a population-centric view (versus an enemy-centric) and divided the population into three bins: Neutral Populace, Support for the Insurgency, and Support for the Host Nation Government. From there, they added the effects that external funding can have on the economic conditions, leading to greater economic investment, further strengthening the economy, and hopefully influencing movement from insurgents back to either neutrals or supporters of the government. Other areas that they address in their model are psychological operations and security. Psychological operations' major objective is to get the "message" or information out, while security covers acts of violence and military presence. The final key point provided by the authors is the importance of intelligence and information and what drives the appropriate mix of resources needed.

The Conflict Modeling, Planning and Outcome Experimentation (COMPOEX) [36] Program is a large political, military, economic, social, information, and infrastructure (PMESII) simulation. Each PMESII area is represented by a set of models, where each set is divided by the level of granularity. The models communicate via a publish and subscribe architecture, posting their output/result to a backplane at the end of each time step, while reading in their inputs at the beginning of the next time step. COMPOEX allows for adaptation to a variety of different countries.

From our understanding of this class of tools/approaches, there are still many open challenges yet to be addressed. These include but are not limited to: model currency, adaptation of the model to other nation-states, sensitivity analysis, visualization of the results and the interaction between physical and social/behavioral models, verification and validation of the model and how stability is measured and defined. Regardless of either modeling class, it is imperative that users (researchers, analysts, or decision makers) understand how their model works, which indicators will provide the most value for what they seek, and which conditions (input variables) one must modify to achieve desirable values for those indicators (outputs). An indicator is defined as "A statistic or parameter measure that, tracked over time, provides information on trends in the condition of a phenomenon and has significance extending beyond that associated with the properties of the statistic itself [13]."

What indicators are of interest to us? What indicators have been studied to provide us with insight as to whether a nation-state is failing? Before we settle on a set of indicators, let us continue to review some of the previous work in this area. We can divide this work into two categories. The first set attempts to answer the question as to *whether the nation-state is moving toward instability or conflict*. In these studies, the researchers were attempting to develop a minimum set of indicators to answer a specific question from a macroscopic view. As an example, Fund for Peace [43], a not-for-profit organization, is attempting to measure the overall stability of every country. As stated on their website, (http://www.fundforpeace.org/), "The Mission of The Fund for Peace is to prevent war & alleviate the conditions that cause war." In support of this endeavor, the site publishes the "Failed State Index Score." This score is based on a number of indicators—some are measureable, while others seem to be subjective. Taken from their website, each country is assessed based on twelve indicators that are divided amongst three categories: Social, Economic and Political.

A second example of defining indicators comes from Nysether [39]. As part of his research, Nysether provided an excellent summary of previous work concerning indicators, as well as a list of indicators used in an attempt to determine which inputs are most significant in determining stability. Nysether used both Barnett's definition of core/gap countries and the Fund for Peace stability index.

An alternative approach in defining a set of indicators looks at key decision-makers' interests regarding the determination of success or failure of stability operations. This can best be shown by highlighting the stability operations currently being pursued in Iraq. By law (Section 9010 of the Department of Defense Appropriations Act 2007, Public Law 109-289 as amended by Section 1308 of Public Law 110-28), the DoD must provide to Congress a report measuring the Operations progress being made. This report includes specific performance indicators and measures of progress toward political, economic, and security stability. This report entitled "Measuring Stability and Security in Iraq [44]" is more often referred to as the "9010 Report."

Yet another way to look at stability indicators is to break them out temporally. One could argue that nation-state stability metrics could be decomposed into three time periods: (1) Near Term—example is PITF which looks at violence; (2) Mid Term—example is Fund for Peace which looks at factors such as GDP; and (3) Long Term Prospects—examples are indicators such as quality of life, happiness, or sustainment.

Which indicators should be supported? The answer to this question is both problem centric and a research question. The community of practitioners (e.g., State Department, USAID, etc.) continues to discuss what these measures should be and no single set has been agreed to. The approach we have taken with the NOEM is to allow the user to choose what they would like to monitor until there becomes an agreed upon set of stability measures. This approach is not without some concerns. If there is no agreed upon set of measures and one allows a user to choose the measures, then the assessment itself might be biased or slanted to what the user believes is important. This is true, but at the present time, we have no better answer and we wait for those more knowledgeable in the area to provide the solution.

1.3 The National Operational Environment Model

There have been a number of contributors to the field of Operations Research as it can be applied to Stability Operations, most notably from Richardson [14], Robbins [15], and [16]. Richardson demonstrated the viability of applying system dynamics modeling techniques to a Stability and Reconstruction (S&R) effort, allowing the statistical analysis of various potential macro-level policy choices. Robbins, under CENTCOM sponsorship, designed and developed a deeper, broader model: the Stabilization and Reconstruction Operations Model (SROM), which allows for a more detailed analysis of the S&R effort than the earlier work done by Richardson.

The SROM is a sub-national, region-based construct, as opposed to the nation-based model put forth by Richardson. This allows the end-users of the model to identify potential problem regions within a nation-state, to test a wide variety of policy options on a national or regional basis, to determine suitable courses of action (COAs) given a specified set of initial conditions, and to determine resource allocation methods that best improve overall national or regional stability. The different policy options are simulated at the sub-national, regional level, providing higher fidelity solutions to the end-user. Analysis of S&R operations via the SROM provides decision makers insight into policy-driven scenarios, affording them with the opportunity of having different policy options modeled and simulated with the outcomes analyzed.

The SROM can be used to analyze the importance of the controllable factors affecting ongoing S&R operations. By changing different factor levels such as troop numbers, the region of troop deployments, troop deployment schedules, indigenous security force training schedules, levels of indigenous popular support, initial number of insurgents, civil-military operations in terms of aid-money disbursed, and other such values included within the model, the factors' relative significance and impact to the success or failure of the particular instance of nation-building may be investigated. By providing decision-makers and analysts a tool that can "perturb current or future conditions," they are able to assist in determining appropriate COAs in stabilization and reconstruction at a regional level within a nation.

The SROM was designed to use two metrics (aka indicators) to determine stability: Coalition Losses and a 90-day moving average for violent deaths. Losing a given percentage of coalition forces will lead to a projected failure in operations, while achieving a violent death rate below a given threshold for each region within the nation-state will indicate stability. Are these the right metrics (indicators) and do they even support the decision-makers requirements? SROM focused mostly on Insurgents and Blue factors. In order to support the tenets from Major General Flynn's recent paper (Flynn et al. 2010), we need to also take into account the operational environment and the residing populace.

While Robbins provided us with a model as described above, his approach was more from a bottom-up, data-driven perspective. Capt. Fensterer, in his thesis, looked at the problem from the reverse: top-down, question-centric. Capt. Fensterer's research [16] provides a decision analysis technique — Value Focused Thinking (VFT) to capture all the important values of nation state stability. Using the DoD Directive 3000.05 as high-level guidance, VFT decomposition of the information within that document produced five fundamental values of stability. They are Economy, Governance, Rule of Law, Security, and Social Well-Being. These constitute the first tier of values in the stability value hierarchy (Fig. 2); referred to within this document as the "Pillars of stability."

The sub-objectives of each of the fundamental values were determined by a further detailed VFT analysis of three accepted nation state stability documents:

- "Beyond Declaring Victory and Coming Home: The Challenges of Peace and Stability Operations" [17]
- "Winning the Peace: An American Strategy for Post-Conflict Reconstruction by Center for Strategic and International Studies" [18]
- "The Quest for Viable Peace: International Intervention and Strategies for Conflict Transformation" [19]

This further analysis decomposed each of the fundamental values of stability and developed a robust set of sub-objectives for each pillar. Later research has suggested that there be four pillars, combining the Security and Rule of Law pillars into a single pillar and calling it Security. Subsequently, the four pillars in which the NOEM will adhere are: Security, Economy, Governance and Social Well-being.

Based on Robbins [15] and the lessons learned from his prototype, SROM, AFRL/RI re-engineered the basic model and have called it the National Opera-

Security: Protecting lives of citizens from immediate and large-scale violence and restoring the state's ability to maintain territorial integrity

Economy: System comprised of policy, macroeconomic fundamentals, free market, and international trade that exchanges wealth, goods, and resources in an environment mostly free of economic criminal activity

Governance: Public management process that involves a constituting process, governmental capabilities, participation (or lack thereof) of citizens, and administrative structure

Rule of Law: Comprehensive, six-element justice and reconciliation effort that involves law enforcement, judicial system, constitution and body of law, corrections system, and past abuse reconciliation mechanisms

Social Well-Being: Sustenance of life and relief of suffering by way of humanitarian aid, best practices, human rights, essential services, and emergency response systems

Fig. 2 Five pillars (fundamental values) as proposed by Fensterer

tional Environment Model (NOEM). Like SROM, NOEM starts with a number of modules based on system dynamics and adds other computational theories, such as agent-based models and game theory. System dynamics is ideally suited for modeling critical infrastructures and similar macroscopic concepts; however, systems dynamics is poorly suited to describe group interactions or behaviors of individuals. In this case, we rely on Agent-Based Models (ABMs). The turn toward ABMs is due to the recognition of the importance of "bottom-up" causal processes in the development of civil unrest [20–23]. ABMs permit researchers to move beyond the limitations of the aggregate approach underlying system dynamics models, encouraging researchers to identify and experiment with the micro-processes involved in the production of emergent macro-scale social patterns of distributions [11]. Neither SD nor ABMs models provide the complete solution, however. To investigate the various strategies that each group can take towards the general populace, we need to look at Gaming Theory and Methodologies. In combination, the three techniques provide the NOEM with the basic building blocks to achieve its goals.

A hierarchical approach has been implemented for the structure of the NOEM. The national environment is modeled as a conglomeration of interconnected regional entities. A sub-national, regional module typically will represent a geopolitical area of interest (governorate, province, etc.) within the environment. A region can be as simple as the entire nation itself, a set of one or more politically divided components (for example, one can divide the nation state into three parts: northern, central or southern regions) or along political boundaries (e.g., in Iraq we can equate each region with a province). The decision as to how to divide the nation-state into regions can be based on how granular you wish the model to be or simply based on the availability or lack of the data to support the individual regions. Each region is modeled as the composite of a number of modules representing the numerous interdisciplinary and interdependent systems existing within a geopolitical region. Each module interacts with other modules within the region, but also with appropriate modules residing in other region modules. Modules exist at a regional level, however many of the modules are so tightly interconnected to modules in other regional modules that a national level abstraction of the module is maintained (e.g., economic, power, etc.).

2 NOEM Overview

The NOEM infrastructure consists of three main components the: (1) Model Development Environment (MDE), (2) Baseline Forecaster (BF), and (3) the Experiment Manager (EM). The MDE facilitates the creation and management of the model and their components. The MDE's Model Builder is used to create a new nation-state model or edit an existing one. Users may create new regions within a nation using the Model Builder and populate those regions with various, existing system modules. The BF continually and automatically provides an updated set of projected futures for any nation-state model. A Baseline Forecast tells us where a nation of interest is headed over time if the nation remains on its current course. The EM enables intelligence analysts and decision makers to perform "What If" analyses in order to compare policy sets and their effect on a nation or region's stability over time, as well as reveal any potential short- to long-term ramifications of pursuing any one policy set. The EM allows one to examine/assess a multitude of policy options while the BF supports the assessment of whether the actions proposed are moving the situation in the desired direction (Fig. 3).

The Decision Support Tools allow one to define an objective, investigate policy set alternatives and obtain English-like explanations of the resulting policy sets. It sits alongside the NOEM components, operates concurrently and interacts with them during the course of operations. We will be discussing this capability in much more detail in the following sections; however before we do, let us provide an overview of the NOEM model.

2.1 The Model

The NOEM model supports the simulation and the analysis of a nation-state's operational environment. Within the MDE, the nation models can be configured for one or

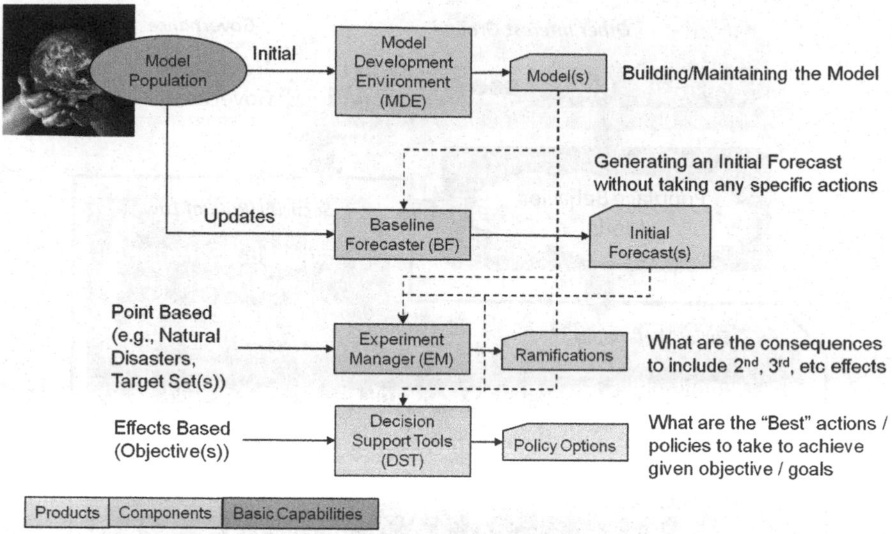

Fig. 3 Overview of the major components that form the NOEM

more regions, where each region is composed of a group of highly interconnected modules which simulate subsystems such as a region's demographics, economy, or critical infrastructure. These modules, in essence, relate to the major pillars of a nation-state based on stability operations theory (Governance, Security/Rule of Law, Economy and Social Well-Being). Figure 4 provides a representative view of the overall model. It is envisioned that additional modules can be added as needed. Such additional modules can include: education, health infrastructure, mining, etc.

Security/Rule of Law is defined as protecting the lives of the populace from immediate and large-scale violence, and ensuring the state's ability to maintain territorial integrity. The Security/Rule of Law pillar is currently comprised of three modules: Indigenous Security Institutions (ISI), Crime, and Police. The ISI module is divided into: Border Patrol, Civil Defense, Facility Protection Services, and the Indigenous Military. Economy is defined as a system made up of various economic policies, macroeconomic fundamentals, free market, and international trade that exchanges wealth, goods, and resources mostly free of economic criminal activity. The Economy pillar is composed of two modules: Economics and Finance & Debt. Governance is defined as a public management process that involves a constituting process, governmental capabilities, and participation of citizens. Social Well-Being is defined as sustenance of life and relieving of suffering by way of humanitarian aid, best practices, human rights, essential services, and emergency response systems. The Social Well-Being pillar is composed of the majority of the modules and includes: Demographics, Health, Migration, Food, and fundamental Utilities (Electric Power, Telecommunications, Natural Gas, Oil, Transportation, and Water & Sanitation).

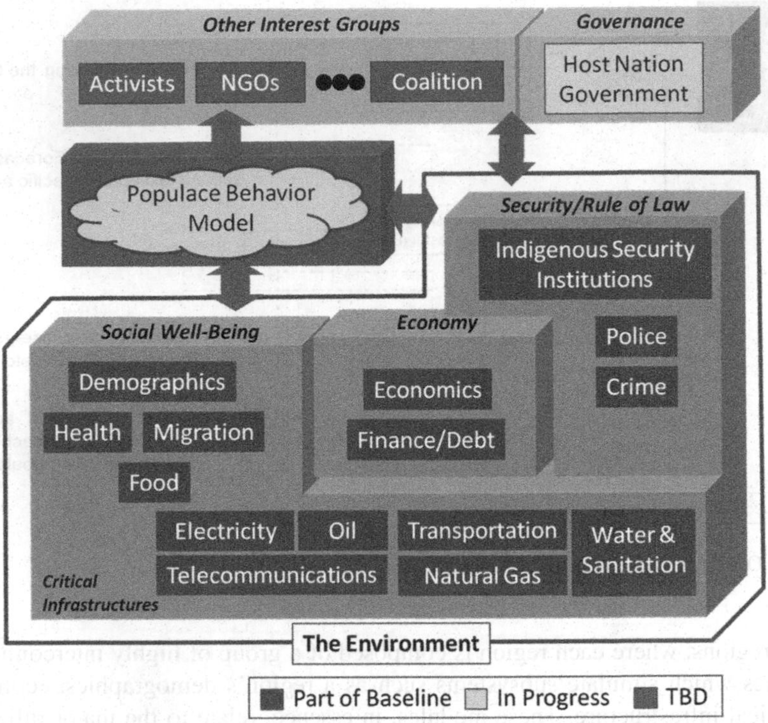

Fig. 4 NOEM model into its constituent modules

The agent-based Populace Behavior Module forms the heart of the NOEM model in the sense that all other modules (resources and security) are in place to support the populace. If the populace is not happy or satisfied to a certain degree of expectation, they could become activists and rebel against the host-nation government. Whether or not segments of the populace become activists depends on many factors, including their perceived hardship, the legitimacy of (or belief in) their government, their level of risk aversion, and the amount/visibility of security forces. Insurgents, Coalition forces, NGOs, and Host Nation Governments within the NOEM are not modeled as agents, but are characterized by the policies or strategies that they implement. Policies implemented by such groups will affect either the overall security within the environment or the services/resources provided to the people. As such, the NOEM team is researching the use of various Gaming Engine Techniques/Technologies that will allow one to play off these various strategies.

3 Using the NOEM Tools

In this section, we discuss the use of the NOEM using two different use cases. Our two use cases will be centered around a fictitious country called "ImagiNation". The island nation of ImagiNation has a history of poor rulers. Recently, ImagiNation's leadership changed and a new ruler has taken power. While this is good news for ImagiNation, its aging infrastructure and tumultuous history have made its populace somewhat unsettled, and anti-government demonstrations are likely. Oil and gas infrastructure exist, but are underutilized, and the economy is stagnant. In the first use case (Point or Event Based Analysis), we are looking at the results of an action or event taking place. In this scenario, we use an earthquake. In the second use case (Prospective Analysis), we are capitalizing on the ability of the NOEM to forecast where the country is headed and make use of this information in an attempt to develop policies that will stop or minimize the potential unstable condition from happening.

3.1 *Point or Event Based Analysis*

In our first use case, an event either has occurred or may occur—an earthquake, for instance. As an example, consider a scenario with a city, Springfield, in the country we call ImagiNation. Springfield is the largest city in ImagiNation and has an estimated population between 7 and 7.5 million. An earthquake in the region with an epicenter at the center of the city will have an impact on a large portion of the population. The question is, then, what this impact will be, and how will it change if the magnitude of the earthquake is 4.5, 7.5, or 8.8. We begin by defining the impact rings for both the environment (destruction of the critical infrastructure) and on the people. In Fig. 5, we define the location (by entering a latitude and longitude in the NOEM's geographical input interfaces) and the day that the event will occur. The next step then is to define a set of concentric circles that define the varying degrees of damage (Infrastructure) and populace affected (People).

After a set of concentric circles are defined, the NOEM will identify any/all entities whose coordinates fall within the rings and degrade them according to the damage range defined by each ring along with the number of people that are affected. Figure 6 provides a sample result.

The next step is to define the outputs or a set of indicators that we wish to monitor. To accomplish this one selects the output tab and the "Associate" tool (Fig. 7). At this point the user selects an input from the input set. The tool examines the model and provides the user with a ranked list of potentially affected outputs, i.e., a suggestion as to what they should be monitoring. Clicking on the item(s) will add it to the list of outputs. One can then select another input and corresponding set of outputs. All outputs can then be saved in a single output file.

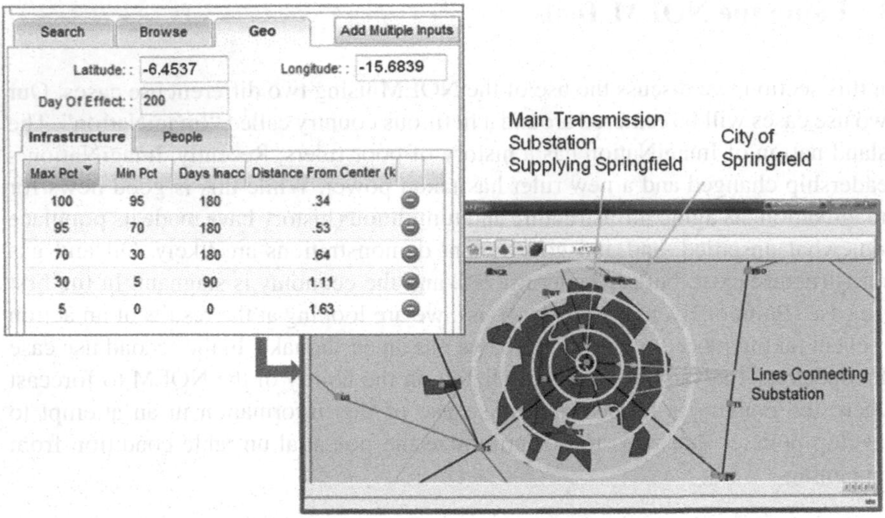

Fig. 5 NOEM geo tool

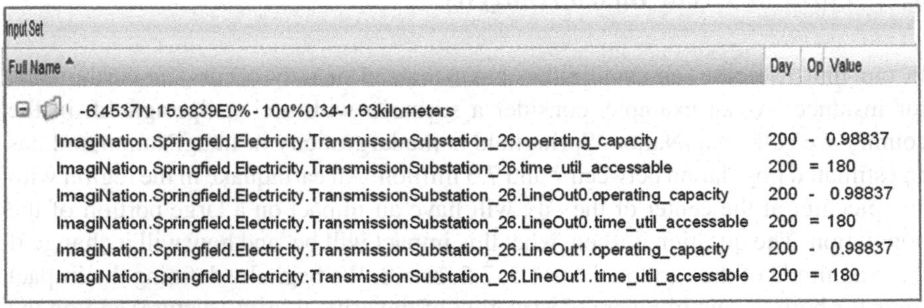

Fig. 6 NOEM generated inputs

At this point, a user would define a set of outputs to monitor, conduct the simulations, and analyze the results through the creation of charts (example charts shown in Fig. 8). The "Blue" arrows depict the interdependencies between the outputs. For example the electricity production (upper-left chart) is needed for both the pumping of potable water (lower-left chart) and for running the industries. If power is down for a sufficient period of time, then there will be an impact on the ability to run the factories, people will be laid off, and unemployment will increase (chart top center). High unemployment and lack of manufacturing will then proceed to impact both the crime rate (lower center chart) and overall GDP (top right most chart). Many of these factors will then impact the general populace (e.g., lack of potable water, lack of income, etc.).

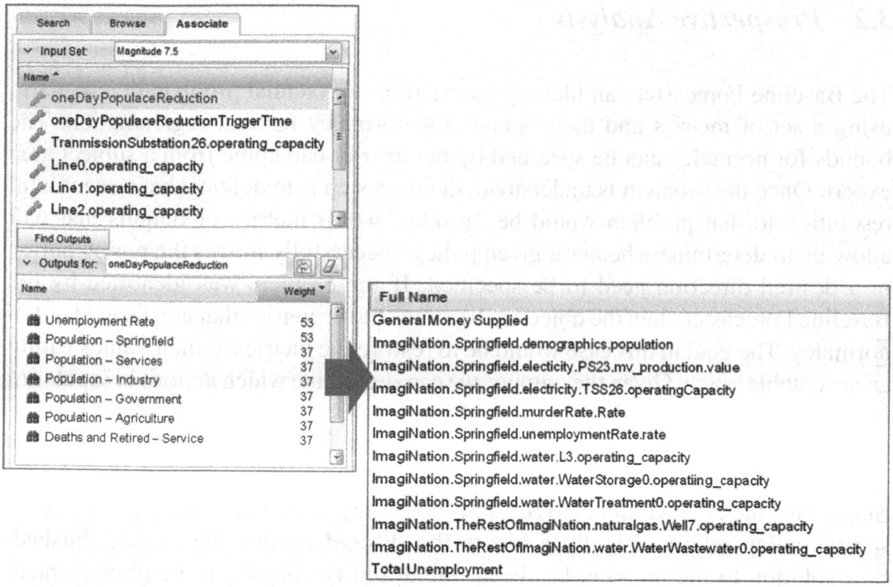

Fig. 7 Generating the output set

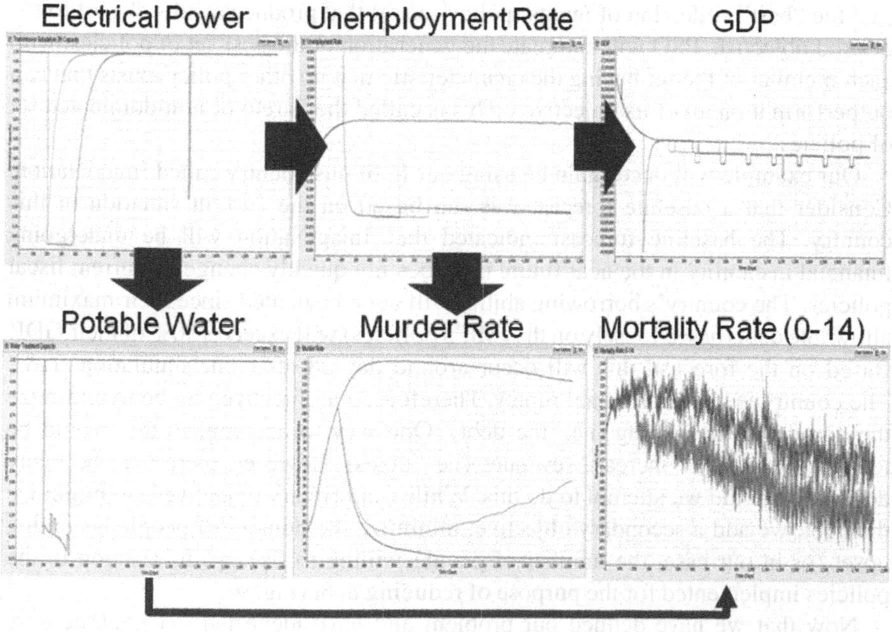

Fig. 8 Output results

3.2 Prospective Analysis

The Baseline Forecaster can identify an existing or potential problem in a scenario using a set of metrics and their bounds for normalcy for that region/nation. The bounds for normalcy can be specified by the user or can come from a subject area expert. Once the problem is understood, the next step is to define what a successful resolution to that problem would be. In other words metrics or outputs that will allow us to determine whether a given policy successfully moves the region/nation in a desired direction need to be specified. If the problem was identified by the Baseline Forecaster, then the objective becomes those metrics that are deemed out of normalcy. The goal in this case would be to restore the metrics to their values within an acceptable range. Given the outputs, we can determine which actions or inputs are necessary to modify in order to achieve the desired objective. These inputs therefore become the initial policy sets. Various combinations of changes to these inputs (which can be generated through the NOEM's Sensitivity Analysis (SA) tool) can be run against the NOEM model to see how close they come to meeting the objective. If any policy set successfully achieves the desired results, the user is finished, as a solution to the problem has been found and needs only to be implemented. Otherwise, we can revise the input values and develop additional/revised policy sets. This can be done through the NOEM's 'Policy Set Optimization' tools (or PSO, currently under development). This process of generate-test-refine can be iterated to find the "best" collection of inputs (policy set(s)) that produce results closest to the desired objective. PSO will automate the generation of the best set of policies, with each member of the set having the characteristic that no other policy exists that can outperform it on all of its objectives. This is called the Pareto or non-dominated set of policies.

Our example will once again be using our fictitious country called ImagiNation. Consider that a baseline forecast was run based on the current situation in this country. The baseline forecast indicated that ImagiNation will be undergoing financial instability in the near future if it does not quickly change its current fiscal policies. The country's borrowing ability will soon be denied since their maximum allowable debt (based loosely on the IMF's policies) will exceed 150% of their GDP. Based on the forecast, this will occur around day 28 after our simulation starts. The country is headed for bankruptcy. Therefore, our objective will be to minimize the amount of borrowing (i.e., the debt). One way to accomplish this would be to stop spending or increase revenues (i.e., taxes). However, there may be many downfalls should we attempt to do this. While our primary objective is to minimize the debt, we add a secondary objective: minimize the number of people becoming upset (or in our case, the number of people willing to rise up) in reaction to the policies implemented for the purpose of reducing debt (Fig. 9).

Now that we have defined our problem and have identified our goal/measure of effectiveness, we next generate an initial set of policies that we believe would likely provide us the "best" solution. Typically, these potential solutions or policy sets are debated within a group, revised, and then an agreed upon direction forward

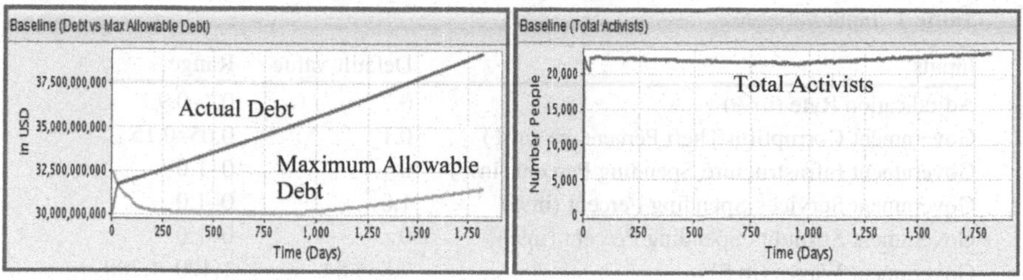

Fig. 9 Baseline forecasts

is adopted. However, how do we know whether any given policy set chosen was the best one (having taken the right actions), the most robust, or the most resilient? How does one know that it is the lowest cost, and how long could it be until one begins to see appreciable results? How long until the full effects can be realized? How does one even know what actions one should take to achieve the most effective possible solution? These are the types of questions policy makers have to answer each day—questions that the NOEM and its suite of Decision Support Tools strive to support.

In our example, our primary objective is to minimize the amount of long term debt accrued by our country, ImagiNation. Another objective is to avoid inciting a riot or revolt; i.e., any actions we take should not increase the discontent of the populace (measured in terms of the number of populace willing to become activists). Thus, our initial goals had been to minimize debt and the number of activists. We found that what we needed to do was to maximize the difference between the maximum allowable debt and the actual debt. This is due to the fact that the maximum allowable debt is a function of the GDP and only provides an upper ceiling as to the amount of money a country can borrow. Solely driving the debt lower does not provide us an insight into whether we are below the maximum allowable debt, since the GDP could decrease and in such a case so would the maximum allowable debt. We weight both the number of activists and the distance between maximum debt and actual debt equally. The next question is to determine to what extent it is possible to achieve this objective. Of course, we can set the goal for both the debt and the number of activists to be zero, but this is unrealistic. With that in mind, what are the realistic values? One way is to simply state a number for each based on expert opinion. A second option might be to derive them based on a set of runs to determine what is achievable. We opted to employ the second approach, and do not specify the actual values for our objective function, which can be written as:

GOAL = MAXIMIZE (MAXIMUM ALLOWABLE DEBT - ACTUAL DEBT)
AND MINIMIZE (TOTAL ACTIVISTS)

After identifying the outputs of interest, we next need to define what inputs or actions we believe are necessary to achieve our desired objective. We again use the associate tool to identify the corresponding inputs based on a given output. The number of inputs displayed will be based on the degree of significance that the

Table 1 Input variables

Inputs	Default value	Range
Adjudication Rate (in%)	0.7	0.1–0.9
Government Corruption Theft Percentage (in%)	0.1	0.05–0.15
Government Infrastructure Spending Percent (in%)	0	0–1.0
Government Services Spending Percent (in%)	1.0	0–1.0
Government Stimulus Spending Percent (in%)	0	0–1.0
Government Wages (in $'s)	2,835.84	1,400–4,200
Initial Police Forces (in Number People)	22,000	112,500–137,500
Interest Rate (in%)	0.045	0.0225–0.0675
Jail Term (in Months)	30	3–100
Long Term Government Share Of Employed (in%)	0.001	0.001–0.05
Mean Adjudication Processing Time (in%)	0.5	0.1–0.9
Military Acting As Police Percent (in%)	0.1	0.0–0.2
Police Forces Goal (in Number People)	22,000	112,500–187,500
Stimulus (in $'s)	10,000,000	0–6,000,000
Tax_Rates (in%)	0.0176	0.03–0.50

input contributes to the selected output (user selectable). Highlighting the inputs will automatically select them for the next step in the process—creating an initial set of policies. For our example, fifteen inputs have been identified. Table 1 provides this list along with their default values and ranges.

This next step is to create an initial set of policies. This is accomplished by using the inputs/outputs defined in the previous steps, their specified ranges, and space filling techniques such as Latin Hypercubes Sampling (LHS) or Leap Halton. The goal is to create an initial set of policies that uniformly cover the design space. This design space is based on the number of inputs, their acceptable ranges, and the total number of allowable runs, which is user selectable. Each simulation run consists of one input or policy set along with the list of desired outputs. These are sent to the NOEM, which executes the given simulation and returns the results. The results of each run can now be compared to the objective function to examine how close to "optimal" the given policy set has come. If one or more of the results or policy sets are within a given tolerance, we are done. Otherwise, we can use the closest policy set so far and evolve or mutate it to generate a series of new policies that can be run and evaluated. This iterative process can be continued until the desired results are found or after a specified maximum number of iterations.

For our example, a total of 1,000 runs were conducted. In this case, since we are minimizing only two values, we can simply plot them as a two dimensional scatter plot—debt versus total activists. Figure 10 provides this plot for 1-year and 5-year projections. We can see that, as time progresses, the dispersion of the results increases.

Based on these runs, we can now find our goal values (the best that can be done independently) for both the delta between the maximum allowable debt and the actual debt, and for total activists. This is accomplished by searching through each run for a given day and finding the minimum (or maximum) value for that output. In our example we have (Table 2).

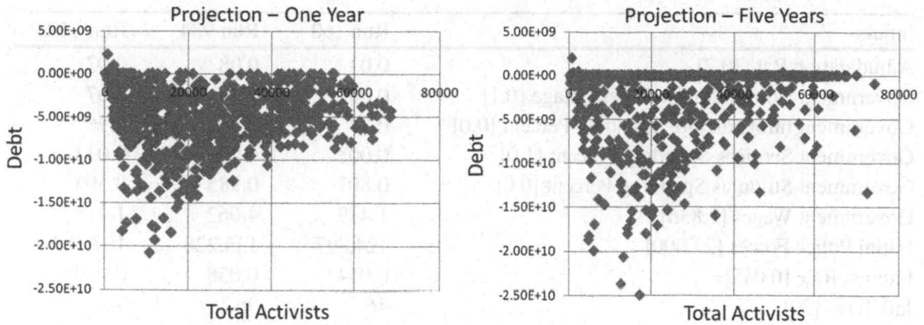

Fig. 10 Results of 1,000 runs at the end of 1 and 5 years

Table 2 Goal values

Output	1 Year	5 Years
Maximum allowable debt – actual debt	2.3E + 09	2.00E + 09
Minimum number of total activist	622	716

Table 3 "Best" policies

	1 Year		5 Years	
Rank	Run No.	Distance	Run No.	Distance
0	160	0.006	160	0.007
1	984	0.063	984	0.062
2	142	0.087	142	0.064
3	145	0.095	49	0.067
4	411	0.096	630	0.070
5	214	0.096	867	0.070
6	810	0.096	500	0.070
7	581	0.096	389	0.070
8	419	0.096	766	0.070
9	515	0.096	681	0.070

In a two variable world, we can simply find the policy set that is closet to our objective. In a multi-dimensional world we can still use the concept of closeness, but generalized for an n-dimensional Euclidean space. We use the goal as the target point and compute a distance between it and each policy set result. Table 3 provides a listing of the closest ten policy sets after 1- and 5-year projections.

Comparing the two projections, we can see that the top three runs are the same, Run 160, 984, and 142. We next examine these top policies for each of our fifteen variables. Table 4 provides these values.

We can now display the results of our policy sets through the NOEM and see how well we did. Figure 11 provides the results for runs 160, 984, and 142. The two smooth lines represent the maximum allowable debt (upper) and the actual debt (lower). The third curve (purple) provides the number of activists.

Table 4 "Best" policies (input values)

Inputs		Run 160	Run 984	Run 142
Adjudication Rate [0.7]		0.03	0.08	0.07
Government Corruption Theft Percentage [0.1]		0.06	0.08	0.07
Government Infrastructure Spending Percent [0.0]		0.24	0.75	0.96
Government Services Spending Percent [1.0]		0.007	0.009	0.033
Government Stimulus Spending Percent [0.0]		0.807	0.985	0.503
Government Wages [2,836]		1,479	4,062	1,417
Initial Police Forces [22,000]		124,897	136,228	107,183
Interest Rate [0.045]		0.034	0.058	0.0398
Jail Term [30]		46	96	74
Long Term Government Share Of Employed [0.001]		0.0457	0.0372	0.0079
Mean Adjudication Processing Time [0.5]		0.23	0.48	0.57
Military Acting As Police Percent [0.1]		0.17	0.03	0.095
Police Forces Goal [22,000]		113,976	168,146	142,586
Stimulus [10,000,000]		471,071	42,325	565,747
Tax_Rates [0.0176]		0.4549	0.1699	0.42498
Total Activist	**[End of Year 1]**	**622**	**2057**	**872**
[End of Year 5]	**716**	**2299**	**985**	
Debt	**[End of Year 1]**	**3.47E + 10**	**3.15E + 10**	**3.37E + 10**
[End of Year 5]	**5.21E + 10**	**4.18E + 10**	**5.05E + 10**	
Maximum Allowable Debt	**[End of Year 1]**	**3.70E + 10**	**3.25E + 10**	**3.40E + 10**
[End of Year 5]	**5.41E + 10**	**4.36E + 10**	**5.08E + 10**	

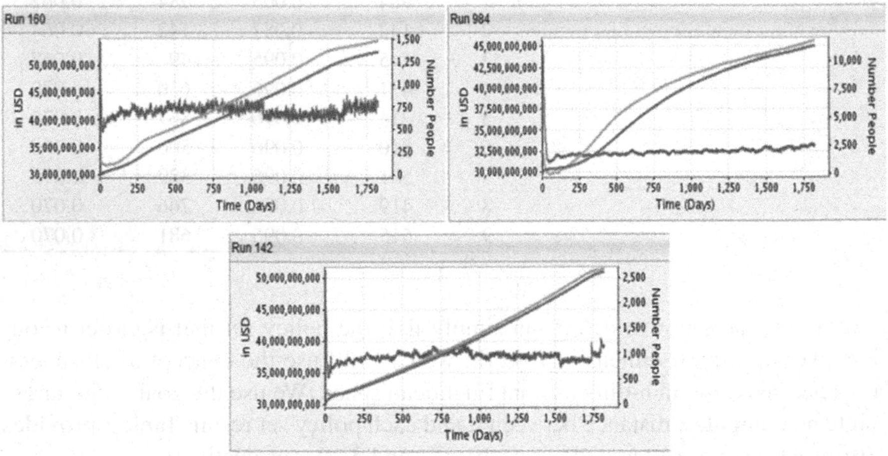

Fig. 11 Results for runs 160, 984 and 142

A second way to view the results is through the use of a Decision Tree. A decision tree is a tree like graph of decisions and their possible consequences. To illustrate, we used the 1-year data and labeled the top ten policies as good (Table 3) and all of the other policies as bad. Figure 12 is an image generated using the rapidminer

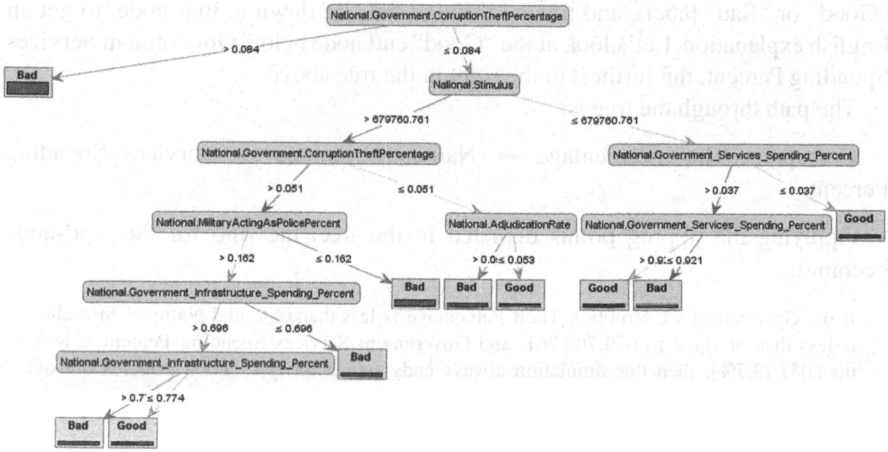

Fig. 12 Decision tree

National.Government.CorruptionTheftPercentage > 0.084: Bad {Good=0, Bad=658}
National.Government.CorruptionTheftPercentage ″ 0.084
| National.Stimulus > 679760.761
| | National.Government.CorruptionTheftPercentage > 0.051
| | | National.MilitaryActingAsPolicePercent > 0.162
| | | | National.Government_Infrastructure_Spending_Percent > 0.696
| | | | | National.Government_Infrastructure_Spending_Percent > 0.774: Bad {Good=0, Bad=12}
| | | | | National.Government_Infrastructure_Spending_Percent ″ 0.774: Good {Good=2, Bad=1}
| | | | National.Government_Infrastructure_Spending_Percent ″ 0.696: Bad {Good=0, Bad=42}
| | | National.MilitaryActingAsPolicePercent ″ 0.162: Bad {Good=0, Bad=243}
| | National.Government.CorruptionTheftPercentage ″ 0.051
| | | National.AdjudicationRate > 0.053: Bad {Good=0, Bad=4}
| | | National.AdjudicationRate ″ 0.053: Good {Good=2, Bad=0}
| National.Stimulus ″ 679760.761
| | National.Government_Services_Spending_Percent > 0.037
| | | National.Government_Services_Spending_Percent > 0.921: Good {Good=2, Bad=0}
| | | National.Government_Services_Spending_Percent ″ 0.921: Bad {Good=1, Bad=30}
| | National.Government_Services_Spending_Percent ″ 0.037: Good {Good=3, Bad=0}

Fig. 13 Text view of the decision tree in Fig. 12

package (see www.rapidminer.com). The tree had been constructed using the information gain criterion (branching based on most information). Based on this decision tree, rules can be derived and English-like explanations can be generated, essentially describing how a good policy for the given scenario is composed.

Figure 13 is a text view of the same graph. The text view provides a set of rules which can be used to form English-like explanations.

If we follow the left branch of the tree we see that if National Government's Theft due to corruption is over a threshold of 0.084, then we will have loss of money and likely a higher debt. The people will see problems with their government, so there will be a high number of activists and hence we always end up with either bad debt or high activists and hence a bad policy (0 good policies out of 658 cases). For deeper ending nodes, it is only a matter of finding an end node (one with a

"Good" or "Bad" label), and tracing through the tree down to that node, to get an English explanation. Let's look at the "Good" end node below Government Services Spending Percent, the furthest to the right in the tree above.

The path through the tree is:

Corruption Theft Percentage → National Stimulus → Services Spending Percent.

Applying the tipping points depicted in the tree, the rule for this end-node becomes:

> If the Government's Corruption Theft Percentage is less than.084, and National Stimulus is less than or equal to 679,760.761, and Government Services Spending Percent is less than.037 (3.7%), then the simulation always ends satisfactorily (3 good policies out of 3 cases).

3.3 Model Validation

How do we know the conclusions or results arrived at through NOEM are correct, or that we should have any confidence in them? This concern brings us to the area of Verification and Validation (V&V). The NOEM is a simulation of real world processes designed to enhance our understanding of the mechanisms, applications and management of those processes in selected world regions. In general, models enable understanding by trading complexity for clarity. Yet, clarity of the processes as captured by the model cannot be assumed, no matter how intuitive [24]. At the same time, we recognize that the content and structure of a model must be assessed in light of its intended use or role in policy development. This requires us to examine the interaction between model builder and model user.

Verification and validation (V&V) of models are well known and well used terms within many scientific fields. Yet, these concepts are difficult to apply to the modeling of Human Social Cultural Behavior (HSCB). Following arguments of Lakatos [25], Sargent [26], and Zacharias [27], the process of *verification* is concerned with answering the question, "was the model built correctly?" That is, verification assesses if the model implementation accurately reflects the developer's conceptual model. By contrast, the process of *validation* is concerned with answering the question, "was the correct model built?" This assessment reflects how accurately the model captures the real world. Since the notion of "correct" might change over time, the validation processes themselves will need to be revisited occasionally to reassess a model's validity. Thus, validation is an ongoing process. For social models, some argue that validation will always be incomplete [28]. Petty [29] states that there are over ninety different V&V methods that he has summarized into four categories: Informal, Static, Dynamic and Formal Methods. All 90 are listed and described in Balci [30]. Informal Methods consist of Inspection, Face Validation and the Turing Test. Static Methods involve Data Analysis and Cause-Effect Graphing, while Dynamic Methods include sensitivity analysis comparison testing, regression analysis and hypothesis testing. The fourth category, Formal Methods, include: Inductive Assertions and Predicate Calculus.

HSCB models present unique challenges for V&V in general, and for operational validation in particular. Unlike models of physical systems, HSCB models cannot be robustly tested through experiments in the real world nor fully verified or validated using historical data [31]. In fact, Sallach [31] indicates that, "It is possible that, for any specific problem type (war, insurgency, state collapse, etc.), all of the relevant events in recorded history may constitute only a small sliver of the possible events that could occur. From the perspective of probability theory, the available *sample* of events is not and cannot be representative of the *population* of possible events." Thus, for historical testing, there may not be enough data for meaningful training and test data sets. So the bottom line is how do we provide the highest level of confidence to the user that we can?

The NOEM takes the concept and ability to perform V&V to the extreme. The model simulates physical systems (which in many cases are well understood and classical V&V can be performed), social systems and behavior systems. As we move up in the model, abstraction becomes the norm. Not only are we contending with the models themselves but we also need to deal with the computational methods employed. The majority of the physical and social systems within the NOEM are implemented using system dynamics while the behavioral components rely on agent-based techniques. We are currently working on two separate, but complimentary approaches to validate the NOEM Model: Face Validation and Inverse V&V.

3.3.1 Verification and Face Validation

The verification of NOEM is being approached in a bottom up fashion. This means, that because NOEM is actually a composite of different modules that represent different interdependent systems, the base modules are verified first on an individual basis. Then, after the individual modules are verified, the two-way interdependency between all modules is verified, followed by the three-way interdependency between all modules, and so on and so forth until all of the interdependencies between modules have been verified. The actual verification of a NOEM module is facilitated by creating a series of policy sets that can be executed in NOEM, called test threads. Test threads are designed to exercise the functionality of an isolated module and the results of these test threads can be analyzed to verify that the NOEM modules are operating as described in the design document. An example of this will be discussed for the electricity module.

The electricity module provides components and procedures to model a system capable of satisfying the electrical demand of different NOEM modules. The basic components of this module are power stations to produce electricity, transmission stations to serve as junctions for incoming and outgoing transmission lines, distribution stations to serve as end point components that demand electricity, and transmission lines to move electricity from one component to another. Of course,

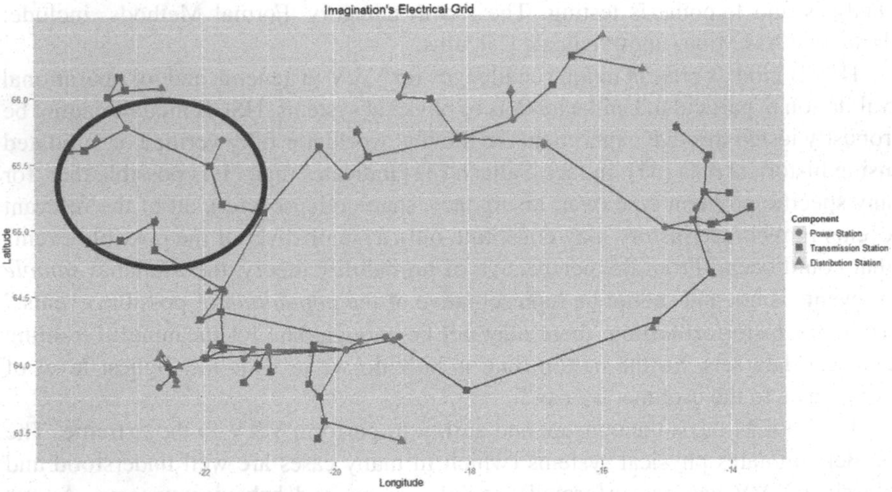

Fig. 14 ImagiNation power grid

the electricity module interacts with many other modules and creates interactions that need to be verified in addition to the electricity module's base functionality. For example, power plants that use natural gas as a fuel require natural gas from the natural gas module, and water treatment facilities from the water and sanitation module require electricity to produce potable water. Following the previously described bottom up approach to verification, the electricity module must first be verified in a standalone environment, and then in conjunction with its other dependent NOEM modules.

To verify the standalone electricity module, we again visit the nation of ImagiNation. ImagiNation provides a robust example for verification, because the nation's critical infrastructure exhibits all of NOEM's critical infrastructure components. This allows all of the interdependencies between the electricity module and the other NOEM modules to be verified using the same model. To verify the functionality of the electricity module specifically, the interdependencies between the electrical grid and the other modules of ImagiNation were removed. Test threads for the module's base functionalities were then developed for an isolated region of ImagiNation's electrical grid, after which, subject matter experts could develop test threads for the whole of ImagiNation's electrical grid to verify larger scope functionalities. The isolated region of ImagiNation's electrical grid is circled in Fig. 14. This region has one power station, four transmission stations, three distribution stations, and eight connecting transmission lines. A graphical representation of this region can be seen in Fig. 15.

Test threads for this isolated region of ImagiNation are designed to test the base functionality of the electricity module. In particular, test threads verify that electricity is generated at the power station, that electricity flows from the power station to the transmission stations, and that electricity then flows from

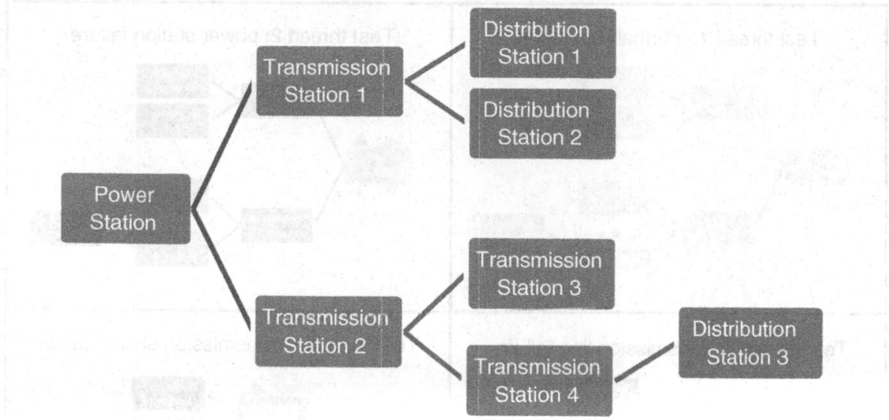

Fig. 15 Graphical representation of power system of interest

the transmission stations to the distribution stations, as the model designers have intended. Test threads also verify that a disruption in the electricity network results in disrupted electricity flow.

Example test threads for the base functionality of the electricity module can be viewed in Fig. 16. Test thread 1 is used to verify that all components function as expected under normal operation. Test thread 2 is used to verify that if a power station fails then its associated electrical components lose electricity. Test thread 3 is used to verify that the loss of a transmission line inhibits the ability of a power source to transfer electricity along that line. Finally, test thread 4 is used to verify that the loss of a transmission station causes a loss in electricity to the outgoing transmission lines and the components to which those are associated. These test threads illustrate scenarios where various components fail and indicate which components should be affected by these failures. Policy sets for these test threads are constructed and executed in NOEM, and then outputs are defined that record the supply of electricity at each component. These outputs are used to verify that the expectations are fulfilled.

After verifying the base functionality of the NOEM electricity module, the map of ImagiNation's electrical grid could be provided to subject matter experts for face validation. These experts could suggest their own test threads on the larger standalone model, the results of which would provide face validation that the electricity module is operating realistically. That is, the experts would use their knowledge and experience to determine whether the model appears to be accurately capturing the desired aspects of the real world. After having verified the standalone electricity module, the basic functionalities of the module do not need to be reconfirmed when the module is joined to its interdependent modules. However the interdependencies created by this joining do need to be confirmed, which can again be done with test threads to verify that the module was built correctly, and use

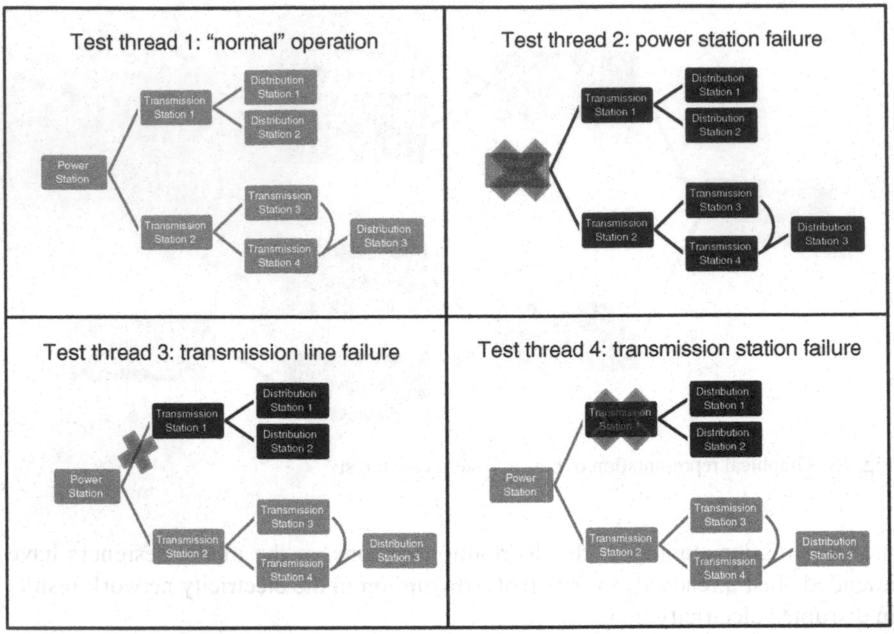

Fig. 16 Sample test threads

of additional expert-developed threads for face validity to determine that the correct model is being built.

Each of the Critical Infrastructures (CIs) will be approached in a similar manner. Once completing the individual CIs, the dependencies amongst them will then be addressed. This will be done by developing a cross-matrix identifying each of the dependencies between the CIs. Some connections might be uni-directional, while others bi-directional but for different reasons. For example the water infrastructure, specifically the pumping station (a node within the water infrastructure module) requires electricity to operate. The electricity CI could in some cases require water for cooling especially if the power station is run on nuclear energy. Once the CIs have been tested, the social components will be added one at time. Interdependencies similar to the CIs will be identified and tested. In this manner we will work our way up the model. We do realize that the CI models are somewhat easy compared to that of the social and behavioral models, but we need to start somewhere—which leads us to more advanced methods of V&V.

3.3.2 Inverse V&V

Lockheed Martin Advanced Technology Laboratories (LM ATL) and Oak Ridge National Laboratory (ORNL) have been working together to develop a technique called Inverse Verification and Validation (IV&V). This re-usable, semi-automated technique can be used for operational verification and validation and can be used both where models can support historical testing, as well as for more exploratory models such as the NOEM. This technique enables one to: (1) build a picture of model verification and validation as it relates to a particular region; and (2) to better understand which relationships represented in the model are also applicable to the region.

IV&V provides the capability to perform multiple verification and validation tests using one technique including Extreme Condition Tests, Event Validity, and Degenerate Tests. It also implements a robust form of Face Validity that enables the model itself to suggest multiple solutions to problems proposed by a subject matter expert (SME) by using a genetic algorithm to perform a search of the solution space. The SME is asked to consider those solutions and determine whether the behavior is reasonable for the particular region of interest.

The IV&V technique is quite straightforward. Consider a model that was built specifically for a region and for which there is a significant amount of historical data. A subset of these historical data points are isolated for training, making a "historical training set", and another subset is isolated for testing, the "historical testing set". For the model under test, the historical training data is used to initialize or calibrate the model to the data set. Models will vary in how the calibration process is performed. For validation, the "historical testing set" is used to develop a fitness function that is suitable for use by a genetic algorithm. This fitness function can be described as the 'target outcome' for the model. The genetic algorithm is guided by the fitness function to find one or more input sets to the model that generates outputs that are similar to the target outcome. The system (or a person) then compares these model-generated input sets to the actual historical inputs states from the "historical testing set" to determine how well the model-generated inputs correspond with the historical data inputs.

Now let us consider this technique in light of the challenges presented by the NOEM, an exploratory, multi-disciplinary model. The issue with not having enough data to train and test a model is particularly pertinent to models using agent-based components like the NOEM where there often are many parameters such that exhaustively exploring the parameter space for verification purposes may be intractable. In addition, the NOEM is an exploratory model that was not designed to represent the specific relationships in a specific region, yet, nonetheless, it enables a user to explore relationships that are in the model and are applicable to a region. Validation of agent-based models such as the NOEM Behavior Module are also challenging because agent-based models trade complexity for clarity and as such may not be very accurate point predictors of future events. Nonetheless, agent-based simulations have been shown to be very useful for exposing and understanding relationships between macro phenomenon and micro behaviors. This kind of modeling becomes useful in analysis such as counter-factuals, what-if hypothesizing, and drill-down to lower level behaviors. These analyses represent plausible situations in the real world, but are not necessarily observed in the real world. In such cases, it is useful to think of the NOEM not as a model of what will happen in the real world, but rather as a model of how experts understand how the world works, taking into account its potential states as well as its realized states. At this point it is useful to return to the definition of validation and repose the question, “was the correct model built?” as, “is the model thinking about this problem the way the expert is thinking about this problem?” This restatement leads us to consider techniques like the IV&V that assist the SME in his role as a proxy for the real world data. Several subject matter experts can be employed in this role, in order to be able to reduce (or at least expose) bias.

We defined two experiments to begin the process of verifying and validating the NOEM model. We chose an application of the NOEM model that characterizes the complex relationships among a Government’s policies and the health of its country as measured by its economy. We apply our approach with the intention of determining both a target and an optimal debt value by manipulating components that comprise a policy consisting of government spending, stimulus into the economy, tax rates, and police forces. The experiments are:

(1) Validation using IV&V and Subject Matter Experts. AFRL has asked three test subjects (SMEs) to use the NOEM to identify optimal solutions as they would be applied to the Democratic Republic of Congo. At the same time, the LM ATL team will use the criteria set forth by AFRL to select an appropriate outcome set and use IV&V to find the top three input sets. The SMEs will evaluate those inputs sets and identify why they did or did not select solutions that are in the input sets found by the IV&V. The result will be descriptions/grading by each SME as to how closely the model is thinking about the posed problem to how the SME is thinking about it.
(2) Verification using IV&V and the modeler or an independent evaluator. In this experiment, there is no expert involved. We are trying to find out if the model is working the way it was designed. A condition of interest (and corresponding fitness function) is determined by someone who is familiar with the model. The IV&V generates sets of inputs that generate those outputs. The modeler grades the model based on whether the inputs that were found would be expected based on the specifications and design of the model.

In these experiments, it is anticipated that we will find multiple input sets that generate the desired outputs for both verification and validation. Some of these will make sense, while others will likely be surprising. In previous experiments using IV&V on the model of ImagiNation, Wedgwood and Schlicher [32] were able to verify a particular emergent mechanism. In the course of that experiment, they uncovered verification errors that were found when the experiment produced surprising results.

Based on our experience, IV&V can be a powerful tool to elicit significant thought processes on the part of the SME for validation purposes and the modeler for verification purposes as they examine both the anticipated and the surprising results. IV&V will help to build confidence in the NOEM model by providing insight into the model verification and validation as it relates to a particular region while helping to better indentify and understand the boundaries of the relationships and mechanisms in the model of a particular region.

4 Conclusion

In this paper, we have described a new and novel capability that will provide not only our Intelligence analysts, but Command & Control (C2) operators and the Decision Makers that they support a greater insight into potential problems before they happen, as well as a better understanding of the ramifications (unintended consequences in the short/long term) that might occur as the result of implementing potential planned actions. The powerful, adaptive, and customizable suite of tools within the NOEM will ultimately bolster the community's ability to continually gauge the stability/health of nations and to more extensively explore a wider range of impacts of implementing, in tandem, a collection of possible policy decisions. The NOEM will change the way the community can effectively convey and share the reasoning behind the evaluation and generation of potential policy sets. Once the complex interrelations within the environment and the effects-over-time of combined actions upon the environment are better grasped, and the capability to evaluate a significantly higher volume of policy permutations is attained, more optimal real-world decisions can be achieved in less time and with reduced surprise.

5 Disclaimer

The views expressed in this article are those of the author and do not reflect the official policy or position of the United States Air Force, Department of Defense or the U.S. Government.

Acknowledgements The authors would like to thank the remainder members of the NOEM development team (from ITT: Chris Kies, Brian Sullivan, Roger Nestler, Andrew Allen; from CUBRIC: Jeff Spaulding) for their dedication and support. We would also like to share our appreciation with Dr. Dennis Leedom (EBR Inc.), Janet Wedgwood (Lockheed Martin) and Bob Schlicher (Oakridge National Laboratory) for their support and collaboration in performing V&V on the NOEM. This work has been partially supported by Dr Joe Lyons, Air Force Office of Scientific Research (AFOSR) under a Laboratory Research Imitation Request (LRIR) effort.

References

1. Army Field Manual 3-24.2, Tactics in counterinsurgency. Headquarters Department of the Army, Washington DC, 21 April 2009, 3–1, see also 1–22 – 1–23
2. Barnett TPM (2004) The pentagon's new map: war and peace in the twenty-first century. G.P. Putnam's Sons, New York
3. The National Security Strategy of the United States of America, September 2002
4. Tyson AS (2008) Standard warfare may be eclipsed by nation-building. By Washington Post Sunday, October 5, 2008
5. Gates RM (2008) Speech to the National Defense University National Defense University (Washington, D.C.) As Delivered by Secretary of Defense Robert M. Gates, Washington, D.C., Monday, September 29, 2008

6. Department of Defense Directive Number 3000.05, November 28, 2005, paragraph 4.2
7. Dixon J (2009) What causes civil wars? Integrating quantitative research findings. International Studies Review 11:707–735
8. Goldstone JA, Bates RH, Epstein DL, Gurr TR, Lustik MB, Marshall MG, Ulfelder J, Woodward M (2010) A global model for forecasting political instability (19 pages; b/w). American Journal of Political Science 54(1):190–208
9. Abdollahian M, Baranick M, Efird B, Kugler J (2006) Senturion a predictive political simulation model. *Defense & Technology Paper*, National Defense University, Washington DC, July, vol. 32
10. Anderson EG (2009) Modeling insurgencies and counterinsurgencies. Proceedings of the *2009 International System Dynamics Conference* (Albuquerque, NM)
11. Epstein JM, Axtell R (1996) Growing artificial societies, social science from the bottom up. Brookings Institution Press, 1775 Massachusetts Avenue, N.W., Washington DC; The MIT Press, Cambridge, Massachusetts & London, England
12. Pierson B, Barge, W, Crane C (2005) The hairball that stabilized Iraq: modeling FM 3–24 National Defense University HSCB Conference. http://www.ndu.edu/ctnsp/HSCB/HSCB COIN - CDRBrett Pierson et al - The Hairball that Stabilized Iraq -Modeling FM 3-24.pdf
13. Kerr A (1995) National environmental indicators: program summary. Environment Canada, State of the Environment Directorate, Indicators Branch, Ottawa. Unpublished
14. Richardson DB (2004) Modeling and analysis of post-conflict reconstruction. MS Thesis, United States Air Force Institute of Technology
15. Robbins MJ (2005) Investigating the complexities of nation building: a sub-national regional perspective. MS Thesis, School of Engineering and Management, Air Force Institute of Technology, Wright Patterson AFB, OH
16. Fensterer GD (2006) Planning and assessing stability operations: a proposed value focus thinking approach. MS Thesis, School of Engineering & Management, Air Force, Institute of Technology, Wright Patterson AFB, OH
17. Manwaring MG, Joes AJ eds (2000) Beyond declaring victory and coming home: the challenges of peace and stability operations. Greenwood Publishing Group, ISBN 0275967689, 9780275967680
18. Orr RC (ed) (2004) Winning the peace: an american strategy for post-conflict reconstruction by center for strategic and international studies. Center for Strategic & International Studies (CSIS), ISBN 0-89206-444-7
19. Covey J, Dziedzic Michael J, Hawley LR (eds) (2005) The quest for viable peace: international intervention and strategies for conflict transformation. United States Institute for Peace, ISBN-10: 1929223676, ISBN-13: 978-1929223671
20. Cederman L-E, Girardi L (2007) Beyond fractionalization: mapping ethnicity onto nationalist insurgencies. American Political Science Review 101:173–185 Copyright © 2007 by the American Political Science Association. DOI: 10.1017/S0003055407070086, Published online: 05 February 2007
21. Karl R, de Rouen JR, Sobek D (2004) The dynamics of civil war duration and outcome. Journal of Peace Research 41:303–320
22. Fearon JD, Laitin DD (2003) Ethnicity, insurgency, and civil war. American Political Science Review 97(1):75–90
23. Kalyvas S (2006) The logic of violence in civil war. Cambridge University Press, New York, 508 p
24. Ruvinsky A, Wedgwood WJ (2011) V&V Techniques. Unpublished – Office of Naval Research, Model Evaluation and Application (MESA)
25. Lakatos I (1965) Falsification and the methodology of scientific research programmes. In: Lakatos I, Musgrave A (eds) Criticism and the growth of knowledge. Cambridge University Press, Cambridge
26. Sargent R (2004) Validation and verification of simulation models. Proceedings of the 2004 Winter Simulation Conference

27. Zacharias G et al (2008) Behavioral modeling and simulation: from individual to societies. The National Academies Press, Washington, DC
28. Hartley D, Starr S (2010) Verification and validation. In: Kott A, Citrenbaum G (eds) Estimating impact: a handbook of computational methods and models for anticipating economic, social, and security effects in international interventions. Springer, New York, pp 311–336
29. Petty M. (2010) Chapter 10: Verification, validation, and accreditation. In: Sokolowski JA, Banks CM (eds) Modeling and simulation fundamentals: theoretical underpinnings and practical domains. John Wiley & Sons, Inc., Hoboken, New Jersey. ISBN-10: 0470486740, ISBN-13: 978-0470486740
30. Balci O. (1998) Verifi cation, validation, and testing. In: Banks J (ed) Handbook of simulation: principles, methodology, advances, applications, and practice. Wiley, New York, pp. 335–393
31. Sallach D (2011) Herding concepts: the contextual validation of social agent models. Presented at The Computational Social Science Society of the Americas, 2011
32. Wedgwood J, Schlicher B (2011) Verification and validation of complex models. Presented at NOEM Days conference. AFRL, Rome
33. Choucri N, Electris C, Goldsmith D, Mistree D, Madnick SE, Morrison JB, Siegel MD, Sweitzer-Hamilton M (2006) Understanding & modeling state stability: exploiting system dynamics. ESD-WP-2006-04, February 2006
34. Joint Publication 3-0 (2006) Joint Operations. Chapter V, 17 Sept 2006
35. Joint Publication 3-07. Stability Operations. 29 Sept 2011
36. Kott A, Corpac PS (2007) COMPOEX Technology to Assist Leaders in Planning and Executing Campaigns in Complex Operational Environments, Twelfth International Command and Control Research and Technology Symposium (12th ICCRTS). 19–21 June 2007, Newport, RI
37. Flynn MT, Pottinger Matt, Paul D (2012) Fixing Intel: a blueprint for making intelligence relevant in Afghanistan Center for a New American Security, www.cnas.org. January 2010
38. Miller JO, Honabarger Lt, Jason B (2006) Modeling and Measuring Network Centric Warfare (NCW) With the System Effectiveness Analysis Simulation (SEAS), Eleventh Annual International Command and Control Research and Technology Symposium (11th ICCRTS), 26–28 Sept 2006, Cambridge, UK
39. Nysether NE (2007) Classifying failing states. MS Thesis, School of Engineering and Management, Air Force Institute of Technology, Wright Patterson AFB, OH
40. O'Brien S (2010) Analyzing Complex Threats for Operations and Readiness (ACTOR). Center For Army Analysis, Fort Belvoir Va. Sep 2001, DTIC Accession Number: ADA399370
41. Pierce BM, Zanol J (2011) MANeuver in N-Dimensional Terrain (MAN^N), A Full Spectrum Maneuver Concept. http://smallwarsjournal.com/jrnl/art/maneuver-in-n-dimensional-terrain-mann, Authors' notes from visit to National Training Center, Ft Irwin, CA, 11–12 Feb 2008
42. Taylor G, Frederiksen R, Vane RR III, Waltz E (2004) Agent-based simulation of geo-political conflict. Presented at Innovative Applications of Artificial Intelligence (IAAI), July 2004
43. The Fund for Peace (www.fundforpeace.org) (FFP)
44. United States Department of Defense. Measuring Stability and Security in Iraq. Quarterly reports to Congress submitted pursuant to the section entitled "Measuring Stability and Security in Iraq" of House Conference Report 109-72 accompanying H.R. 1268, Emergency Supplemental Appropriations Act for Defense, the Global War on Terror, and Tsunami Relief, 2005, Public Law 109-13. http://www.defense.gov/home/features/iraq_reports/index.html
45. United States Government Accountability Office Military Operations: Actions Needed to Improve DoD's Stability Operations Approach and Enhance Interagency Planning. Report to the Ranking Member, Subcommittee on National Security and Foreign Affairs, Committee on Oversight and Government Reform, House of Representatives, May 2007
46. Wagenhals, L, Levis, A (2001), Modeling Effects-Based Operations in Support of War Games, in A. Sisti & D. Trevisani, eds, Enabling Technology for Simulation Science V, vol. 4367, International Society for Optical Engineering, pp. 365–376
47. Del Valle S, Stroud P, Mniszewski S (2008) Dynamic Contact Patterns and *Social Structure in Realistic Social Networks, Social Networks: Development, Evaluation and Influence.* Nova Science Publishers, 2008.

A Multi-Method Approach for Near Real Time Conflict and Crisis Early Warning

Sean P. O'Brien

1 Introduction

In 2010, I wrote an article [8] for *International Studies Review* that described the objectives and emerging results of a project I led at the Defense Advanced Research Projects Agency (DARPA) which attempted to develop an Integrated Crisis Early Warning System (ICEWS). That piece reported on our progress to integrate conflict and instability forecasting models developed by some of the most prominent scholars in the field to generate forecasts of a variety of events of interest (i.e., international crises, insurgencies, rebellions, domestic political crises, and ethnic/religious violence) with higher degrees of accuracy and precision than any one model alone could generate. By the time I left DARPA in April 2011, we had integrated these models, along with real time event data collection and visualization technologies, into a system that was transitioning for use within the US Department of Defense. In this chapter, I illustrate the components of this system and describe the challenges we faced in our attempt to transform academic analyses and techniques into a powerful capability that can be applied by a diverse set of users to a variety of contemporary foreign policy challenges.

1.1 Building on Previous Research

Schrodt and Gerner trace the roots of contemporary early warning research to programs funded by the US Department of Defense (particularly DARPA) in the 1960s [1, 11]. With the US and Soviet Union in the throes of the Cold War, scholars

S.P. O'Brien (✉)
College of William and Mary,
e-mail: seanob88@gmail.com

V.S. Subrahmanian (ed.), *Handbook of Computational Approaches to Counterterrorism*,
DOI 10.1007/978-1-4614-5311-6_18,

and policymakers were eager to identify conditions and sequences of interactions between the US and its principal adversary that might provide early warning of the use of military force that could quickly escalate out of control. By identifying the seeds of such crises in their earliest stages, the hope was that policymakers might heed warnings of the potentially catastrophic consequences of their actions and take deliberate measures to step back from the abyss.

The scholars involved in these programs identified the universe of potential nation-state "event interactions", across the entire continuum of the conflict-cooperation spectrum, and formalized them into various taxonomies, chief among them being the Conflict and Peace Databank (COPDAB) and World Event Interaction Survey (WEIS) frameworks [6]. Hordes of graduate students and professors were funded to sift through news reports (primarily the *New York Times*) to extract data that reflected *who* was doing *what* to *whom, when, where,* and *how*. These data sets were then analyzed with a variety of analytic techniques to identify patterns that might prove instructive for identifying and validating early warning indicators of impending military conflict. Unfortunately, the volume of news reports quickly overwhelmed these "human coders," who struggled to consistently maintain the currency of the data, imperiling the objectives of the research programs to generate timely and accurate crisis early warnings. COPDAB and WEIS data were not extended beyond 1978.

Schrodt himself would later go on to make a lasting contribution to this research agenda, and brought a new level of sophistication to the field of political science in general with his development of a suite of automated tools to automatically extract these "events data" from media reports with speeds and volume for the first time approaching what would be required for near real time monitoring of event interactions. His KEDS (Kansas Event Data System) research program at the University of Kansas generated dozens of doctoral dissertations and hundreds of conference papers and peer-reviewed journal articles. It also influenced the development of other automated event data coders including those developed by Shellman [12] and Bond [2]. One of Schrodt's former students [9] was among the first to demonstrate the potential of the automated events data approach for generating timely, policy-relevant warnings of major conflict events, with their application of these tools to an analysis of events associated with the Serbia-Kosovo War. Schrodt applied Hidden Markov Models (HMMs) to various event data series and demonstrated their potential for forecasting temporal variations in high levels of conflict during the Balkan Civil War [10]. Schrodt's TABARI/CAMEO taxonomy and event data protocols were further extended and improved upon in DARPA's Integrated Crisis Early Warning System (ICEWS) program.

While Schrodt and his colleagues were refining their automated data extraction framework, other scholars were drawn to identifying the seeds of crises using other types of data and analysis techniques. In 1994, the genocide in Rwanda—in which 500,000 Tutsi and Hutus were rapidly and systematically slaughtered—shocked the world's conscience. In response, Vice President Al Gore commissioned the CIA that same year to develop and validate an analytic model that would provide early warning indications of when a state might "fail" two years in advance [3].

The CIA contracted with Science Applications International Corporation (SAIC) who recruited some of the most prominent scholars (primarily social scientists) to generate the data and identify and evaluate research methodologies to achieve these objectives. The State Failure Task Force (SFTF), later renamed the Political Instability Task Force (PITF), identified four events that they argued, reflected state failure:

- Genocides and Politicides
- Ethnic Wars
- Abrupt regime transitions
- Revolutions

The Task Force collected thousands of time series indicators at the country-year level that measured a wide variety of societal macro-structural characteristics, including measures of broad trends in economics, demographics, politics, environment, and social characteristics. The scholars evaluated dozens of techniques and hundreds (if not thousands) of model specifications and concluded that their best model of state failure was a logistic regression model that included only three variables: a country's infant mortality rate, regime type, and trade openness (a ratio of its imports/exports relative to the size of its economy). In retrospective assessments, these three factors were able to discriminate between failed and non-failed states two years in advance with about 70% accuracy (though with somewhat less precision). King and Zeng would later criticize the Task Force methodology on several technical details, and offered their own model that included two additional factors (effectiveness of legislature and fraction of population in military), which they claimed could forecast state failure with 80% accuracy [4].

O'Brien extended the PITF methodology by evaluating other potential crisis antecedents, research methodologies, and dependent variables that reflected not just *whether* a country would become unstable, but also the level of intensity of that instability [7]. Using a split-sample research design, with training and test sets of various lengths, O'Brien's ACTOR (*Analyzing Complex Threats for Operations Readiness*) model was able to correctly forecast the level of intensity of instability in 200 countries, 5 years in advance with about 80% accuracy (albeit, with performance degrading over time). The research team forecasted macro-structural factors for each country 15 years into the future, and used the forecasts of these macro-structural factors in each country to compute the likelihood that each country would experience a particular level of intensity of instability in any given year through 2015 [7]. Doing so allowed one to identify those countries likely to continue stabilizing or de-stabilizing, assuming they continued along their macro-structural trajectories. These forecasts were used in a variety of US Army analytic studies to, for instance, determine the potential demand for US military deployments, evaluate the Army's portfolio of Theater Security Cooperation (TSC) activities and future plans, and were included in several "watch list reports" produced by the Intelligence Community.

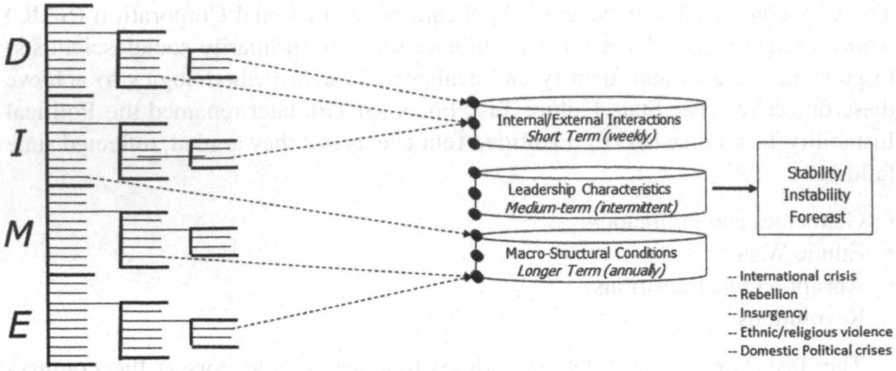

Fig. 1 Overview of the ICEWS system. DIME is the acronym the US military often uses to categorize the Diplomatic, Information, Military, and Economic (DIME) activities it conducts in various Areas of Responsibility (AORs) as it strives to achieve its intermediate- and end-state objectives

1.2 DARPA's ICEWS Program

Scholars who study crisis and instability typically examine the phenomenon from one, and usually only one, of three different perspectives. The different perspectives can be distinguished by the types of factors they consider, and their corresponding forecast horizons (reflected in the cylinders in Fig. 1). As noted above, some instability forecasting models use only macro-structural factors, neglecting the dynamic interactions that take place within and between countries, as well as the characteristics of a country's leadership and decision making style, which can decisively influence the trajectory of a crisis (witness Saddam Hussein's miscalculations in the run up to the most recent US-Iraq War, Hosni Mubarak's decision to stand down rather than repress Egyptian protestors during the Arab Spring, or Hugo Chavez's bombastic approach to foreign relations which has earned him friends and enemies in equal measure). By the same token, scholars who focus on the influence of decision maker styles, cognitive frameworks, personality, and psychological pre-dispositions often ignore the macro-structural environment that enables or constrains a leader's ability to react to internal and external pressures. And, finally, those who examine sequences of internal and external events often neglect the importance of leaders, their decision calculus, and the environment within which they contemplate their actions.

The goal of DARPA's Integrated Crisis Early Warning System (ICEWS) was to leverage the best of these forecasting models and data collection and analysis techniques, integrate them within the context of an intuitive, user-friendly software system, and demonstrate the utility of that system to automatically and consistently generate forecasts of a variety of events of interest, in near real time. This was to be accomplished in a way that could inform decisions by senior leaders about how to allocate resources to mitigate the undesirable consequences. The ICEWS program kicked off in 2007 with three teams. Each team was led by a large defense contractor and consisted of a mix of small businesses and several of the most prominent scholars in the field. Incentivized by performance metrics that increased in complexity over time, the three teams competed to develop alternative ICEWS systems using different perspectives, data sources, and modeling techniques. Lockheed Martin-Advanced Technology Lab (LM-ATL) was the most successful contractor, and the only one funded to produce the final ICEWS system.

In phase 1 of the program (October 2007–December 2008), we focused on integrating multiple forecasting models across the three levels of analysis. In phase 2 (January 2009–December 2011), the objective was to transition from retrospective, "proof of concept" forecasting to forecasting real world events in near-real time. An important secondary objective in phase 2 was to develop the decision support framework and methodology for evaluating alternative courses of action that could mitigate de-stabilizing events. After all, it is one thing to forecast the occurrence of a de-stabilizing event, along with the key factors driving it. It is quite another to link those crisis drivers to sets of resources and courses of action that could be brought to bear to mitigate their undesirable consequences. If one can demonstrate an ability to forecast events of interest with high degrees of accuracy and precision then one has identified *causal structure* in the factors that generate these crises [4]. If reducing the likelihood of insurgency from 90% to 60% is an important *end-state objective*, the societal factors that create that potential—the internal and external dynamic events that generate pressure, a country's leadership characteristics, and the macro-structural environment—are the *intermediate objectives* that need to be affected or influenced with policy interventions in order to achieve the desired end-state objective. Thus, our intent in phase 2 of the program was to solicit from senior decision makers their stability objectives for countries in regions for which they were responsible. Because we had identified causal structure in the factors that lead to or were associated with a variety of events of interest, we can identify constellations of these factors (i.e., the independent variables) and the extent to which they would need to be influenced in order to achieve particular stability objectives. These *independent* variables in our forecasting models would become *dependent* variables in analyses designed to determine how they could be impacted by various Diplomatic, Information, Military, or Economic (DIME) activities.

Finally, in phase 3 of the ICEWS program (originally scheduled to last from January 2011 to April 2012) the original intent was to subject the ICEWS system to an independent test and evaluation with potential users in the field. The US government continuously deploys resources as it responds to humanitarian disasters, engages in military exchanges and exercises, hosts foreign delegations, sells military equipment, and assists partner nations in developing the capacity to respond to internal and external challenges. These activities almost always have short-, medium-, and long-term objectives. And the achievement of those objectives is measurable (albeit with some difficulty). This parameter, which we denote as O_{us} in Fig. 2, could be measured and averaged across all objectives and countries in a given region over any time period. In addition, we could incorporate these resource allocations within our decision support framework and estimate their *likely*

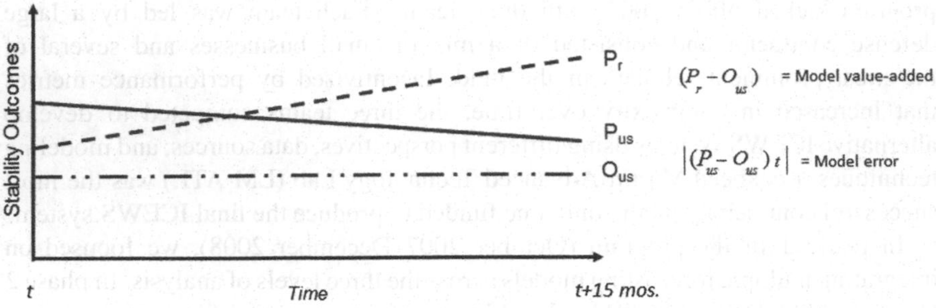

System provides value-added over current approaches if the model value-added is larger than model error and model error is bounded, i.e.,

$$\frac{\Sigma\,(P_r - O_{us})\,t}{\Sigma\,\left|(P_{us} - O_{us})\,t\right|} > 1 \text{ AND } \Sigma\,\left|(P_{us} - O_{us})\,t\right| < 25\%$$

Fig. 2 ICEWS Phase III test and evaluation metrics. Although part of the original research plan, the experiment was never conducted

impact on those same objectives (denoted P_{us} in Fig. 2). This parameter measures model error relative to "ground truth", and our goal was to minimize the error to the extent possible over the 15 month test period. Finally, we could calculate a third parameter, P_r. P_r measures the impact that ICEWS-derived "optimal" resource packages might have had on the end-state objectives had those courses of action been taken in response to the forecasted crisis. The difference between P_r and O_{us} is the valued-added by the ICEWS system. Unfortunately, the program ended at DARPA and portions transitioned for use before we had an opportunity to conduct the experiment.

1.3 Adjusting to Operational Reality: Lessons Learned from the ICEWS Program

ICEWS was coordinated with a major regional military command from the beginning of the program. Representatives from several regional commands had been lobbying DARPA for some time to invest more research funding in social science tools and techniques that they could use to better monitor, anticipate, and respond to challenges in their area of responsibility. Those challenges were largely rooted in social and behavioral disruptions generated by ideological extremism, drug-trafficking, lack of government performance and capacity, government corruption, and human rights abuses, among others. Once the four-year ICEWS program was officially approved, and our research funding secured (at least for phase 1), we regularly briefed our plans and emerging results to key military stakeholders.

From the beginning, we were immediately presented with several challenges that needed to be overcome:

- Senior military leadership is subject to high turnover rates, in which some general and flag officers remain in their positions for relatively short periods of time, often two years or less. This meant that the leaders who lobbied DARPA for greater investment in social science systems like ICEWS had moved on to their next assignments by the time the program was approved. In such instances, we were required to "re-socialize" the program with the new leadership to seek their continued support and participation in shaping the program in a way that maximized its potential usefulness. This cycle repeated itself with each new milestone, typically every six months. Indeed, by the time the research program transitioned for use four years after it began, we were working with our fourth generation of senior leaders.
- Policy makers and analysts are very busy people, responsible for dealing with multiple, high-consequence issues under immense pressure and shifting deadlines. Although they were very interested in shaping the trajectory of the ICEWS program to maximize its benefits, they had little time or patience for "science experiments" or to participate in series of workshops in which tangible progress was difficult for them to discern. They needed reliable tools to make their jobs easier, and they needed them immediately. These requirements were somewhat inconsistent with a long-term, scientific research and development program. We did, though, manage to adapt. Specifically, we agreed to provide access to individual components of the ICEWS system to analysts and decision makers as they matured. As we made progress, we presented analysts with prototype "mockups", semi-functioning systems that were more sophisticated than a set of working papers, PowerPoint slides, or screenshots, but less than a full functioning prototype. These experiences provided us with invaluable feedback about the kinds of data, analyses, functions, and visualizations users found useful or meaningless for accomplishing their daily tasks. After two or three iterations, we were often able to return with fully functioning prototype systems that users were able to apply, experimentally, to real world issues. By taking this approach, our research benefited from the user feedback and our access to real world scenarios, and the users had some familiarity with each system component by the time the next component was ready for evaluation. Nevertheless, we learned that a new system component needed to be introduced for initial test and evaluation about every six months in order to maintain momentum along with the attention of potential users in the ICEWS program. To operate effectively in such an environment, we needed to carefully and continuously monitor and adjust our schedule and funding priorities, in light of all the bureaucratic and political pressures that constrained our ability to do so.
- Most senior decision makers and analysts in US military commands have high standards of analytic rigor and closely scrutinize data, methods, and tools for their accuracy, relevance, and reliability. At the same time, the standards by which they judge the adequacy and validity of new models differ markedly from the way in

which most scholars evaluate their wares. The standard statistics and criteria that most political scientists use to judge their models—R^2, F statistics, Z-scores, among a host of others—are next to meaningless to users in the field. Military analysts typically possess an enormous reservoir of knowledge on a wide range of issues and conditions in the countries they monitor. As a result, they examine new tools or data sources in light of their ability to capture, track, and illuminate key features of these countries or regions with which they are quite familiar. Thus, we placed a heavy premium on transparency and the replicability of all of the data, tools, and models that we introduced for evaluation and consideration for operational transition, which were evaluated from both a qualitative and quantitative perspective.

2 Components of the ICEWS System

The research team developed three key components of the ICEWS system:

- iTRACE: ICEWS Trend Recognition and Assessment of Current Events. iTRACE used event data collected in near real time from over 100 international, regional, and country-specific news media sources covering events occurring in most major countries in South and Southeast Asia, and Central and South America. iTRACE leverages an improved version of the TABARI system to transform this unstructured text into structured indices that reflect *who* is doing *what*, to *whom*, *when*, *where*, and *how*. These data can be customized into an array of event interaction indices that can be monitored over time, or they can be fed into any number of computational and statistical forecasting models.
- iSENT: ICEWS Sentiment Analysis. iSENT uses a very similar approach as iTRACE, but rather than coding who is doing what, to whom, we code who is *saying* what about whom, and what issues or activities. These data are extracted from blogs and various social media platforms. iSENT serves a role similar to public opinion polls in the analysis stream, but it is designed to automatically, efficiently, and inexpensively generate near real time data that is more timely and focused than most public opinion polling firms can provide.
- iCAST: ICEWS Forecasting. ICAST uses data from iTRACE, iSENT, and other sources (i.e., country macro-structural conditions) to forecast a variety of events of interest including rebellions, insurgencies, domestic political crises, international crises, and ethnic/religious violence.

2.1 *iTRACE*

Figure 3 displays a screenshot from the iTRACE system. Using an enhanced version of the TABARI system, we coded over 12 million reports from over 100 major

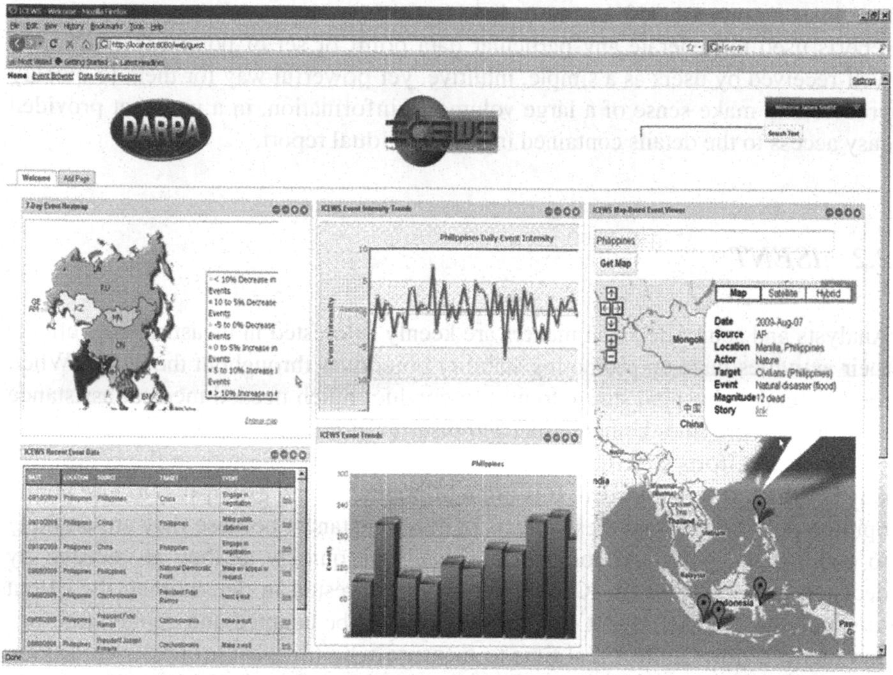

Fig. 3 Screenshot from the iTRACE system

international and regional news sources over the period 1998 to present. Initially, our aim was to generate quantitative data to use in a variety of individual models, and aggregate the output of those models into a combined forecast for each event of interest. We had no use for the stories themselves. However, in discussions with key stakeholders, we learned that one of the biggest challenges facing defense and intelligence analysts as they try to monitor and assess events around the world is the sheer volume of information. What they needed were tools that could automatically process documents, display relationships among actors and events on the fly, and the ability to drill down on trend lines into the corpus of news articles that contained additional details and context on events of interest. Thus, rather than having to read through thousands or tens of thousands of news reports to identify details of interest, inflections in the time series data plots provide analysts with cues and a means to access that subset of reports they need for any particular time period and activity of interest. The iTRACE system provides a near real time, 360° view of *who* is doing *what* to *whom, when, where,* and *how*. It provides an intuitive layering of charts, graphs and storyboards that can quickly point the analyst to that set of reports most relevant to an issue, region, or time period of interest.

The iTRACE framework has features similar to iGoogle and allows users to fully customize the addition and layering of charts, graphs, and alerts. The system is also designed to be fully transparent and traceable: From any graph or chart, the user is never more than five mouse clicks away from the actual time-stamped news reports used to generate any particular data point or set of points. iTRACE was well-received by users as a simple, intuitive, yet powerful way for them to quickly process and make sense of a large volume of information, in a way that provided easy access to the details contained in any individual report.

2.2 *iSENT*

Analysts and senior decision makers are keenly interested in measuring the effects their activities have on promoting stability objectives throughout the world. When the US sends a hospital ship into port to conduct much needed medical assistance to people, or provides humanitarian and disaster aid in general, the hope is that the people these efforts were designed to help will view the gesture in the benevolent way in which it was intended. It is often impractical (and expensive) to hire public opinion polls to measure these effects in every instance, because they are unlikely to conduct the polls and compile the results as quickly as they are needed by decision makers. Thus, in ICEWS, we were interested in determining the extent to which the TABARI event data *approach* could be adapted and applied to blogs and media reports in near real time to measure who (or what groups) were saying what, about these activities, or any other personality or issue of interest. If we could demonstrate that an automated sentiment analysis capability generated data that was consistent with public opinion polls, such a capability would provide analysts with user-friendly tools that could be applied across a wide range of issues in a way that was inexpensive and timely relative to standard public opinion polls.

We conducted a pilot study to assess the feasibility of developing such a capability. We chose to limit the pilot study to an examination of Taiwanese sentiment toward presidential candidate Ma Ying-Jeou over the period April 2007–July 2008, just prior to, and just after, the Presidential election in Taiwan. The goal was to determine the extent to which various sentiment measures correlated with public opinion polls, and the extent to which popular sentiment correlated to various political behaviors, both cooperative and hostile [15].

Two approaches were taken in this pilot study. The first was a "Bag of Words" (BoWs) approach. The BoWs begins by scoring an array of sentiment words according to the extent to which their connotation is positive or negative. For any text that is analyzed, one simply calculates the number of positive and negative words contained in the text, which gives an overall impression of the sentiment expressed within it. Such an approach provides useful information, but can be quite crude, in that it is unable to distinguish between the source and the target of the sentiment expression. Further, if an actor were to express strong, positive sentiment toward some aspects of a person, group, or issue, and negative expressions toward other aspects, the end result would show neutral sentiment. In fact, in many cases it is useful—even crucial—to know specifically those aspects toward which one has positive feelings and those for which sentiment is more negative, so that the negative aspects can be addressed when desired or necessary (consider, for instance, the importance of doing so for a new product introduction, like an Apple iPhone 4s). It is in that vein that we sought, for the second approach, to adapt the TABARI event data framework toward measuring the expression of various sentiments expressed by one actor toward another (which we refer to as *dyadic sentiment speech acts*).

The primary difference between the events and sentiment frameworks is that, in the latter, we only use words identified as conveying sentiment. As such a new sentiment verb dictionary had to be created. This new dictionary had over 800 verbs and verb phrases. Each word was rated by a group of linguists on a scale from −1 to 1 based on their implied sentiment. For instance, words such as *excellent* and *great* received values close to 1, while words such as *despair* and *revolting* received values close to -1. Actors making the statements were included in an actor dictionary developed specifically for Taiwan. Targets included individuals as well as terms focusing on the economy, countries, organizations, and security. There were over 200 Taiwanese-specific actors alone. The advantage of speech acts is that one can observe who is directing sentiment towards whom, and any events or political implications that sentiment might produce.

In order to perform the analysis, over 1 GB of text was downloaded. The following Taiwanese news publications were used for this analysis: *China Post, Taipei Times, Central News Agency, Kuomintang (KMT) News Network, and Taiwan Review*. These account for the most popular sources of general Taiwanese news available in English. Among the blogs downloaded include *Forumosa: Taiwan Politics, That's Impossible: Politics from Taiwan, Far Eastern Sweet Potato, Taiwan Matters, Sun Bin, Rank, Only Red Head in Taiwan, Jerome F. Keating's Writings, and It's Not a Democracy it's a Conspiracy.* These were found to be the most popular and most accessible blogs emanating from Taiwan. Public opinion polls were gathered from a variety of poll-administrating agencies namely *United Daily News, United Evening News, Global Views Magazine, TVBS News, China Times, Taiwan Apple Daily News, ERA Television, and The Executive Yuan Research, Development and Evaluation Commission*. Although hundreds of questions were administered in these polls, only 33 were relevant and used for this analysis. Questions covered topics on the economy, culture, international relations, and support for various leaders within the government, among many others. That said the polling data were very spotty in the sense that the polls were not conducted with any regular frequency. Many of the polling questions were repeated only periodically, every few months or weeks. Around the time of the election, the questions regarding Ma became more frequent.

Figure 4 shows the relationship between sentiment data generated using the dyadic sentiment and BoWs approaches relative to public opinion polls measuring popular support for presidential candidate Ma Ying-Jeou. Both the BoWs and the dyadic sentiment approaches track the polling data rather nicely, though dyadic sentiment appears to be a closer fit. We were curious to assess the extent to which

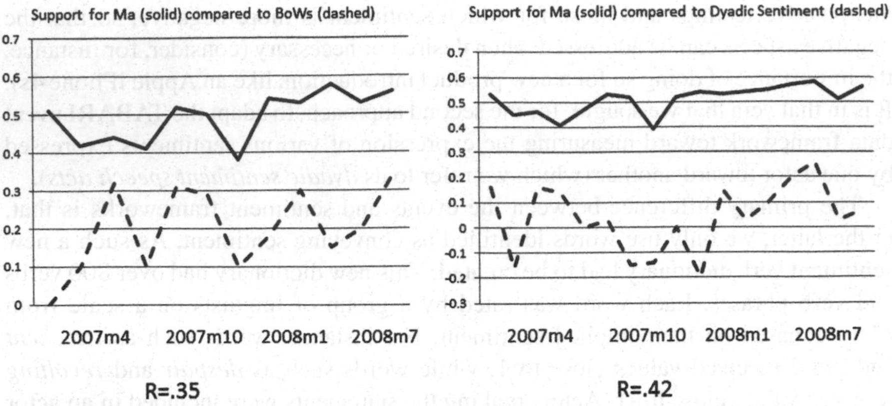

Fig. 4 Comparison of public opinion polls and different sentiment analysis approaches

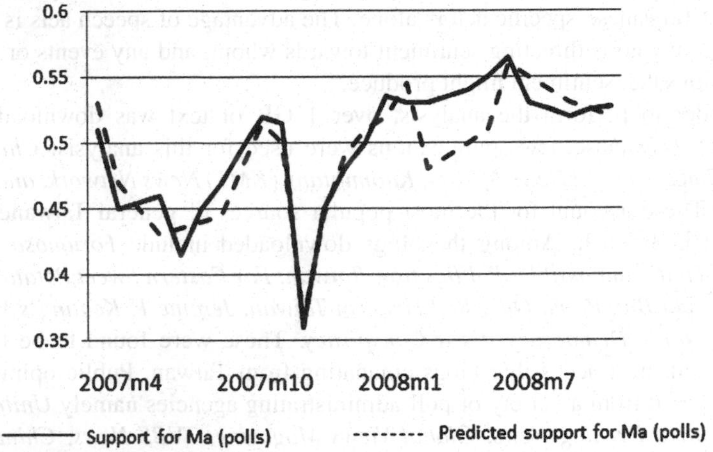

Fig. 5 A model of Taiwanese polling data using sentiment measure (monthly)

we could predict the polling data with our measure of sentiment. If we could predict the polls using our automatically generated sentiment data, than sentiment might be a suitable alternative to polling data.

To conduct the analysis, the polling data was placed on the left side of the equation as the dependent variable. On the right side of the equation, lags of the sentiment measures were used as independent variables. Figure 5 plots the actual polling data against the predicted polling data generated from the sentiment model. As one can see, the model fitted values closely resemble the actual polling data. The two series correlate at .92. As such, automatically generated sentiment analysis may be a timelier, less expensive substitute for more traditional public opinion polls.

2.3 *iCAST*

One of the first objectives of the ICEWS program was to demonstrate our ability to forecast 5 events of interest: domestic political crises, ethnic/religious violence, rebellions, insurgencies, and international crises. We also developed an index that was used to measure the intensity of violence experienced by countries on a quarterly basis. Historical data on these events were collected on a quarterly basis from a variety of open sources [8]. Initially, the data were collected for 29 major countries in the US Pacific Command (USPACOM) Area of Responsibility (AOR). Because our aim was to field a *generalizable* capability to forecast progress toward or away from stability, a viable solution involved the ability to forecast country instability using a consistent set of factors applicable to all countries in the PACOM AOR. A portion of the historical data (1998–2004) was provided to the ICEWS performers to use as training in model development; the remaining data (2005–2006) were withheld for testing the modeling solutions according to the following performance metrics:

Accuracy:	*# of correct predictions*
# of predictions made	
Recall:	*# of correctly predicted conflicts*
	# of conflicts that occurred
Precision:	*# of correctly predicted conflicts*
	# of conflicts predicted to occur

The recall and precision scores measure the (inverse) of false positives and false negatives respectively. Ideally, we strive to develop models that produce precision *and* recall scores as close to 1 as possible.

The ICEWS performers comprised three competing teams of experts drawn from industry and the social science academic community. Though each team was provided with the same, focused set of outputs for model training, it was each team's responsibility to collect and evaluate their own input data for its predictive potential. The most successful team was led by Lockheed Martin-Advanced Technology Laboratories (LM-ATL), assisted by several prominent scholars and industry partners. The team successfully integrated and applied six different conflict modeling approaches, including agent based models [5], logistic regression [13, 14], Bayesian models, and geo-spatial network models [16, 17]. A final model was developed by aggregating the forecasts from the above mentioned models using Bayesian techniques. The Bayesian aggregator ascribed more or less confidence to any individual model forecast, according to the countries and events of interest for which that model had historically demonstrated high performance.

The results of the evaluation are reported elsewhere [8], so I will not dwell on them here. The team led by Lockheed Martin did demonstrate an ability to forecast insurgencies, rebellions, ethnic/religious violence, and their levels of intensity three months in advance with accuracy exceeding 80%. The team fell well short of that

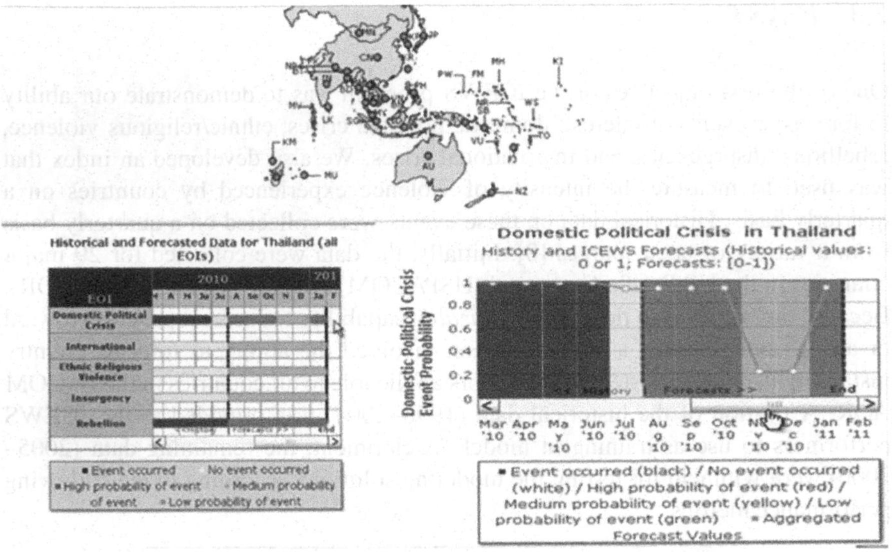

Fig. 6 Screenshot from the iCAST forecasting module in ICEWS

threshold for international crises and domestic political crises. The former could possibly be explained by our unit of analysis (country-quarter, country-month), which did not adequately allow for explicit consideration of dyadic or multi-adic features of groups of countries. The disappointing results for domestic political crises may be driven in part by the fact that these crises are less intense, less violent forms of both rebellions and ethnicreligious violence. In other words, rebellions and ethnicreligious violence often begin as domestic political crises, and we have yet to identify the factors that would allow us to detect these events in their earlier stages.

Figure 6 shows a screenshot from the iCAST component of the ICEWS software system. Charts and analyses are layered, providing users with the flexibility to generate executive summary views at the regional or country-specific level, or drill into the forecasts all the way down to the raw historical data used to generate them. Users can customize their dashboard displays, adding or deleting alternative views or layers, and arranging them in any configuration they desire. A "what if" analysis module allows users to assess how increasing or decreasing individual or groups of indicators (i.e., levels of government repression, economic development, democracy, levels of hostile rhetoric) could change the forecast for any events of interest in any particular country.

Senior decision makers, analysts, and planners, understandably, were skeptical of the veracity of the forecasts when the forecasts were first presented to them for review. Specifically, they demanded transparency into the mechanics of the forecasting models—for instance, what caused a probability of insurgency to increase or decrease at any particular time? Figure 7 illustrates an example that I found useful in trying to foster an understanding and appreciation for the

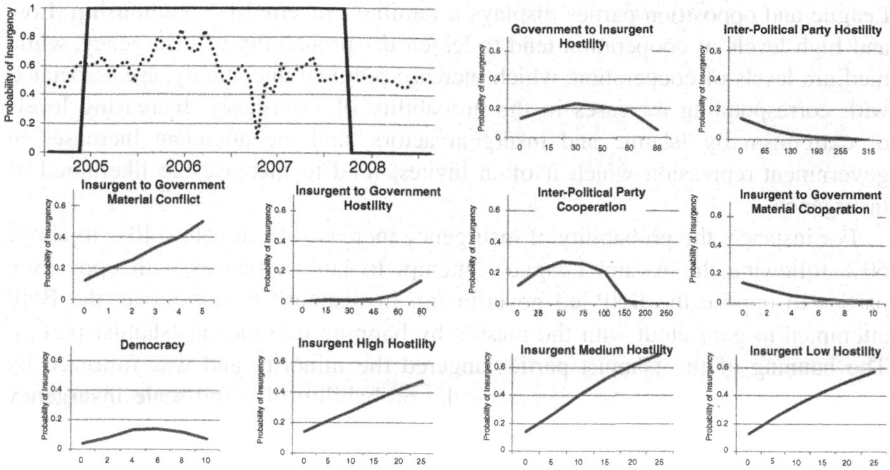

Fig. 7 Retrospective forecast of insurgency in Bangladesh (*top left corner*). Solid line indicates the beginning and ending of the insurgency. The *dashed line* is the monthly probability forecast (3 month look-ahead) that an insurgency would occur, generated from the ICEWS system. The remaining charts show the conditional probability of insurgency given different values on the independent variables

model-generated forecasts. Figure 7 depicts a monthly, retrospective forecast of insurgency in Bangladesh. The insurgency began in late 2004 and lasted into late 2007, as indicated by the solid line in the top left corner of the figure. The dashed line in that figure indicates the probabilities of insurgency ICEWS generated, given the variables identified in the retrospective analyses as being important drivers of insurgency. A subset of those key variables, extracted from one of the logistic regression models used in ICEWS, are also displayed in Fig. 7 as conditional probabilities of insurgency given different values on the dependent variable.

Taken together, the subset of 11 variables measure the character and intensity of interactions between the Bangladeshi government, civil society, major political parties, and insurgent groups, along with macro-structural factors such as the level of government repression (political terror) and regime type (democracy). As the values of the independent variables change over time, as a function of structural shifts and dynamic interactions between groups and individuals, they are associated with corresponding changes in the probability of insurgency. The probability of insurgency increases as the events and conditions in Bangladesh moved along the more de-stabilizing parts of the curve; the probability of insurgency declines as these events and conditions occur along the more stabilizing parts of the curves.

The variation in the probability of insurgency in Bangladesh can be attributed to three key factors: inter-party cooperation between the Awami League and opposition parties to oppose the Bangladesh National Party (BNP) led government; government repression of radical Islamic parties and insurgent groups; and the hostility of these groups toward the BNP-led government. Cooperation between the Awami

League and opposition parties displays a familiar "inverted-U" relationship. Low and high levels of cooperation tend to lessen the probability of insurgency, while medium levels of cooperation, which increase political uncertainty, are associated with corresponding increases in the probability of insurgency. Increasing levels of extremism by Islamic and insurgent actors, and the attendant increases in government repression which it often invites, tend to increase the likelihood of insurgency.

For instance, the probability of insurgency increased from below 40% to above 60% following the Awami League's attempt to build alliances with opposition parties to oppose the BNP-led government. In response to this move, the BNP attempted to gain clout with the masses by banning two radical Islamist parties. The banning of the Islamist parties angered the minority and was followed by a series of Islamist attacks, increasing the probability of a full-scale insurgency from 60% to 82%. As the level of cooperation between the Awami League and the opposition parties subsided, the series of insurgent attacks continued through April 2006. The government responded to the attacks with increased repression, detaining hundreds of members of the two banned parties. The probability of insurgency thus remained high as repression and dissent continued through the middle of 2006. The probability of insurgency began to wane as the Awami League, which had previously demanded political reforms as a pre-condition for its participation in elections, agreed to do so without any of its demands being met. A caretaker government was then formed with the cooperation of the military, which arrested several corrupt politicians. As a result, the probability of insurgent attacks decreased below 50%.

3 Summary and Conclusion

The ICEWS system is the culmination of 40 years of significant, incremental advancements by dozens of scholars and industry partners, as they pursued their ambitious objective to develop a near real time, automated crisis early warning and assessment system. We were able to extend the system to South and Southeast Asia, and Central and South America before the program formally ended at DARPA. Several research challenges remain for future researchers to contemplate:

1. *Decision support: linking courses of action to crisis drivers.* Because we rapidly transitioned from R&D to technology transition in the final year, and because we were unable to procure data in sufficient quantities, we jettisoned the decision support objective. With respect to data, there was no centralized, historical data set that we could draw upon that provided an accounting for all the different activities and resource allocations the US conducted as it engaged with various countries abroad. Our attempts to generate these data ourselves through open sources proved woefully unsatisfactory. Even had we obtained the necessary data, we faced a daunting challenge trying to link so many activities and resource allocations to so many possible crisis drivers, not to mention accounting for short

and long-term time lags. If sufficient data could be located, quasi-experimental designs might prove to be a useful way to establish these linkages and effects. This is an important research objective because policymakers will remain hard-pressed to design effective and appropriate crisis responses until they have some confidence in the likely effects of different courses of actions.

2. *Policy-relevant antecedents vs. contemporaneously measured correlates.* Ideally, the goal of a crisis early warning system is to identify those seemingly benign features, conditions, and events that, in certain configurations provide sufficient early warning of major stability disruptions. Many factors identified in the literature, and confirmed in ICEWS, as valid "predictors" of crisis and instability are static, policy-irrelevant features that are not amenable to change through policy interventions. These include the presence of mountainous terrain, aggregate size of population, and even, I would submit, religious and ethnic homogeneity. Even in ICEWS, many of the correlates we validated as reliable early warning indicators were, in some sense, contemporaneously measured features of the dependent variable. For instance, a key determinant of insurgency is high level of insurgent hostility; a key driver of ethnic/religious violence is the level of hostility expressed by certain ethnic and religious organizations. These antecedents are among the very types of phenomenon we seek to anticipate. By the time we begin observing them, it may well be too late. Future researchers might find that a fruitful next step would be to search for and validate new candidate antecedents at earlier stages along the crisis early warning timeline.
3. *Geo-specificity of events.* Contemporary events data frameworks that rely on news media reports will often fail to identify the specific location (i.e. city) of reported events up to 80% of the time, either because the location from which the report is generated is different from the location of the event itself, or because the specific location of the event is not mentioned at all. Lacking means to reliably and precisely geo-locate events, our ability to more specifically identify localized vulnerabilities is severely blunted. The research community would benefit from the development of new and novel techniques to better infer or triangulate on the specific location of event occurrences.

References

1. Andriole SJ, Hopple GW (1984) The rise and fall of event data: from basic research to applied use in the US department of defense. International interactions 10:293–309
2. Bond D, Bond J, Oh C, Jenkins JC, Taylor, CL (2003) Integrated data for events analysis (IDEA): an event typology for automated events data. J of peace research 40:733–745
3. Esty DC, Goldstone J, Gurr TR, Harff B, Levy M, Dabelko GD, Surko PT, Unger AN (1998) The state failure task force report: Phase II findings. http://globalpolicy.gmu.edu/pitf/SFTF%20Phase%20II%20Report.pdf. Accessed 8 June 2009
4. King G, Zeng L (2001) Improving forecasts of state failure. World politics 53:623–658
5. Lustick IS, Miodownik D, Eidelson RJ (2004) Secessionism in multicultural states: does sharing power prevent or encourage it. American political science review 98: 209–230

6. McClelland CA, Hoggard G (1969) Conflict patterns in the interactions among nations. In: Rosenau JN (ed) International politics and foreign policy, Revised edn.Free press, New York
7. O'Brien SP (2002) Anticipating the good, the bad, and the ugly: an early warning approach to conflict and instability analysis. J of conflict resolution 46:791–811
8. O'Brien SP (2010) Crisis early warning and decision support: contemporary approaches and thoughts on future research. International studies review 12:87–104
9. Pevehouse JC, Goldstein JS (1999) Serbian compliance or defiance in Kosovo? statistical analysis and real-time predictions. J of conflict resolution 43:538–546
10. Schrodt PA (2000) Forecasting conflict in the Balkans using hidden markov models. http://web.ku.edu/ keds/papers.dir/KEDS.APSA00.pdf, Accessed 2 March 2012
11. Schrodt PA, Gerner DJ (1997) Empirical indicators of crisis phase in the Middle East, 1979–1995. J of conflict resolution 41:529–552
12. Shellman SM (2004) Measuring the intensity of intranational political interactions event data: two interval-like scales. International interactions 30:109–141
13. Shellman SM (2008) Machine coding non-state actors' behavior in intrastate conflict. Political analysis 16:464–77
14. Shellman SM, Hatfield C, Mills M (2010) Disaggregating actors in intrastate conflict. J of peace research 47:83–90
15. Shellman S, Covington M (2009) Automated sentiment analysis. http://www.dtic.mil/cgi-bin/GetTRDoc?AD = ADA532194. Accessed 2 March 2012
16. Ward MD, Gleditsch KS (2002) Location, location, location: an MCMC approach to modeling the spatial context of war and peace. Political analysis 10:244–260
17. Ward MD, Weidmann NB (2010) Predicting conflict in space and time. J of conflict resolution 54:883–901

A Realistic Framework for Counter-terrorism in Multimedia

Durat-ul-Ain Mirza and Nasrullah Memon

1 Introduction

The presence of Web 2.0 has increased the percentage of users accessing video sharing sites. Regardless of its advantages like ease of use and access, speed and security; the technology also manifests some of its emergent plights [1, 2]. Terrorists adopted this technology as an effective means to achieve their goals. Investigations reveal that young terrorists are computer sawy and make use of multimedia to deliver their message to their potential audience [3], for example, the homegrown terrorists and negative radicalization spreading in the United States of America and the Europe. Numerous incidents [3, 4] of such terrorists and extremists, sharing videos and radicalizing text via blogs, forums, articles, emails, etc., for their message proclamation, have occurred.

Due to the controversial understanding of the term, we mention the most common definition of 'terrorist'; "the person or group that is involved in some unlawful, violent or criminal activities for some political or religious ideology" [5]. Nonetheless the means that they adopt for spreading terrorism is also worth noticeable. 'Terrorism' is interpreted as "violent events that are intended to deliver fear and terror to both its near and far victims" [6], and broadcasting videos of catastrophic events covering the agony and sufferings of the nearby victims, terrifies and radicalizes far located masses directly. Terrorists use videos for broadcasting their message of spreading fear, and to some extent, to gain sympathy. In an investigative study [4], authors explain how videos on the internet play a vital role in spreading terrorism. The authors analyzed YouTube videos and the corresponding comments and feedback of viewers, and observed auto-radicalization springing from those extremist videos. While studying viewership tendency, they revealed that the *'footage of suicide bombing'* was largely viewed as compared to other violent

D. Mirza • N. Memon (✉)
The Maersk McKinney Moller Institute University of Southern Denmark, Denmark
e-mail: memon@mmmi.sdu.dk

V.S. Subrahmanian (ed.), *Handbook of Computational Approaches to Counterterrorism*, DOI 10.1007/978-1-4614-5311-6_19,

videos. Several incidents of arrested terrorists later revealed how they got trapped-in through radicalization and extremism after watching violent content on the internet.

Internet security provides some solutions to block and filter inappropriate content (regardless of being terrorist-induced only) and among some popular systems are Net Nanny, BESS (web filtering software operating in majority in the US), I-Gear, CYBER sitter, CYBER Patrol, etc. All of them are textual keyword or URL based scanners, usually with an encrypted or unencrypted (Net Nanny) list of blocked sites [7]. They turn inefficient when inappropriate multimedia content appears which is not captioned or addressed (in the URL) with any of the suspicious textual tags existing in the pre-defined list of these system. This is also obvious for video sharing sites that get updated every minute on an average and avail a silent escape from these systems.

The main challenge remains to scan and search the video databases for inappropriate and particularly violent content triggered by terrorists particularly. Like terrorism, 'violence' is also a broad term and has various aspects; violence with: physical objects, persons or masses, domestic or public, and similar others. However, there are some severe forms of violence which terrorists use and which moves the viewers more intensely than others. We explain these in detail in the next section.

Automatic recognition of violence in the videos is a vast challenge, broadly in the artificial intelligence industry and precisely within the computer vision community. We propose to make use of existing knowledge of video content analysis techniques to develop a framework for detection and recognition of violence found in terrorist videos on the World Wide Web. The system can also serve for filtering violence in all other videos including, random videos available on hundreds of websites, movies, news' segments, documentaries, etc. However, the strategic aim is to counter tactics that terrorists make use of. To the best of our knowledge, no system exists that scans the broad scope of violent content from the videos. As in the context of deployment, the servers of video sharing websites as well as the individuals at the user end can chose to avail from this system to restrict the malicious video content on their systems.

This paper presents some novel contributions. Considering violent videos as the most effective means of terrorists, we first explain the elements or indicators that define 'violence in videos' explicitly. Additionally we categorize these elements into four streams and present a novel, unified framework to detect violence without audio assistance. Furthermore, we briefly compare existing techniques in literature for various functions that participate in the framework. Section 2 defines the term violence in videos and its various dimensions and levels of abstraction. It further explains the handling of semantics in videos in the computer vision industry to give the reader a glimpse of its use in our framework. Lastly, it accounts for existing systems that have been used to detect violence in videos and their limitations. Section 3 details the proposed framework and discusses the role of the four streams participating in the system to block radicalized videos based on violence. Section 4 outlines the performance of the system and concludes the chapter with some future directions in this area of research.

2 Violence in Videos

Violence in videos spans a broad scope with respect to its meaning. Training machines to identify and interpret violence in videos completely is a challenging task. In this regard, some systems exist but they are limited to one aspect of violence while ignoring the other. We elaborate these in following sub-sections.

2.1 Violence Identification in Videos

Violence in videos can be characterized somewhat differently than its actual dictionary definition. The common perception of violence needs to be understood clearly before we proceed towards developing annotations for the concept of violence. Existing systems are based on the discrete definitions of violence assumed by the authors. We discuss the systems in Sect. 2.3.

As a broad term, violence is a concept built upon several objects or entities, their states and actions in videos. It can also be viewed as a 'process' with respect to the dynamics of video sequences. It initiates with some indicators that lead to some subsequent process involving actions and events. To better understand the concept, we analyzed a database of videos that contained violence and observed the various contexts of violence found in these videos. The database was collected from online videos of violent events taken place around the globe and several presentation videos (available on the internet) prepared by extremists showing violent content. These videos were observed to sprout negative radicalization that we learnt from the feedback in the corresponding comments. We also added some movies that are age-restricted due to violence, to this database. After careful examination of the violent scenes in this dataset, we were able to classify violent scenes into several categories tabulated in Table 1. We divide the entities with respect to the video dynamics; image indicators and video sequence entities. The image indicators and their corresponding states are shown. For automatic recognition systems, the measurement metrics for these indicators are also mentioned herein. The video sequence entities are related to actions and are discussed in Sect. 3.1. However, the entity definition is related to the levels (discussed below) and to a greater category of human action recognition.

We further observed the perception levels of violent scenes with respect to actions and events. Coherent to the perception at different age groups and levels of nerves to digest it, we recorded the levels of violence from a deep analysis of the responses of the NC-17[1] and R-rated[2] movies and from the comments on violent internet videos[3].

[1]No children under 17 (NC-17), www.en.wikipedia.org/wiki/List_of_NC-17_rated_films.

[2]Movies that are restricted for children or require parental check/guideline.

[3]Many of these videos were later removed from the video sharing sites after user complaints and flagging.

Table 1 Classification of objects, states and actions found in violent scenes

Dynamics	Entities			
2D Image Indicators	Blood	Fire	Skin	Facial expression
Measure	Ratio	Volume	Bruises Presence of face	Without smile
State of the indicators	Person in a hitting state, dead or injured state	Dark/light	Age and gender Man/woman/child	Horror, surprise, crying
Video sequence	Levels of violence Various approaches of human action recognition.			

Table 2 Levels of violence for various actions and events

Level	Action	Event
Intense	Evisceration (human, dog, etc.)	Explosion, bombing Wounded person in dead state and people crying around
	Sword killing Human burning Beheading, hitting with weapon	Munitions dropping from aircraft and resulting in explosion
	Hitting women, children, old or handicapped people Shooting Torture, extreme pain	Person to person fighting with weapons
Moderate	Person to person fighting without weapons	Military outrage or firing
	Injured persons	Soldiers, gunned men
Mild	Other demeanors	Riots, rebellion, insurgency

Named as Intense, Moderate and Mild, we explain the levels of abstraction, with respect to its affect and inline to age statistics, in Table 2. These actions and events can be overlapping, isolated or static regarding their appearance in the story-line of the video.

Based on the information in Tables 1 and 2, we can define four streams of violence according to the types of indicators, actions and events. Streams (or paths) A, B, C and D serve as the main control structures in the proposed framework. This was very important because in existing systems, indicators and actions are not associated to each other and the idea of having a defined path assures the patterns of events or states followed in development of a violent scene. For example, the most common indicator associated with violence is blood, identified by the red pixels in a video frame, and associated with fast motion as suggested by the authors in [8]. The red pixels could be the red clothing of a sports team in the ground or a girl dancing in a red outfit. They both contain red pixels and fast motion, but are not violent scenes. Similarly, fire detected in a scene followed by screams in video could be a festival with fireworks with music and screams in the background.

Following are the streams of violence; they all may or may not have relation or dependency in a violent scene. The indicators Blood, Fire, Face and Crowd are the fundamental entities; the system traces frames prior to these as well as forthcoming ones to ensure the occurrence of a complete stream. Details of these streams have been discussed in Sect. 3.1, where they appear as the core functions of the proposed framework.

Stream A: Blood → skin → face → unhappy expression?
Stream B: Fire → explosion → smoke → weapon/gunned men? injured/dead person(s)?
Stream C: Face → facial expression → masked? full body → action? weapon?
Stream D: Crowd → faces→expressions →agitating? calmer?

To understand how a semantic concept as violence can be actually detected in the digital videos, we explain the content analysis procedure for videos in the next section.

2.2 Semantics Extraction in Videos

Extracting the semantics from visual data is a critical goal as compared to the self-evident textual semantics; and this is known as the 'semantic gap'. The content analysis of videos functions according to the video dynamics and occurs in the following arrangement:

1. Video content analysis begins with preprocessing the video file in order to transform the input video to some suitable form (changing file formats, containers, compressing/decompressing) that is compatible for efficient performance of the system.
2. Video shots/clips collect to form a video sequence [9] and a video story [10]. Shot boundary detection (SBD) is an important process that identifies the temporal positions where a scene or shot transits into the next one. This task requires highly efficient features for differentiating the frames from each other whenever a shot boundary occurs. The three most common transitions are: the hard or abrupt transition, known as 'cut'; the soft or gradual transition, identified as 'dissolve'; and the special effects, commonly having 'fades' and 'wipes'. SBD algorithms search for the maximum visual discontinuities between two consecutive frames for the hard transition detection and between temporally distant frames for the other types of transitions. Noise produced due to camera motions is also handled therein. The state-of-art in SBD algorithm [11] is an ensemble of features extracted from the fundamental descriptors of video frames (or images) like color, shape, motion, etc. They use machine learning algorithms for training large video sets.

3. Once a video clip has been retrieved, the next step is to figure out the frame (or set of frames) that maximally represent any particular video shot. This process is called keyframe extraction. Generally, two aspects are considered for this task; first is to minimize the redundancy in the visual content within a shot, second is that the keyframe should capture maximum relevant aspects of the visual content. There are several methods to identify the keyframes from video shots. The simplest one is to select the most redundant ones from the beginning and ending frames of a shot [12]. More sophisticated techniques make use of clustering methods [13, 14], by grouping similar visual content based frames into spectral clusters and selecting the most representative frame from each cluster. A video clip can have a single or multiple keyframes according to the complexity of the content. However, multiple keyframes make up a good representation for large motion objects and rapid camera zooming [15].
4. The keyframe, detected to represent a video clip, is used for information retrieval using image mining. It includes segmentation and identification of objects, persons, parts of person like 'pose estimation', 'hand gesture' or 'face recognition' from the 2D keyframe. Moving deeper, 'a pointing finger, a strong fist' and 'a happy, sad or angry face' depict strong emotions supporting the identification of affect in image. The image segmentation algorithms for foreground, background [16], objects and persons identification are utilized. These include detection of edges, corners, blobs, shapes, faces and much more. Most of these make use of colors, shape and textural image descriptors like SIFT, SURF, LESH, HoG,[4] template and pattern matching algorithms. Known-object detection use template matching methods. Over the last few years, the supervised learning methods, especially the bag-of-visual words [17] approach has proven promising results in object recognition.
5. A set of keyframes participate in the action recognition process. Based on motion, the process of action recognition begins with some initial identification of interest points and follows that main person or object movement over the set of keyframes [18]. Early research in action recognition focused simple actions like clapping, walking and jogging and less literature is found for violent and aggressive actions [13]. Those methods used silhouettes based recognition of actions [19] and later the bag of visual code book or visual word histograms appeared for describing an action. During the last decade, TRECVID[5] accelerated the progress in all these areas of video content analysis research. The target objects, actions and relevant parameters combine to detect events, usually assisted with the support of audio and text indicators [20].

---

[4]See www.wikipedia.org for explanation of these terms.

[5]TRECVID is a worldwide competition organized annually by the National Institute of Standards and Technology, USA. It provides large video datasets to participants for performing experiments for various tasks of the video analysis process. It has served as a platform leading to state of the art technologies. See more at: http://trecvid.nist.gov/

6. Once objects and actions have been identified, they both contribute to the process of 'content' modeling. Content depicts the semantics and can be modeled in several ways. Consequently, there are three levels of the semantic concepts hierarchy [10]. The top level is the 'topic' of the video clip; in our case it is an 'action' or 'violent' video. It is called simply the semantics or 'concept' of a video. The middle level is the 'events' built upon the dynamics of the videos' low level features. For instance, an 'action' video clip topic may comprise of events like explosion, fighting, gunshots, etc. They make up the highest semantic concept for the video genres. The third level contains the 'objects' and 'sites or locations'.
7. Finally the semantics get annotated. Early video semantic models were manually annotated; a successful application of text annotation of videos is ANVIL[6] used for video retrieval research. However, it became unsuccessful while handling large video archives. Other notable methods for searching a particular concept include the query methods or code-books. In Query-by-example, the user choses an image as a query and the system returns images found most similar to it. Second, the query-by-sketch method allows the user to draw a rough sketch of the image they want to find [21, 22]. For an automatic system to detect violence, the tedious task is to collect images or sketches for all the entities identified in Table 1.

The process is illustrated in Fig. 1. Preprocessing and shot boundary detection result in a set of keyframes. A single keyframe is used for object detection and a set of keyframes helps extracting the regions around the object using motion analysis. Actions are detected from objects and elements in the keyframes. Events are detected from actions, objects and regions around objects. Finally, the concept is modelled for the entire content. Semantics is associated at each level; with objects, regions around the objects, actions, events and the concepts. Several models have been proposed to reduce the semantic gap at various levels.

For object level semantics, a real-time algorithm called 'Automated Linguistic Indexing of Picture—Real time' (ALIPR) [23] defines the models of image features for each concept by generating a signature of each image in a concept and building a generative model from it.

For action level semantic modeling, bag-of-visual words (BoVW) approach was proposed [17]. The BoW features are constructed by clustering spatio-temporal interest points, followed by assignment of each of these interest points to its closest vocabulary word (cluster) and finally the histogram of video words is computed over a space-time volume describing an action. An interesting bag-of-words model called the 'Semantics Preserving BoW' (SPBoW) model [24] considered the distance between the semantically identical features as a measurement of the semantic gap and learned a codebook for minimizing the semantic gap.

---

[6] Available at: http://www.anvil-software.de

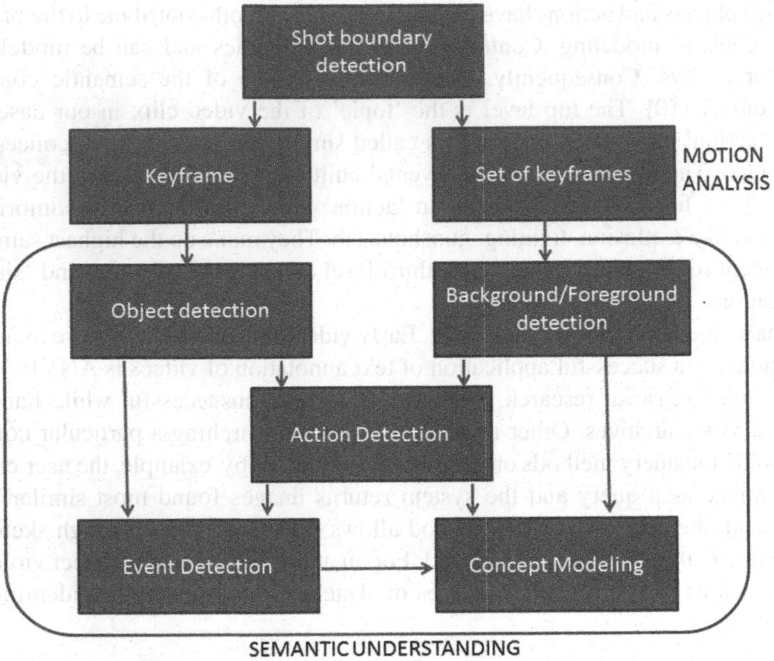

Fig. 1 Video content analysis for extracting semantics. Semantics is based on objects, regions surrounding objects, actions and events

Table 3 Some concepts related to violence from the LSCOM concept collection. They are built for the Afghan battles' news reports [25]

Battles/violence in mountains
Landmines exploding in barren landscapes
Funeral procession of young victims of bombing
Man firing at soldier
Series of explosions in hilly terrain
Masked gunmen
Group of people with pile of weapons
Dead and injured people

However, BoVW faced some problems due to the visual vocabulary construction, therefore a standard vocabulary development project was organized by Smith et al. [25]. Named as Large Scale Concept Ontology or LSCOM, the project focused the news reports and managed to list 449 concepts annotated for the TRECVID 2005 video database. Some of these concepts related to violence are listed in Table 3. These entities appear from the use case of news footage of Afghan battles.

Meanwhile, video event representation recieved new solutions [26]. VERL emerged as a 'video event representation language' and was meant for annotating video content including objects, states and events; further classifying these into types and subtypes. These VERL expressions are based on types and subtypes of events, their temporal relations, with respect to some inference rules and control structures. The semantics of these expressions is obtained by translating them into equivalent first order logic sentences. Video event mark-up language (VEML) is built for specific instances of objects and events detected in video streams for exchange between research groups and analysis techniques. The work was expected to extend and improve but many hidden subtleties appeared behind its failure [26].

Hence, semantics extraction from videos is a challenging procedure and requires good representation of features at all stages. Our purpose to develop a system to detect violence from videos has an underlying structure similar to one described above. However, we first have a look at the existing methods and solutions for violence detection in videos in the following section.

2.3 Existing Methods

Researchers have explored violence detection in videos using single and multi-modalities. The audio assistance in videos accelerates the arousal of emotions and grabs more attention; and is found in both movies and terrorist videos. Violence detection research has been limited due to the lack of a commonly accepted definition of violence in videos and existing methods are incomparable due to this assumption. Section 2.1 gave a clear concept of violence. However, in this section, we discuss the existing methods and their shortcomings in contrast to the streams of violence discussed earlier.

Violence in videos has been detected using either visual data, audio data or both visual and audio data. The audio features for gunshots, explosions and screams can strongly aid in decision making but it also tends to be the computationally expensive and appears to fail in situations where the background sound is not natural or when the audio is not present. Radicalizing violence containing videos have been noticed to usually contain some fairly irrelevant audio. We argue that audio can be misleading despite its supportive role in some situations. Having said this, we discuss the research that has not considered audio in their systems. However interested readers may look into references [27, 28, 29] for audio assisted violence detection approaches.

Early systems in violence detection include the well-known two person fight action detection by Datta and Shah [19]. The authors focused human violence like fist fighting and kicking between two persons. Their method begins with subtraction of background from the persons that are in motion. Silhouettes are fitted for the moving person followed by the identification of neck, shoulder and a head tracking box (which later is used to detect the jerk in the person being hit). This is done using sum of squared differences of color and measuring the acceleration vector for motion. Orientation map for arms and legs is computed next. In case the persons do not point fingers at each other neither do they hit the other, violence is not detected. The method set building block for future action recognition approaches but limits to controlled environment and fails when more than two persons appear or when any one of the persons fall down.

Violence detection using blood and skin identification followed by a motion intensity metric looking for violent scene in images has been reported by Clarin et al. [30]. The system, called DOVE (Detection of Movie Violence using Motion Intensity Analysis on Skin and Blood), is based on a modular approach with first detecting scene boundary by thresholding color RGB histograms; second module detecting skin and blood as indicators from the rest of the scene; the third module tracking the connected component or blob candidates for violence that are related to skin and blood; the last module analyzes the motion intensity describing the violent sequences by very fast motion. The authors calculate the pixel change ratio of skin and blood pixels for the last module. The authors tested their system on American good quality movies and reported long processing duration for their system. Another apparent constraint is the indicator specific association with motion intensity which limits the system to fewer violent scenes detection.

Unrestricting the blood and skin indicators approach and with the improvement in semantic extraction approaches, Souza et al. [31] tested the bag of visual words based on image descriptors like scale invariant feature transform (SIFT) and spatio-temporal interest points (STIP), which is an extended operator from the Harris corner operator in space-time and is identified by high variation of intensity in motion and space. Histograms of oriented gradients and that of optical flow combine to form HNF features that are extracted for 3D patches of video around the STIPs that are detected. The authors describe actions that characterize violent demeanors involving face punch, fast movements of limbs, kicking etc. Considering video shots as an input to STIP and the keyframe as an input to SIFT respectively, visual words are developed from STIP using clustering and local interest points are detected from SIFT. A histogram of these bags of visual words is generated. Linear support vector machines (SVMs) are trained on these representations for classifying violent and non-violent scenes. Since SIFT is a local descriptor it performed significantly poor than STIP which handles 3D structure of video and is suitable for motion. The method was later compared with motion SIFT (MoSIFT) by Bermejo et al. [32]. They also considered violence as fight scenes and developed their own fight dataset from hockey videos. Despite good performance, MoSIFT appeared to be computationally more expensive than STIP because it consists of two portions: a standard SIFT descriptor extracting histograms of oriented gradients and a parallel histogram of optical flows that represent local motion.

Recently, Chen et al. proposed to detect violence in movies defining violence as action scene with a bloody frame [33]. Observing filming rules for action scenes to consist of fast motion and fast edits, they used SVMs to classify action scenes from the video sequences based on fast camera motion and average shot length. Subsequently, they detect the 'bloody frame' from these action scenes based on bright and dark red color pixels, the blood blobs using connected component analysis and face detection using the well-known Viola Jones face detector. An action scene containing atleast one bloody frame is considered to be violent. The system faces the same drawback like other red pixel indicator-based systems.

Generally, these systems were tested on violent movies [30, 33] while some authors collected their own datasets [31] and still others used only surveillance videos that have least camera motion [19]. All the above mentioned systems have the common drawback of assuming a constrained definition of violence. The scope of these systems hence becomes very limited and inefficient for real-time violence detection from large video databases. Table 4 summarizes the shortcomings of these systems that manifest the challenge of a generally accepted definition of violence and particularly those which radicalize greatly.

The shortcomings of existing methods are evident in Table 4; however few important aspects need elaboration. The system [31] depends on the role of motion in detecting violence; motion does not necessarily have to play a role always. Images showing dead persons with heads and arms/legs slaughtered are more violent even without motion. Presentation style videos also tend to have sequence of slides rather than continuous motion. Moreover, the fast motion assumption restricts the system only to movies and action scenes. Our target is not only to identify violent movies or scenes automatically but also to detect violent images that terrorists use for radicalization. The next section presents the proposed framework that provides a complete and high-level solution to these problems.

3 Proposed Methodology

In this section, a novel framework is proposed to serve for the broad definition of violence in videos. As a comprehensive framework, the system is capable of those characteristics that a good violent detection system can have; being fully automatic, real-time, capable of working with video feeds as well as images, without audio aid, able to understand the sequence of violent events and accordingly capable to recognize the intensity of violence.

3.1 A Realistic Framework

The indicators used for violence detection in existing literature are quite useable like blood pixels, face and fire. The proposed framework is large and is built upon several existing systems (especially for these indicators) for which we give a high-level overview, detailing enough to give the reader some perspective about their functionality.

Figure 2 shows the block diagram of the framework. The first module is pre-processing. An incoming video feed is preprocessed for detecting the shot boundaries and keyframes are extracted subsequently using methods explained in Sect. 2.3. Image mining for the indicators begins here and proceeds towards four respective modules according to the violence streams described in Sect. 2.1. The streams are predefined functions that control flows for various sub-functions explained below.

Table 4 Existing violence detection systems and their pros and cons

System	Pros	Cons
Person to person violence detection [19]	Calculate the responding 'jerk' of the beaten person.	Not suitable for more than two persons.
	Analyze violence at the object level.	Fails when a person falls down.
		Violence constrained to human kicking and fist fighting only.
		Used videos from surveillance cameras that do not have large camera motion.
Detection of Movie violence using Motion Intensity Analysis on Skin and Blood (DOVE) [30]	Judge the intensity of the violence.	Use only skin and blood. Optimized for only American movies for their good picture quality.
Violence Detection in Video using Spatio-temporal Features [31]	Did not limit to skin and blood colors, instead used actions and demeanors for this. Compared SIFT and STIP testing violent actions.	Only detected aggressive actions (including fights, kicking etc.) Restricted violence is detected.
		STIP had to outrun since actions were considered for violence only.
		Used most videos from surveillance cameras that do not have large camera motion.
Violence Detection in Video Using Computer Vision Techniques [32]	MoSIFT performed better in identifying the fighting actions.	MoSIFT performed better but was computationally expensive.
	Violence is defined as fighting only.	
		Dataset is from hockey league videos having less camera motion.
Violence detection in Movies [33]	Detection of action scenes based on fast motion and edits, bloody frame detection is important but may work only for fast action movies.	Tested on 4 movies: Gladiator, Passion of Christ, Kill Bill, and First Blood-John Rambo. Testing limited to films with assumption of fast motion and fast edits.
		Does not appear to detect animals (esp. dog) or human's (faceless image) evisceration.
		Dependent on red color pixels for blood, so leads to much misdetection.

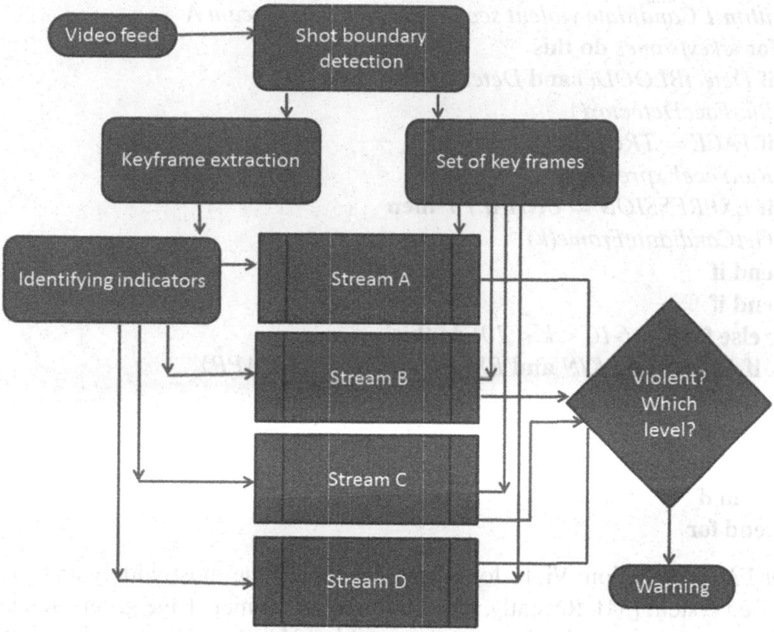

Fig. 2 Flow chart for the entire framework

A set of features appear as an outcome of these modules, which then enters the decision making module that also identifies the level of violence explained in Table 2. For an intense and the moderate level violence, the system generates a WARNING message.

Four primary indicators are searched in a keyframe: BLOOD, FIRE, FACE and CROWD. If any the features is found 'TRUE' the system proceeds to enter one of the corresponding streams. Build upon logical and conditional structures, the streams are the core functions in the framework that serve to analyze the storyline of the scene (discussed later) and relations of the objects and action features. We explain the stream modules below.

Stream A: BLOOD is identified by red pixels and blobs of blood are detected using connected component methods in a frame *k*. The system next searches for SKIN in the same keyframe using method described in [30], and FACE (detection method described below) in frames ranging *(-10 < k < 10)*. If FACE = TRUE the subsequent EXPRESSION is searched. For all true values of BLOOD, SKIN and FACE EXPRESSION = UNHAPPY, the scene is spotted as candidate for VIOLENT scene. We restrict the system to UNHAPPY expression at the moment, however in future we plan to consider other facial expressions as sadness, fear, surprise, anger, disgust. Motion analysis based information can be added to strengthen the decision, otherwise it is not necessary. Algorithm 1 sketches the computational steps required for processing stream A. The feature vector records the true values of blood, skin and face and the candidate frame index.

Algorithm 1 *Candidate violent scene detection with stream A*

for *k keyframes* **do this**
if *DetectBLOOD()* **and** *DetectSKIN()* **then**
RunFaceDetector()
if *FACE = TRUE* **then**
RunFaceExpression()
if *EXPRESSION = UNHAPPY* **then**
GetCandidateFrame(k)
end if
end if
else for *f* = *(-10 < k < 10)* **do this**
if *FACE* **and** *SKIN* **and** *EXPRESSION = UNHAPPY*
GetCandidateFrame(f)
end if
end for
end if
end for

For FACE detection, Viola-Jones face detector is the most widely and successfully used system [34]. Recently, recognition of an estimated age group, gender and mood/expression have also been made possible [35]. Regardless of the results for larger databases and wide range of expressions, our focus is all values of 'mood' that are not 'happy' showing that they can be sad, surprised, crying, disgusting, etc., and a 'smile' value 'false' with a percentage above a threshold. We tested several images with faces from violent videos and the facial expression system generated results similar to those shown in Fig. 3. The picture has poor quality, yet the 'mood' and 'smile' values can be observed. The system is also capable of tracking side view of a face and can be used for detecting multiple faces for crowd detection that is used in stream D.

Stream B: FIRE is identified initially from the color pixels, their optical flow and quasi-periodic behavior in flame behavior from a set of frames in the temporal wavelet transform[7]. The irregular boundary of dancing flame and upward moving smoke motion vectors classify the fire, flame and smoke from the video feed. Subsequently, it can be noticed if some previous frames voted for some rapid upward and outward motion flow indicating an explosion event, then the occurrence of smoke confirms the explosion or bombing. Traditional fire and smoke detectors in videos initially search some interest point and later track these particles. Explosion is simply detected with the occurrence of fast upward motion vectors in less number of frames with a possibility of judder in the scene. Fire and smoke vote for confirmation of a violent scene.

Violent event followed by sufferings have more impact on the viewers. Facial expression detected as crying, shocked or unhappy can appear after some disastrous

[7] http://www.ee.bilkent.edu.tr/~signal/VisiFire/index.html

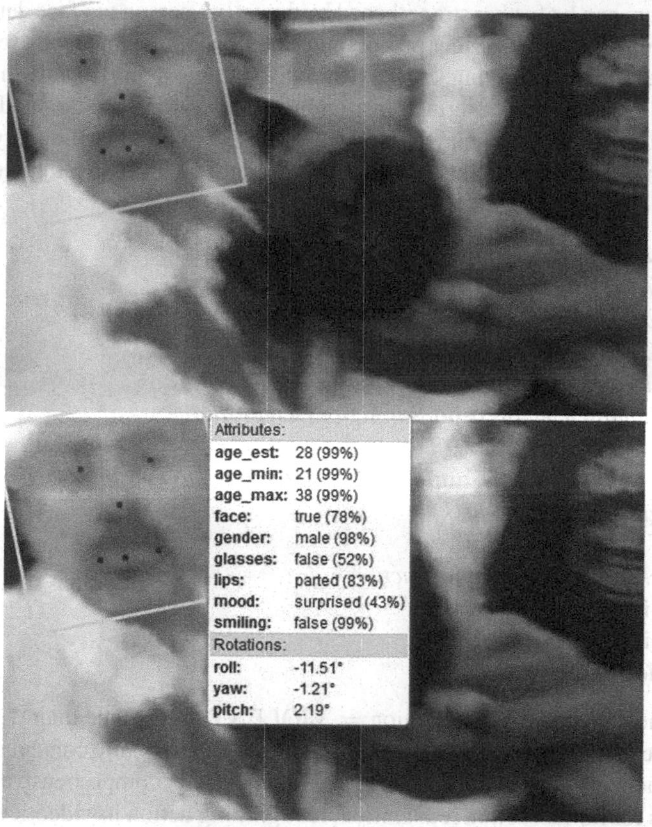

Fig. 3 Face, mood, gender, estimated age range and smile = false (99 %) detected by the facial expression detection system. The picture is of an insurgent event, a man handling a dead person recognized by blood. System used is available online at: http://developers.face.com/tools/faces/detect

event. The stream can verify the violent candidate scene by detecting a face and unhappy expression in latter keyframes of subsequent video shots. Another option is to identify dead persons in later keyframes. Initiating from face again, if the person is motionless, having zero motion vectors in the frame grabber, then s/he is considered dead. Similarly, huge volume of smoke in later frames vote for the candidate scene. Additionally, fast camera motion intensity can aid in stronger decision.

Stream C: As explained earlier, FACE and EXPRESSION are detected from a single keyframe. The AGE and GENDER attributes are also important as minors and women are more vulnerable to gather sympathy of viewers, if the EXPRESSION is found UNHAPPY. The MOOD and SMILE attributes show null values in two cases; either if the face is visible from a side pose or if the face is masked with a veil. In the latter case, two eyes coordinates have some values, instead of NULL.

The stream can be further strengthened by detecting person via face detection and can run into stream C for further processing. The concept behind this stream relates directly to actions like fighting, hitting, with/without weapons. Usually movies have this type of violence, hence termed 'action movies,' comprising the moderate level of violence. Fighting itself may not appear intense violent, but the story followed enhances the effect of violence. Algorithm 2 sketches the main functionality for stream C violence.

***Algorithm 2** Candidate set of frames for stream C violence*

for *f frames* **do this**
RunFaceDetector()
if *FACE* = *TRUE* **and** *Detectedfaces* > *1* **then**
for *all faces* **do this**
RunHumanDetector() **and** *RunActionDetector()*
end for
if *Action* = *VIOLENT* **and** *EXPRESSION(k + 1to k + 5)* = *UNHAPPY* **then**
GetCandidateKeyframe(k)
else
check next shot for violent action
end if
end if
end for

The conditional statement, Action = VIOLENT, is another main function described here. Action recognition is an essential task of automatic content recognition in 3D sequences. According to authors [36] of a very comprehensive survey on approaches for human action recognition, recognizing action in videos comprises of three main steps:

1. Representation extraction—extraction of low level features like color, texture and optical flow,
2. Action modeling—representations are mapped into different action categories, ranging from templates mapping to direct comparison with sophisticated machine learning based modeling schemes, and
3. Action recognition—matching unlabeled video data to the action models to classify the type of actions.

The authors [36] explain action recognition approaches for videos from a single camera; this is suitable for our target actions that include fighting, aggression movements/states of single/multiple persons, beheading, person throwing something at somebody (stones, fire grenade, etc.) or hitting action, static person shooting while targeting something and similar demeanors.

Human actions can be broadly categorized as fast or slow and single or multiple persons. Motion analysis is essential to identify motion intensity and is computed with the motion vectors. Not being necessarily related to human actions, most of these videos contain fast motion and juddering in the scenes. Juddering is an intense camera motion and can be detected using motion vectors ratio between two frames judged with respect to a threshold.

Based on the three steps mentioned above, we follow the part based approach in contrast to the global appearance description for action recognition. Part-based approaches consist of identification of small regions around the spatio-temporal interest points (STIP) for a fraction of the complete action spanning few frames. STIPS are extracted by detecting some corners in a frame first. Small cuboids of regions are extracted around these corners, and clustered to form meaningful groups. For example, a man raising his hand to hit someone, is one state for which the STIPs generate a bag of clustered cuboids, also called the BoVW. Further a histogram from this BoVW along time domain is generated. These BoVWs are labeled for specific action states. For matching to another BoVWs for the same action, clustering will collect the BoVW for the previous action state. Hence, action representation extraction is performed with the corners and STIPs extraction. Action modeling is performed during BOVW and its histogram generation. Finally, matching of these histograms with another one for a maximum matching performance identifies the required action.

The computer vision community has widely worked to test several actions, the STIP based BOVW approach is relatively easy to calculate and train. So we adapt our system according to this approach. A 'batting' action is analogous to the 'beheading'. Both have the same state of representation: a person raises the weapon or bat in his hands, moves it towards the target or ball and hits the object that is in front of it. STIP based BoVW are well suited for this task. Fighting between two persons also follows similar method with a different vocabulary.

We categorize the various actions according to the levels identified in Table 2, to have a good representation for every action or behavior. A sample view of fight scene identified using STIP based BoVW [32] is shown in Fig. 4 below.

Stream D: CROWD can be detected in several ways. Two types of crowds can be observed, a moving crowd and a static crowd. The moving crowd is identified by motion patterns at a slow speed. While the static crowd can be identified using the number of faces in that crowd. If more than p faces appear in a crowd for a longer range of frames, then it is a static crowd. Next the EXPRESSIONS of the detected faces vote for the MOOD of the crowd. The more the number of UNHAPPY faces, the less the chances of a happy crowd which is commonly found in footages of concerts or television reality shows.

In case the crowd scene has no faces and motion is the element of signaling a crowd, then the subsequent scenes are observed for FACEs and their EXPRESSIONs are recorded. This stream of violence is mostly dependent on the storyline discussed below. A CROWD keyframe followed by other violent stream indicates the crowd as agitating, or riots and insurgency found in the video.

Finally, the output feature vectors from these streams, enter the decision box where the sequence of events is computed and evaluated to pass a final decision. The main task of the decision box is to analyze the story-line and degree of violent scene or video respectively. These functions are explained in the following two sections.

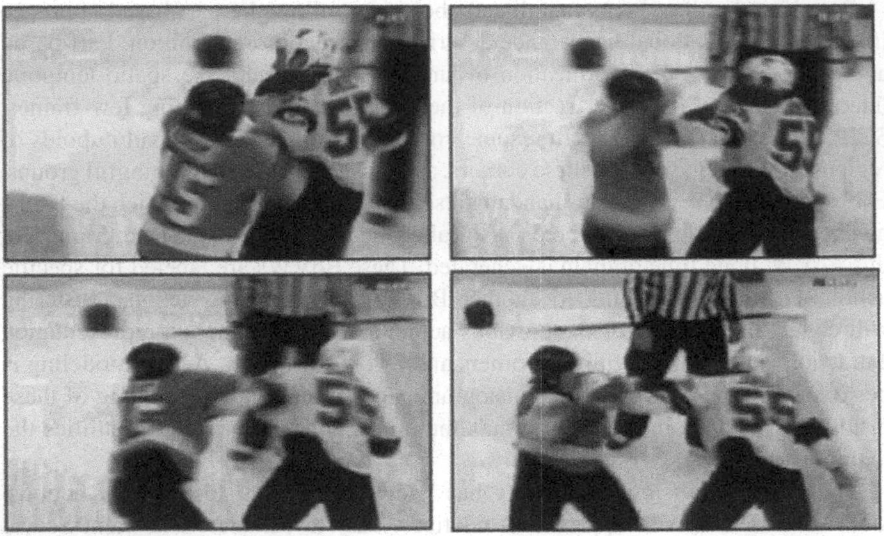

Fig. 4 Sample fight action scene [32]

3.2 Story-Line of Violent Scene

A very sensitive observation of radicalization is related to the sequence of images that appear in videos. After any explosion or fire outrage, fighting or hitting or beheading, one can wisely judge that the scene or the story of this movie was sad or mesmerizing in a negative aspect, sympathizing, heart catching or eventually a threat to social welfare.

Analogous to the scene segmentation in video content analysis, the story-line for violent scenes is translational. Scene segmentation is related to the high level semantic correlation of video content. The methods of video scene segmentation are built upon the principle of 'content coherence' [10]. Boggs and Petrie [37] explain a scene as "a series of shots that communicate a unified action with a common locale and time". Although scene segmentation becomes easier in domain specific segmentation, i.e., scenes can be easily segmented in TV news, sports following the specific structure adapted by such videos [38]. However, this is not the case for terrorism videos [39]. Scene detection in such videos becomes more challenging due to lack of professional video making rules and unstructured story patterns.

An efficient parameter for scene segmentation is the video clip similarity; calculated with respect to time adjacency and visual similarity. Automatic scene segmentation is modeled according to the time information, e.g., a bomb blast clip and another about its aftermath have significant timing information that contributes to the scene segmentation. In the context of visual similarity [14], two consecutive clips (one pass) and the first and third clip (second pass) are usually compared to identify a scene boundary depending on their keyframes. Representation of scenes as variables and their temporal relation as expressions build up an effective story line. Expressions developed for every concept in the video, if collected in a timeline corresponding to the video sequence and shot appearance, can give a fair judgment for violence in the video.

For example, a shot describing stream A as *Sa*, followed by another from Stream B as *Sb*, intervened by multiple non-violent shots can be written as:

$$Sa - T - T - T - Sb - T \qquad (1)$$

where T represents non-violent shot.

3.3 Degree of Violence

A degree of violence in a video is evaluated using a threshold; for m number of total shots in a video feed, and n number of violent shots detected intense and/or moderate, then the degree of violence V_d in the video is calculated as:

$$Vd = \left(\frac{n}{m}\right) * 100 \qquad (2)$$

We can add weights to the expression for the levels of violence with a larger value of α for intense violence and an average value of β for moderate level of violence. Multiplying these weights to the corresponding shots $n_1, n_2, \ldots, n_m$ increase the value V_d for an extremely violent video. For *np,* number of violent shots that were intensely violent and *nq,* number of shots that were moderately violent, the equation for degree of violence V_d is:

$$Vd = \left(\frac{\alpha\,(np) + \beta\,(nq)}{m}\right) * 100 \qquad (3)$$

An important use of measuring the degree of violence is because regular movies do not have a high value of this ratio whereas the terrorists' videos that are meant to attract sympathy contain more violence. News video feed, reporting specifically to some violent event may also have a greater value and gets alert for a precautionary message broadcasting *'Not for Children'*.

4 Discussion

Representaion of a language requires keen details, well developed relations/connections among words which are objects/persons/animals in case of a video/image. The proposed framework is a good representation for violent videos.

Most of the existing appraoches used for building the system are simple and easy to compute; and were hence selected for a reduced processing speed and less computational expense in the framework.

Human analysis of the video datasets lead to the selection of indicators and generation of the corresponding streams. The four streams are interdependent and interelated, removing duplicate computations and memory usage. The functions based on supervised learning methods are promising enough to generate good results. The framework is completely functional and effective. It does not fail in unusual cases; if an indicator is not found, the algorithm searches for the preceeding and proceeding keyframes; secondly, the indicators like face, facial expression and blood are detected in almost every stream ensuring not to miss any detection. Moreover, the story-line generation strengthens the performance by checking the corresponding scene sequence expressions as shown in Eq. 1.

For the first time, we have set blood pixel color ranging from light red to dark red and maroon, to ensure the system does not miss any illuminated blood pixels. Face detector, selected for the framework, is able to handle the noise artifacts successfully; it is incredibly efficient for occluded face images like ones with varying illumination, closed eyes, spectacles and even a masked face.

Despite the powerful techniques, there are some limitations. If the face is covered completely, with even the eyes not visible, then the detector fails. Similarly, the facial expressions have been restricted to check only those that do not lie under the HAPPY category. The other expressions like anger, surprise, sad, crying, etc., are UNHAPPY expressions, but their explicit detection and recognition can improve the performance of the system. Further classification for these expressions can also help to generate new sub-streams for decision making. We adopted the simpler appraoch in order to reduce the computational load.

The current layout of the framework and the streams are highly realistic and effective. However, the overall framework needs to be tested on large video datasets. This would allow a better outlook towards the deficiency in the system and direct towards the areas that need improvement. The future works include these tasks.

5 Conclusion

Extremism, an outcome of radicalization, is a process which changes an individual from passiveness or activism to become more revolutionary, militant or extremist. Radicalization gradually grows from perception after watching violence in videos that leaves a silent, natural atrocity and scarcity of sentiments of civility and humane sentiments. In this chapter, we focused one of the visual means that terrorists use to threaten and/or to deliver their message. The chapter presents a novel framework in a modular fashion, to detect violence in visual data. Regardless of the distinct and restricted definitions of violence, system is designed to accommodate a much broader definition of violence. All modules constituting this framework have compatible representation and comparison features. The system also identifies the 'narrative flow' of violence in addition to the extent/degree of violence in videos that can aid to disproportionate the purpose of terrorists.

Besides the potential of the proposed system, the emerging solutions towards facial expression recognition, action and human behavior recognition remains a promising direction to pursue effectiveness in the current system. As a novel framework, several areas can be further improved and analyzed for future work. Despite the progress made, the workable system is extendable to exercise on much more complex scenes.

References

1. Yariv T, Gabriel W (2002) www.terror.com-Terror on the Internet. Studies in Conflict & Terrorism, 25:317–332
2. Dahl G, Vigna SD (2009) Does movie violence increase violent crime? The Q. J. Econ., MIT Press, 124(2):677–734
3. Using the Web as a Weapon: The Internet as a Tool for Violent Radicalization and Homegrown Terrorism. (2007) (Hearing on Subcommittee on Intelligence, Information Sharing and Terrorism Risk Assessment of the Committee on Homeland Security House of Representatitves, one hundred tenth congress 2007)
4. Conway M, McInerney L (2008) Jihadi video & auto-radicalization: evidence from an exploratory YouTube study. Euro ISI 2008, First European Conference on Intelligent and Security Informatics.
5. Definition of 'Terrorist' taken from www.wikipedia.org
6. The Translational Terrorism, Security and Rule of Law Report (2008) Terrorism and The Media. WorkPackage 4, Deliverable 6. Accessed 25th Jan 2012 http://www.transnationalterrorism.eu/tekst/publications/WP4%20Del%206.pdf
7. List of blocking software: www.peacefire.org/info/blocking-software-faq.html
8. Chen L, Hsu H, Wang L, Su C (2011) Violence detection in movies. Eighth international conference computer graphics, imaging and visualization. IEEE Computer Society, pp 119–124
9. Reisz K, Millar G (1968). The technique of film editing. Focal Press, Oxford UK
10. Hanjalic A (2004) Content based analysis of digital video. Kluwer Academic Publishers, Delft
11. Smeaton AF, Over P. (2010) Video shot boundary detection: seven years of trecvid activity. Computer Vision and Image Understanding 114:411–418
12. Zhang H, Low CY, Smoliar SW, Zhong D (1995) Video parsing, retrieval and browsing: an integrated and content-based solution. Proceedings of the ACM Conference on Multimedia, New York, USA
13. Jain AK, Murty MN, Flynn PJ (1999) Data clustering: a review. ACM Computing Surveys 31(3):264–323
14. Chasanis V, Likas A, Galatsanos N (2007) Scene detection in videos using shot clustering and symbolic sequence segmentation. IEEE Proceedings of Multimedia Signal Processing, Crete, Greece
15. Yuan Y, Pang Y, Pan J, Li X (2009) Scene segmentaion based on IPCA fro visual surveillance. Elsevier Neurocomputing 2450–2454
16. Wang W, Yang J, Gao W (2008) Modeling background and segmenting moving objects from compressed video. IEEE Proceedings on Circuits and Systems for Video Technology, vol. 18, 670–681
17. Wang Y, Xing CX, Zhou L (2006) Video semantic models: survey and evaluation. IJCSNS, vol. 6, 10–20

18. Benetos E, Siatras S, Kotropoulos C, Nikolaidis N, Pitas I (2008) Movie analysis with emphasis to dialogue and action scene detection. Multimedia Processing and Interaction. Springer Science + Business Media, pp 157–177
19. Datta A, Shah M, Lobo NV (2002) Person-on-person violence detection in video data. IEEE 16th International Conference on Pattern Recognition, pp 433–438
20. Bovik A (2000) Handbook of Image and video processing. Academic Publishers, San Diego, CA
21. Ma H, Zhu J, Lyu MR, King I (2007) Bridging the semantic gap between image content and tags. Journal of Latex Files, vol. 12, 1–12
22. Mihling M, Ewerth R, Friesleben B (2009) Improving semantic video retrieval via object-based features. IEEE International Conference in Semantic Computing. IEEE Computer Society, pp 109–115
23. Li J, Wang JZ (2007) Real-time computerized annotation of pictures. Proceedings of the ACM Multimedia Conference, Santa Barbara, CA, October 2006. pp. 911–920
24. Wei L, Hoi SC, Yu N (2010) Semantics-Preserving BoW model and applications. IEEE Transactions on Image Processing, vol. 19, 1908–1920
25. Smith JR (2006) Large-scale concept ontology for multimedia. IEEE Computer Society 86–91
26. Nevatia R, Hobbs J, Bolles B, (2004) An ontology for video event representation. Conference on Computer Vision and Pattern Recognition Workshop, p 119
27. Giannakopoulos T et al (2006) Violence content classification using audio features. Fourth Hellenic conference on Artificial Intelligence, pp 502–507
28. Zajdel W et al (2007) Cassandra: audio-video scene fusion for aggression detection. IEEE Conference on Advanced Video and Signal Based Surveillance, pp 200–205
29. Lin J, Wang W (2009) Weakly supervised violence detection in movies with audio and video based co-training. Lecture Notes in Computer Science 5879:930–935
30. Clarin C et al (2006) DOVE: detection of movie violence using motion intensity analysis on skin and blood. 6th Philippine Computing Science Congress, Computing Society of Phillipines, pp 150–156
31. Souza FD et al (2010) Violence detection using spatio-temporal features. Proceedings SIBGRAPI 10, Proceedings of the 2010 23rd SIBGRAPI Conference on Graphics, Patterns and Images. ISBN: 978-0-7695-4230-0 doi:10.1109/SIBGRAPI.2010.38, pp. 224–230
32. Bermejo E, Deniz O, Bueno G, Sukthankar R (2011) Violent action recognition using computer vision techniques. 14th International Conference on Computer Analysis of Images and Patterns, Seville, Spain
33. Chen L, Hsu H, Wang L, Su C (2011) Violence detection in movies. IEEE Eigth International Conference on Computer Graphics, Imaging and Visualization, pp 119–124
34. Viola P, Jones MJ (2004) Robust real-time face detection. IJCV 57:137–154
35. Vinnay Bettadapura (2012) Face expression recognition and analysis: the state of the art. Tech Report, arXiv:1203.6722
36. Lopes APB, Valle EA Jr, de Almeida JM, de Araújo AA (2010) Action recognition in videos: from motion capture labs to the web. arXiv:1006.3506v1. ACM classes, CVIU, I.4.8; I.4.10
37. Boggs J, Petrie DW (2000) The art of watching films. Mountain View, CA, USA
38. Choros K, Pawlaczyk P. (2010) Content-based scene detection and analysis method for automatic classification of TV sports news. RSCTC, LNAI 6086. Springer-Verlag Berlin Heidlberg, pp 120–129
39. Salem A, Reid E, Chen H (2008) Multimedia content coding and analysis: unraveling the content of jihadi extremist groups' videos. Studies in Conflict and Terrorism, pp 605–626. Routledge Taylor and Francis Group

PROTECT in the Ports of Boston, New York and Beyond: Experiences in Deploying Stackelberg Security Games with Quantal Response

Eric Shieh, Bo An, Rong Yang, Milind Tambe, Craig Baldwin, Joseph DiRenzo, Ben Maule, Garrett Meyer, and Kathryn Moretti

1 Introduction

The global need for security of key infrastructure with limited resources has led to significant interest in research conducted in multiagent systems towards game-theory for real-world security. As reported previously at AAMAS, three applications based on Stackelberg games have been transitioned to real-world deployment. This includes ARMOR, used by the Los Angeles International Airport to randomize checkpoints of roadways and canine patrols [16]; IRIS, which helps the US Federal Air Marshal Service [22] in scheduling air marshals on international flights; and GUARDS [17], which is under evaluation by the US Transportation Security Administration to allocate resources for airport protection. We as a community remain in the early stages of these deployments, and must continue to develop our understanding of core principles of innovative applications of game theory for security.

To this end, this chapter presents a new game-theoretic security application to aid the United States Coast Guard (USCG), called *Port Resilience Operational/Tactical Enforcement to Combat Terrorism* (PROTECT), which originally appeared [20]. The USCG's mission includes maritime security of the US coasts, ports, and inland waterways; a security domain that faces increased risks in the context of threats

E. Shieh (✉) • B. An • R. Yang • M. Tambe
University of Southern California
e-mail: eshieh@usc.edu; boa@usc.edu; yangrong@usc.edu; tambe@usc.edu

C. Baldwin • J. DiRenzo • B. Maule • G. Meyer • K. Moretti
United States Coast Guard
e-mail: craig.w.baldwin@uscg.mil; joseph.direnzo@uscg.mil; ben.j.maule@uscg.mil; garrett.r.meyer@uscg.mil; kathryn.a.moretti@uscg.mil

V.S. Subrahmanian (ed.), *Handbook of Computational Approaches to Counterterrorism*, DOI 10.1007/978-1-4614-5311-6_20,

such as terrorism and drug trafficking. Given a particular port and the variety of critical infrastructure that an adversary may attack within the port, the USCG conducts patrols to protect this infrastructure; however, while the adversary has the opportunity to observe patrol patterns, limited security resources mean that USCG patrols cannot be at every location 24/7. To assist the USCG in allocating its patrolling resources, similar to previous applications [16, 17, 22], PROTECT uses an attacker-defender Stackelberg game framework, with USCG as the defender against terrorist adversaries that conduct surveillance before potentially launching an attack. PROTECT's solution is to typically provide a mixed strategy, i.e. randomized patrol patterns taking into account the importance of different targets, and the adversary's surveillance and anticipated reaction to USCG patrols.

While PROTECT builds on previous work, this chapter highlights five key innovations. The first and most important is PROTECT's departure from the assumption of perfect rationality on the part of the human adversaries. While appropriate in the initial applications (ARMOR, IRIS, GUARDS) this assumption of perfect rationality is well-recognized as a limitation of classical game theory, and bounded rationality has received significant attention in behavioral game-theoretic approaches [4]. Within this behavioral framework, quantal response equilibrium has emerged as a promising approach to model human bounded rationality [4, 14, 24] including recent results illustrating the benefits of the quantal response (QR) model in security games contexts [25]. Therefore, PROTECT uses a novel algorithm called PASAQ [26] based on the QR model of a human adversary. To the best of our knowledge, this is the *first time that the QR model has been used in a real-world security application*.

Second, PROTECT improves PASAQ's efficiency via a compact representation of defender strategies exploiting dominance and equivalence analysis. Experimental results show the significant benefits of this compact representation. Third, PROTECT addresses practical concerns of modeling a real-world maritime patrolling application in a Stackelberg framework. Fourth, this chapter presents a detailed simulation analysis of PROTECT's robustness to uncertainty that may arise in the real-world. For various cases of added uncertainty, the chapter shows that PROTECT's quantal-response-based approach leads to significantly improved robustness when compared to an approach that assumes full attacker rationality.

PROTECT has been in use at the port of Boston since April 2011 and been evaluated by the USCG. This evaluation brings forth our final key contribution: for the first time, this chapter provides real-world data comparing human-generated and game-theoretic schedules. We also provide results from an Adversarial Perspective Team's (APT) analysis and comparison of patrols before and after the use of the PROTECT system from a viewpoint of an attacker. Given the success of PROTECT in Boston, we are now extending it to the port of New York, which is a much larger, more complex security environment. Based on the outcome there, it may potentially be extended to other ports in the US.

Fig. 1 USCG boats patrolling the ports of Boston and New York. (**a**) PROTECT is being used in Boston. (**b**) Extending PROTECT to New York

2 Background

In this section, we explain the framework for Stackelberg security games, for which PROTECT is modeled. Previous deployed security applications that are based on Stackelberg games are then presented.

2.1 Stackelberg Security Game

Recent security games have increasing relied on a Stackelberg framework as Stackelberg games are based on a model of a leader and follower. In a security domain, the defender has a set of targets that must be defended with only a limited number of resources. The attacker is able to observe the strategy of the defender and utilize this information in planning an attack. This fits the description of a Stackelberg game with the defender in the role of the leader and the attacker in the role of the follower. In the security domain, an action, or *pure strategy*, for the defender is an allocation of resources on a set of targets or patrols, e.g., list of checkpoints at the LAX airport or flight schedule for federal air marshals. The pure strategy of an attacker is a target, e.g., an airport terminal or a flight. The strategy for the leader, or defender, is a *mixed strategy*, i.e., a probability distribution over the pure strategies. The strategy for the follower, or attacker, is also a mixed strategy. However, dependent on the model of the attacker, the mixed strategy for the attacker may be equivalent to a pure strategy where a particular pure strategy, or target,

Table 1 Payoff table for example Stackelberg security game

	Defender		Attacker	
Target	Covered	Uncovered	Covered	Uncovered
t_1	5	−3	−4	4
t_2	10	−7	−1	2

is selected with a probability of 1 while all other strategies have a probability of 0. For each target, there exists payoff values that represents the utility for both the attacker and defender in the scenario of either a successful or failed attack.

In a security game, each target has a set of four payoff values: the reward and penalty for both the defender and attacker in the case of a successful or failed attack. These four payoff values are enough for all possible outcomes in the security domain. Table 1 gives an example of a Stackelberg security game composed of two targets, t_1 and t_2, and the four corresponding payoffs for each one. In this example, if the defender chose to cover or protect target t_1 and the attacker attacked target t_1, then the defender would receive a reward of 5 while the attacker would receive a penalty of −4.

An assumption made by security games is that the defender's payoff for covering a target is greater than the payoff for leaving a target uncovered, and that the attacker's payoff for attacking an uncovered target is greater than the payoff for attacking a target that is covered. This assumption holds in the real-world. Another aspect of security games is that the payoff of a scenario is dependent only on the target that is attacked and whether or not it is covered by the defender [10]. The payoff does not rely on the defender coverages on the targets that are not attacked. For example, if the attacker decides to attack target t_1, whether or not the defender covers target t_2 does not impact the final payoff value. This aids in computing the defender's optimal strategy because many resource allocations of the defender are identical.

2.2 Deployed Security Applications

The Stackelberg security game framework has been used and in the following deployed security applications: ARMOR [16], IRIS [22], and GUARDS [17]. The ARMOR system (Assistant for Randomized Monitoring over Routes) has been at use in the Los Angeles International Airport (LAX) since 2007 in scheduling checkpoints on inbound roads into the airport and patrol routes for bomb-sniffing canine units in the eight terminals of LAX. The terminals in LAX have various characteristics such as passenger population, physical size, and international or domestic flights. This impacts the risk assessment and subsequently the payoffs for each individual terminal. The LAX personnel are limited in both the human resources available to conduct checkpoints and the number of canine units. ARMOR is used to optimally allocate the resources in such a way as to improve effectiveness of the checkpoints and canine patrols while also avoiding patterns that an adversary may exploit. It uses a Bayesian Stackelberg solver known as DOBSS (Decomposed Optimal Bayesian Stackelberg Solver) [15] to compute the optimal mixed strategy for both where to set up checkpoints and the terminals for the canine units.

A second application is called IRIS (Intelligent Randomization In Scheduling) and has been used by the US Federal Air Marshals Service (FAMS) since 2009 in scheduling air marshals to flights that originate in and depart from the United States. The purpose of the air marshals is to both dissuade potential adversaries and prevent possible attacks. Flights have different values based on various factors such as thc number of passengers, the source/destination countries and cities, and special events that can impact the risk for certain flights. IRIS is used to schedule tours of flights that not only obey different constraints (e.g., boarding time or flight time), but also must take into account the limited number of air marshals and varying values of each flight. Due to the exponential increase in the possible number of schedules based on the number of flights and resources, DOBSS was not applicable and thus the faster ASPEN [8] solver was used to generate the thousands of flights per day.

The third security application that is based on the Stackelberg security game framework is GUARDS (Game-theoretic Unpredictable and Randomly Deployed Security). GUARDS is used by the United States Transportation Security Administration (TSA) in scheduling resources to conduct various security activities. While many people are aware of common security activities, such as passenger screening, this is just one of many security layers TSA personnel implement to help prevent potential threats. These layers can involve hundreds of heterogeneous security activities executed by limited TSA personnel leading to a complex resource allocation challenge where the TSA is unable to run every security activity all the time [17]. To address the many heterogeneous security activities, GUARDS created a new game-theoretic framework that takes into account heterogeneous defender activities along with a compact modeling of the large number of threats.

3 USCG and PROTECT's Goals

The USCG continues to face challenges with evolving asymmetric threats within the maritime environment not only within the maritime global commons, but also within the ports and waterways that make up the United States Maritime Transportation System. The former Director of National Intelligence, Dennis Blair noted in 2010 a persistent threat "from al-Qa'ida and potentially others who share its anti-Western ideology. A major terrorist attack may emanate from either outside or inside the United States" [3]. This threat was reinforced in May of 2011 following the raid on Osama Bin Laden's home, where a large trove of material was uncovered, including plans to attack an oil tanker. "There is an indication of intent, with operatives seeking the size and construction of tankers, and concluding it's best to blow them up from the inside because of the strength of their hulls" [6]. These oil tankers transit the

U.S. Maritime Transportation System. The USCG plays a key role in the security of this system and the protection of seaports to support the economy, environment, and way of life in the US.

Coupled with challenging economic times, USCG must operate as effectively as possible, achieving maximum benefit from every hour spent on patrol. As a result, USCG is compelled to re-examine the role that optimization of security resource usage plays in its mission planning—and how innovations provided by game theory can be effectively employed.

The goal of PROTECT is to use game theory to assist the USCG in maximizing its effectiveness in its Ports, Waterways, and Coastal Security (PWCS) Mission. PWCS patrols are focused on protecting critical infrastructure; without the resources to provide 100 % on-scene presence at any, let alone all, parts of the critical infrastructure, optimization of security resource is critical. Towards that end, unpredictability creates situations of uncertainty for an enemy and can be enough to deem a target less appealing.

The PROTECT system, focused on the PWCS patrols, addresses how the USCG should optimally patrol critical infrastructure in a port to maximize protection, knowing that the adversary may conduct surveillance and then launch an attack. While randomizing patrol patterns is key, PROTECT also addresses the fact that the targets are of unequal value, understanding that the adversary will adapt to whatever patrol patterns USCG conducts. The output of PROTECT is a schedule of patrols which includes when the patrols are to begin, what critical infrastructure to visit for each patrol, and what activities to perform at each critical infrastructure. While initially pilot-tested in the port of Boston, the solution technique was intended to be generalizable and applicable to other ports.

4 Key Innovations in PROTECT

The PWCS patrol problem was modeled as a leader-follower (or attacker-defender) Stackelberg game [7] with USCG as the leader (defender) and the terrorist adversaries in the role of the follower (attacker). The choice of this framework was supported by prior successful applications of Stackelberg games [21]. In this Stackelberg game framework, the defender commits to a mixed (randomized) strategy of patrols, whereas the attacker conducts surveillance of these mixed strategies and responds with a pure strategy of an attack on a target. The objective of this framework is to find the optimal mixed strategy for the defender.

Stackelberg games are well established in the multi-agent systems literature [5, 11, 13, 21]. The last several years have witnessed the successful application of multi-agent systems in allocating limited resources to protect critical infrastructures [2, 9, 12, 17]. The framework of game-theory, and more specifically, of Stackelberg games, is well suited to formulate the strategic interaction in security domains in which there are usually two players: the security force (defender) commits to a security policy first and the attacker (e.g., terrorist, poacher and smuggler) conducts surveillance to learn the policy and then takes his best attacking action.[1] Stackelberg games have been widely used for modeling/reasoning complex security problems and a variety of algorithms have been proposed to efficiently compute the equilibrium strategy, i.e., the defender's best way of utilizing her limited security resources. (There is actually a special class of Stackelberg games that often gets used in these security domains, and this class is referred to as security games.) In the rest of this section, we describe the application of the Stackelberg game framework in multiple significant security domains.

We will now discuss three of PROTECT's key innovations over previous such works. We begin by discussing how to practically cast this real-world maritime patrolling problem of PWCS patrols as a Stackelberg game (Sect. 4.1). We also show how to reduce the number of defender strategies (Sect. 4.2) before addressing the most important of the innovations in PROTECT: its use of the quantal response model (Sect. 4.3).

4.1 Game Modeling

To model the USCG patrolling domain as a Stackelberg game, we need to define (i) the set of attacker strategies, (ii) the set of defender strategies, and (iii) the payoff function. These strategies and payoffs center on the targets in a port, for example pieces of critical infrastructure. In our Stackelberg game formulation, the attacker conducts surveillance on the mixed strategies that the defender has committed to, and can then launch an attack. Thus, the attacks an attacker can launch on different possible targets are considered as his/her pure strategies.

However, the definition of defender strategies is not as straightforward. Patrols last for some fixed duration during the day as specified by USCG, e.g. 4 h. Our first attempt was to model each target as a node in a graph and allow patrol paths to go from each individual target to (almost all) other targets in the port, generating an almost complete graph on the targets. This method yields the most flexible set of patrol routes that would fit within the maximum duration, covering any permutation of targets within a single patrol. This method unfortunately faced significant challenges: (i) it required determining the travel time for a patrol boat for each pair of targets, a daunting knowledge acquisition task given the hundreds of pairs of targets; (ii) it did not maximize the use of port geography whereby boat crews could observe multiple targets at once and; (iii) it was perceived as micromanaging the activities of the USCG boat crews, which was undesirable.

Our improved approach to generating defender strategies therefore grouped nearby targets into patrol areas. The presence of patrol areas led the USCG to redefine the set of defensive activities to be performed on patrol areas to provide a more accurate and expressive model of the patrols. Activities that take a longer time

[1]Or the attacker may be sufficiently deterred and dissuaded from attacking the protected target.

Table 2 Portion of a simplified example of a game matrix

Patrol Schedule	Target 1	Target 2	Target 3	Target 4
(1:k_1), (2:k_1), (1:k_1)	50,−50	30,−30	15,−15	−20,20
(1:k_2), (2:k_1), (1:k_1)	100,−100	60,−60	15,−15	−20,20
(1:k_1), (2:k_1), (1:k_2)	100,−100	60,−60	15,−15	−20,20
(1:k_2), (2:k_1), (1:k_2)	100,−100	60,−60	15,−15	−20,20
(1:k_1), (3:k_1), (2:k_1), (1:k_1)	50,−50	30,−30	15,−15	10,−10
(1:k_1), (2:k_1), (3:k_1), (1:k_1)	50,−50	30,−30	15,−15	10,−10

provide the defender a higher payoff compared to activities that take a shorter time to complete. This impacts the final patrol schedule as one patrol may visit fewer areas but conduct longer-duration defensive activities at the areas, while another patrol may have more areas with shorter-duration activities.

To generate all the permutations of patrol schedules, a graph $\mathcal{G} = (\mathcal{V}, \mathcal{E})$ is created with the patrol areas as vertices $\mathcal{V}$ and adjacent patrol areas as edges $\mathcal{E}$. Using the graph of patrol areas, PROTECT generates all possible patrol schedules, each of which is a closed walk of $\mathcal{G}$ that starts and ends at the patrol area $b \in \mathcal{V}$, the base patrol area for the USCG. The patrol schedules are a sequence of patrol areas and associated defensive activities, and are constrained by a maximum patrol time τ.

The graph $\mathcal{G}$ along with the constraints b and τ are used to generate the defender strategies (patrol schedules). Given each patrol schedule, the total patrol schedule time is calculated (this also includes traversal time between areas, but we ignore this for expository purposes); we then verify that the total time is less than or equal to the maximum patrol time τ. After generating all possible patrol schedules, a game is formed where the set of defender strategies is composed of patrol schedules and the set of attacker strategies is the set of targets. The attacker's strategy was based on targets instead of patrol areas because an attacker will choose to attack a single target.

Table 2 gives an example, where the rows correspond to the defender's strategies and the columns correspond to the attacker's strategies. In this example, there are two possible defensive activities, activity k_1 and k_2, where k_2 provides a higher payoff for the defender than k_1. Suppose that the time bound disallows more than two k_2 activities (given the time required for k_2) within a patrol. Patrol area 1 has two targets (numbered 1 and 2) while patrol areas 2 and 3 each have one target (numbered 3 and 4, respectively). In the table, a patrol schedule is composed of a sequence of patrol areas and a defensive activity in each area. The patrol schedules are ordered so that the first patrol area in the schedule denotes which patrol area the defender needs to visit first. In this example, patrol area 1 is the base patrol area, and all of the patrol schedules begin and end at patrol area 1. For example, the patrol schedule in row 2 first visits patrol area 1 with activity k_2, then travels to patrol area 2 with activity k_1, and finally returns to patrol area 1 with activity k_1. For the payoffs, if a target i is the attacker's choice and is also part of a patrol schedule, then the defender would gain a reward R_i^d while the attacker would receive a penalty P_i^a, else the defender would receive a penalty P_i^d and the attacker would gain a reward R_i^a. Furthermore, let G_{ij}^d be the payoff for the defender if the defender chooses patrol j and the attacker chooses to attack target i. G_{ij}^d can be represented as a linear combination of the defender reward/penalty on target i and A_{ij}, the effectiveness probability of the defensive activity performed on target i for patrol j, as described by Eq. 1. The value of A_{ij} is 0 if target i is not in patrol j.

$$G_{ij}^d = A_{ij} R_i^d + (1 - A_{ij}) P_i^d \tag{1}$$

For instance, suppose target 1 is covered using k_1 in strategy 5, and the value of A_{15} is 0.5. If $R_1^d = 150$ and $P_1^d = -50$, then $G_{15}^d = 0.5(150) + (1-0.5)(-50) = 50$. ($G_{ij}^a$ would be computed in a similar fashion.) If a target is visited multiple times with different activities, only the highest quality activity is considered.

In the USCG problem, rewards and penalties are based on an analysis completed by a contracted company of risk analysts that looked at the targets in the port of Boston and assigned corresponding values for each one. The types of factors taken into consideration for generating these values include economic damage and injury/loss of life. Meanwhile, the effectiveness probability values A_{ij} for different defensive activities are decided based on the duration of the activities. Longer activities lead to a higher probability of capturing the attackers. While Table 2 shows a zero-sum game, the algorithm used by PROTECT is *not limited to a zero-sum game*; the actual payoff values are determined by the USCG.

4.2 Compact Representation

In our game, the number of defender strategies, i.e. patrol schedules, grows combinatorially, generating a scale-up challenge. To achieve scale-up, PROTECT uses a compact representation of the patrol schedules using two ideas: (i) combining equivalent patrol schedules, and (ii) removal of dominated patrol schedules.

With respect to equivalence, different permutations of patrol schedules provide identical payoff results. Furthermore, if an area is visited multiple times with different activities in a schedule, only the activity that provides the defender the highest payoff requires attention. Therefore, many patrol schedules are equivalent if the set of patrol areas visited and defensive activities in the schedules are the same, even if their order differs. Such equivalent patrol schedules are combined into a single compact defender strategy, represented as a set of patrol areas and defensive activities (and omitting any ordering information). Table 3 presents a compact version of Table 2, with the game matrix simplified using equivalence. For example, the patrol schedules in the rows 2–4 from Table 2 are represented as a compact strategy $\Gamma_2 = \{(1,k_2), (2,k_1)\}$ in Table 3.

Next, the idea of dominance is illustrated in Table 3. Note the difference between Γ_1 and Γ_2. Since activity k_2 gives the defender a higher payoff than k_1, Γ_1

Table 3 Example compact strategies and game matrix

Compact Strategy	Target 1	Target 2	Target 3	Target 4
$\Gamma_1 = \{(1 : k_1), (2 : k_1)\}$	50,−50	30,−30	15,−15	−20,20
$\Gamma_2 = \{(1 : k_2), (2 : k_1)\}$	100,−100	60,−60	15,−15	−20,20
$\Gamma_3 = \{(1 : k_1), (2 : k_1), (3 : k_1)\}$	50,−50	30,−30	15,−15	10,−10

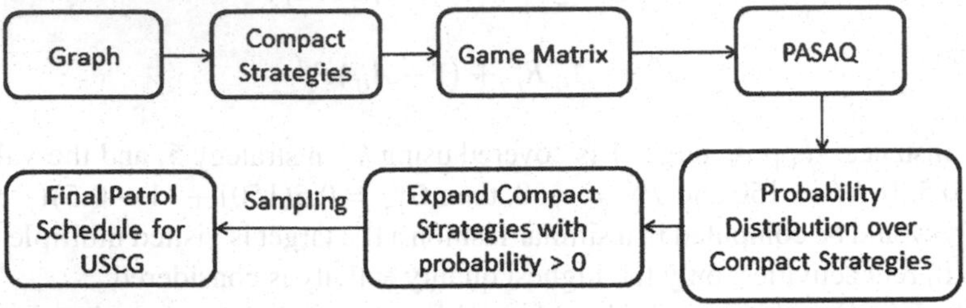

Fig. 2 Flow chart of the PROTECT system

can be removed from the set of defender strategies because Γ_2 covers the same patrol areas while giving a higher payoff for patrol area 1. To generate the set of compact defender strategies, a naive approach would be to first generate the full set of patrol schedules and then prune the dominated and equivalent schedules. Instead, PROTECT uses three ideas to quickly compute the compact strategies: (i) computation of a starting point for compact strategy generation; (ii) computation of a stopping point and; (iii) verification of feasibility in compact strategies.

While generating compact strategies, we first generate compact strategies containing $\hat{n}$ patrol areas, then $\hat{n} - 1$ patrol areas and so on down to $\check{n}$ patrol areas. $\hat{n}$ is called the starting point and is defined as τ/ρ where τ is the maximum patrol time and ρ shortest duration of a defensive activity. The maximum number of areas in any compact strategy must be less than or equal to $\hat{n}$. For example, if there are 20 patrol areas, $\tau = 100$ min and $\rho = 10$ min, then the algorithm will start by generating compact strategies with ten patrol areas. It must be verified that a feasible patrol schedule can be formed from each compact strategy. This is achieved by constructing the shortest patrol schedule that is equivalent to the compact strategy, and comparing the patrol travel time against τ.

Let $S(n)$ represent all the compact strategies that contain n patrol areas. If $S(\check{n})$ contains all the compact strategies that are covered with the highest quality defensive activity at each patrol area, the process of generating compact strategies will terminate and $\check{n}$ is called the stopping point of enumeration. Any compact strategy that contains fewer than $\check{n}$ patrol areas will be dominated by a compact strategy in $S(\check{n})$.

Figure 2 shows a high-level view of the steps of the algorithm using the compact representation. The compact strategies are used instead of full patrol schedules to generate the game matrix. Once the optimal probability distribution is calculated (as explained in Sect. 4.3) for the compact strategies, the strategies with nonzero probability are expanded to a complete set of patrol schedules.

In this expansion from a compact strategy to a full set of patrol schedules, we need to determine the probability of choosing each patrol schedule, since a compact strategy may correspond to multiple patrol schedules. The focus here is on increasing the difficulty for the attacker of conducting surveillance by increasing unpredictability,[2] which we achieve by randomizing uniformly over all expansions of the compact defender strategies. The uniform distribution provides the maximum entropy (greatest unpredictability). Thus, all the patrol schedules generated from a single compact strategy are assigned a probability of v_i/w_i where v_i is the probability of choosing a compact strategy Γ_i and w_i is the total number of expanded patrol schedules for Γ_i. The complete set of patrol schedules and the associated probabilities are then sampled and provided to the USCG, along with the start time of the patrol generated via uniform random sampling.

4.3 *Human Adversary Modeling*

While previous game-theoretic security applications have assumed a perfectly rational attacker, PROTECT takes a step forward by addressing this limitation of classical game theory. Instead, PROTECT permits a boundedly rational adversary, using a quantal response (QR) adversary reasoning model, which has shown to be a promising model of human decision-making [14, 18, 25]. A recent study demonstrated the use of QR as an effective prediction model of humans [24]. An even more relevant study of the QR model was conducted by Yang et al. [25] in the context of security games where this model was shown to outperform competitors in modeling human subjects. Based on this evidence, PROTECT uses a QR model of a human adversary. (Aided by a software assistant, the defender still computes the optimal mixed strategy.)

The QR model adapts ideas from the literature which presumes that humans will choose better actions at a higher frequency, but with noise added to the decision-making process following a logit distribution:

$$q_i = \frac{e^{\lambda G_i^a(x_i)}}{\sum_{j=1}^{T} e^{\lambda G_j^a(x_i)}} \tag{2}$$

The parameter $\lambda \in [0, \infty]$ represents the amount of noise in the attacker's strategy, with a value of 0 indicating a uniform random probability over attacker strategies and a value of ∞ indicating a perfectly rational attacker. q_i is the probability that the attacker chooses target i; $G_i^a(x_i)$ is the attacker's expected utility of attacking target i given x_i, the probability that the defender covers target i; and T is the total number of targets.

[2]Creating optimal Stackelberg defender strategies that increase the attacker's difficulty of surveillance is an open research issue in the literature; here we choose to maximize unpredictability as the first step.

Table 4 PASAQ notation as applied to PROTECT

t_i	Target i
R_i^d	Defender reward for covering t_i if attacked
P_i^d	defender penalty for not covering t_i if attacked
R_i^a	Attacker reward for attacking t_i if not covered
P_i^a	Attacker penalty for attacking t_i if covered
A_{ij}	Effectiveness probability of compact strategy Γ_j on t_i
a_j	Probability of choosing compact strategy Γ_j
J	Total number of compact strategies
x_i	Marginal coverage on t_i

To apply the QR model within a Stackelberg framework, PROTECT employs an algorithm known as PASAQ [26]. PASAQ computes the optimal defender strategy (within a guaranteed error bound) given a QR model of the adversary by solving the following nonlinear, non-convex optimization problem P (see Table 4 for notation):

$$P : \begin{cases} \max\limits_{x,a} \dfrac{\sum_{i=1}^{T} e^{\lambda(R_i^a-(R_i^a-P_i^a)x_i)}((R_i^d - P_i^d)x_i + P_i^d)}{\sum_{i=1}^{T} e^{\lambda(R_i^a-(R_i^a-P_i^a)x_i)}} \\ x_i = \sum\limits_{j=1}^{J} a_j A_{ij}, \quad \forall i \\ \sum\limits_{j=1}^{J} a_j = 1 \\ 0 \le a_j \le 1, \quad \forall j \end{cases}$$

The first line of the problem corresponds to the computation of the defender's expected utility resulting from a combination of Eqs. 1 and 2. Unlike previous applications [11, 21], x_i in this case not just summarizes presence or absence on a target, but also the effectiveness probability A_{ij} on the target as well.

The key idea in PASAQ is to use a piecewise linear function to approximate the nonlinear objective function appearing in formulation P, and thus convert it into a Mixed-Integer Linear Programming (MILP) problem, which can then be solved in a reasonable amount of time. Such a problem can easily include assignment constraints giving an approximate solution for a Stackelberg game against a QR-adversary with assignment constraints, as is the case in this setting. See [26] for details.

As with all QR models, a value for λ is needed to represent the noise in the attacker's strategy. Based on discussions with USCG experts about the attacker's behavior, a λ value of 0 (uniform random) and ∞ (fully rational) were ruled out. Under the payoff data for Boston, an attacker's strategy with $\lambda = 4$ starts approaching a fully rational attacker—the probability of attack focuses on a single target. It was determined from the information gathered from the USCG that the attacker's strategy is best modeled with a value $\lambda \in [0.5, 4]$. A discrete sampling approach was used to determine a λ value giving the highest average expected utility across attacker strategies within this range, yielding $\lambda = 1.5$. Selecting an appropriate value for λ remains a complex issue, however, and it is a key agenda item for future work.

5 Evaluation

This section presents evaluations based on (i) experiments completed via simulations and (ii) real-world patrol data along with USCG analysis. All scenarios and experiments, including the payoff values and graph (composed of nine patrol areas), were based on the port of Boston. The defender's payoff values lie in the range $[-10, 5]$ while the attacker's payoff values lie in the range of $[-5, 10]$. The game was modeled as a zero-sum game[3] in which the attacker's loss or gain is balanced precisely by the defender's gain or loss. For PASAQ, the defender's strategy uses $\lambda = 1.5$ as mentioned in Sect. 4.3. All experiments are run on a machine with an Intel Dual Core 1.4 GHz processor and 2 GB of RAM.

5.1 Memory and Run-time Analysis

This section presents the results based on simulation to show the efficiency in memory and run-time of the compact representation versus the full representation (Sect. 4.2). In Fig. 3a, the x and y axes indicate the maximum patrol time allowed and the memory needed to run PROTECT, respectively. In Fig. 3b, the x and y axes indicate the maximum patrol time allowed and the run-time of PROTECT,

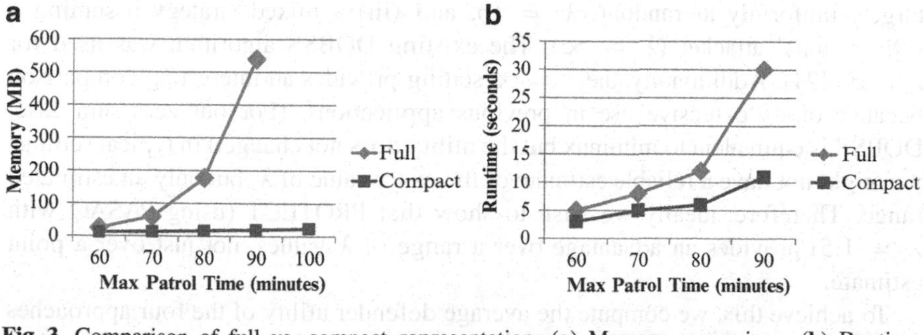

Fig. 3 Comparison of full vs. compact representation. (**a**) Memory comparison. (**b**) Runtime comparison

[3]In general these types of security games are non-zero-sum [21], though for Boston as a first step it was decided to cast the game as zero-sum.

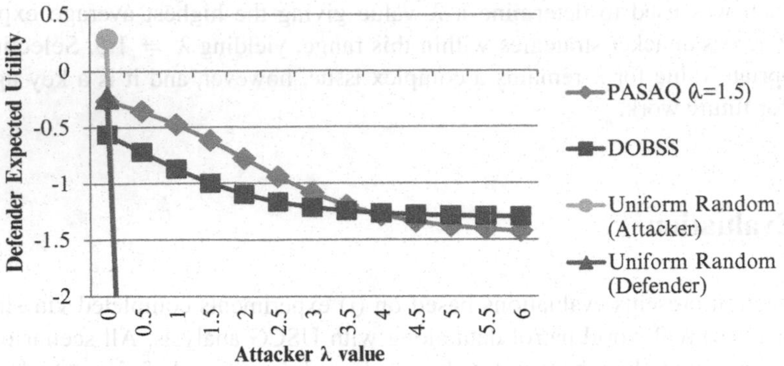

Fig. 4 Defender's expected utility when varying λ for attacker's strategy

respectively. The maximum patrol time allowed determines the number of combinations of patrol areas that can be visited; that is, the x-axis indicates a scale-up in the number of defender strategies. When the maximum patrol time is set to 90 min, the full representation takes 30 s and uses 540 MB of memory while the compact representation takes 11 s to run and requires 20 MB of memory. Due to the exponential increase in the memory and run-time for the full representation, it cannot feasibly be scaled up beyond 90 min in our simulation setting.

5.2 Utility Analysis

Since we are working with real data, it is interesting to understand whether PROTECT using PASAQ with $\lambda = 1.5$ provides an advantage over (i) a uniform random defender's strategy; (ii) a mixed strategy assuming the attacker chooses targets uniformly at random ($\lambda = 0$), and (iii) a mixed strategy assuming a fully rational attacker ($\lambda = \infty$). The existing DOBSS algorithm was used for $\lambda = \infty$ [21]. Additionally, the $\lambda = \infty$ setting provides an interesting comparison because of its extensive use in previous applications. (For our zero-sum case, DOBSS is equivalent to minimax but the utility does not change.) In typical settings we might not have a reliable estimate of the exact value of λ, but only an estimated range. Therefore, ideally we wish to show that PROTECT (using PASAQ with $\lambda = 1.5$) provides an advantage over a range of λ values, not just over a point estimate.

To achieve this, we compute the average defender utility of the four approaches above as λ varies over the range [0, 6], which is a more conservative range than [0.5, 4]. In Fig. 4, the y-axis indicates the defender's expected utility; the x-axis indicates the λ value used for the attacker's strategy. Both uniform random strategies perform well when the attacker's strategy is based on $\lambda = 0$. However, as λ increases, both strategies quickly drop to a very low defender expected utility. In

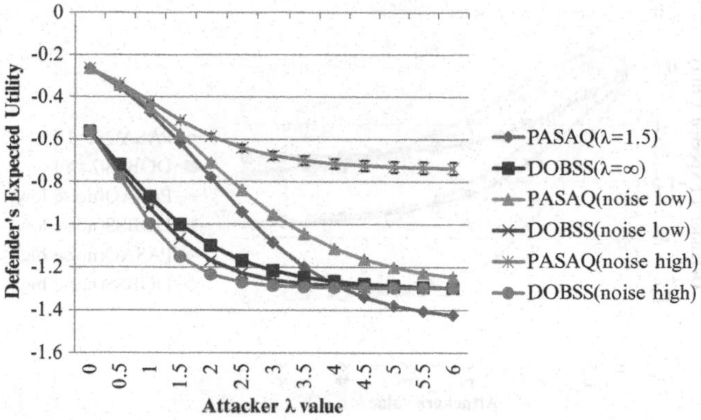

Fig. 5 Defender's expected utility: observation noise

contrast, PASAQ with $\lambda = 1.5$ provides a higher expected utility than that it does assuming a fully rational attacker over a range of attacker λ values (and indeed over the range of interest), not just at $\lambda = 1.5$.

5.3 Robustness Analysis

In the real world, observation, execution, and payoffs are not always perfect due to various causes: noise in the attacker's surveillance of the defender's patrols, the many tasks and responsibilities of the USCG, whose crews may be pulled off a patrol, and limited knowledge of the attacker's payoff values. Our hypothesis is that PASAQ with $\lambda = 1.5$ is more robust to such noise than a defender strategy (such as DOBSS [21]) that assumes full attacker rationality; that is, we believe PASAQ's expected defender utility will be more robust than DOBSS' over the range of attacker λ of interest. This is illustrated by comparing both PASAQ and DOBSS against observation, execution, and payoff noise [11, 13, 27]. (A comparison of the uniform random strategies was omitted due to its poor performance shown in Fig. 4.) All experiments were run generating 200 samples with added noise, averaging over all the samples. In Figs. 5, 6, and 7, the y-axis indicates the defender's expected utility, and the x-axis indicates the attacker's λ value, with error bars depicting the standard error.

The first experiment considers observational noise, which means the attacker has noise associated with observing the defender's patrol strategy, as shown in Fig. 5. In this scenario, if the defender covers a target with probability p, the attacker may perceive the probability to be uniformly distributed in $[p - x, p + x]$ where x is the noise value. The low observation error corresponds to $x = 0.1$, and the high error to $x = 0.2$. Contrary to expectation, we find that observation error leads to an increase in defender expected utility in PASAQ, but a potential decrease (or no change) in

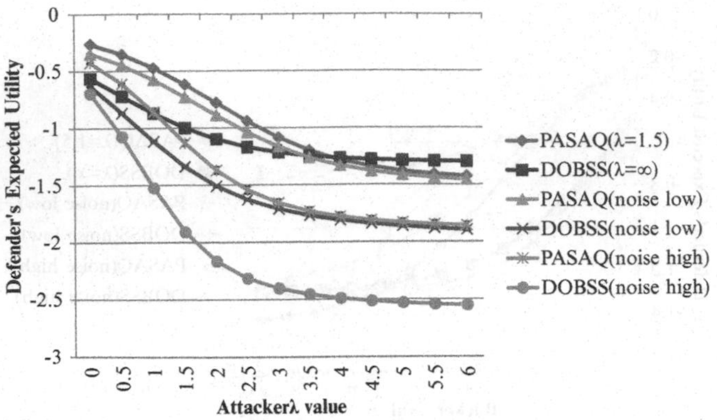

Fig. 6 Defender's expected utility: execution noise

DOBSS. Thus PASAQ ends up dominating DOBSS by a larger margin over bigger ranges of λ values, further strengthening the case for using PASAQ rather than a full-rationality model.

An example illustrates PASAQ's unexpected behavior. Suppose the defender's strategy is **c** and there are two targets t_1 and t_2 with defender expected utilities of $U_1^d(\mathbf{c}) = -2$ and $U_2^d(\mathbf{c}) = -1$ (with the attacker's expected utility $U^a(\mathbf{c})$ being the opposite, since the game is zero-sum). For an attacker strategy with a larger λ, the adversary will choose to attack t_1 and the defender would receive a utility of -2. When observation noise is added, increases in the coverage of t_1 results in decreases in $U_1^a(\mathbf{c}')$, so the attacker might choose to attack t_2 instead, giving the defender a greater utility. If the coverage of t_1 decreases, $U_1^a(\mathbf{c}')$ will increase and the attacker will still choose to attack t_1, but $U_1^d(\mathbf{c})$ will remain the same as when there was no noise.

DOBSS exhibits a different trend because DOBSS minimizes the maximum attacker's expected utility or, in our zero-sum setting, maximizes the minimum defender's expected utility. This results in multiple targets with the same minimum defender's utility, which are referred to as an *attack set* [21]. Typically, when the coverage over the attack set varies due to observation error, some of the targets have less coverage and some have more, but the attacker ends up attacking the targets in the attack set regardless, giving the defender almost no change in its expected utility.

For the second experiment, noise is added to the execution phase of the defender, as shown in Fig. 6. If the defender covered a target with probability p, this probability now changes to be uniformly distributed within $[p - x, p + x]$, where x is the noise value. The low execution error considered is $x = 0.1$, and the high error is $x = 0.2$. The key takeaway here is that execution error leads to PASAQ dominating DOBSS over all tested values of λ, further strengthening the reason to use PASAQ rather than a full-rationality model. When execution error is added, PASAQ dominates DOBSS because the latter seeks to maximize the minimum defender's expected utility, and so multiple targets will have the same minimum

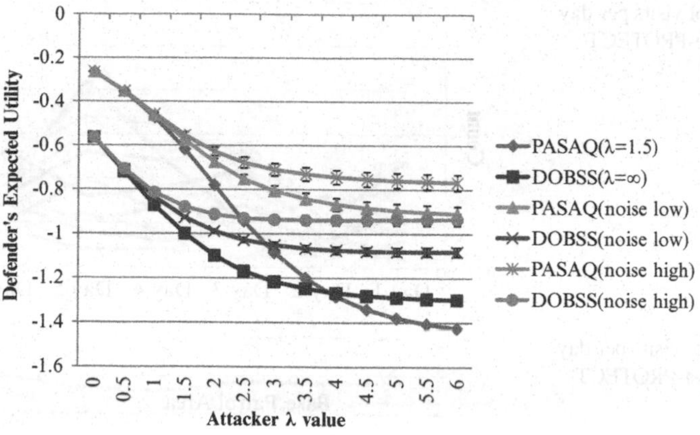

Fig. 7 Defender's expected utility: payoff noise

defender utility. For DOBSS, when execution error is added there is a greater probability that one of these targets will have less coverage, resulting in a lower defender's expected utility. For PASAQ, typically only one target has the minimum defender expected utility. As a result, changes in coverage do not impact it as much as DOBSS. As with observation error, as execution error increases, the advantage of PASAQ over DOBSS for the defender's expected utility grows even greater.

In the third experiment, shown in Fig. 7, payoff noise is added by aggregating mean-0 Gaussian noise to the attacker's original payoff values (similar to [11]). As more noise is added to the payoffs, both defenders' strategies result in an increase in the defender's expected utility because the game is no longer zero-sum. The low payoff noise corresponds to a standard deviation of 1 while a high payoff noise corresponds to a standard deviation of 1.5. Similar to the previous experiments, when payoff noise is added, PASAQ again dominates DOBSS. As noise is added to the attacker's payoff but *not* the defender's payoff, the attacker's strategy may no longer yield the lowest possible defender expected utility. For example, with no payoff noise, target t_1 gives the attacker the highest utility and the defender the lowest utility. When noise is added to the attacker's payoffs, t_1 may no longer give the attacker the highest utility; instead, he/she will choose to attack target t_2, and the defender receives a higher utility than t_1. In essence, with a zero-sum game the defender plans a conservative strategy, based on maximin, and as such any change in the attacker is to the defender's benefit.

5.4 USCG Real-World Evaluation

In addition to the data made available from simulations, the USCG conducted its own real-world evaluation of PROTECT. With permission, some aspects of the evaluation are presented in this chapter.

Fig. 8 Patrol visits per day by area – pre-PROTECT

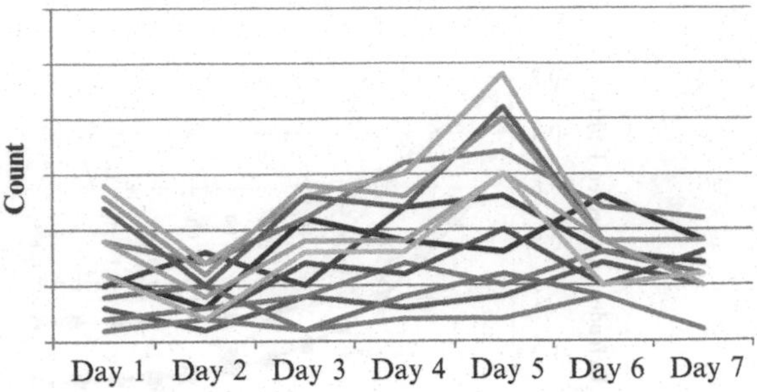

Fig. 9 Patrol visits per day by area – post-PROTECT

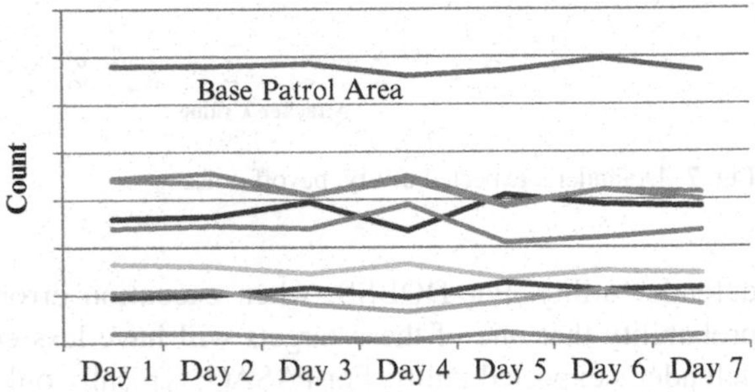

Real-world scheduling data: A key novelty of this chapter is the inclusion of actual data from USCG patrols before and after the deployment of PROTECT at the port of Boston. Only recently has there been real-world data used in game theoretical security applications. A previous application by Shakarian et al. [19] used real data about Improvised Explosive Device (IED) attacks, to predict locations that the adversary may try to hide. Our application and use of real world data is different in that it shows and compares actual patrols conducted by USCG personnel before and after PROTECT.

Figures 8 and 9 present the frequency of visits by USCG to different patrol areas over a number of weeks. The x-axis indicates day of the week, and the y-axis indicates the number of times a patrol area is visited on that day. The y-axis is intentionally blurred for security reasons, since this is real data from Boston. There are more lines in Fig. 8 than in Fig. 9 because during the implementation of PROTECT, new patrol areas were formed, containing more targets and thus fewer patrol areas. Figure 8 depicts a definite pattern in the patrols. While there is a spike in patrols executed on Day 5, there is a dearth of patrols on Day 2. Besides this pattern, the lines in Fig. 8 intersect, indicating that on some days a higher value target was visited more often while on other days it was visited less often, even though the value of a target does not change day-to-day. This means that there was not a consistently high frequency of coverage of higher value targets before PROTECT.

In Fig. 9, we notice that the pattern of low patrols on Day 2 (from Fig. 8) disappears. Furthermore, lines do not frequently intersect, that is, higher-value targets are visited consistently across the week. The top line in Fig. 9 is the base patrol area and is necessarily visited more often than the other patrol areas.

Adversary Perspective Teams (APT): To obtain a better understanding of how the adversary views the potential targets in the port, the USCG created an Adversarial Perspective Team (APT), a mock attacker team. The APT provides assessments from the terrorist perspective and as a secondary function, assesses the effectiveness of the patrol activities before and after deployment of PROTECT. In their evaluation, the APT incorporates the adversary's known intent, capabilities, skills, commitment, resources, and cultural influences. In addition, it screens attack possibilities and assists in identifying the level of deterrence projected at and perceived by the adversary. For the purposes of this research, the adversary is defined as individuals with ties to al-Qa'ida or its affiliates.

The APT conducted a pre- and post-PROTECT assessment of the system's impact on an adversary's deterrence at the port of Boston. This analysis uncovered a positive trend in which the deterrence effectiveness increased between the pre- to post- PROTECT observations.

Additional Real-world Indicators: The use of PROTECT and the APT's improved guidance given to boat crews on how to conduct patrols jointly provided a noticeable increase in the quality and effectiveness of the patrols. Prior to implementing PROTECT, there were no documented reports of illicit activity. After implementation, USCG crews, reported more illicit activities within the port and provided a noticeable "on the water" presence with industry port partners commenting, "the Coast Guard seems to be everywhere, all the time." With no actual increase in the number of resources applied, and therefore no increase in capital or operating costs, these outcomes support the practical application of game theory in the maritime security environment.

5.5 *Outcomes Following the Boston Implementation*

After evaluating the performance and impact of PROTECT at Boston, the USCG viewed this system as a success. As a result, PROTECT is now being deployed in the port of New York, a much larger and more complicated security environment. We were presented with an award for the work on the PROTECT system for Boston Harbor, which reflects the USCG's recognition of PROTECT's impact and value.

6 Lessons Learned: Putting Theory into Practice

The development of the PROTECT model was a collaborative effort involving university researchers and USCG personnel, including decision-makers, planners and operators. Building on the lessons reported in [21] for working with security organizations, we informed the USCG of (i) the assumptions underlying the game-theoretic approaches, e.g. full adversary rationality, and strengths and limitations of different algorithms, rather than pre-selecting a simple heuristic approach; (ii) the need to define and collect correct inputs for model development and; (iii) a fundamental understanding of how the inputs affect the results. We gained new insights on real-world applied research, in particular involving the following issues: (i) unforeseen positive benefits that can occur when security agencies are compelled to reexamine their assumptions; (ii) the requirement to work with multiple teams in a security organization at multiple levels of their hierarchy and; (iii) the necessity of preparing answers to practical end-user questions not always directly related to "meaty" research problems.

The first insight came about when the USCG had to reassess their operational assumptions as a result of working through this research problem. A positive result of this reexamination prompted the USCG to develop new PWCS mission tactics, techniques, and procedures. Through an iterative development process, the USCG reassessed the reasons why boat crews performed certain activities and whether they were sufficient. For example, instead of "covered" vs. "not covered" as the only two possibilities at a patrol point, there are now multiple sets of possible activities for each patrol point.

The second insight is that applied research requires the research team to collaborate with planners and operators on the multiple levels of a security organization to ensure that the model accounts for all aspects of a complex real-world environment. Initially when we started working on PROTECT, the focus was on patrolling each individual target. This appeared to micromanage the activities of boat crews, and it was through their input that individual targets were grouped into patrol areas associated with a PWCS patrol. On the other hand, input from USCG headquarters and the APT mentioned earlier led to other changes in PROTECT, e.g. departing from a fully rational adversary model to a QR model.

The third insight involves the need to develop answers to end-user questions that are not always related to "meaty" research questions but are related to the larger knowledge domain on which the research depends. One example of this involved the user citing that one patrol area was being repeatedly visited, which seemed to suggest the schedules were not genuinely random. After assessing this concern, we determined that the cause of the repeated visits to the patrol area was its high reward, an order of magnitude greater than those of the rarely visited patrol areas. PROTECT correctly assigned patrol schedules that covered the more "important" patrol areas more frequently. In another example, the user noted that PROTECT did not assign any patrols to start at 4:00 AM or 4:00 PM over a 60 day test period. They expected patrols would be scheduled to start at any hour of the day, prompting them to ask if there was a problem with the program. This required us to develop a layman's briefing on probabilities, randomness, and sampling. With 60 patrol schedules, a few start hours may not be chosen given our uniform random sampling of the start time. These practitioner-based issues demonstrate the need for researchers to not only be conversant in the algorithms and mathematics underlying the research, but also be able to explain from a user's perspective why solutions make sense. An inability to address these issues would result in a lack of real-world user confidence in the model.

7 Summary and Related Work

This chapter reports on PROTECT, a game-theoretic system deployed by the USCG in the port of Boston since April 2011 for scheduling their patrols. USCG has deemed the deployment of PROTECT in Boston a success and efforts are underway to deploy PROTECT in the port of New York, and to other ports in the United States. PROTECT uses an attacker-defender Stackelberg game model, and includes five key innovations.

First, PROTECT moves away from the assumption of perfect adversary rationality used in previous work, relying instead on a quantal response (QR) model of the adversary's behavior. While the QR model has been studied extensively in behavioral game theory, to the best of our knowledge this is its first real-world deployment. Second, to improve PROTECT's efficiency, we generate a novel compact representation of the defender's strategy space, exploiting equivalence and dominance. Third, the chapter shows how to practically model a real-world (maritime) patrolling problem as a Stackelberg game. Fourth, we provide experimental results illustrating that PROTECT's QR model of the adversary is better able to handle real-world uncertainties than a perfect rationality model can. Finally, for the first time in a security application evaluation, we use real-world data, providing (i) a comparison of human-generated security schedules and those generated via a game-theoretic algorithm, and (ii) results from an APT's analysis of the impact of the PROTECT system. We also outlined insights gained from the project, which include the ancillary benefits due to a review of assumptions made by security agencies and the importance of answering questions not directly related to the research problem.

As a result, PROTECT has advanced the state of the art beyond previous applications of game theory for security. Prior applications mentioned earlier, including ARMOR, IRIS or GUARDS [21], have each provided unique contributions in applying novel game-theoretic algorithms and techniques. Interestingly, these applications have revolved around airport and air-transportation security. PROTECT's novelty is not only its application domain in maritime patrolling, but also in the five key innovations mentioned above, particularly the moving away from the assumption of perfect rationality by using the QR model.

In addition to game-theoretic applications, the issue of patrolling has received significant attention in the multi-agent literature. These include patrol work done by robots primarily for perimeter patrols that have been addressed in arbitrary topologies [2], maritime patrols in simulations for deterring pirate attacks [23], and in research on the impact of uncertainty in adversarial behavior [1]. PROTECT differs from these approaches in its use of a QR model of a human adversary in a game-theoretic setting, and in being a deployed application. Building on this initial success of PROTECT, we hope to deploy it at more and much larger-sized ports. In so doing, in the future, we will consider significantly more complex attacker strategies, including potential real-time surveillance and coordinated attacks.

Acknowledgements We thank the USCG offices, and particularly Sector Boston, for their exceptional collaboration. Thanks to Matt Johnson for technical assistance in the preparation of this chapter. The views expressed herein are those of the author(s) and are not to be construed as official or reflecting the views of the Commandant or of the U.S. Coast Guard. This research was supported by the United States Department of Homeland Security through the Center for Risk and Economic Analysis of Terrorism Events (CREATE) under award number 2010-ST-061-RE0001.

References

1. Agmon N, Kraus S, Kaminka GA, Sadov V (2009) Adversarial uncertainty in multi-robot patrol. In IJCAI
2. Basilico N, Gatti N, Amigoni F (2009) Leader-follower strategies for robotic patrolling in environments with arbitrary topologies. In AAMAS
3. Blair D (2010) Annual threat assessment of the US intelligence community for the senate select committee on intelligence. http://www.dni.gov/testimonies/20100202_testimony.pdf
4. Camerer CF (2003) Behavioral Game Theory: Experiments in Strategic Interaction. Princeton University Press, USA
5. Conitzer V, Sandholm T (2006) Computing the optimal strategy to commit to. In ACM EC
6. Dozier K (2011) Bin laden trove of documents sharpen US aim. http://www.msnbc.msn.com/id/43331634/ns/us_news-security/t/bin-laden-trove-%documents-sharpen-us-aim/
7. Fudenberg D, Tirole J (1991) Game Theory. MIT, USA
8. Jain M, Kardes E, Kiekintveld C, Ordonez F, Tambe M (2010) Security games with arbitrary schedules: A branch and price approach. In AAAI
9. Jain M, Tsai J, Pita J, Kiekintveld C, Rathi S, Tambe M, Ordonez F (2010) Software assistants for randomized patrol planning for the LAX airport police and the federal air marshal service. Interfaces 40:267–290
10. Kiekintveld C, Jain M, Tsai J, Pita J, Tambe M, Ordóñez F (2009) Computing optimal randomized resource allocations for massive security games. In AAMAS
11. Kiekintveld C, Marecki J, Tambe M (2011) Approximation methods for infinite bayesian Stackelberg games: modeling distributional uncertainty. In AAMAS
12. Korzhyk D, Conitzer V, Parr R (2010) Complexity of computing optimal Stackelberg strategies in security resource allocation games. In Proceedings of The 24th AAAI conference on artificial intelligence. Atlanta, GA, USA, pp 805–810
13. Korzhyk D, Conitzer V, Parr R (2011) Solving Stackelberg games with uncertain observability. In AAMAS
14. McKelvey DR, Palfrey TR (1995) Quantal response equilibria for normal form games. Games Econ Behav 10(1):6–38
15. Paruchuri P, Pearce JP, Marecki J, Tambe M, Ordonez F, Kraus S (2008) Playing games with security: An efficient exact algorithm for Bayesian Stackelberg games. In AAMAS
16. Pita J, Jain M, Western C, Portway C, Tambe M, Ordonez F, Kraus S, Parachuri P (2008) Deployed ARMOR protection: the application of a game-theoretic model for security at the Los Angeles International Airport. In AAMAS
17. Pita J, Tambe M, Kiekintveld C, Cullen S, Steigerwald E (2011) GUARDS – game theoretic security allocation on a national scale. In AAMAS
18. Rogers BW, Palfrey TR, Camerer CF (2009) Heterogeneous quantal response equilibrium and cognitive hierarchies. J Econ Theory, 144(4)
19. Shakarian P, Dickerson JP, Subrahmanian VS (2012) Adversarial geospatial abduction problems. ACM Trans Intell Syst Technol 34:1 to 34:35
20. Shieh E, An B, Yang R, Tambe M, Baldwin C, DiRenzo J, Maule B, Meyer G (2012) Protect: A deployed game theoretic system to protect the ports of the united states. In International Conference on Autonomous Agents and Multiagent Systems (AAMAS), Taipei, Taiwan

21. Tambe M (2011) Security and game theory: algorithms, deployed systems, lessons learned. Cambridge University Press, NY
22. Tsai J, Rathi S, Kiekintveld C, Ordonez F, Tambe M (2009) IRIS: a tool for strategic security allocation in transportation networks. In AAMAS. Budapest, Hungary pp 37–44
23. Vanek O, Jakob M, Hrstka O, Pechoucek M (2011) Using multi-agent simulation to improve the security of maritime transit. In MABS
24. Wright J, Leyton-Brown K (2010) Beyond equilibrium: Predicting human behavior in normal form games. In AAAI
25. Yang R, Kiekintveld C, Ordonez F, Tambe M, John R (2011) Improving resource allocation strategy against human adversaries in security games. In IJCAI
26. Yang R, Tambe M, Ordonez F (2012) Computing optimal strategy against quantal response in security games. In AAMAS
27. Yin Z, Jain M, Tambe M, Ordóñez F (2011) Risk-averse strategies for security games with execution and observational uncertainty. In AAAI

Government Actions in Terror Environments (GATE): A Methodology that Reveals how Governments Behave toward Terrorists and their Constituencies

Laura Dugan and Erica Chenoweth

1 Introduction

With the persistent alarm being raised about terrorist violence by the media and government officials it is unsurprising that scholarship in this area has grown well beyond its traditional disciplinary boundaries (i.e., political science and international relations). As scholars from disciplines such as criminology [27, 30], computer science [11, 12, 35], economics [25], and others get more involved, more data sources have become available [1, 19, 28, 50] and more sophisticated analytical methods have been applied to terrorism research [14, 17, 30]. Yet, research on the effectiveness of counterterrorism measures has only incrementally improved in recent years [33].

Since 2004, empirical evaluations have assessed a wide range of efforts to reduce terrorism, such as the installation of metal detectors [16], target hardening and defensive fortification [36], negotiation [2, 5, 7, 37], targeted assassination [6, 22, 41, 54], curfews and containment strategies [29], violent repression or military retaliation [29, 51], and indiscriminate repression [34], among others. While all of these efforts have been informative, their findings are based on government actions that were identified by the authors as worthy of attention. By hand selecting counterterrorism efforts to evaluate, we still are left in the dark on whether other less extraordinary actions by states played any role in reducing (or increasing) terrorist violence. To further complicate this potential bias, the types of actions that have been evaluated by scholars can typically fall under the purview of

L. Dugan (✉)
University of Maryland, College Park, MD, 20742, USA
e-mail: ldugan@umd.edu

E. Chenoweth
University of Denver, Denver, CO, 80208, USA
e-mail: erica.chenoweth@du.edu

V.S. Subrahmanian (ed.), *Handbook of Computational Approaches to Counterterrorism*, DOI 10.1007/978-1-4614-5311-6_21,

coercive or repressive actions. With the exception of negotiations and maybe some target hardening efforts, most of the counterterrorism activities described above are repressive. Thus, in addition to the apparent systematic omission from these studies of less dramatic government actions, we also miss most conciliatory efforts to reduce violence [15].

To be fair, scholars have evaluated almost exclusively highly visible attempts by governments to repress terrorism because those are the only efforts that we know about. Despite the vast amount of attention that has been paid to thoroughly measure the dependent variable (terrorism) (see chapter 1 in this book—LaFree and Dugan), until now little effort has been made to measure a critical independent variable that would systematically capture efforts by governments to stop terrorism. This chapter describes one such effort, the Government Actions in Terror Environments (GATE) database, which methodically captures a full range of reported actions by governments in select countries. In this chapter, we describe the process used to collect the GATE data, provide a summary of its key variables across five countries, and then assess the effects of repressive and conciliatory actions on terrorist violence for two of the countries.

2 What is known about Government Actions to End Terrorism

In 2006 Lum and Kennedy published a review of evaluation studies on counterterrorism. After systematically reviewing 20,000 pieces of scholarship from 1971 through 2004, they only identified 7 research studies that met the criteria of a moderately rigorous evaluation. As a reference point, a similar review conducted on criminal justice practices found more than 500 rigorous and scientifically sound program impact studies [49]. Since then research that empirically assesses the effects of counterterrorist efforts has grown. Conclusions have been mixed, as some find that repressive actions do seem to reduce terrorist violence [29, 34], but with caveats showing a greater increase in terrorist violence after repression [29, 42, 51]. This finding is troublesome for many reasons; especially because these harmful effects are only now being detected after years of practice, highlighting the importance of continued evaluation of ongoing practices.

Further, limitations in data availability lead to other problems that are found in much of the recent scholarship on counterterrorism, including (1) oversimplification of counterterrorism; (2) relying on monadic methods to study relational questions; and (3) neglecting to resolve endogeneity problems between the independent and dependent variables. We discuss each of these problems in detail below.

The first major problem is a tendency to oversimplify measures of state actions, oftentimes representing them with a basic dichotomous variable, such as repression or no repression, targeted killings or no targeted killings, negotiations or no negotiations. But we know that policymakers rarely choose solely between negotiating and not negotiating [3, 10, 38]. Instead, they usually simultaneously consider a wide array of policy choices, such as repression, blockades, limited political concessions, appealing to international partners, etc., oftentimes constructing a mixed bag of policies. Thus, when scholars ask questions like, "Do targeted killings work?" they often neglect to consider targeted killings within the context of other concurrent or even potential strategies [10]. For example, when LaFree and colleagues evaluated whether internment, criminalization/Ulsterization, Falls Curfew, Operation Motorman, and two other military interventions worked at reducing Republican terrorism in Northern Ireland between 1969 and 1992, they ignored all other efforts that were also being implemented during that period—possibly confounding their conclusions with the effects of these unmeasured efforts [29]. Because of this tendency, it is difficult to evaluate the value of particular strategies relative to other choices that might have been more effective. Furthermore, mischaracterizing the choice set as a series of dichotomous choices also reduces the relevance of such research to policymakers. Thus a more informative question might be, "Do targeted killings work compared with all of the other potential strategies that states have at their disposal?", and "With what other combination of actions do targeted killings reduce terrorist violence?"

The second problem in the literature is that scholars often study terrorism (and, by extension, counterterrorism) from a monadic perspective [9, 53]. In other words, tests on the effects of counterterrorism look for a decrease in the number of terrorist attacks in general, without considering the target of the specific action. For example, if we were interested in the question of whether a U.S. counterterrorism policy against a particular terrorist group resulted in terrorist retaliation against the U.S., we would require information about the identities of the terrorist perpetrators to determine whether the relationship existed. Most current studies use monadic analyses, meaning that they ignore the identities of the perpetrators, and instead count the number of terrorist attacks against the U.S. following the counterterrorist action regardless of the attacks' origins. Monadic relationships are represented in Fig. 1. As this Figure illustrates, there is no assumed relationship between the counterterrorism actions and who they might be directed toward. Instead, tests assess general changes in terrorist behavior after actions are taken [40, 51, 55]. In contrast, Fig. 2 depicts a dyadic relationship between state A and terrorist organization X. Here we see that the counterterrorist actions are directed toward a specific perpetrator in time t, and its effects are measured by changes in terrorist attacks by that perpetrator in time t+1.

In order to better demonstrate these criticisms of typical counterterrorism research, we present two hypothetical datasets. The first, shown in Table 1, contains an example of what a country-year data set looks like when the unit of analysis is monadic and the intervention is operationalized as an indicator variable is valued at 1 on during the years of operation and as 0 during the other years. The focus of this dataset is clearly on the country and the number of attacks per year. This is the most common approach in terrorism studies at the moment, as scholars often use an indicator of whether a specific intervention is at play, in additional to possible aggregate measures of regime capabilities, political repression, or human

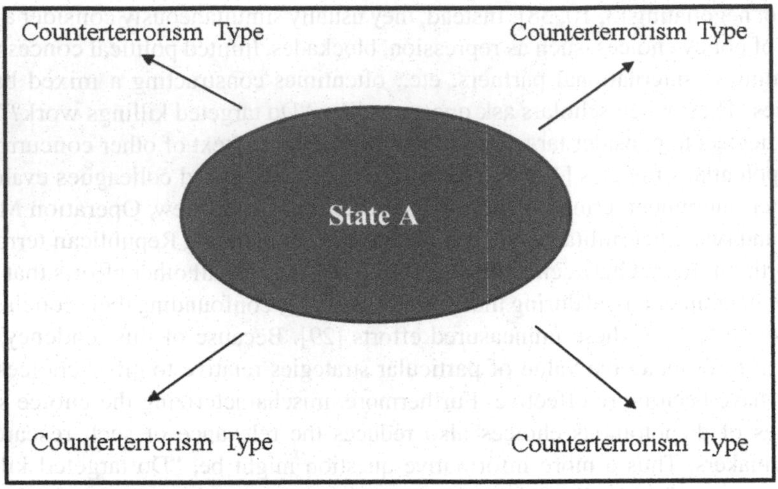

Fig. 1 Monadic relationship between state A and terrorist attacks

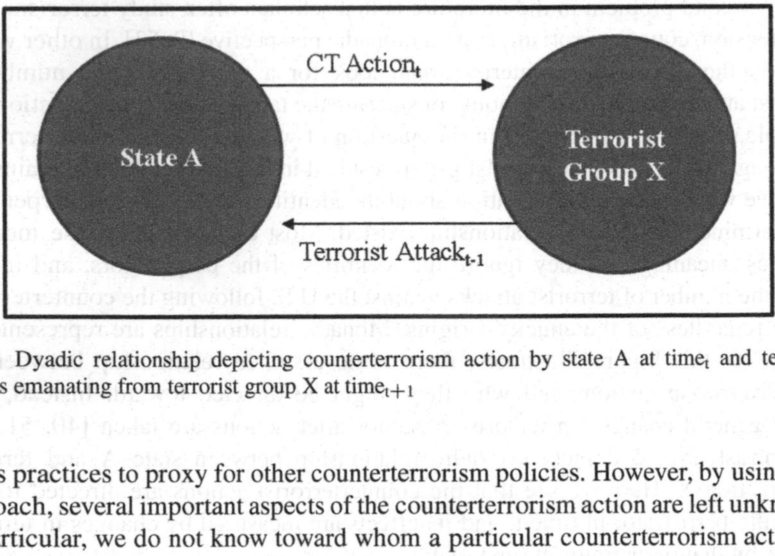

Fig. 2 Dyadic relationship depicting counterterrorism action by state A at time$_t$ and terrorist attacks emanating from terrorist group X at time$_{t+1}$

rights practices to proxy for other counterterrorism policies. However, by using this approach, several important aspects of the counterterrorism action are left unknown. In particular, we do not know toward whom a particular counterterrorism action is directed, nor do we know anything about the perpetrators of the terrorist attacks. Scholars ignore whether the 19 attacks against Country A in 2000 were perpetrated by different organizations or just a single group. There is no way to account for whether these attacks perpetrated by an organization toward whom Country A is actively implementing a counterterrorism campaign or by a different group. It is possible to make some educated guesses based on what is known about Country A, but without additional information about the specific attacks and intentions of the interventions, causal inference is impossible.

Table 1 Hypothetical example of monadic country-year data structure with a simplistic intervention

Country	Year	Intervention	Number of attacks in next year
A	2000	0	19
A	2001	1	8
A	2002	1	12
A	2003	0	19
A	2004	0	15
B	2000	1	2
B	2001	0	0
B	2002	1	4
B	2003	0	4
B	2004	0	3
C	2000	0	0
C	2001	0	0
C	2002	0	0
C	2003	1	0
C	2004	1	1

Table 2 Hypothetical example of dyadic country-year data structure with a mixture of contextual interventions

Country	Year	Number of repressive acts	Number of conciliatory acts	Target of intervention	Number of attacks in next year
A	2000	3	2	Group X	1
A	2001	4	1	Group X	4
A	2002	1	0	Group X	1
A	2003	0	0	Group X	6
A	2004	2	1	Group X	0
A	2000	5	5	Constituency of Group X	2
A	2001	7	6	Constituency of Group X	0
A	2002	3	7	Constituency of Group X	0
A	2003	4	3	Constituency of Group X	0
A	2004	2	8	Constituency of Group X	0
A	2000	1	1	Group Y	6
A	2001	0	0	Group Y	4
A	2002	0	0	Group Y	0
A	2003	1	2	Group Y	0
A	2004	1	4	Group Y	0

A dyadic data set that includes more contextual measures of counterterrorism would look like Table 2, which disaggregates the Country A data shown in Table 1 using hypothetical organizations and their constituencies. Notice that the data are now relational: the unit of analysis is the yearly interaction between the country and target over the time period of interest. The target could also be a foreign organization, so that we can measure the effects of Country A's counterterrorism on transnational terrorist incidents. By considering relational aspects of counterterrorism, including both the state and the target, scholars can advance more empirically accurate assumptions, causal inferences, and predictions about the conditions under which counterterrorist tactics reduce attacks by certain types of targets [13]. Indeed, such a link seems necessary to make causal inferences about the effects of counterterrorism policies on terrorist activity.

While qualitative research on counterterrorism often uses dyadic analysis in practice, most large-sample studies of counterterrorism use monadic analysis. In fact, Young and Findley find that since 2001, nearly 80 articles have been published on terrorism in the top nine general political science and international relations peer-reviewed journals.[1] Among those employing statistical methods to explain terrorist attacks, all use monadic analysis. To date, we are only aware of three studies [9, 26, 53] that incorporate dyadic analysis of terrorist events. Further, only three published studies [11, 20, 29] apply quantitative, dyadic analysis to questions about counterterrorism effectiveness; although all of these are single-country case studies.[2]

Finally, studies of terrorism and counterterrorism rarely account for endogeneity between the independent and dependent variables. As Piazza and Walsh [40: 413) point out, the relationship between social dissent is reciprocal, meaning that "dissent leads to repression, which in turn fuels dissent." One way to address this endogeneity problem is to use instrumental variables to proxy for the endogenous covariates; but instrumental variables are very difficult to locate. Several studies from the late 1980s and early 1990s accounted for simultaneity by incorporating relational data and Vector-Autoregression (VAR) models in studies of counterterrorism, which helped to account for dyadic interactions and endogeneity [4, 8, 18, 31, 32]. VAR is an estimation method that allows the researcher to account for the evolution of a relationship given interdependencies across multiple time series. However, the inferences available from these studies were necessarily constrained by the reliance on the ITERATE data set, which focuses only on incidences of transnational terrorism and state response. As a consequence, few studies have explored the effects of different counterterrorism strategies on domestic terrorism, which represents the vast majority of terrorist activity [28].

Thus, the goal of developing the GATE data has been to address all prior conceptual and empirical limitations by creating a comprehensive event-based dataset that documents specific actions by governments with information on who those actions target. In order to avoid oversimplifying the measures of state actions, the GATE

[1] Their sample includes *American Political Science Review, Journal of Politics, American Journal of Political Science, International Organization, International Studies Quarterly, World Politics, Journal of Conflict Resolution, Journal of Peace Research,* and *Conflict Management and Peace Science.*

[2] Network analysis provides another way to incorporate relational data into the field. In this case, the relationships are between and among terrorist groups, rather than between the groups and their targets. There have been some recent advances with network analysis, such as Asal and Rethemeyer's Big, Allied, and Dangerous database [1] and Horowitz and Potter's [24] study on the effects of network centrality on the diffusion of terrorist tactics.

data are not constrained to arbitrary temporal aggregation, allowing for details to be reported on each action. We include a wide range of state actions under the category of "counterterrorism." While many studies have focused exclusively on repression or concessions, we collect data on thousands of types of state actions—from raids and arrests to allowing telephone lines to be built in refugee camps—so that we can explore the relative effects of different types of interactions. This allows us to move away from misleading characterizations of conflict as a series of dichotomous choices and to consider a wider range of conflict actions [10]. By providing a wide range of actions, scholars are free to operationalize counterterrorism according to their theoretical motivations. For example, the hypothetical dataset shown in Table 2 can be constructed from the GATE data by aggregating all repressive and all conciliatory state actions to the year for each country. The GATE data address the second limitation by allowing scholars to incorporate dyadic analysis by only accounting for only those actions that are targeted toward specific organizations, terrorist movements, or terrorist constituencies (again as shown in Table 2). Finally, the data are organized in a way that allows researchers to estimate the effects of tactics on terrorist attacks using methods that account for the reciprocal effects that terrorist attacks have on counterterrorism actions and vice versa.

3 Introducing the GATE Database

Thus far we have developed the GATE database of counterterrorism actions for five countries: Algeria, Egypt, Israel, Lebanon, and Turkey. For the remainder of this chapter we will describe how it was developed; report patterns of counterterrorist activity for all countries; and present findings on the effects of conciliatory and repressive actions on Palestinian terrorism in Israel and Kurdish terrorism in Turkey from June 1987 through December 2004.

We collected the GATE data using Textual Analysis by Augmented Replacement Instructions (TABARI), which searches Reuters news articles and identifies observations that match the criteria of an extensive set of dictionarics designed to capture international and domestic activity [43, 44].[3] We preferred Reuters over other wire services because of its consistent editorial control and its tendency to use a simpler sentence structure and vocabulary than alternative news sources [45, 46]. TABARI is an automated text-coding program that codes the lead sentence of news articles based on noun and verb pattern recognition. This method is surprisingly accurate and considerably more efficient than coding entire stories [43, 44].[4] For other recent applications of TABARI, see Clauset et al. [11]; Shellman [47]; and Shellman et al. [48].

[3]In our case, we used the CAMEO coding scheme.

[4]Schrodt estimates that TABARI codes 33 million times faster than the average human coder (2006).

Table 3 Seven-point guide for the conciliatory-repression scale [15]

1 = accommodation	
	appeasing or surrendering to adversary
	making full concessions according to opponent's demands
	action required
2 = conciliatory action	
	making material concessions
	taking action that signals intention to cooperate or negotiate with opponent
3 = conciliatory statement or intentions	
	expressing intention to cooperate or showing support
	verbal expression short of physical action
4 = neutral OR ambiguous	
	no clear moves toward or away from resolution of conflict
	includes all attempts to ask for help from a third party to resolve the conflict
	requires more context to determine whether it is conciliatory or repressive
	includes all infighting over Palestinians within the Israeli government
5 = verbal conflict	
	express intent to engage in conflict or threaten
	decline to cease ongoing conflict; maintain the status quo during conflict
	short of physical action
6 = physical conflict	
	physical or violent action aimed at coercing opponent
	no apparent intention to kill
7 = extreme deadly repression	
	physical action exhibiting intent to kill
	torture or severe violence (such as severe beatings), which could easily kill someone

In order to get a pool of potential stories relevant to the criteria of GATE, we downloaded Reuters articles from Factiva using country-specific search terms (52,575 for "Algeria*"; 109,694 for "Egypt*"; 243,448 stories for "Israel*"; 67,107 for "Leban*"; and 152,998 for "Turk*") as the sole search criterion for the period January 1, 1987 to December 31, 2004. Reuters starts its Factiva archive in June 1987, limiting the beginning of GATE to that date. The end date was selected due to the constraints of the original grant that supported this research. Future versions of GATE will include more recent events.

After TABARI identified politically relevant news articles, we filtered the output to keep only those actions that government actors implemented toward terrorist-relevant sub-state targets. We chose to not select on the specific types of verbs to assure that all unexpected actions would be captured. This method ensured that we captured a wide range of actions that may not immediately seem like counterterrorism but may be relevant to the overall conflict (e.g., allowing developers to build better water wells in the Palestinian territories). We further wrote a program that automatically coded each action according to several additional dimensions. First, we established a Conciliatory-Repression scale for each action, illustrated in Table 3. The scale features distinctions in the intensity of the action as well as its

Table 4 Examples of common actions for each scale item [15]

Accommodation/Full concessions	**Verbal conflict**
Withdrew from town	Make pessimistic comment
Signed peace accord	Dismissed
Handed town to Palestinians	Blame for attack
Conciliatory action	Deny responsibility
Met to discuss	Threaten military force
Released	**Physical conflict**
Lifted curfew	Demolished
Pulled out	Barred
Investigate abuse	Sealed off
Conciliatory statement or intensions	Imposed Curfew
	Arrested
Expressed optimism	**Extreme deadly repression**
Agreed to hold talks	Shot dead
Praised Palestinians	Fired missiles
Expressed desire to cooperate	Clashed with
Admitted mistake	Raided
Neutral or ambiguous	Helicopter attack
Infighting over	
Failed to reach agreement	
Host a visit	
Appealed for third party assistance	
Investigating	

relative placement of the action on a conciliation-repression spectrum, similar to the Goldstein scale [21]. Examples of actions that commonly fell under each code in the GATE data are listed in Table 4.

Second, we autocoded each observation for whether the event was directed toward a discriminate or indiscriminate target based on the nature of the verb pattern. Discriminate actions are those that attempt to single out "guilty" or "suspected" parties from uninvolved parties (e.g., made an arrest). Indiscriminate actions are those that directly affect uninvolved people (i.e. those that are not suspected of involvement in terrorist activity; e.g., raided a town). Finally, we autocoded for whether the action was material or nonmaterial. Material actions are those that involve physical contact between state and nonstate actors, whereas nonmaterial actions are not physical (i.e. they are typically verbal actions, such as a decision or a sentencing).

Following the autocoding stage, research assistants hand-checked each observation (as many as 10,000 cases for one country) to ensure that TABARI coded each story correctly, that the autocoding was appropriate to the context of the lead sentence, and to mark for removal any cases that were irrelevant for the project or duplicative of other actions. During this cleaning process, we also attributed each government action to politicians, the military, the judiciary, the police, and for

Table 5 Codebook for GATE

Variable	Definition
ID	Unique action identifier.
Date	Date of the news article.
Leadsent	Entire lead sentence.
Campyes	1: Part of multiple related actions; 0: Not part of multiple related actions.
Campid	Unique identifier for all related series of actions.
Camptype	Specific type of related series of actions (e.g. criminal justice, military operation, negotiations).
Campspec	Specific details of related series of actions.
Actor	Entity that initiated the action. Typically this is a politician, the military, the police, or the judiciary.
Action	This is a simple phrase that describes the action (e.g., negotiated, killed dead).
Target	This is the entity toward which the action was directed. It uses TABARI codes (e.g., PSEREB equals Palestinian rebels).
Conc_repr	This is the seven-point scale described in Table 3.
Discrim	1: The action affected only a guilty target (e.g., terrorist); 0: The action affected a general group of people (e.g., Palestinian civilians).
Material	1: The action was physically tangible; 0: The action was an intangible statement or gesture.

Turkey, the militia. This process revealed a high degree of error, particularly with the conciliatory-repressive, indiscriminate-discriminate, and material-nonmaterial scales. Research assistants corrected these errors, and we checked their changes to ensure a high degree of intercoder reliability. Finally, we linked all related actions through unique identifiers. For example, multiple actions can capture the investigation, arrest, trial, conviction, and release of a specific terrorist. The GATE data are coded so these related actions are all linked. The resulting files contain more than 9,500 actions for all five countries. Table 5 lists all relevant variables.[5]

[5]There were 9,530 actions detected for these countries from 1988 through 2004. As the authors publish studies using the data, they will become available to other researchers. The monthly version of GATE-Israel is available on the American Sociological Review's website in an on-line appendix associated with their article [15].

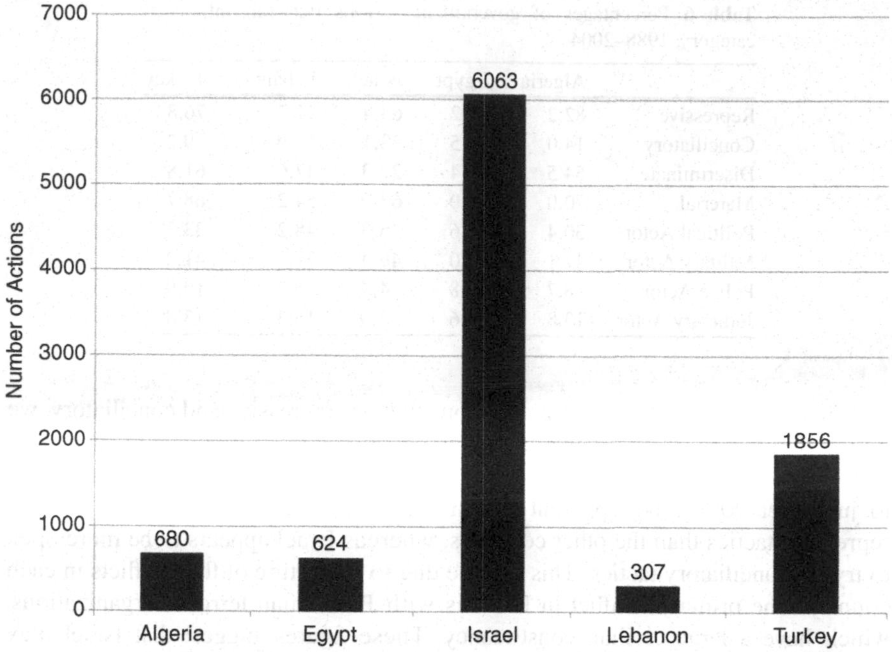

Fig. 3 Total number of government actions by country, 1988–2004

3.1 The Data Described

Figure 3 shows the number of government actions relevant to terrorism for each of the five countries in GATE for the years 1988 through 2004. As we can see, Israel initiated more than three times the number of actions that Turkey initiated; and the other three countries initiated fewer than 1,000 reported actions over the seventeen year period cover by the data. Despite the large differences across these countries, we point out that the documented 307 actions by the Lebanese government are still more than what has been previously available for a single country.

We now look within the government actions and examine the percent distributions based on the type of actions across several dimensions: repressive vs. conciliatory, discriminate vs. indiscriminate, material vs. nonmaterial, and by the type of actor. Table 6 presents the percent of actions that fall into each category. Actions are marked as conciliatory when their code in the Conciliatory-Repressive Scale (in Table 2) is rated as 1, 2, or 3. Conversely, actions are marked as repressive when their code is rated as 5, 6, or 7. When considering the percent of discriminate cases versus indiscriminate and material versus nonmaterial, we only present the percent discriminate and material because each only has two possible outcomes (i.e., percent indiscriminate = 100 - percent discriminate). We include both the percent repressive and conciliatory because a subset of cases are considered neutral (item 4 in Table 2), making the summation of the percent repressive and the percent conciliatory less than 100.

Table 6 Percentages of government actions that fall into each category, 1988–2004

	Algeria	Egypt	Israel	Lebanon	Turkey
Repressive	82.2	90.2	63.4	71.7	76.3
Conciliatory	14.0	8.5	28.1	16.9	9.2
Discriminate	54.5	66.4	22.3	47.6	61.8
Material	70.0	84.0	63.9	54.2	68.7
Political Actor	30.4	21.6	45.3	48.2	33.3
Military Actor	47.1	13.0	46.4	31.3	41.2
Police Actor	8.7	50.8	4.4	5.2	10.9
Judiciary Actor	13.8	14.6	3.9	15.3	13.8

When we examine the percent of actions that are repressive and conciliatory, we notice two things. First, all five countries are more likely to resort to repressive actions than conciliatory actions. Percent repression ranges from just over 63% to just over 90%. Also apparent is that Egypt is much more likely to rely on repressive tactics than the other countries, whereas Israel appears to be more open to trying conciliatory tactics. This may be due to the nature of the conflicts in each country. The primary conflict in Israel is with Palestinian terrorist organizations, which have a large civilian constituency. These figures suggest that Israel may have made explicit efforts to accommodate their needs through concessions. In contrast, in Egypt the primary terrorist activity occurred in the mid-1990s, when a variety of Islamist groups launched an internal war against Hosni Mubarak's regime. These acts were accompanied by massive increases in repression, while conciliatory acts remained virtually nonexistent. Indeed, when they did occur, conciliatory acts generally consisted of prisoner releases of accused Islamists or members of the Muslim Brotherhood. However, the Egyptian regime and Egyptian terrorists generally met violence with violence during this period of high conflict.

When we consider the distribution of discriminate actions across countries, some interesting differences emerge. Once again, Israel and Egypt mark the endpoints on the range, with Israel having the fewest relative discriminatory acts (22.3%) and Egypt having the most (66.4%). The low percentage of discriminate actions by the Israeli government suggests that it did not distinguish between Palestinian civilians and Palestinian terrorists. In contrast, Algeria, Turkey and Egypt seemed to target their actions toward specific suspects without involving innocent civilians. The distribution of material actions ranges from 54.2% in Lebanon to 84.0% in Egypt. While all countries are more likely to take tangible actions over intangible gestures, the distribution, once again suggest that Egypt is the most aggressive.

Finally, when we consider the primary actors of the government actions we see that Egypt once again stands out. For the other four countries, either the military or politicians implemented around 80% of the actions. In Egypt, nearly 60% of terrorist relevant actions are perpetrated by police. This makes sense in the Egyptian system, because the 500,000-strong police force was the primary body charged with maintaining internal security under Hosni Mubarak. In Israel, Turkey, and Algeria,

Algeria	Egypt	Israel
Muslim Rebels (67.6%)	Islamic Terrorists (73.1%)	Palestinian Civilians (74.4%)
Muslim Extremists (12.5%)	Islamic Extremists (8.5%)	Palestinian Terrorists (17.7%)
Others (6.3%)	Egyptian Civilians (8.0%)	Israeli Civilians (5.8%)
Algerian Civilians (5.7%)	Leftists (5.3%)	Lebanese Terrorists (1.4%)
Muslim Civilians (3.2%)	Others (1.4%)	Lebanese Civilians (0.4%)
FLN (1.6%)	Islamic Civilians (1.3%)	Other (0.3%)
The Berbers (1.3%)	Christians (1.3%)	
Algerian Rebels (1.0%)	International Terrorists (1.1%)	
International Terrorists (0.6%)		

Lebanon	Turkey
Lebanese Terrorists (19.2%)	Kurdish Rebels (47.7%)
Palestinian Terrorists (15.3%)	Islamic Civilians (21.6%)
Palestinian Civilians (9.1%)	Kurdish Civilians (20.6%)
Moslem Civilians (7.5%)	Turkish Civilians (4.4%)
International Terrorists (6.5%)	Other (2.3%)
Others (5.2%)	Islamic Rebels (1.9%)
Moslem Terrorists (4.6%)	Left-Wing (1.5%)
Israelis (1.6%)	

Fig. 4 Targets of government actions, 1988–2004 (terrorist targets in gray)

the military represented the plurality of actions, with politicians following close behind. In Lebanon this trend is reversed: politicians claimed the most actions, with the military following close behind.

Figure 4 presents the distributions of targets of the government actions from 1988 through 2004, ordered by most frequently targeted to least frequently targeted. In order to more easily distinguish between terrorists and their constituencies the terrorists are marked in gray. Most obvious is that terrorist relevant actions can target both civilians and terrorists. In the Israeli case, nearly 75% of the actions are directed toward Palestinian civilians, whereas all of the other countries most frequently target a terrorist movement. This figure also shows that more than 90% of the actions in Israel were directed toward Palestinians civilians or terrorists. Nearly 6% were directed toward Israeli civilians, and nearly 2% were directed against Lebanese civilians or terrorists. In Algeria, more than 80% of actions targeted Muslims (civilians, extremists, or rebels)—a natural representation of the fact that the primary terrorist threat in Algeria emerged from Islamic groups during the series. In the Egyptian case, the primary targets were Islamic terrorists and extremists, representing more than 80% of the targets. The targets of Lebanese counterterrorism are more diverse, with Lebanese civilians representing the largest target category (at 30.9%), followed by Lebanese and Palestinian terrorists (19.2 and 15.3%, respectively). In Turkey, close to 70% of the actions targeted Kurdish civilians or terrorists, whereas 21.6% targeted Islamic civilians. Only a very small percentage of actions targeted Islamic terrorists (nearly 2%) or left-wing terrorists (1.5%).

4 Exploring Counterterrorism Effectiveness Using GATE Data

We now briefly demonstrate the usefulness of the GATE data to explore the effectiveness of government actions on terrorism. Using the classification of conciliation or repression described above, we first compare annual trends of counterterrorism actions with annual trends of terrorist attacks in each country in Fig. 5. This figure presents five bar graphs—one for each country—that depict repressive actions with the black bars and conciliatory actions in the gray bars; all measured by the scale for government actions on the left side of the graph. Terror attacks are shown by the solid black line; and its scale is shown in the right axis of each graph.

Turning first to the Algeria case (a. Algeria) of this figure, one can see that the vast majority of repressive actions accompanied the Algerian Civil War through the 1990s, with repressive policies declining in the late 1990s as (primarily Islamist) terrorist attacks increased. Although some concessions were offered to Islamists in the late 1980s and early 1990s, there were virtually no conciliatory actions during or since the civil war, at which point the government adopted a near-universal repressive stance against Islamic fundamentalists. The relatively close tracking of government action and terror makes it difficult to draw any conclusions about the importance of repression for stopping terror.

In Egypt (b. Egypt), the primary terrorist activity occurred in the mid-1990s, when a variety of Islamist groups launched an internal war against Hosni Mubarak's regime. These acts were accompanied by massive increases in repression, while conciliatory acts remained virtually nonexistent in Egypt. Indeed, when they did occur, conciliatory acts generally consisted of prisoner releases of accused Islamists or members of the Muslim Brotherhood. However, the Egyptian regime and Egyptian terrorists generally met violence with violence during this time period. Clearly government action and terrorism track each other rather closely. More sophisticated analysis can better discern whether the terrorists respond to the government action, or whether the government is actually responding to the terrorism.

In the Israeli case (c. Israel), we see that that during the First Intifada in Israel (1987–1993), the Israeli government employed persistent repressive action, which it later combined with conciliatory action culminating in the Oslo accords (September 1993). It seems that terrorist violence dropped when both types of actions were

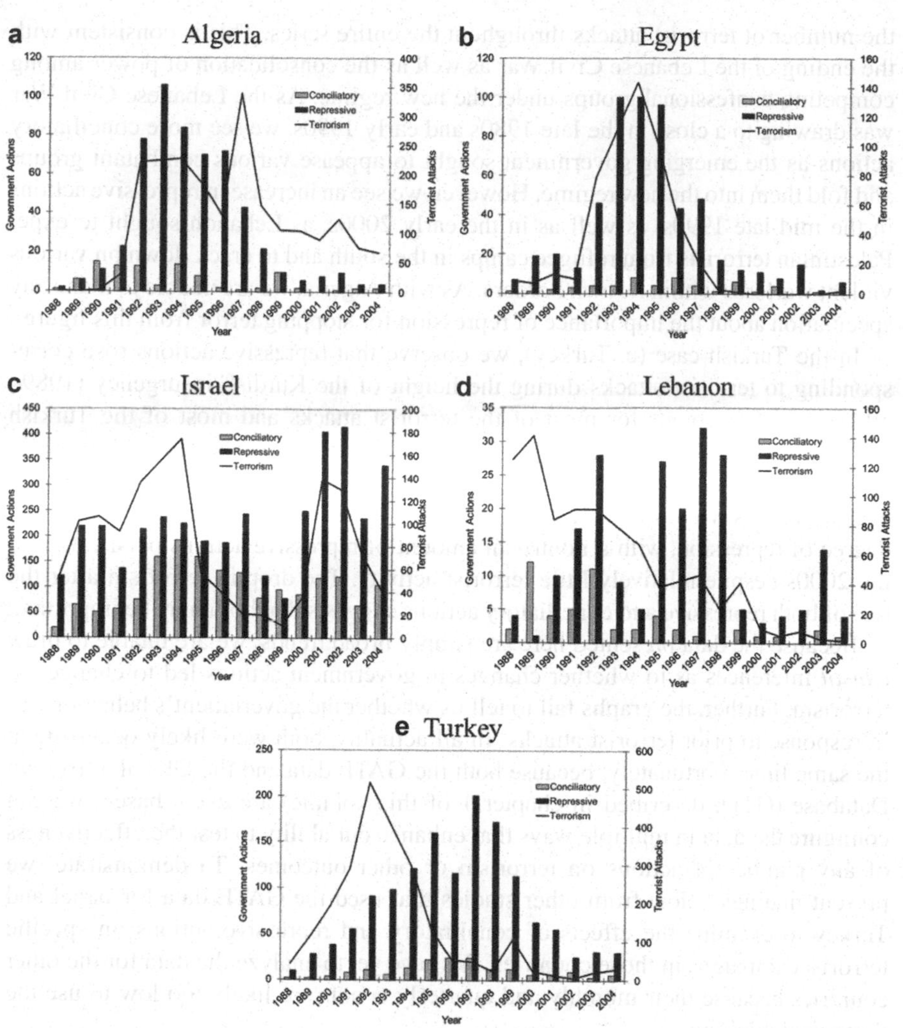

Fig. 5 Conciliatory and repressive government actions and terrorist attacks, 1988 through 2004

being taken. During the Second Intifada (2000–2005), the Israeli government acted with a record high frequency of repressive actions. In fact, when we look closer to the types of actions during that period, we discover that most actions were extremely repressive with the intent to kill (scale 7). While this figure does not tell us whether governments must use repression to stop terrorism, it does suggest that less repressive means, as were used during the First Intifada, might also be promising in reducing terror attacks.

In Lebanon (d. Lebanon), we see more of a mixed bag, which is likely due to the relatively low number of terrorist-relevant government actions compared to the other countries. We also see that Lebanon has experienced a gradual downward trend in the number of terrorist attacks throughout the entire series. This is consistent with the ending of the Lebanese Civil War as well as the consolidation of power among competing confessional groups under the new regime. As the Lebanese Civil War was drawing to a close in the late 1980s and early 1990s, we see more conciliatory actions as the emerging government sought to appease various combatant groups and fold them into the new regime. However, we see an increase in repressive actions in the mid-late 1990s, as well as in the early 2000s, as Lebanon sought to expel Palestinian terrorists from refugee camps in the south and to crack down on various violent warlords within its own borders. As with Algeria, we are unable to make any speculation about the importance of repression for stopping terror from this figure.

In the Turkish case (e. Turkey), we observe that repressive actions rose corresponding to terrorist attacks during the height of the Kurdish insurgency (1989–1995), which accounts for most of the terrorist attacks and most of the Turkish government's actions. It is also clear that conciliatory actions began only to rise in the mid-late 1990s, near the termination of the PKK-led Kurdish insurgency. Yet even as terrorist violence subsided, the Turkish government maintained a fairly high degree of repression, with a nontrivial amount of repressive actions persisting into the 2000s despite relatively little terrorist activity. The drop in terrorism after the rise of both repressive and conciliatory actions suggests that both may be important.

Because the data presented here are simply broad trends, we are unable to draw *causal* inferences as to whether changes in government actions led to changes in terrorism. Further, the graphs fail to tell us whether the government's behavior was in response to prior terrorist attacks. In all actuality, both were likely occurring at the same time. Fortunately, because both the GATE data and the Global Terrorism Database (GTD; described in chapter 1 of this volume) are event-based, we can configure the data in multiple ways that enhance our ability to test the effectiveness of any number of actions on terrorism or other outcomes. To demonstrate, we present findings below from other studies that used the GATE data for Israel and Turkey to examine the effects of conciliatory and repressive actions on specific terrorist campaigns in those countries. We have yet to analyze the data for the other countries because their monthly and quarterly counts are likely too low to use the same methodology.

4.1 The Effectiveness of Israeli Actions on Palestinian Terrorist Violence

For the following analysis, we ask the simple question: Does the number of repressive or conciliatory actions by Israel in one month affect the number of terrorist attacks the following month? To address this question, we combine the GTD and GATE-Israel data into a monthly time-series dataset that records the total number of government actions for each month and the total number of Palestinian

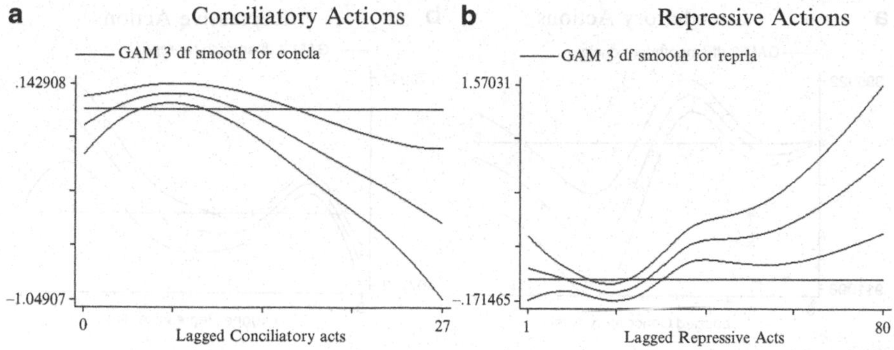

Fig. 6 Partial prediction of conciliatory and repressive actions by Israel from last month on terrorist attacks in the current month [15]

terrorist attacks for each month.[6] Only those government actions that were directed toward Palestinian civilians or terrorists were selected in order to assure our study conforms to the logic of dyadic analysis.

To better estimate the relationship between government actions and terrorist attacks, we use a nonparametric graphical approach, Generalized Additive Models (GAM), which allows us to visually examine the relationships between government actions and the number of terrorist attacks during the next month or week, while controlling for period of data collection in the GTD and the number of previous terrorist attacks lagged back to four months (or weeks), reducing problems of endogeneity described above. This strategy uses a smoothing function to isolate the relationship between actions and attacks without imposing assumptions about linearity [23]. The method produces graphs that show the partial predictions of our independent variables with confidence bands, allowing us to visually examine the nature of the relationships for consistency with our hypotheses [52]. The resulting graphs show the predicted number of attacks in the current month based on the number of conciliatory or repressive actions in the previous month. If both types of actions are effective, we would expect the line for each graph to be sloping downward, indicating that more actions lead to less terrorism.

The results are presented in Fig. 6, which show the partial prediction of conciliatory (a) and repressive actions (b) in the previous month on terrorist attacks in the current month. The horizontal line in both graphs shows the location of zero. This figure shows that a low number of conciliatory actions may lead to a slight increase in terrorist attacks (all three lines are above zero); however, as Israel initiates more conciliatory actions, the expected number of attacks in the next month is more likely to drop. This finding suggests that conciliatory actions do indeed

[6] A much more elaborate analysis is found in Dugan and Chenoweth [15]. Also see that article for details about how the time-series dataset was constructed. Further, we also include in this analysis all terror attacks with an unknown perpetrator.

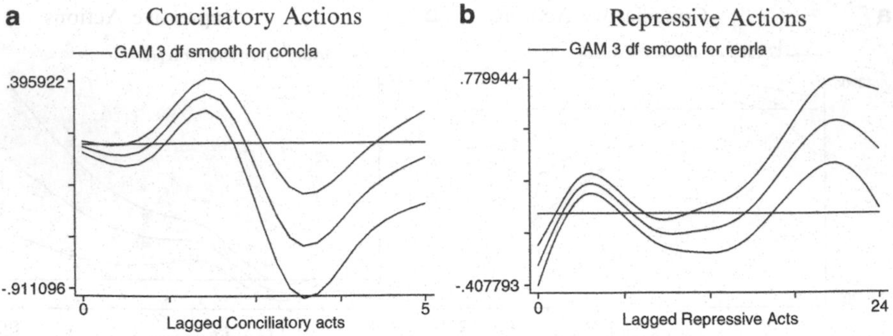

Fig. 7 Partial prediction of conciliatory and repressive actions by Turkey from last quarter on terrorist attacks in the current quarter

lower Palestinian terrorist attacks, especially if Israel initiates many conciliatory actions within a month.

As we turn to repressive actions (Fig. 6b), we see that a small number of actions seem to produce no effect on Palestinian terror attacks. However, as the Israeli government initiates more repressive actions, the expected number of attacks rises. This effect is consistent with the argument that repressive actions can produce a backlash, leading to more terrorism.

4.2 The Effectiveness of Turkish Actions on Kurdish Terrorist Violence

The conclusions that we drew from Israel can only be generalized to Israel between June of 1987 and December 2004. They cannot speak to how other terror campaigns respond to conciliatory and repressive actions by the government. Thus, we further investigate the effects of conciliatory and repressive acts by the Turkish government on Kurdish terrorism. As with the Israeli case, we only include in this analysis those actions by Turkey that were directed toward relevant targets (i.e., Kurdish civilians or Kurdish terrorists), making the analysis dyadic. Further, we only included terror attacks by Kurdish organizations (e.g., Kurdistan Workers Party, Dev Sol).[7] However, because there are much fewer Turkish than Israeli actions (1,856 versus 6,063) and because we have very few conciliatory acts in Turkey (173), we conduct the analysis on quarterly (or three month) intervals.

Figure 7 shows the graphical relationship between conciliatory (a) and repressive (b) government actions directed toward Kurdish civilians or terrorists on terror attacks by Kurdish militants in the following quarter. The relationship between

[7] As with Israel, we also include terror attacks by unknown perpetrators.

conciliatory acts and terror (shown in Fig. 7) is difficult to interpret. Notice that the horizontal axis shows that the largest number of conciliatory acts in any given quarter is only 5. This is very low when we compare it to Israel, which initiated as many as 27 conciliatory actions in one-third the time (i.e., a month as opposed to a quarter). In fact, a quick scan of the data for Turkey shows that in 36% of the quarters, Turkey made no conciliatory gestures toward the Kurds. When they did, in nearly 80% of the remaining quarters they initiated less than three conciliatory actions. If we take this information and again consider Fig. 7, then we can see that the most relevant part of the graph is its far left portion when conciliatory actions are relatively few. Here we do observe a downward slope as these acts move from zero to 1 or 2, before it gets very unstable (by going up, down, and up again), suggesting that there might be some value for Turkey to explore more conciliatory measures.

The effects of repressive actions against Kurds on Kurdish terrorism are shown in Fig. 7b. Here we see that the effects are more stable than the conciliatory relationship, although the story is still somewhat complicated. The initial drop, followed by a rise (and then a drop again) suggests that a little repression might be useful, but more repression might lead to increases in Kurdish terror. The final drop on the left portion of the screen was calculated from less than three quarters of the data and therefore should be interpreted with caution. Further, the top line never drops below zero suggesting further caution is needed. However, as with the Israeli case, repression may have further exacerbated the Kurdish terrorist campaigns.

5 Conclusion

In this paper, we have identified several problems in existing counterterrorism studies that we hope to have remedied with the GATE database. Future articles will apply this data set to hypotheses commonly articulated in the literature, while disaggregating the independent variables; incorporating the relative and simultaneous effects of a variety of policies; and structuring the data to include dyadic relationships. Moreover, to address the problem of endogeneity, future analyses will incorporate the mutual dependence between these major trends to disentangle relationships and identify possible causality. We will employ numerous estimation procedures to determine the conditions under which certain counterterrorism policies succeed.

We expect to use the data from the GATE database to address a wide range of questions including estimating determinants of specific types of government actions, such as upcoming elections. Also, the amount of detail available in the lead sentences alone allows us to study the effectiveness of verbal messages on a host of outcomes beyond terrorist violence, such as Palestinian elections. Furthermore, we are currently linking the tactical actions that are tied to the same strategic campaign in order to more specifically evaluate the broader strategies and their components. Finally, we plan to match the lead sentence to the entire news article to glean details (such as casualties, GIS coordinates, etc.) on each counterterrorism action to provide important context for more thorough analysis. In sum, we expect to eventually produce a global series of counterterrorism databases to help researchers finally assess the effectiveness of different strategies to reduce terrorism for the full range of perpetrators.

References

1. Asal V, Rethemeyer K (2008) The nature of the beast: organizational structures and the lethality of terrorist attacks. J Polit 70:437–449
2. Bapat N (2006) State bargaining with transnational terrorist groups. Int Stud Quart 50:213–229
3. . Birnir JK, Baranov O, Perkoski E, Chenoweth E (2010) The turnaround of terror: constituent populations, terrorists, and counter-terror. Unpubl Manuscr, U Md
4. Brophy-Baermann B, Conybeare JA (1994) Retaliating against terrorism: rational expectations and the optimality of rules versus discretion Am J Polit Sci 38:196–210
5. Browne J, Dickson E (2010) "We don't talk to terrorists": on the rhetoric and practice of secret negotiations. J Confl Resolut 54:379–407
6. Byman D (2006) Do targeted killings work? Foreign Aff (85)95–107
7. Byman D (2009) Talking with insurgents: a guide for the perplexed. Wash Quart 32:125–137
8. Cauley J, Iksoon Im E (1988) Intervention policy analysis of skyjackings and other terrorist incidents. Am Econ Rev 78:27–31
9. . Chenoweth E (2009) War initiation and transnational terrorism: is there a causal connection? Paper presented at the annual meeting of the International Studies Association, New York, USA, February
10. Chenoweth E, Lawrence A (2010) Rethinking violence: states and non-state actors in conflict. MIT Press, Cambridge
11. Clauset A, Heger L, Young M, Skrede Gleditsch K (2010) The strategic calculus of terrorism: substitution and competition in the Israel-Palestine conflict. Coop Confl 45:6–33
12. Clauset A, Young M, Skrede Gleditsch K (2007) On the frequency of sever terrorist events. J Conf Resolut 51:58–87
13. Cunningham D, Skrede Gleditsch K, Saleyhan I (2009) It takes two: a dyadic analysis of civil war duration and outcome. J Conf Resolut 53: 570–597
14. Dugan L (2011) The series hazard model: an alternative to time series for event data. J Quant Crim 27: 379–402
15. Dugan L, Chenoweth E (2012) Moving beyond deterrence: the effectiveness of raising the benefits of abstaining from terrorism in Israel. Am Soc Rev 77:forthcoming 597–624.
16. Dugan L, LaFree G, Piquero AR (2005) Testing a rational choice model of airline hijackings. Crim 43:1031–1066
17. Dugan L, Yang SM (2011) Introducing group-based trajectory analysis and series hazard modeling: Two innovative methods to systematically examine terrorism over time. In: Lum C, Kennedy L (eds) Evidence-based counterterrorism policy. Springer, New York
18. Enders W, Sandler T (1993) The effectiveness of antiterrorism policies: a vector-autoregression-intervention analysis. Am Polit Sci Rev 87:829–844
19. Freilich JD, Chermak SM, Belli R, Gruenewald J, Parkin WS (2012) Introducing the United States Extremist Crime Database (ECDB). Unpubl Manusc, John Jay Coll Crim Justice, CUNY
20. Gil-Alana L, Barros C (2010) A note on the effectiveness of national anti-terrorist policies: evidence from ETA. Confl Manag Peace Sci 27:28–46
21. Goldstein JS (1992). A conflict-cooperation scale for WEIS event data. J Confl Resolut 36: 369–385
22. Hafez M, Hatfield J (2006) Do targeted assassinations work? A multivariate analysis of Israel's controversial tactic during the Al-Aqsa uprising. Stud Confl Terror 29:359–382

23. Hastie TJ, Tibshirani RJ (1990) Generalized additive models. Chapman and Hall, New York
24. Horowitz M, Potter P (2009) The network of terrorist organizations: the link between effectiveness and tactics. Unpubl Manuscr, U Pa
25. Krueger AB (2007) What makes a terrorist: economics and the roots of terrorism (New Edition). Princeton University Press, Princeton
26. Krueger AB, Laitin DD (2008) Kto-Kogo?: A cross-country study of the origins and targets of terrorism. In: Keefer P, Loayza N (eds) Terrorism, economic development, and openness. Cambridge University Press, New York
27. LaFree G, Dugan L (2004) How does studying terrorism compare to studying crime? In DeFlem M (ed) Terrorism and counter-terrorism: criminological perspectives. Elsevier, New York
28. LaFree G, Dugan L (2007) Introducing the global terrorism database. Terror Polit Violence 19:181–204
29. LaFree G, Dugan L, Korte R (2009) The impact of British counterterrorist strategies on political violence in Northern Ireland: Comparing deterrence and backlash models. Crim 47:501–530
30. LaFree G, Freilich JD (2012) Editors introduction: quantitative approaches to the study of terrorism (special issue). J Quant Crim 28:1–5
31. Landes W (1978) An economic study of U.S. aircraft hijacking, 1961–1976. J Law Econ 21: 1–31
32. Lapan HE, Sandler T (1988) To bargain or not to bargain: that is the question. Am Econ Rev 78:16–21
33. Lum C, Kennedy LW, Sherley A (2006) Are counter-terrorism strategies effective? The results of the Campbell systematic review on counter-terrorism evaluation research. J Exp Crim 2:489–516
34. Lyall J (2009) Does indiscriminate violence incite insurgent attacks? Evidence from Chechnya. J Confl Resolut 53:331–362
35. . Mannes A, Shakarian J, Sliva A, Subrahmanian VS (2011) A computationally-enabled analysis of Lashkar-e-Taiba attacks in Jammu & Kashmir. Intelligence and Security Informatics Conference (EISIC), 2011 European, 224–229, 12–14 Sept. 2011
36. Markovsky D (2004) How to build a fence. Foreign Aff 83:50–64
37. Neumann P (2007) Negotiating with terrorists. Foreign Aff 86 128–138
38. . Perkoski E, Chenoweth E (2010) The effectiveness of counterterrorism in Spain: a new approach. Paper presented at the International Studies Association annual meeting, New Orleans, LA, USA, March 15–17
39. Piazza JA, Walsh JI (2009) Transnational terrorism and human rights. Int Stud Quart 53: 125–48
40. Piazza JA, Walsh JI (2010) Physical integrity rights and terrorism. Polit Sci Polit 43:411–414
41. Plaw A (2008) Targeting terrorists: a license to kill? Ashgate, London
42. Rasler K (1996) Concessions, repression, and political protest in the Iranian revolution. Am Soc Rev 61:132–152
43. Schrodt PA (2001) Automated coding of international event data using sparse parsing techniques. Unpubl Manusc U Kans
44. Schrodt PA (2006) Twenty years of the Kansas Event Data System project. Unpubl Manusc, U Kans
45. Schrodt PA, Gerner DJ (1994) Validity assessment of a machine-coded event data set for the Middle East, 1982–1992. Am J Polit Sci 38:825–854
46. Schrodt PA., Davis SG, Weddle JL (1994) Political science: KEDS: a program for machine coding events data. Soc Sci Comput Rev 12:561–588
47. Shellman SM (2008) Coding disaggregated intrastate conflict: machine processing the behavior of substate actors over time and space. Polit Anal 16:464–477
48. Shellman SM, Hatfield C, Mills MJ (2010) Disaggregating actors in intranational conflict. J Peace Res 47:83–90

49. Sherman LW, Gottfredson D, MacKenzie DL, Eck J, Reuter P, Bushway S (1997) Preventing crime: what works, what doesn't, what's promising: a report to the United States congress. National Institute of Justice, Washington, DC
50. . Smith BL, Damphousse KR (2002) American Terrorism Study: Patterns of Behavior, Investigation and Prosecution of American Terrorists, Final Report to the National Institute of Justice, Award Number: 1999-IJCX-0005
51. Testas A (2004) Determinants of terrorism in the Middle East. Terror Polit Violence 16: 253–273
52. Xiang D (2001) Fitting generalized additive models with the GAM procedure, SUGI Proceedings. SAS Institute, Inc., Cary NC
53. . Young J, Findley M (2009) Promise, problems, and pitfalls of terrorism research. Paper presented at the Midwest Political Science Association annual meeting, Chicago, IL, USA, March 10–12
54. Zussman A, Zussman N (2006) Assassinations: evaluating the effectiveness of an Israeli counterterrorism policy using stock market data. J Econ Perspect 20:193–206
55. Walsh JI, Piazza JA (2010) Why respecting human rights reduces terrorism. Comp Polit Stud 43:551–577

Part V
New Directions

A CAST Case-Study: Assessing Risk in the Niger Delta

Nate Haken, Patricia Taft, and Raphaël Jaeger

1 Introduction

The first step in any effective counterterrorism effort is a robust risk assessment of the socio-political landscape. Clearly, the social science behind any such assessment must be sound. So much the better if the tools used to support the assessment allow the user to track trends on a real-time basis, to drill down from the national to the sub-national level, and to manipulate the data so as to unpack the specific combinations of risk factors most salient to the particular area of concern.

Three key interlocking components are equally critical for a good assessment of the risk of terrorism: (1) A sound theoretical framework, (2) relevant and timely data at varying levels of spatial-temporal granularity, and (3) meaningful aggregation, integration, and interpretation of the data for a good understanding of the risk profile of the location or locations in question. If any one of these three components is lacking, the risk assessment will likely be weakened, depending on the research question. Technologists frequently overlook the importance of theory. Academics sometimes ignore the value that computers and information technology can bring to the data question. Subject matter experts notoriously disregard both theory and data, in favor of context-specific qualitative analysis.

Perhaps, aside from the bias of specialists, one reason it is difficult to bring the three components together for a robust risk assessment is that the data challenges are so steep in this field, especially in those cases where terrorism emanates from under governed spaces with forbidding terrain and limited access. Even with a

N. Haken (✉) • P. Taft
The Fund for Peace, Washington, DC, USA
e-mail: nhaken@fundforpeace.org; ptaft@fundforpeace.org

R. Jaeger
The Fund for Peace, Washington, DC,
e-mail: rjaeger@fundforpeace.org

V.S. Subrahmanian (ed.), *Handbook of Computational Approaches to Counterterrorism*, 489
DOI 10.1007/978-1-4614-5311-6_22,

solid theory underpinning the research where ideal metrics have been carefully vetted and categorized, finding data to map to those indicators can be prohibitive. Scaling and normalizing the data to generate actionable insights presents additional methodological conundrums.

Given these challenges, and the enormous financial and human capital required to tackle them, such work has traditionally been the exclusive purview of governments and large international organizations. But recent developments in technology and computational methods offer a great deal of promise, and have rendered conflict assessment of this kind within the reach of nongovernmental organizations such as The Fund for Peace.

There are various models and systems that have been developed in recent years to forecast the probability of conflict[1,2]. Leaning less heavily on quantitative methods, The Conflict Assessment System Tool (CAST) also takes advantage of these developments, not so much for forecasting as to track and understand current trends in conflict risk. This chapter will outline the three components (theory, data, analysis) as incorporated by CAST and illustrate its use for the assessment of risk in the Niger Delta region of Nigeria, specifically the states of Delta, Bayelsa, Rivers, Akwa Ibom, and Abia in the year 2011, where there was significant levels of militancy, criminality, and political violence.

1.1 Component 1: Theory

CAST was created by Pauline H. Baker, former President of the Fund for Peace, and measures the social, economic, and political/military pressures on the state as well as the state's capacity to withstand those pressures. It is relevant to counterterrorism efforts in that terrorism is a tactic frequently used in the context of asymmetric and irregular warfare, often in an environment where the hegemonic role of the state is challenged. A state can fail absent the emergence of irregular warfare. And irregular warfare can be waged without the use of terrorism. However, there is an increased risk of terror when the state is very fragile. The CAST framework is used in The Fund for Peace's annual Failed States Index, which is published in Foreign Policy Magazine. It is also used for a wide range of conflict assessments at the national, provincial, and local levels.

---

[1]Brandt, Patrick T., John R. Freeman and Philip A. Schrodt. 2011. "Real Time, Time Series Forecasting of Inter- and Intra-state Political Conflict." Conflict Management and Peace Science 28(1):41–64.

[2]Weidmann, Nils B. and Michael D. Ward. 2010. "Predicting Conflict in Space and Time." Journalof Conflict Resolution 54(6):883–901.

Drawing from the scholarship on state failure[3,4], CAST considers a failed state to be one that cannot or will not fulfill basic functions, such as the provision of essential services and human security. When the state weakens, social, economic, political, and security pressures can become unmanageable. Thus, CAST takes an "immunological approach," whereby a failing state has recognizable clusters of symptoms. CAST tracks 12 broad social, economic, political, and security indicators and a total of about 200 measures and sub-measures that are monitored for assessing risk of internal conflict or state failure. The specific menu of measures and sub-measures depends upon the parameters of the project and are selected on the basis of relevance to the context and research question as well as availability of data.

A complete list of the 12 indicators is as follows:

1. Mounting Demographic Pressures
2. Massive Movement of Refugees or Internally Displaced Persons
3. Legacy of Vengeance-Seeking Group Grievance or Group Paranoia
4. Chronic and Sustained Human Flight
5. Extreme Uneven Economic Development Along Group Lines
6. Sharp and/or Severe Economic Decline
7. High Criminalization and/or Delegitimization of the State
8. Progressive Deterioration of Public Services
9. Suspension or Arbitrary Application of the Rule of Law and Widespread Violation of Human rights.
10. Security Apparatus Operates as a "State Within a State"
11. Dramatic Rise of Factionalized Elites
12. Intervention of Other States or External Political Actors

In CAST, juxtaposed against the 12 indicators listed above are the state institutions whose role is to manage the pressures on the state. A country may experience high pressures, as measured in the indicators, but may have relatively strong state institutions. Colombia is an example of this type of country, which may be more resilient to collapse than a country with comparable pressures and weaker state institutions. Nigeria, however, is a country with weak state institutions. As measured by The Fund for Peace in their 2011 Country Profile of Nigeria, the institutions were assessed as follows, illustrating the vulnerability of Nigeria to pressures and shocks to the 12 CAST indicators. When all 177 countries were plotted on a graph where the vertical axis was capacity and the horizontal axis was pressure, Nigeria fell on the bottom right quadrant with high pressure and low capacity (Fig. 1).

According to CAST, for the pathology of conflict to emerge, there must be deep structural imbalances which give rise to immediate triggers. Depending on

[3]Fearon J, Laitin D. 2003. "Ethnicity, insurgency and civil war". Am. Polit. Sci. Rev. 97:75–90.

[4]Rotberg, Robert I. When States Fail: Causes and Consequences. Princeton: Princeton University Press, 2004.

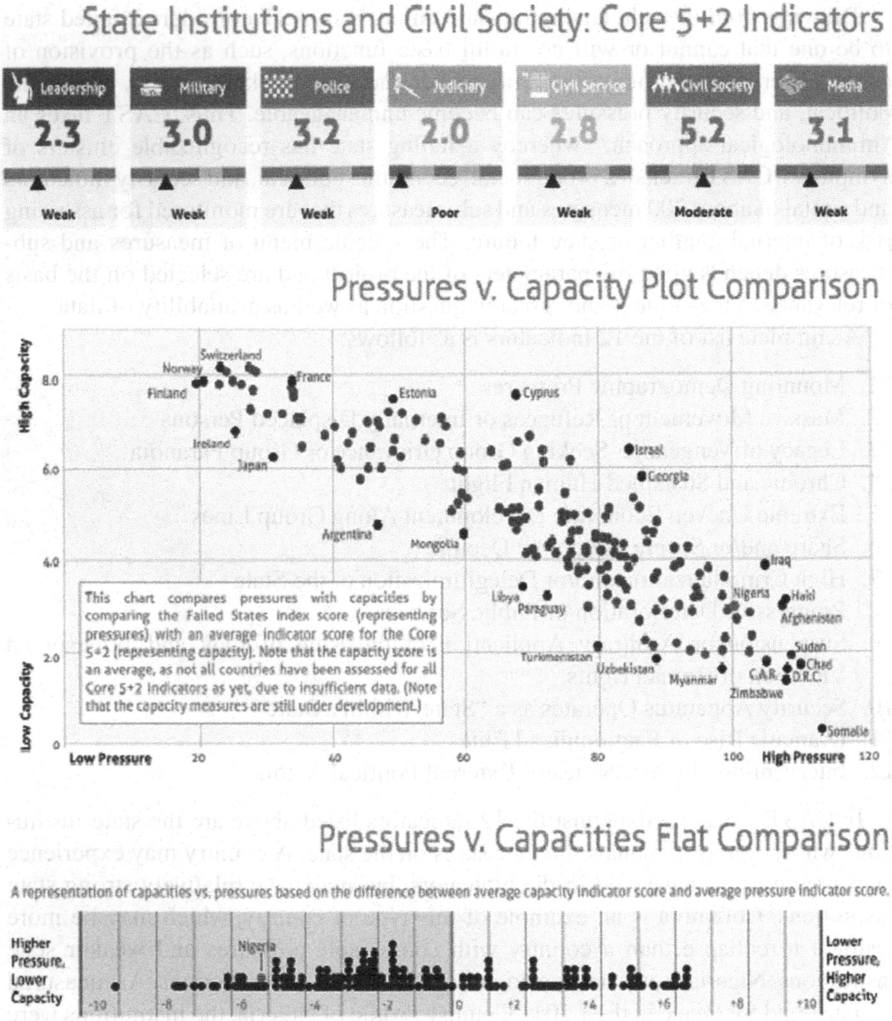

Fig. 1 Nigeria's pressures v. capacities, courtesy of the fund for peace

the effectiveness of the state institutions, these events can lead to conflict and state collapse on the one hand, or reconciliation and the rule of law on the other. Policy makers must reduce the pressures, as measured by the CAST indicators, and build up the institutions, while preparing for and mitigating shocks. Thus CAST provides a map to guide policy makers through the murky waters of conflict and can provide a better understanding of the vulnerability of a particular area to the risk of terrorism.

1.2 Component 2: Data

The data piece is perhaps the most difficult aspect to a good risk assessment, considering the fact that in places of high risk, data is especially difficult to find. Economic and development statistics may be dated or questionable. Conflict data is typically aggregated annually, which does not allow for an assessment of a rapidly changing risk environment. Data also frequently depends on statistics provided by parties to conflict, raising questions of reliability. Recent developments in computational and information technology, however, allow analysts to quantify socio-political risk factors at a level of spatial-temporal granularity that were previously impossible to measure.

Two methods that have shown particular promise as applied to the CAST framework, are 1) automated content analysis of internet news and media sources, and 2) participatory conflict assessment and early warning networks in conflict-affected countries using old and new media to collect and map incident reports of risk factors to track trends at the local level. These two methods are useful for measuring very dynamic, event-driven risk factors.

These event-driven risk factors are embedded in a more static socio-political context. It is important to measure both the dynamic factors and the contextual factors. Layering content analysis and incident data over macroeconomic and health statistics renders a rich profile of risk.

1.2.1 Context

To assess the contextual factors, CAST draws on pre-existing datasets such as those produced by the United Nations and the World Bank and maps those variables to the 12 indicators. For this case study on Nigeria for the year 2011, CAST pulled data on 84 measures from the UN Millennium Indicators, the World Development Indicators, the Economist Intelligence Unit, the Swiss Federal Institute of Technology, Polity IV, Freedom House, Transparency International, Uppsala University's Department of Peace and Conflict Research, the World Food Program, Standard and Poor's Transfer and Convertibility Assessment, and the University of Kentucky's Coup d'état dataset. Those 84 measures were normalized against a global sample of 178 countries and scaled from 0–10 with 10 being the most pressure and 0 being the least pressure. They were then grouped according to 40 Groups (Table 1). *Note once again (and it cannot be emphasized enough) that this is not a forecasting tool.* The groups are all weighted equally, and are aggregated to the indicator level. The indicators are all weighted equally and are aggregated to an overall CAST score. Upon identifying spikes in the trends, a qualitative review of the raw data provides interpretive texture to the analysis. Absent this qualitative review (in Sect. 1.3 of this chapter) the numbers would be merely notional, suggestive of trends but not predictive of conflict onset. There is an assumption that when the pressures on the state go up, as measured by the CAST indicators, that it is more difficult for the state to manage those pressures, thus putting a strain on the social contract between the government and the governed. In the table below, the categories with CAST scores of over 8.0 have relatively severe pressure, as normalized against a wider sample of 178 countries (Table 1).

These 40 Groups were then aggregated to the Indicator level. There was no data at this stage for measuring Factionalized Elites. This indicator level data was then averaged with the Failed States Index score for the previous year in order to factor in reverberations from the events of 2010. Below is a chart depicting the integration of the most recently available quantitative data with the previous year's FSI score to arrive at a baseline score measuring the socio-political context pressures. The next step will be to layer in the event-driven pressures of 2011 as measured by content analysis (Table 2).

1.2.2 Events

Now that we have arrived at an index that estimates the contextual pressures faced by Nigeria, we need to layer in the events of 2011 to see how they impacted the overall risk environment. CAST uses automated content analysis of electronic documents from over 30,000 media sources to identify major events that, in combination with the socio-political context, can further exacerbate state weakness and instability.

Particular strengths of this tool are:

- Information can be updated on a daily basis (for the sake of this case study, data were aggregated on a monthly basis.)
- Analysis can be undertaken at any level: regional, national, provincial, local
- Automation greatly reduces labor costs
- Standardization reduces human bias and increases the ability to compare and contrast across space and time.
- It can be tailored to fit the needs of a particular research or policy question.

For this CAST study, 2,565 Boolean phrases were used to scan 7,452,053 documents from 2011 about the 55 countries most at risk of State Failure as identified in the Failed States Index of the previous year. Phrases were also used to scan 29,352 articles about the five Nigerian states of interest in the Niger Delta Region. The monthly tally of articles about a particular type of pressure in a particular country was divided by the sample size for that country-month to control for media drift and uneven coverage.[5]

The graph below (Fig. 2) shows the percentage of articles in a given month about the five Niger Delta states of interest that deal with eight CAST indicators as measured by content analysis. For the parameters of this research, content analysis was not used to measure four of the 12. Indicator 4 (Human Flight),

[5]For more information about content analysis as employed by FFP, please read "Casting Globally: Using Content Analysis for Conflict Assessment and Forecasting," Nate Haken, Joelle Burbank, and Pauline H. Baker, *Military Operations Research Society Journal*, Vol. 15, No. 2, 2010.

Table 1 CAST scores by group and indicator (contextual factors)

Indicator	Group	CAST score
Demographic Pressures	Population pressures	8.82
Demographic Pressures	Malnutrition and Hunger	4.97
Demographic Pressures	Disease (Malaria and HIV)	9.31
Demographic Pressures	Life Expectancy and Mortality	9.32
Demographic Pressures	Pollution	6.81
Demographic Pressures	Drinking Water	8.99
Refugees and IDPs	Absorption Capacity	8.23
Refugees and IDPs	Refugees/IDPs	5.86
Group Grievance	Discrimination	3.96
Group Grievance	Aggrieved Population	8.15
Human Flight	Brain Drain	5.11
Human Flight	Physicians per capita	6.72
Human Flight	Net Migration	5.14
Uneven Development	Income Distribution	6.52
Uneven Development	Services Distribution	7.26
Economic Pressures	Fiscal Pressure	2.02
Economic Pressures	Employment Pressures	4.74
Economic Pressures	Poverty Pressures	8.40
Economic Pressures	GDP per capita growth	2.16
Economic Pressures	Inflation, consumer prices (annual %)	9.59
State Legitimacy	Corruption	8.10
State Legitimacy	Representativeness in Government	7.59
State Legitimacy	Functioning of Government	7.63
Public Services	Policing	6.92
Public Services	Education	7.38
Public Services	Water and Sanitation	8.79
Public Services	Health System	8.65
Public Services	Phones and Internet	7.40
Public Services	Power	8.01
Public Services	Roads	6.99
Human Rights	Press and Civil Liberties	6.16
Human Rights	Political Liberties	7.18
Human Rights	Gender	9.30
Human Rights	Political Terror	9.00
Security Apparatus	Small Arms	8.00
Security Apparatus	Deaths from Internal Conflict	2.00
Security Apparatus	Demonstrations	8.00
Security Apparatus	One-sided and non-state Fatalities	8.99
External Intervention	Economic and Development Assistance	1.97
External Intervention	Credit Rating	6.07

[a]Each group is made up of multiple measures. In the case of Population Pressures, the measures comprised Adolescent fertility rate births per 1,000 women (15–19); Birth rate, crude (per 1,000 people); Population growth (annual %); and Population ages 0–14 (% of total). The measure breakdown of each group is not publically available

Table 2 Nigeria CAST indicator scores (baseline)

	Pre-existing data sets	FSI (from 2010)	Baseline score (context)
Demographic Pressures	8.15	8.3	8.22
Refugees/IDPs	7.04	6	6.52
Group Grievance	6.05	9.6	7.83
Human Flight	5.66	7.7	6.68
Uneven Development	6.89	9	7.95
Economic Pressures	7.78	7.3	7.54
State Legitimacy	7.78	9	8.39
Public Services	7.73	9	8.37
Human Rights	7.71	8.6	8.15
Security Apparatus	6.75	9.1	7.92
Factionalized Elites	NA	9.5	9.5
External Intervention	3.33	6.9	5.12

5 (Uneven Development), and 6 (Poverty/Economic Decline) are inherently more contextual than the others, making them less amenable to content analysis as a viable method for assessing risk. In addition, Indicator 12 (External Intervention) was not particularly relevant to a study of pressures in the Niger Delta during this period.

Note that in this graph you cannot compare the overall intensity of one indicator to another, given unevenness in media coverage by issue. However, you can see very clearly that most of the indicators are strongly correlated and there were spikes in April (around the time of the election) and in November (when President Goodluck Jonathan declared the removal of the fuel subsidy) (Fig. 2).

Before we could integrate the dynamic content analysis data with the contextual socio-economic factors, we normalized the data as follows. For each indicator, the top ten percentile of scores (for the 55 most at-risk counties) in each month was considered to have a HIGH event-driven impact on the risk profile. The top twenty and thirty percentile was considered to have a MEDIUM impact. The bottom 70 percentile were considered to have no event-driven impact on the scores. Thus for the five Niger Delta states of interest content analysis found especially high levels of impact in indicators 3 (Group Grievance), 7 (State Legitimacy), and 11 (Factionalized Elites), particularly in the month of April (Table 3).

Taking the 2011 context score for Nigeria as a whole as the baseline, we then layered in the salience of the events such that if in a given month there was a HIGH event salience, the baseline score was multiplied by 120%. If there was a Medium event salience, the score was multiplied by 110%. This rendered the following integrated CAST scores (Table 4).

Taking the sum total of all 12 indicators and placing that total on a graph (on a scale of 0 to 120), we could track trends in overall pressure, which spiked in April and again in November (Fig. 3).

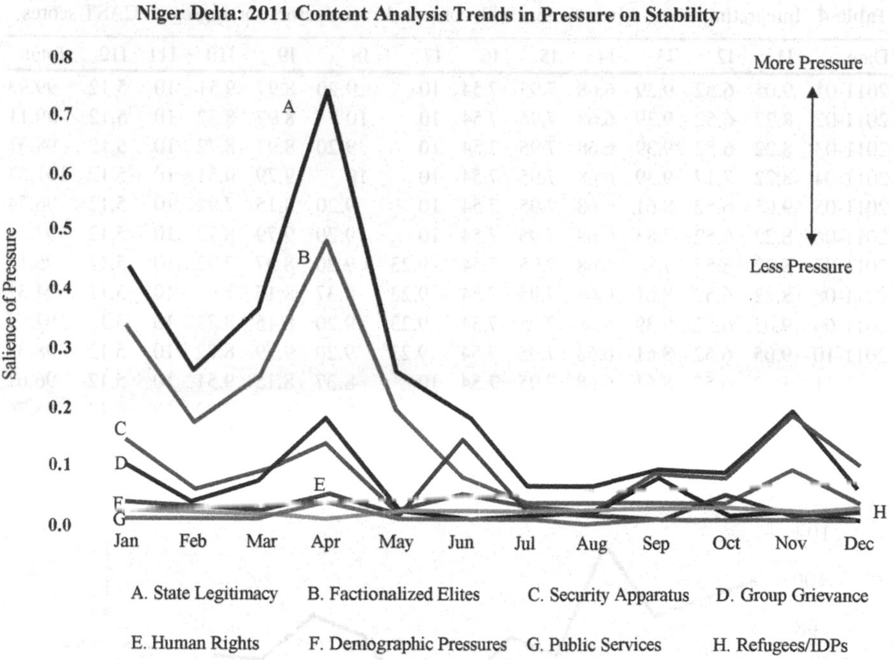

Fig. 2 Measuring trends in event-driven factors using content analysis

Table 3 Measuring the impact of event-driven factors using percentiles of content analysis data

Date	I1	I2	I3	I4	I5	I6	I7	I8	I9	I10	I11	I12
2011-01	M		H	NA	NA	NA	H	M	M	H	H	
2011-02			H	NA	NA	NA	H	H	M	M	H	
2011-03			H	NA	NA	NA	H	M	M	M	H	
2011-04		M	H	NA	NA	NA	H	H	H	H	H	
2011-05	M		M	NA	NA	NA	H	M			H	
2011-06				NA	NA	NA	H	M	H	M	H	
2011-07				NA	NA	NA	M	M	M		M	
2011-08			M	NA	NA	NA	M				M	
2011-09	M		H	NA	NA	NA	M	M		M	M	
2011-10	M		M	NA	NA	NA	M	M	H	M	M	
2011-11			M	NA	NA	NA	H			H	M	
2011-12			M	NA	NA	NA	M	M	M		M	

1.2.3 Participatory Early Warning and Conflict Mapping

The final step in the data piece is to incorporate information generated by networks of local stakeholders based in the conflict-affected communities being assessed. This information, mapped using geo-based technology, can then be juxtaposed with the CAST trends for a high textured evaluation of risk. Local knowledge

Table 4 Integrating event-driven factors with context-driven factors for aggregate CAST scores

Date	I1	I2	I3	I4	I5	I6	I7	I8	I9	I10	I11	I12	Total
2011-01	9.05	6.52	9.39	6.68	7.95	7.54	10	9.20	8.97	9.51	10	5.12	99.93
2011-02	8.22	6.52	9.39	6.68	7.95	7.54	10	10	8.97	8.72	10	5.12	99.11
2011-03	8.22	6.52	9.39	6.68	7.95	7.54	10	9.20	8.97	8.72	10	5.12	98.31
2011-04	8.22	7.17	9.39	6.68	7.95	7.54	10	10	9.79	9.51	10	5.12	101.37
2011-05	9.05	6.52	8.61	6.68	7.95	7.54	10	9.20	8.15	7.92	10	5.12	96.74
2011-06	8.22	6.52	7.83	6.68	7.95	7.54	10	9.20	9.79	8.72	10	5.12	97.56
2011-07	8.22	6.52	7.83	6.68	7.95	7.54	9.23	9.20	8.97	7.92	10	5.12	95.18
2011-08	8.22	6.52	8.61	6.68	7.95	7.54	9.23	8.37	8.15	7.92	10	5.12	94.31
2011-09	9.05	6.52	9.39	6.68	7.95	7.54	9.23	9.20	8.15	8.72	10	5.12	97.55
2011-10	9.05	6.52	8.61	6.68	7.95	7.54	9.23	9.20	9.79	8.72	10	5.12	98.39
2011-11	8.22	6.52	8.61	6.68	7.95	7.54	10	8.37	8.15	9.51	10	5.12	96.67
2011-12	8.22	6.52	8.61	6.68	7.95	7.54	9.23	9.20	8.97	8.72	10	5.12	96.76

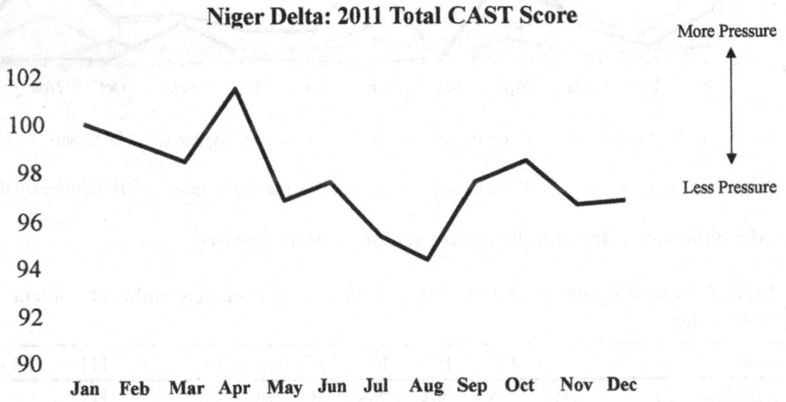

Fig. 3 Aggregating the integrated indicator scores for an overall trend of risk

and perspectives on the conflict risk factors are frequently overlooked in risk assessments. Given the ubiquity of social media and web-mapping technology, filling this gap can be relatively cost-effective. FFP, in partnership with the Port Harcourt-based Institute of Human Rights and Humanitarian Law (IHRHL) has cultivated a network of local participants in the five Niger Delta states of interest who have been thoroughly trained in the CAST framework. These participants regularly enter reports of conflict risk factors into a database which is uploaded to a map produced by The Gadfly Project[6] (www.camp.thegadflyproject.org). The raw data, with the full description of each incident or issue is password protected. But the wider public can query the database to populate the map with markers according

[6]A formation of volunteers dedicated to providing technology solutions to NGO's and economically emerging governments. Their website can be found at http://www.thegadflyproject.org/.

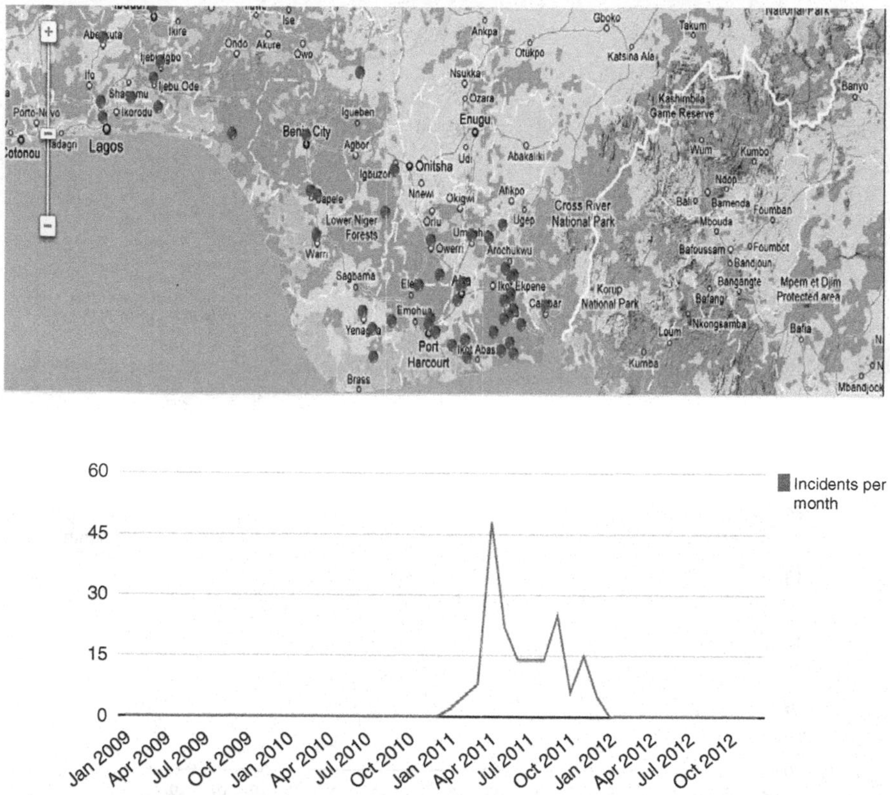

Fig. 4 All risk factors-2011

to date, location, and category. The project is called the Universal Network of Local Knowledge (UNLocK)[7] and is currently operating in Liberia, Uganda, and Nigeria.

For the year 2011 there were 180 incidents uploaded to the map, each of which could fall into as many as three separate categories. Included are four screenshots of the web-map. The first screenshot (Fig. 4) shows all conflict risk factors reported by the early warning network participants during the year. The second screenshot drills down to show only the risk factors relating to pressures on the Security Apparatus indicator. The two categories within the Security Apparatus indicator most pertinent in this context are Organized Crime and Shootings/Killings. Organized crime is particularly relevant in the Niger Delta as a signal of risk because since the 2009 Amnesty program, much of the militancy has partially morphed into criminality including abductions, oil bunkering, and armed robbery. Abductions, for instance, often had a criminal dimension (ransom) as well as a political dimension (intimidation of a candidate or party). Thus the last two screen shots depict the distribution of reports broken out by those two categories (Figs. 4, 5, 6, and 7).

[7] http://www.fundforpeace.org/global/?q=unlock

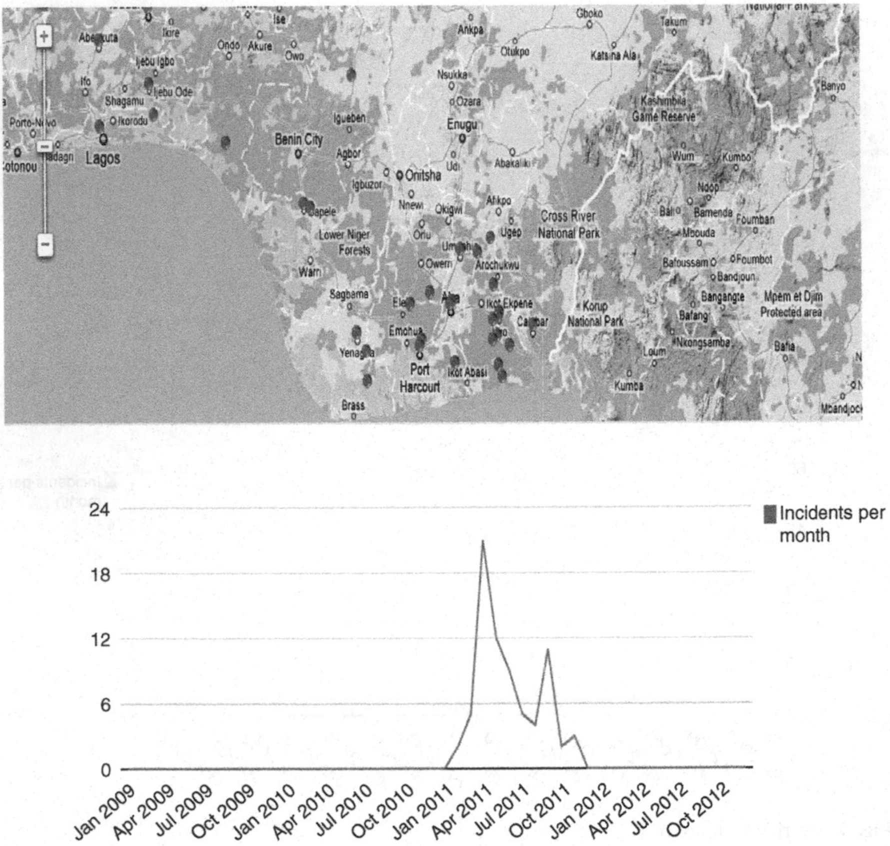

Fig. 5 Security apparatus risk factors-2011

Using this mapping tool, in addition to the CAST trend lines, one can identify areas of high risk, periods of high risk, and analyze the types of risk factors that are most salient in different times and locations. For instance, if you search for only the incidents that relate to Organized Crime, within CAST Indicator 10 (Security Apparatus), you get a more granular picture, whereby 14 out of the 23 incidents reported took place in Akwa Ibom state. Six of those took place in the month of May, 2011. It is important to note, however, that the distribution of incident reports reflects the distribution and priorities of the UNLocK participant on the ground more than it does the distribution of risk factors by location or indicator, highlighting the importance of ensuring a wide network and triangulation of this data against all the other data-streams described in this paper.

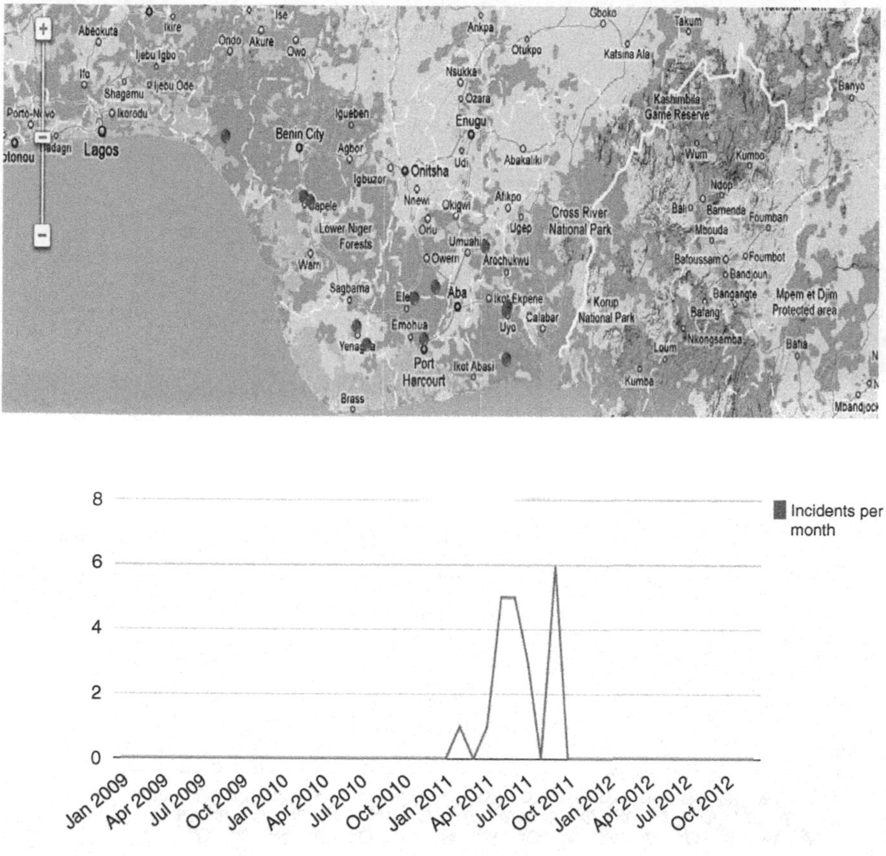

Fig. 6 Security apparatus/organized crime-2011

1.3 *Component 3: Analysis*

1.3.1 Background: The Origins of Conflict in the Niger Delta

Integrating, juxtaposing, and visualizing data will only take you so far. At the end of the day, there needs to be a robust analysis of that information, put in the proper historical context for actionable insight into the trends being identified. This is the point at which innovative technologies and sound theory must give way to the subject matter experts.

The roots of the ongoing crisis in the Niger Delta can be traced much further back than the start of oil production in the 1950s. The Niger Delta is indeed a microcosm of greater Nigeria, with many of the same unique attributes that make it at once a showcase of cultural diversity as well as a tinderbox of explosive social, political, economic and security factors.

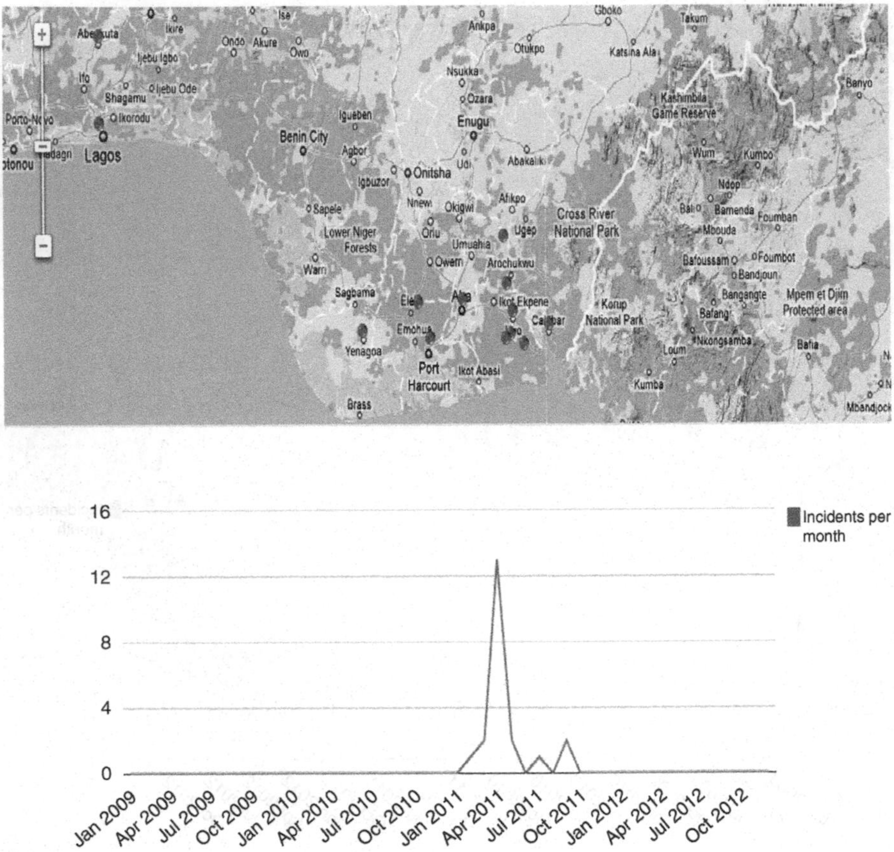

Fig. 7 Security apparatus/shootings and killings-2011

Nigeria is the most populous country in Africa, with an estimated 162,000,000 people living within its borders. It is also the most heterogeneous country in sub-Saharan Africa, with nearly 250 ethno-linguistic groups, of whom approximately 50% identify as Muslim, 40% Christian, with the remaining 10% adhering to traditional practices or a blend thereof. Topographically, it is as diverse as its inhabitants, with the north of the country being arid in contrast to the rich mangroves of the tropical south. Demographically, Nigeria is also a very young country, with nearly half of the population under the age of fourteen. However, despite its large, young population and relative wealth, the average life expectancy is only 47 years of age and the country has one of the highest infant mortality rates in the world.[8]

Nigeria is also emblematic of many social and political extremes, with decades of military rule and war giving way to the proliferation of states which was

[8]CIA World Factbook, 2011.

initially an attempt to ensure equal ethnic and religious representation in order to discourage secessionist movements such as the one which led to the 1967 Biafra War. Additional mechanisms have also been established to balance the interests of the major socio-political constituencies, which have had the perverse effect of further polarizing communities as well as contributing to the sprawl of bureaucracy, and institutionalizing the power of individuals at the state level, rather than those of state institutions such as the legislature and the treasury. This, in turn, has created highly corrupt patronage networks and a political system that was centered on competing ethno-religious identities rather than focusing on individual merit and past performance as qualifications for political leadership. Despite the informally mandated "power shifting" system, which requires the presidency to rotate between the Muslim North and the Christian South every two terms, or eight years, there continues to be high levels of religious tension that has given rise to acts of violent extremism, particularly in the north.[9]

Since the beginning of civilian rule in 1999, Nigeria's social and political landscape has been dominated by decades of a de-facto one-party political system that has further strained national unity and fostered widespread corruption, with the dominant political party, the PDP, continuing to rely on informal power brokers, or "godfathers," to maintain control using complex systems of patronage and oil revenues. Thus, even at the state level, local governors from both the PDP and the opposition parties have a keen economic interest in ensuring their political survival by assimilating into the current system.[10]

The system of "haves" and "have-nots" is most glaring in the sheer economic disparities plaguing Nigeria. In a country highly dependent on a single resource, oil, that has allowed for the proliferation of billionaires, 70% of Nigerians live in poverty with 35% classified as living in absolute poverty.[11] It is this grinding poverty, exacerbated by corrupt and negligent governance and institutional decline that has kept Nigeria on the brink of disaster since its independence in 1960. And this is nowhere as obvious as in the Niger Delta.

The root causes of the Niger Delta's cycle of violence are linked to the vastly uneven distribution of resources that come from the oil industry, coupled with environmental degradation and long-simmering minority tensions. The oil-rich region is located in southern Nigeria at the delta of the Niger River. There are nine oil producing states in the Niger Delta, including the five states covered in this chapter's analysis. All of Nigeria's oil comes from the Delta region which has a population of approximately 32 million, or 20% of the overall population of the country. The demographics of the Niger Delta mirror those of larger Nigeria, with

[9] Baker, Pauline, et al. "Can Nigeria Achieve Unity in Diversity? A Case Study on the Management of Identity." Fund for Peace Publications, December 2010.

[10] Ibid.

[11] Francis, Paul, et al. "Securing Development and Peace in the Niger Delta: A Social and Conflict Analysis for Change." Woodrow Wilson International Center for Scholars Publication. January 2011.

over 62% of the population under the age of thirty and more than forty different ethnic groups, with the Ijaw comprising the largest group. Social indicators such as access to education and health care are poor and infant mortality rates are high. Unemployment is acute despite the booming oil economy and this, when factored with a large youth population with little education and few opportunities, means a restive population easily mobilized for violent aims.[12]

The first significant incidence of violent conflict in the Niger Delta presaged the 1967 Biafra War when an Ijaw-led insurgency movement, known as Niger Delta Volunteer Force, claimed independence for the Niger Delta People's Republic, which consists of both Rivers and Bayelsa States. Although the twelve-day uprising eventually ended unsuccessfully, it represented the first onset of widespread unrest that would eventually culminate in a cycle of violence that continues to the present day. Both the conflict in the Niger Delta and the Biafra War were resource-driven, with local constituencies demanding that oil companies negotiate directly with them, rather than through the federal authorities. In 1967, in an attempt to quell the rising tensions in the Delta, the government broke the East into three states, initially raising hopes for more direct local control of the majority of the oil-producing areas. Optimism was quickly dashed, however, by the 1969 Petroleum Decree that gave the federal government control and exclusive ownership over the petroleum resources. This law was followed up with the 1978 Land Use Decree, which nationalized all state and locally administered land, essentially ending the claims of minority tribes to their land or the resources taken from it.[13]

Throughout the 1980s and into the 1990s tensions about the disbursement of oil revenue seethed alongside growing group grievance surrounding what minority groups perceived as an unresponsive and increasingly corrupt government. In 1990, the grievances of local communities were put forth in the Ogoni Bill of Rights which called for "true" federalism along with greater resource control, self-determination, social justice and calls to end the irresponsible practices that led to the degradation of Ogoni land. This movement, which also held non-violent action as its centerpiece, was largely ignored by the federal government. In the same year, an unrelated oil protest triggered the first significant military action against civilians, further heightening already tense relations between the national government and local communities. Also during this time, relationships continued to fracture between and within ethnic groups and communities as well as between oil companies and communities. The federal government's unwillingness or inability to adequately

---

[12]Baker, Pauline, et al. "Can Nigeria Achieve Unity in Diversity? A Case Study on the Management of Identity." Fund for Peace Publications, December 2010.

[13]Baker, Pauline, et al. "Can Nigeria Achieve Unity in Diversity? A Case Study on the Management of Identity." Fund for Peace Publications, December 2010; Francis, Paul, et al. "Securing Development and Peace in the Niger Delta: A Social and Conflict Analysis for Change." Woodrow Wilson International Center for Scholars Publication. January 2011.

address local grievances and provide basic human services fueled deep resentments that were often vented against the oil companies, triggering further repressive action from the state.[14]

Following the assassination of the Ogoni leader, Ken Saro-Wiwa in 1995, violent uprisings began to take hold throughout the region, as more minority groups demanded ownership of land and direct access to oil revenues. This resulted in a hardening of social identities and increased violence as weapons flowed into the region and the military began to employ increasingly violent tactics in order to repress the uprisings. By 2005, the UNDP estimated that 20,000–25,000 armed youths comprising about fifty groups were operating in the oil-producing regions and violent community conflicts in Rivers, Bayelsa and Delta states had increased to over 120 per year[15]. During this time period, hostage taking, oil-bunkering and direct attacks on oil facilities and the pipelines became common occurrences. In 2003, the government deployed 15,000 Army, Navy and Air Force troops to the region as part of the Joint Task Force (JTF) in a campaign known as "Operation Restore Hope" which resulted in the deaths of hundreds of civilians, many of them unarmed, and the wide-scale destruction of villages and farmland. In 2005, an umbrella group called Movement for the Emancipation of the Niger Delta (MEND), comprised of various armed groups, began a violent campaign against the state, abandoning any remaining initiatives at a peaceful resolution to the conflict after the failure of a government-brokered peace agreement in 2004.[16]

MEND proved to be a formidable force in the region, able to reach offshore oil platforms and wreak havoc that hindered production, falling from a peak production rate of 2.6 million barrels per day (mbpd) to 1.3 mbpd in 2009. In May of 2009, the government launched an all-out military operation against MEND in the western Delta, including aerial bombardment campaigns that razed entire villages and sent thousands fleeing the area, creating an internal displacement crisis. Finally, later that year, President Yar'Adua was able to negotiate an amnesty program with senior MEND commanders, and over 26,000 combatants disarmed and demobilized. Beyond the monetary incentive—each former combatant received a monthly stipend of $433, more than three times the minimum wage— the Amnesty program also provided skills training for social reintegration and employment. As of August 2011, it was estimated that 19,000 men and women had gone through a non-violence and social reintegration course while approximately 3,200 had been matched to training programs at various Nigerian institutions.[17]

[14]Asuni, Judith Burdin. "Understanding the Armed Groups of the Niger Delta, A Working Paper" Council on Foreign Relations Publication, September 2009.

[15]UNDP. 2006. "Niger Delta Human Development Report." United Nations Development Program, Abuja. http://web.ng.undp.org/reports/nigeria_hdr_report.pdf.

[16]Francis, Paul, et al. "Securing Development and Peace in the Niger Delta: A Social and Conflict Analysis for Change." Woodrow Wilson International Center for Scholars Publication. January 2011.

[17]Ibid.

While the military crackdown and the Amnesty program were largely successful in quelling the large-scale violent attacks in the Delta, human insecurity remains a major problem as significant social, political, and economic challenges remain. Among the key social issues affecting the Delta overall, and the five states chosen for this analysis in particular, are ethnic, communal, and intra-communal tensions that have sprung up around real and perceived economic disparities. An inter-generational division has resulted from the unraveling of traditional societal and familial norms in the face of decades of violent conflict and trans-communal challenges that the traditional institutions are unequipped to deal with effectively. Demographically, despite Nigeria's oil-producing states receiving large revenue allocation, these funds have not translated to addressing core drivers of conflict emanating from a basic lack of access to social goods and public services like healthcare or education. Environmental degradation continues to be a huge driver of instability in the region, with oil spills and ilegal oil bunkering contaminating farmland and water sources and causing myriad health problems in addition to already staggering rates of HIV and Malaria infections. In addition, outbreaks of violence as well as environmental factors also influence the flow of internally displaced persons, taxing land and limited resources in already population-dense areas, greatly increasing the chance of conflict.[18]

Severe human rights abuses have been a factor throughout the lifetime of the conflict in the Delta, with multiple atrocities committed by the government, various armed groups and gangs, private security forces, as well as overall ethnically- and politically- motivated violence. Although kidnapping and hostage taking of foreign nationals from oil facilities and offshore platforms has dropped significantly over the past two years, it continued to pose a threat to local communities and is often motivated by criminal gains and political intimidation, particularly around election periods. Lack of development in the region, both in terms of public services and infrastructure, as well as minimal job opportunities, also fuels a high degree of criminality and corruption. In this environment, the informal sector dominates resulting in a lack of transparency and accountability, further fueling distrust of the government as well as among various ethnic groups. There is a high degree of factionalization among political elites in the five states analyzed in this chapter, emanating from competition for oil revenue and resources that drive identity and personality-based politics that foment violence encourage a "winner takes all" political culture.[19]

Group grievance, between and among the multitude of minority tribes that occupy the Delta, as well as between these groups and the larger Hausa, Yoruba, and Ibo tribes, is easily manipulated for both political and economic gains. Particularly acute in the Delta, identity-based grievances stem from both the under-representation of minority groups at the federal level and the real and perceived

[18] Asuni, Judith Burdin. "Understanding the Armed Groups of the Niger Delta, A Working Paper" Council on Foreign Relations Publication, September 2009.

[19] Baker, Pauline, et al. "Can Nigeria Achieve Unity in Diversity? A Case Study on the Management of Identity." Fund for Peace Publications, December 2010.

exploitation of land and resources. Absent adequate mechanisms for conflict resolution or the airing of group grievances at the state level, perceptions of group exclusion and exploitation have been deepened by the erosion of traditional dispute resolution mechanisms amidst the culture of violence and conflict that have plagued the region. In this atmosphere, violent protests are a common form of social unrest and often further solidify group consciousness.[20]

Because the Niger Delta, and larger Nigeria as a whole, has inadequate systems or processes by which to deal with group grievances before they become violent, the traditional response has been the employment of an overwhelming use of force by the security apparatus, culminating in a wave of arrests, assaults, displacement and gender-based violence. As noted by Francis et al. in "Securing Peace and Development in the Niger Delta," following waves of violent protest and the responding overwhelming use of force by the state security apparatus—sometimes coupled with monetary and political concessions—what ensues is a 'soft stalemate,' where both sides agree to an informal cession of hostilities. However, when the promised concessions are slow to come, perceived as inadequate by the protestors, or never materialize, the cycle of violence begins anew, often drawing in new groups. The 2009 Amnesty is viewed by many scholars and experts to be just such an arrangement, albeit a moderately to highly successful one. When the program of monetary pay-outs and other benefits comes to an end, many fear that without any real progress at addressing the core structural and developmental issues that plague the Delta, there will be little incentive for armed groups not to return to violence.[21]

In addition to the myriad social and political factors that fuel the cycle of violence and conflict in the Niger Delta, the levels of instability in the region have led to the proliferation of multiple groups with access to the use of violent force. The JTF, which has multiple bases throughout the Delta states, is widely employed by the government to quell both violent and non-violent protests, often using overwhelming force. Human rights abuses are frequent and, as demonstrated at various times over the past decade, the state security apparatus occasionally employs "scorched earth" tactics, razing villages in an attempt to root out militia leaders and quell any future acts of subversion. The training of these forces is not uniform and some are reportedly complicit in criminal activities such as ilegal oil bunkering or the sabotage of oil infrastructure. In addition to public security forces, multiple armed groups are present throughout the region, coalescing around various political, social and criminal interests. These militia and vigilante groups are sometimes used as "guns for hire" by various entities, including the state, oil companies, and local political and economic power brokers. Finally, due to the high levels of insecurity in

---

[20]Asuni, 2009; International Crisis Group: "Nigeria's Federal Experiment" Africa Report Number 119, October 2006.

[21]Baker, Pauline, et al. "Can Nigeria Achieve Unity in Diversity? A Case Study on the Management of Identity." Fund for Peace Publications, December 2010; Francis, Paul, et al. "Securing Development and Peace in the Niger Delta: A Social and Conflict Analysis for Change." Woodrow Wilson International Center for Scholars Publication. January 2011.

the region, outside companies often employ private security forces to protect their personnel and economic interests. These private contractors can include elements linked to local armed groups, who are paid to protect oil infrastructure, or ex-combatants who have gone through government-sanctioned rehabilitation programs. In each case, they may have a vested interest in fomenting instability in order to ensure their job security.[22]

Importantly, as noted above, with the shifting political and social dynamics in the Delta, coupled with the widespread availability of weapons and easy mobilization of fighters, there has been a sharp rise in the level of organized crime. Militant groups, frustrated with slow or non-existent political, economic, and social gains, have increasingly turned to abductions for ransom, oil bunkering, armed robbery, weapons trafficking, and other illicit activities. Despite the 2009 Amnesty Law, high levels of unemployment and underemployment and a growing youth population with few opportunities for education have led to a burgeoning and increasingly sophisticated criminal network. While the amnesty did initially draw in many of MEND's top level commanders and a militants, it was not uniformly accepted by all groups nor did it address the core structural and economic disparities that fueled the conflict from the onset. There is a perception among some, for example, that the Ijaw ethnic group benefited disproportionately from the Amnesty Program, exacerbating the potential for ethnic conflict down the road. While oil companies and the government have made strides in putting in place more development-related programs and community assistance packages, the lure of profits from the criminal sphere remains too lucrative to effectively combat the level of organized crime present in the region. The complicity of local level political figures and the armed forces in such activities further complicates efforts to dismantle the system.[23]

1.3.2 The Niger Delta in 2011

Digging into the raw data (particularly the UNLocK incident reports, and the documents highlighted by content analysis) behind the charts and figures above, the dynamics of risk in the year 2011 can be unpacked, for a deeper understanding of recent trends. The complex and interrelated patchwork of social, economic, political and military factors that have fueled unrest in the Niger Delta for decades continued to influence the stability of the region throughout the year. However, with the continuing amnesty, and an election widely regarded as the most free and fair in Nigeria's history, there were areas of improvement.

---

[22]Francis, Paul, et al. "Securing Development and Peace in the Niger Delta: A Social and Conflict Analysis for Change." Woodrow Wilson International Center for Scholars Publication. January 2011.

[23]Baker, Pauline, et al. "Can Nigeria Achieve Unity in Diversity? A Case Study on the Management of Identity." Fund for Peace Publications, December 2010; International Crisis Group, "Ending Unrest in the Niger Delta," Africa Report Number 135, December 2007.

Of the social and political indicators analyzed here, demographic-related pressures and access to public services did not show significant signs of change. As indicated in the previous section on data collection and context, certain indicators are less amenable to content analysis as a means to measure risk and, short of a catastrophic event, demographic pressures and public services tend to change at a much slower rate than other indicators. Similarly, for the indicator that measures human flight, or "brain drain," positive and negative trending tends to occur over longer periods of time and change is not as easily captured in a twelve-month time period. Despite these caveats, it is clear from the data that there were no significant improvements in those indicators in 2011, with poverty, disease, poor sanitation, environmental degradation caused by oil spills, and overall population pressures remaining high.

The economic indicators, which capture uneven economic development and overall economic pressures, also require a higher degree a contextual analysis (rather than an analysis of event-driven pressures) and were not, on their own, useful in assessing the risk of terrorism during this time period. However, taken in combination with the other indicators and as a general barometer of the overall risk environment, it is clear that poverty and corruption remain critical underlying factors of social and political unease and drivers of conflict. While the continued Amnesty pay-outs to rebel leaders and other militants continued throughout the time period analyzed and certainly helped to ease economic tensions in some respects, they cannot be considered as influencing long-term economic and social development.

The indicators that proved most salient for a risk assessment during this time period were political and security-driven, although they certainly fueled group grievance. At the national level, following the death of President Yar'Adua in late 2010, the country was thrown into a political and ethno-religious crisis as the ruling party's policy of rotating the presidency between the Muslim North and Christian South again came to the forefront. Yar'Adua, only two years into his first term, was a northerner while his vice president, Goodluck Jonathan, was an ethnically Ijaw son of the south from the state of Bayelsa in the Niger Delta region. Upon President Yar'Adua's death, Jonathan was sworn in temporarily until the elections slated for April of 2011. Controversy immediately arose surrounding the question of whether only northerners should contend for the PDP nomination, given that the presidency was technically to be held by one group or the other for a full eight years. Naturally, debates surrounding the issues greatly heightened group grievances as well as factionalized political elites around ethnic and religious identities.

Within the states analyzed in this paper, Jonathan's position as PDP standard bearer was not controversial, however. Support of Goodluck Jonathan was widespread. But group grievances did escalate during this time at the local level. Sharp political rhetoric and attempts to secure votes by the candidates through a combination of bribery and intimidation brought these grievances to the surface. During the early part of 2011, as well as in the months following the election, conflict rippled throughout the region between, pastoralists and farmers, indegenes and non-indegenes, political parties, rival gangs, and state security forces and local communities. Protests frequently turned violent and there were multiple reports of violence and killings targeted against specific groups. During the year, the Delta also witnessed such violence that triggered the excessive use of force by the JTF and local police, including one incident in Delta state where a JTF operation reportedly resulted in the death of 51 civilians, including women and children. Overall, throughout the year, in election-related violence, general protests, and state-led action against armed groups and gangs, human rights violations occurred widely. Such violations included reports of rape, torture, assassination, indiscriminate killings, and other campaigns of political terror.

In the pre-and post-election environment in the Niger Delta, there was a sharp increase in the political factionalization of elites. As noted earlier in the analysis, the country and region have both been virtually captive to the dominate PDP party since the re-instatement of civilian rule in 1999. Due to the exceptionally lucrative economic driver of state control over resource revenue, the political landscape in the Niger Delta, and in the states analyzed, has been driven by zero-sum political tactics. These tactics have historically included the use of intimidation and violence against opposition parties, journalists, and civil society groups attempting to challenge the status quo or expose corrupt practices. During 2011, there was a strong spike in political factionalization that can be correlated with a rapid deterioration of perceived state legitimacy. Although by most expert accounts the freeness and fairness of this election was historical, there still remained numerous incidents of ballot-box snatching, vote-buying, and other corrupt practices. In one instance in Akwa Ibom State, the local governor had reportedly promised certain constituents monetary remuneration through "solidarity packages," which, when they were perceived as inadequate, quickly turned to violence. In other incidences reported, by the UNLocK network rural women were bribed with rice, industrial salt, and money in exchange for their vote.

The proliferation of armed gangs, rival militias, and weapons in the region also fueled violence and greatly exacerbated the risk potential of terrorist action as these groups were mobilized and manipulated for political and economic gains. The employment of gangs by both the dominant and opposition parties to intimidate both voters and rival politicians was widespread. In a landscape dominated by extreme social and economic disparities, and with the ready availability of a disillusioned and an unskilled young labor pool, criminality and violence are rife. For the time period analyzed, particularly around the election, these gangs and militias were mobilized along ethnic and political lines, together with multiple widespread and lucrative opportunities to engage in criminal enterprises. While oil-bunkering, attacks on oil facilities and the targeting of foreign nationals for kidnap and ransom have long been drivers of violence and conflict in the Delta, the added dimension of turning additional profit through the use of political terror was acute during 2011. Pressures on the indicator that measures the security apparatus, and the correlation to political factionalization, group grievance, and state legitimacy peaked in April 2011 and, although decreased thereafter. This represented the employment of various armed factions and gangs to intimidate local populations as well as minority groups through many of the same tactics of kidnapping for ransom, torture, assassination, and gender-based violence.

Finally, although the violence that peaked in the Delta was not as acute as was experienced in the North during this period, and while the elections were largely deemed fair, this analysis proved useful in digging into the data for a deeper understanding of factors behind the risk of terror in the region. These factors include the continued social and economic disparities that are unique to the region and fuel instability and violence. The overall reliance of the Nigerian economy on the oil sector already predisposes it for wider instability due to the lack of diversification of economic revenue. In the Niger Delta, where high levels of wealth and power coexist along with extreme levels of poverty and social fragmentation, this gives rise to severe political cleavages and criminality that are easily manipulated. As social and identity-based movements for greater self-determination and political recognition continue to be frustrated at the local level, the lure of the illicit economy and illegal criminal enterprise are heightened. Adding to an already dangerous mixture of social and political ideologies, easily exploitable economic resources, and a growing youth population with few opportunities for employment or skilled labor, a culture of violence and impunity prevails.

1.3.3 Conclusion and Outlook for the Future: Mitigating Terrorism Risks

Overall, the continuing cycle of violence in the Niger Delta over the past two decades has hindered both social and economic development and given rise to an environment of criminality and impunity. It has also devastated local communities and hindered the delivery of basic public services deemed critical to reverse the trend toward continued instability. Moreover, despite the recent Amnesty Program and the suppression of violence in the region, the root causes that have fueled the cycle of violence and instability have not been fundamentally addressed and continue to simmer beneath the surface. Deep-seated group grievances that date back to the colonial era have not been sufficiently addressed and, more often, are exploited at the national and local level in today's high-stakes political and economic environment. Amid the cyclical violence, a breakdown in the traditional fabric of local communities has resulted in fewer indigenous efforts to mediate conflict than at any time in the past and compromised the success of such initiatives. Compounding these factors is a culture of violence that prevails following decades of armed insurgencies and opportunities for armed groups to be co-opted by a variety of actors for political and economic purposes.

Although the outlook for the Niger Delta in the coming years is precarious, there have been successful initiatives to address the security-development vacuum that exists in the region. Oil companies have increasingly filled the space through multiple community development and enrichment programs that have had varying degrees of success but demonstrate a developing level of company-community cooperation that have proven useful in averting outbreaks of violence.[24] Additionally, companies have also recognized the long-term benefits of employing a systematic approach to the local hiring of security personnel by relying increasingly on trusted community organizations that provide trained and vet candidates. Increasingly, such organizations are subsidized by both companies and the local government through both direct and indirect support for community development, discouraging the practice of hiring short-term, one-off contractors who may have a vested interest in contributing to instability to ensure employment opportunities.

National efforts at coordinating development programs and providing opportunities for gainful employment have been slower to materialize, due largely to the highly centralized, top-down, nature of such programs. However, the creation of development agencies at the state level and an increase in government-funded studies aimed at finding innovative ways to discourage violence and promote social and political inclusion are promising signs. Increases in public-private partnerships and better donor coordination at the international level have also been positive trends over the past decade. Perhaps more than the rest of Nigeria, the Niger Delta has a vibrant and diverse civil society that is increasingly viewed by both outside actors and the state as fundamental to the determination of the future of the region.

Nevertheless, core weaknesses at the federal and state levels, coupled with knee-jerk policies of violent repression and short-term pay-offs, continue to pose significant obstacles to progress. Despite multiple initiatives aimed at tackling the environment of violence and impunity that plague the Delta, an overall national and regional strategy is lacking. Moreover, as long as the short-term economic and political pay-offs to using violence and inciting instability remain, the potential risk factor for terrorism will prevail. Only through a comprehensive strategy that includes both top-down and bottom-up initiatives that address the multitude of social, political, economic and security-related drivers that fuel terrorism and violent conflict, will the region become stable and prosperous. In the end, such a state would benefit all interests by promoting the well-being and safety of the Niger Delta's fragile human and ecological systems. Until such coordination for the overall benefit of all actors is realized, the root causes driving violence and upheaval will similarly present the opportunity for terrorist exploitation.

References

1. Asuni JB (2009) Understanding the armed groups of the Niger delta, a working paper. Council on Foreign Relations Publication. http://www.cfr.org/nigeria/understanding-armed-groups-niger-delta/p20146

[24]Francis, Paul, et al. "Securing Development and Peace in the Niger Delta: A Social and Conflict Analysis for Change." Woodrow Wilson International Center for Scholars Publication. January 2011.

2. Baker PH, Ferroggiaro W, Haken N (2010) Can Nigeria achieve unity in diversity? A case study on the management of identity. Fund for Peace Publications. http://www.fundforpeace.org/global/?q=cpgpr1012
3. Brandt PT, Freeman JR, Schrodt PA (2011) Real Time, Time Series Forecasting of Inter- and Intra-state Political Conflict. Conflict Management and Peace Sci 28(1):41–64
4. Fearon J, Laitin D (2003) Ethnicity, insurgency and civil war. Am Polit Sci Rev 97:75–90
5. Francis P, LaPin D, Rossiasco P (2011) Securing development and peace in the Niger delta: a social and conflict analysis for change. Woodrow Wilson International Center for Scholars Publication. http://www.wilsoncenter.org/sites/default/files/AFR_110929_Niger%20Delta_0113.pdf
6. Haken N, Burbank J, Baker PH (2010) Casting globally: using content analysis for conflict assessment and forecasting. Mil Operations Rcs Soc J 15(2):5–19
7. International Crisis Group (2007) Ending unrest in the niger delta. Africa Report Number 135, December
8. International Crisis Group (2006) Nigeria's federal experiment. Africa Report Number 119, October 2006
9. Rotberg RI (2004) When states fail: causes and consequences. Princeton University Press, Princeton
10. UNDP (2006) Niger delta human development report. United Nations Development Program, Abuja. http://web.ng.undp.org/reports/nigeria_hdr_report.pdf
11. Weidmann NB, Ward MD (2010) Predicting conflict in space and time. J Conflict Resolution 54(6):883–901

Policy Analytics Generation Using Action Probabilistic Logic Programs

Gerardo I. Simari, John P. Dickerson, Amy Sliva, and V.S. Subrahmanian

1 Introduction

Action probabilistic logic programs (*ap*-programs for short) [15] are a class of the extensively studied family of probabilistic logic programs [14, 21, 22]. *ap*-programs have been used extensively to model and reason about the behavior of groups and an application for reasoning about terror groups based on *ap*-programs has users from over 12 US government entities [10]. *ap*-programs use a two sorted logic where there are "state" predicate symbols and "action" predicate symbols[1] and can be used to represent behaviors of arbitrary entities (ranging from users of web sites to institutional investors in the finance sector to corporate behavior) because they

[1] Action atoms only represent the fact that an action is taken, and not the action itself; we assume that effects and preconditions are generally not known.

G.I. Simari (✉)
Department of Computer Science, University of Oxford, Oxford OX1 3QD, UK
e-mail: gerardo.simari@cs.ox.ac.uk

J.P. Dickerson
Gates-Hillman Center, School of Computer Science, Carnegie Mellon University, Pittsburgh, PA 15213, USA
e-mail: dickerson@cs.cmu.edu

A. Sliva
College of Computer and Information Science, Northeastern University, Boston, MA 02115, USA
e-mail: asliva@ccs.neu.edu

V.S. Subrahmanian
Department of Computer Science, University of Maryland College Park, College Park, MD 20742, USA
e-mail: vs@cs.umd.edu

V.S. Subrahmanian (ed.), *Handbook of Computational Approaches to Counterterrorism*, DOI 10.1007/978-1-4614-5311-6_23,

r_1▷clashCas(*jk*‹ 1) : [0▷85‹0▷91] ← socStrife(*pak*‹ 1) ∧ allianceNSAG(1)▷
r_2▷clashCas(*jk*‹ 1) : [0‹0▷03] ← socStrife(*pak*‹ 0) ∧ allianceNSAG(0)▷
r_3▷clashCas(*jk*‹ 1) : [0▷42‹0▷48] ← socStrife(*pak*‹ 0) ∧ allianceNSAG(1)▷
r_4▷murder(*jk*‹ 1) : [0▷52‹0▷58] ← trainCamp(*pak*‹ 1) ∧ relOrg(1)▷
r_5▷murder(*jk*‹ 1) : [0‹0▷03] ← trainCamp(*pak*‹ 0) ∧ relOrg(0)▷
r_6▷murder(*jk*‹ 1) : [0▷20‹0▷26] ← trainCamp(*pak*‹ 0) ∧ relOrg(1)▷
r_7▷fedayeenAtt(*jk*‹ 1) : [0▷59‹0▷65] ← govMilSupp(*pak*‹ 1) ∧ relOrg(1)▷
r_8▷fedayeenAtt(*jk*‹ 1) : [0‹0▷03] ← govMilSupp(*pak*‹ 0) ∧ relOrg(0)▷
r_9▷fedayeenAtt(*jk*‹ 1) : [0▷22‹0▷28] ← govMilSupp(*pak*‹ 0) ∧ relOrg(1)▷

Fig. 1 A small set of rules modeling Lashkar-e-Taiba

consist of rules of the form "if a conjunction C of atoms is true in a given state S, then entity E (the entity whose behavior is being modeled) will take action A with a probability in the interval $[\ell, u]$."

In such applications, it is essential to avoid making probabilistic independence assumptions, since the approach involves *finding out* what probabilistic relationships exist and then exploiting these findings in the forecasting effort. For instance, Fig. 1 shows a small set of rules automatically extracted from data gathered in the *Computational Modeling of Terrorism* (CMOT) project[2] about Lashkar-e-Taiba's past (referred to from now on as LeT), where predicates correspond to rules in the data set and in general a value of 0 indicates that the action is not performed or the condition does not hold.[3] Rule 1 says that LeT engages in clashes in Jammu and Kashmir (J&K, from now on), inflicting casualties in security forces (action `clashCas`), with probability between 0.85 and 0.91 whenever there is social strife in Pakistan (condition `socStrife`), and LeT engages in alliances with non-state armed groups (condition `allianceNSAG1`). Rules 2 and 3, also about such clashes, refer to the probabilities when these conditions are slightly altered. The rest of the rules refer to murders committed in J&K, and Fedayeen attacks carried out in J&K, involving the following conditions: `trainCamp`, which means that LeT maintains training camps; `relOrg`, referring to the condition of LeT as a religious group, and `govMilSupp`, referring to the Pakistani government giving military support. *ap*-programs have been extensively (and successively) used by terrorism analysts to make predictions about terror group actions [10, 19]. The analysis of LeT from which these rules were taken is described in depth in [20].

Suppose, rather than predicting what action(s) a group would take in a given situation or environment, we want to determine what *we can do* in order to induce a given behavior by the group. For example, a policy maker might want to understand what we can do so that a given goal (e.g., the probability of LeT engaging in clashes causing casualties is below some percentage) is achieved, given some constraints on what is feasible. The *basic abductive query answering problem* (BAQA) deals

[2] http://www.umiacs.umd.edu/research/LCCD/projects/let.jsp

[3] Note that variables can have more than two possible values; therefore, even though `murder`(1) is equivalent to ¬`murder`(0) because `murder` is a binary variable, this does not hold in general.

with finding how to *reach* a new (feasible) state from the current state such that the *ap*-program associated with the group and the new state jointly entail that the goal will be true within a given probability interval.

We will also take the problem one step further by reasoning about how the entity being modeled *reacts* to our efforts. We are interested in identifying the *best* course of action on our part, given some additional inputs regarding the cost of exerting influence in the environment and how desirable certain outcomes are; this is called the *cost-based query answering problem* (CBQA). We describe a heuristic algorithm based on probability density estimation techniques that can be used to tackle CBQA with large instances, and then present parallel algorithms capable of solving these problems faster. Finally, we describe a prototype implementation and experimental results showing that our algorithms scale well, and achieve results that are useful in practice.

2 Preliminaries

We now overview the syntax and semantics of *ap*-programs from [15].

2.1 Syntax

We assume the existence of a logical alphabet that consists of a finite set $\mathscr{L}_{cons}$ of constant symbols, a finite set $\mathscr{L}_{pred}$ of predicate symbols (each with an associated arity) and an infinite set $\mathscr{L}_{var}$ of variable symbols; function symbols are not allowed. Terms, atoms, and literals are defined in the usual way [17]. We assume that $\mathscr{L}_{pred}$ is partitioned into disjoint sets: $\mathscr{L}_{act}$ of *action symbols* and $\mathscr{L}_{sta}$ of *state* symbols. Thus, if $t_1, \ldots, t_n$ are terms, and p is an n-ary action (resp. state) symbol, then $p(t_1, \ldots, t_n)$, is called an *action (resp. state) atom*.

Definition 1 (Action formula). A (ground) action formula is defined as:

- A (ground) action atom is a (ground) action formula;
- If F and G are (ground) action formulas, then $\neg F$, $F \wedge G$, and $F \vee G$ are also (ground) action formulas.

The set of all possible action formulas is denoted by $formulas(B_{\mathscr{L}_{act}})$, where $B_{\mathscr{L}_{act}}$ is the Herbrand base associated with $\mathscr{L}_{act}$, $\mathscr{L}_{cons}$, and $\mathscr{L}_{var}$.

Definition 2 (*ap*-formula). If F is an action formula and $\mu = [\alpha, \beta] \subseteq [0, 1]$, then $F : \mu$ is called an *annotated action formula* (or *ap*-formula), and μ is called the *ap*-annotation of F.

In the following, we will use APF to denote the set of all possible *ap*-formulas.

Definition 3 (World/State). A *world* is any finite set of ground action atoms. A *state* is any finite set of ground state atoms.

It is assumed that all actions in a world are carried out more or less in parallel and at once, given the temporal granularity adopted along with the model. Contrary to (related but essentially different) approaches such as stochastic planning, we assume here that it is not possible to directly reason about the effects of actions. One reason for this is that in many applications (e.g., counter-terrorism), there are many, many variables, and the effects of our actions are not well understood. We now define *ap*-rules.

Definition 4 (*ap*-rule). If F is an action formula, $B_1,\ldots,B_n$ are state atoms, and μ is an *ap*-annotation, then $F:\mu \leftarrow B_1 \wedge \ldots \wedge B_m$ is called an *ap-rule*. If this rule is named r, then $Head(r)$ denotes $F:\mu$ and $Body(r)$ denotes $B_1 \wedge \ldots \wedge B_n$.

Intuitively, the rule specified above says that if $B_1,\ldots,B_m$ are all true in a given state, then there is a probability in the interval μ that the action combination F is performed by the entity modeled by the *ap*-rule.

Definition 5 (*ap*-program). An *action probabilistic logic program* (*ap*-program for short) is a finite set of *ap*-rules. An *ap*-program Π' such that $\Pi' \subseteq \Pi$ is called a *subprogram* of Π.

Figure 1 shows a small portion of an *ap*-program we derived automatically to model LeT's actions. On the average, we have derived *ap*-programs consisting of approximately 11,500 *ap*-rules per terror group.

Henceforth, we use *Heads*(Π) to denote the set of all annotated formulas appearing in the head of some rule in Π. Given a ground *ap*-program Π, we will use *sta*(Π) (resp., *act*(Π)) to denote the set of all state (resp., action) atoms that appear in Π.

Example 1 (Worlds and states). Coming back to the *ap*-program in Fig. 1, the following are examples of worlds:

$$\{\texttt{clashCas}(jk,1)\},\{\texttt{clashCas}(jk,1),\texttt{fedayeenAtt}(jk,1)\},\{\}$$

The following are examples of states:

$$\{\texttt{socStrife}(pak,0),\texttt{trainCamp}(pak,1)\},$$
$$\{\texttt{socStrife}(pak,1),\texttt{relOrg}(1)\}.$$ ■

2.2 Semantics of ap-Programs

We use $\mathscr{W}$ to denote the set of all possible worlds, and $\mathscr{S}$ to denote the set of all possible states. It is clear what it means for a state to satisfy the body of a rule [17].

Definition 6 (Satisfaction of a rule body). Let Π be an *ap*-program and s a state. We say that s *satisfies* the body of a rule $F : \mu \leftarrow B_1 \wedge \ldots \wedge B_m$ if and only if $\{B_1, \ldots, B_M\} \subseteq s$.

Similarly, we define what it means for a world to satisfy a ground action formula:

Definition 7 (Satisfaction of an action formula). Let F be a ground action formula and w a world. We say that w *satisfies* F if and only if:

- If $F \equiv a$, for some atom $a \in B_{\mathscr{L}_{act}}$, then $a \in w$;
- If $F \equiv F_1 \wedge F_2$, for formulas $F_1, F_2 \in formulas(B_{\mathscr{L}_{act}})$, then w satisfies F_1 and w satisfies F_2;
- If $F \equiv F_1 \vee F_2$, for formulas $F_1, F_2 \in formulas(B_{\mathscr{L}_{act}})$, then w satisfies F_1 or w satisfies F_2;
- If $F \equiv \neg F'$, for some formula $F' \in formulas(B_{\mathscr{L}_{act}})$, then w does not satisfy F'.

Finally, we will use the concept of *reduction* of an *ap*-program w.r.t. a state:

Definition 8 (Reduction of an *ap*-program). Let Π be an *ap*-program and s a state. The *reduction of* Π *w.r.t.* s, denoted Π_s, is the set $\{F : \mu \leftarrow Body \mid s$ satisfies $Body$ and $F : \mu \leftarrow Body$ is a ground instance of a rule in $\Pi\}$. Rules in this set are said to be *relevant* in state s.

The semantics of *ap*-programs uses possible worlds in the spirit of [9, 11, 23]. Given an *ap*-program Π and a state s, we can define a set $LC(\Pi, s)$ of linear constraints associated with s. Each world w_i expressible in the language $\mathscr{L}_{act}$ has an associated variable p_i denoting the probability that it will actually occur. $LC(\Pi, s)$ consists of the following constraints.

1. For each $Head(r) \in \Pi_s$ of the form $F : [\ell, u]$, $LC(\Pi, s)$ contains the constraint $\ell \leq \sum_{w_i \in \mathscr{W} \wedge w_i \models F} p_i \leq u$.
2. $LC(\Pi, s)$ contains the constraint $\sum_{w_i \in \mathscr{W}} p_i = 1$.
3. All variables are non-negative.
4. $LC(\Pi, s)$ contains only the constraints described in (1)–(3).

While [15] provide a more formal model theory for *ap*-programs, we merely provide the definition below. Π_s is *consistent* iff $LC(\Pi, s)$ is solvable over $\mathbb{R}$.

Definition 9 (Entailment of an *ap*-formula). Let Π be an *ap*-program, s a state, and $F : [\ell, u]$ a ground action formula. Π_s *entails* $F : [\ell, u]$, denoted $\Pi_s \models F : [\ell, u]$ iff $[\ell', u'] \subseteq [\ell, u]$ where:

- $\ell' =$ **minimize** $\sum_{w_i \in \mathscr{W} \wedge w_i \models F} p_i$ **subject to** $LC(\Pi, s)$.
- $u' =$ **maximize** $\sum_{w_i \in \mathscr{W} \wedge w_i \models F} p_i$ **subject to** $LC(\Pi, s)$.

Note that, even though Definition 9 defines entailment for reduced programs (i.e., w.r.t. a state), the definition contemplates general programs, since given an arbitrary set of rules there always exists a state that makes it relevant.

The quantity ℓ' in the above definitions is the smallest possible probability of F, given that the facts in Π are true. In the same vein, u' is the largest such probability. If the $[\ell', u']$ interval is contained in $[\ell, u]$, then $F : [\ell, u]$ is definitely entailed by Π.

3 Abductive Queries to Probabilistic Logic Programs

The first kind of queries that we will study are called *basic abductive queries*, and the associated problem is called the *Basic Abductive Query Answering Problem* (BAQA for short). Suppose s is a (current) state, G is a goal (an action formula), and $[\ell, u] \subseteq [0, 1]$ is a probability interval. The BAQA problem tries to find a new state s' such that $\Pi_{s'}$ entails $G : [\ell, u]$. However, s' must be *reachable* from s. Reachability expresses the changes that we can make in the environment; for instance, we might be able to relieve social strife in Pakistan and influence the Pakistani government to not provide military support to LeT, but perhaps influencing LeT to not be a religious organization is out of realm of possibilities.

For this, we assume the existence of a reachability predicate *reach* specifying *direct* reachability from one state to another. *reach** is the reflexive transitive closure of *reach* and *unReach* is its complement. For now, we will assume that *reach* is provided and can be queried in polynomial time. However, in order to develop practical algorithms, later we will investigate one way in which *reach* can be specified (called *reachability constraints*).

Example 2 (Reachability between states). Suppose, for simplicity, that the only state predicate symbols are those that appear in the rules of Fig. 1, and consider the set of states in Fig. 2. Then, some examples of reachability are the following: $reach(s_4, s_1)$, $reach(s_1, s_4)$, $reach(s_1, s_3)$, $reach(s_2, s_3)$, $reach(s_3, s_2)$, $reach(s_2, s_5)$ $reach(s_3, s_5)$,, $\neg reach(s_5, s_2)$, and $\neg reach(s_5, s_3)$. ■

We can now state the BAQA problem formally:

BAQA Problem.
Input: An *ap*-program Π, a state s, a reachability predicate *reach* and a ground *ap*-formula $G : [\ell, u]$.
Output: "Yes" if there exists a state s' such that $reach^*(s, s')$ and $\Pi_{s'} \models G : [\ell, u]$, and "No" otherwise.
A solution to a BAQA instance is a sequence of states that ends in a state for which the corresponding subprogram entails the probabilistic goal. Therefore, such a solution corresponds to a *policy* that can be implemented to try and bring about the goal by carrying out the actions prescribed by the sequence.

$s_1 = \{$`socStrife`$(pak, 1)$, `allianceNSAG`(1), `trainCamp`$(pak, 0)$, `relOrg`(1), `govMilSupp`$(0)\}$
$s_2 = \{$`socStrife`$(pak, 1)$, `allianceNSAG`(1), `trainCamp`$(pak, 0)$, `relOrg`(0), `govMilSupp`$(0)\}$
$s_3 = \{$`socStrife`$(pak, 0)$, `allianceNSAG`(0), `trainCamp`$(pak, 0)$, `relOrg`(0), `govMilSupp`$(0)\}$
$s_4 = \{$`socStrife`$(pak, 0)$, `allianceNSAG`(1), `trainCamp`$(pak, 0)$, `relOrg`(1), `govMilSupp`$(0)\}$
$s_5 = \{$`socStrife`$(pak, 1)$, `allianceNSAG`(1), `trainCamp`$(pak, 1)$, `relOrg`(1), `govMilSupp`$(1)\}$

Fig. 2 A small set of possible states

Example 3 (Solution to BAQA*).* Consider once again the program in the running example and the set of states from Fig. 2. If the goal is $clashCas(jk, 1) : [0, 0.3]$ (we want the probability that LeT engages in clashes in J&K causing casualties to be at most 0.3) and the current state is s_4, then the problem is solvable because Example 2 shows that state s_3 can be reached from s_4, and $\Pi_{s_3} \models clashCas(jk, 1) : [0, 0.3]$. ■

There may be costs associated with transforming the current state s into another state s', and also an associated probability of success of this transformation (e.g., the fact that we may try to reduce social strife in Pakistan may only succeed with some probability). We will address this problem in Sect. 4.

The BAQA problem can be shown to be intractable both in the general case as well as in constrained subcases; for a formal treatment of the complexity of BAQA, we refer the reader to [26–28]. It turns out that the complexity of this problem is caused by two factors; specifically, we need to address the following two problems:

(P1) Find a subprogram Π' of Π such that when the body of all rules in that subprogram is deleted, the resulting subprogram entails the goal, and
(P2) Decide if there exists a state s' such that $\Pi' = \Pi_s$ and s is reachable from the initial state.

We now present algorithms and techniques for addressing these problems.

3.1 Algorithms for BAQA over Threshold Queries

In this section, we leverage the above intuition that BAQA can be decomposed into two subproblems to develop algorithms for a special case of BAQA: answering *threshold queries*. These queries are over goals of the form $F : [0, u]$ (ensure that F's probability is less than or equal to u) or $F : [\ell, 1]$ (ensure that F's probability is at least ℓ). This is a reasonable approach, since threshold goals can be used to require that certain formulas (actions) should only be entailed with a certain maximum probability (upper bound) or should be entailed with at least a certain minimum probability (lower bound). We start by inducing equivalence classes on subprograms that limit the search space, helping address problem (P1).

Definition 10 (Equivalence of *ap*-programs). Let Π be a ground *ap*-program and F be a ground action formula. We say that subprograms $\Pi_1, \Pi_2 \subseteq \Pi$ are *equivalent* given $F : [\ell, u]$, written $\Pi_1 \sim_{F:[\ell,u]} \Pi_2$, iff $\Pi_1 \models F : [\ell, u] \Leftrightarrow \Pi_2 \models F : [\ell, u]$. Furthermore, states s_1 and s_2 are *equivalent* given $F : [\ell, u]$, written $s_1 \sim_{F:[\ell,u]} s_2$, iff $reach(s_1, s_2)$, $reach(s_2, s_1)$, and $\Pi_{s_1} \sim_{F:[\ell,u]} \Pi_{s_2}$.

Intuitively, sub-programs Π_1, Π_2 are equivalent w.r.t. $F : [\ell, u]$ whenever they both entail (or do not entail) the annotated formula in question. For clarity, when the probability interval is evident from context, we will omit it from the notation.

Algorithm 1: simpleAnnBAQA($\Pi, s, G : [\ell_G, u_G]$)

1. Return *true* if there exists a consistent subprogram $\Pi' \subseteq \Pi$ such that:
   a. If $u_G = 1$, then at least one rule $r \in \Pi'$ must have head $F : [\ell_F, u_F]$ such that $F \models G$ and $\ell_G \leq \ell_F$; otherwise (*i.e.*, $\ell_G = 0$), at least one rule $r \in \Pi'$ must have head $F : [\ell_F, u_F]$ such that $G \models F$ and $u_G \geq u_F$;
   b. State s' for which $\Pi_{s'} = \Pi'$ is such that $reach^*(s, s')$.
2. **Active rule set initialization** $[active(\Pi, G : [\ell_G, u_G])]$: Initialize this set by selecting rules of the form $r : F : [\ell_r, u_r] \leftarrow s_1 \wedge \ldots \wedge s_n$ such that $F \wedge G \not\models \bot$;
   Passive rule set initialization $[passive(\Pi, G : [\ell_G, u_G])]$: Initialize this set with the rules in Π not identified as active;
   Conflicting rule set initialization $[conf(\Pi, G : [\ell_G, u_G])]$:
   For each rule $r_i : F : [\ell_r, u_r] \leftarrow s_1 \wedge \ldots \wedge s_n$ do:
   a. If $\ell_G = 0$, $F \models G$, and $\ell_r > u_G$ then add r_i to set $conf(\Pi, G : [\ell_G, u_G])$
   b. Otherwise (*i.e.*, $u_G = 1$), if $G \models F$ and $u_r < \ell_G$ then add r_i to the set $conf(\Pi, G : [\ell_G, u_G])$.
3. **Candidate active rule set:** Let $candAct(\Pi, G : [\ell_G, u_G]) = active(\Pi, G : [\ell_G, u_G]) \setminus conf(\Pi, G : [\ell_G, u_G])$;
4. Consider the set $candAct(\Pi, G : [\ell_G, u_G]) \cup passive(\Pi, G : [\ell_G, u_G])$ and, for each pair of rules $r_i : F_i : [\ell_{r_i}, u_{r_i}] \leftarrow s_1^i \wedge \ldots \wedge s_n^i$ and $r_j : F_j : [\ell_{r_j}, u_{r_j}] \leftarrow s_1^j \wedge \ldots \wedge s_m^j$ such that $F_i : [\ell_{r_i}, u_{r_i}]$ and $F_j : [\ell_{r_j}, u_{r_j}]$ are mutually inconsistent, add the pair (r_i, r_j) to a set called $inc(\Pi)$.
5. Return *true* if there exists a set of rules $\Pi' \subseteq candAct(\Pi, G : [\ell_G, u_G]) \cup passive(\Pi, G : [\ell_G, u_G])$ such that no pair $\{r_1, r_2\} \subseteq \Pi'$ belongs to $inc(\Pi)$ and:
   // Iterate favoring subsets of Π that contain rules in candAct($\Pi, G : [\ell_G, u_G]$)
   a. $\Pi' \models G : [\ell_G, u_G]$;
   b. There exists state s' for which $\Pi_{s'} = \Pi'$ such that $reach^*(s, s')$
   // Not all subprograms are feasible; e.g., if two rules have the same state, one cannot be chosen without the other.
6. Return *false*;

Fig. 3 An algorithm to solve BAQA assuming a threshold goal

Example 4 (Equivalence of ap-programs). Let Π be the *ap*-program from Fig. 1, and formula $F = \texttt{clashCas}(jk, 1) : [0.4, 1]$. Intuitively, the definition of equivalence between *ap*-programs w.r.t. an action formula states that rules that don't influence the probability of the formula are immaterial. Therefore, we can conclude, for instance, that $\{r_1, r_4, r_7\} \sim_F \{r_1, r_6, r_9\}$, since the rules that change in these two sets do not influence the probability with which F is entailed. Equivalence of states given a formula is analogous. ■

Relation $\sim$, both between states and between subprograms, is clearly an equivalence relation. The algorithm in Fig. 3 first tries to identify consistent subprograms that contain rules that clearly entail the goal (and are easily identifiable). If this is not possible, the algorithm continues by trying to leverage the presence of equivalence classes in the input *ap*-program Π. We now present an example reviewing how this algorithm works.

Example 5 (simpleAnnBAQA over the running example). Suppose Π is the *ap*-program of Fig. 1, the goal is `clashCas`$(jk,1)$: $[0,0.3]$ (abbreviated with $G:[0,0.3]$ from now on) and the state is s_4 from Fig. 2; note that $\Pi_{s_4}=\{r_3,r_6,r_9\}$ and that clearly $\Pi_{s_4}\not\models$ `clashCas`$(jk,1)$: $[0,0.3]$. The first step checks for possibilities to leverage subprogram equivalence; clearly, rule r_2 satisfies the condition, and we thus only need to verify that some subprogram containing it is reachable. Assuming the same reachability predicate outlined in Example 2, state s_3 is reachable from s_4; this corresponds to choosing subprogram $\Pi'=\{r_2,r_5,r_8\}$.

Finally, to illustrate Step 2 of the algorithm, which looks for rules whose heads involve formulas related to the goal, note that in this case this is simple since all the heads of rules in Π are atomic—therefore *passive*$(\Pi_{s_4},G:[0,0.3])=\emptyset$, and the set of active rules contains all the rules in Π. ■

Next, we will explore a particular way in which *reach* can be expressed, and how this can be leveraged to solve the reachability problem (P2). The key is that the reachability predicate will be expressed through *reachability constraints*:

Definition 11 (Reachability constraint). Let F and G be first-order formulas over $\mathscr{L}_{sta}$ and $\mathscr{L}_{var}$, connectives $\wedge$, $\vee$, and $\neg$, such that the set of variables over F is equal to those over G, and all variables are assumed to be universally quantified with scope over both F and G. A *reachability constraint* is of the form $F\not\rightsquigarrow G$; we call F the antecedent and G the consequent of the constraint, and its semantics is:

$$unReach(s_1,s_2)\Leftrightarrow s_1\models F \text{ and } s_2\models G$$

where s_1 and s_2 are states in $\mathscr{S}$.

Reachability constraints simply state that if the first formula is satisfied in a certain state, then no states that satisfy the second formula are reachable from it. We now present an example of a set of reachability constraints.

Example 6 (Reachability constraints). Consider again the setting and *ap*-program from Fig. 1. The following are examples of reachability constraints:

rc_1 : `relOrg`$(1)\not\rightsquigarrow$ `relOrg`(0)

rc_2 : `govMilSupp`$(pak,1)\not\rightsquigarrow$ `trainCamp`$(pak,0)$

rc_3 : `allianceNSAG`$(1)\not\rightsquigarrow$ `socialStrife`(0)

Constraint rc_1, for instance, states that we are not capable of influencing the group being modeled to not be a religious organization. ■

Algorithm *simpleAnnBAQA-Heur-RC* (Fig. 4) is optimistic and assumes that Step 1a of *simpleAnnBAQA* will yield at least one entailing formula for the goal; furthermore, it takes advantage of the structure added by the presence of reachability constraints. After checking for the simple necessary entailment condition, the algorithm continues by executing the steps of *simpleAnnBAQA* that compute the sets *active*$(\Pi,G:[\ell_G,u_G])$, *passive*$(\Pi,G:[\ell_G,u_G])$, *candAct*$(\Pi,G:[\ell_G,u_G])$,

Algorithm 2: simpleAnnBAQA-Heur-RC($\Pi, s, G : [\ell_G, u_G], RC$)

1. If the condition in Step 1a of *simpleAnnBAQA* does not hold, return *false*;
2. Execute Steps 2, 3, and 4 of *simpleAnnBAQA*;
3. let *goalState, goalStateAct, goalStateConf,* and *goalStateInf* be logical formulas over the sets $\mathscr{L}_{sta}$ and $\mathscr{L}_{var}$;
4. initialize *goalState* to null, *goalStateAct* to $\perp$, and *goalStateConf, goalStateInc* to $\top$;
5. for each rule $r_i \in candAct(\Pi, G : [\ell_G, u_G])$ with $Head(r_i) = F : [\ell_F, u_F]$ do
   if $[u_G = 1, F \models G$, and $\ell_G \leq \ell_F]$ or $[\ell_G = 0, G \models F$, and $u_G \geq u_F]$
   then set $goalStateAct := goalStateAct \vee Body(r_i)$;
6. for each rule $r_i \in conf(\Pi, G : [\ell_G, u_G])$ do
   set $goalStateConf := goalStateConf \wedge \neg Body(r_i)$;
7. for each pair of rules $(r_i, r_j) \in inc(\Pi)$ do
   set $goalStateInc := goalStateInc \wedge \neg(Body(r_i) \wedge Body(r_j))$;
8. set $goalState := goalStateAct \wedge goalStateConf \wedge goalStateInc$;
   // goalState describes the states for which the corresponding set of relevant rules satisfy the input goal
9. return $decideReachability(s, goalState, RC)$;

Fig. 4 A heuristic algorithm, based on simple sufficient conditions of entailment, to solve BAQA assuming that the goal is an *ap*-formula of the form either $G : [0, u]$ or $G : [\ell, 1]$ and that the state reachability predicate *reach* is specified as a set RC of reachability constraints

$conf(\Pi, G : [\ell_G, u_G])$, and $inc(\Pi)$. It then builds formulas generated by reachability constraints that solution states must satisfy (under the optimistic assumption); the algorithm uses a subroutine *formula*(s) which returns a formula that is a conjunction of all the atoms in state s and the negations of those not in s. In Step 5, the formula describes the fact that at least one of the states that make relevant entailing rules (as described in Algorithm *simpleAnnBAQA*) must be part of the solution; similarly, Step 6 builds a formula ensuring that none of the conflicting active rules can be relevant if the problem is to have a solution. Finally, Step 7 describes the constraints associated with making relevant rules that are probabilistically inconsistent. Noticeably absent are the "passive" rules from the previous algorithm; such rules impose no further constraints on the solution space under the assumptions being made by the algorithm. The last two steps put subformulas together into a conjunction of constraints, and the algorithm must decide if there exist any states that model formula *goalState* and are eventually reachable from s.

Deciding eventual reachability, as we have seen, is one of the main problems that we set out to solve as part of BAQA. Though there are many possible ways to implement this subroutine, here we will explore a SAT-based algorithm, which is presented in Fig. 5. This algorithm is simple: if the current state does not satisfy *goalState*, it starts by initializing formula *Reachable* which will be used to represent the set of eventually reachable states at each step. The initial formula describes state s, and the algorithm then proceeds to select all the constraints whose antecedents are entailed by *Reachable*. Once we have this set, *Reachable* is updated to the conjunction of the negations of all the consequents of constraints in the set. We are done if either *Reachable* at this point models *goalState*, or the old version of

Algorithm 3: decideReachability-SAT($s, goalState, RC$)

1. let *Reachable* be a formula initialized to *formula*(s);
2. set Boolean variable *done* := ($Reachable \wedge goalState \not\models \bot$);
3. while not *done* do
   set $Reachable_{old} := Reachable$;
   let $RC_{curr} \subseteq RC$ be the set of constraints $F_i \not\leftrightarrow G_i$ such that $Reachable \models F_i$;
   set $Reachable := (\bigwedge_{F_i \not\leftrightarrow G_i \in RC_{curr}} \neg G_i)$;
   set $done := ((Reachable \wedge goalState) \not\models \bot) \vee (Reachable \models Reachable_{old})$;
4. return ($Reachable \wedge goalState \not\models \bot$);

Fig. 5 An algorithm to decide reachability from a state s to any of the states that satisfy the formula *goalState*, where reachability is expressed as a set RC of reachability constraints; a formula is derived describing the set of all possible states eventually reachable from the initial one

Reachable is modeled by the new one, i.e., no new reachable states were discovered. The following is an example of how *decideReachability* – *SAT* works.

Example 7. Consider the *ap*-program from Fig. 1, along with constraint rc_3 from Example 6. As we saw in Example 5, if the goal is $\texttt{clashCas}(jk, 1) : [0, 0.3]$ and the current state is s_4 from Fig. 2, then rule r_2 needs to be made relevant, while r_1 and r_3 should not be relevant, and the rest do not influence the outcome. This yields the following *goalState* formula:

$$\texttt{socStrife}(pak, 0) \wedge \texttt{allianceNSAG}(0) \wedge \neg \left(\bigvee_{i=1,3} Body(r_i) \right)$$

Reachable starts out with *formula*(s_4) (that is, the conjunction of all atoms in the state) and, as $Reachable \models \texttt{allianceNSAG(1)}$, it gets updated to:

$$\neg \texttt{socStrife}(pak, 0)$$

which is mutually unsatisfiable with *goalState*. In the next iteration, however, as *Reachable* does not entail the antecedent of rc_3, it gets updated to $\top$, which means that there are no constraints regarding the states that can be reached, and therefore the algorithm will answer *true*. ■

4 Cost-Based Abductive Query Answering

In this section, we expand on the basic query answering problem described above and assume that there are *costs* associated with transforming the current state into another state, and also an associated *probability of success* of this transformation; e.g., the fact that we may try to reduce social strife in Pakistan may only succeed with some probability. To model this, we use three functions:

Definition 12. A *transition function* is any function $T : \mathcal{S} \times \mathcal{S} \to [0,1]$, and a *cost function* is any function $cost : \mathcal{S} \to [0,1]$. A *transition cost function*, defined w.r.t. a transition function T and some cost function *cost*, is a function $cost_T : \mathcal{S} \times \mathcal{S} \to [0,\infty)$, with $cost_T(s,s') = \frac{cost(s')}{T(s,s')}$ whenever $T(s,s') \neq 0$, and ∞ otherwise.[4]

The rationale behind the above definition is that transitions with high probability of occurring are considered to be "easy", and therefore have a low associated cost.

Function $cost_T$ describes *reachability* between any pair of states—a cost of ∞ represents an impossible transition. The cost of transforming a state s_0 into state s_n by intermediate transformations through the sequence of states $seq = \langle s_0, s_1, \ldots, s_n \rangle$ can be defined in the following manner:

$$cost^*_{seq}(s_0, s_n) = e^{\sum_{0 \le i < n, s_i \in seq} cost_T(s_i, s_{i+1})} \tag{1}$$

Note that Eq. 1 is only one possible way of computing the cost of transitions through a sequence; the only hard requirement is that the function must be monotonic (the costs could, for instance, be additive instead of multiplicative). One way in which cost functions can be specified is in terms of *reward functions*.

Definition 13 (Reward functions). An *action reward function* is a partial function $R : APF \to [0,1]$. An action reward function is *finite* if $dom(R)$ is finite.

Let R be a finite reward function and Π be an *ap*-program. An *entailment-based reward function* for Π and R is a function $E_{\Pi,R} : \mathcal{S} \to [0,\infty)$, defined as:

$$E_{\Pi,R}(s) = \sum_{F:[\ell,u] \in dom(R) \dot{\wedge} \Pi_s \models F:[\ell,u]} R(F : [\ell,u]) \tag{2}$$

Reward functions are used to represent how desirable it is, from the reasoning agent's point of view, for a given annotated action formula to be entailed in a given state by the model being used. Here, we will assume that all reward functions are finite. We use this notion of reward to define a natural *canonical cost function* as $cost^\circ(s) = \frac{1}{E_{\Pi,R}(s)}$ when $E_{\Pi,R}(s) \neq 0$, and 1 otherwise, for each state s. From now on, we assume that all transition cost functions are defined in terms of a canonical cost function.

Example 8. An example of an entailment-based reward function is as follows. Consider state s_2 from Fig. 2, and annotated formulas $F_1 =$ `clashCas`$(jk,1) \wedge$ `murder`$(jk,1) : [0.5,1]$, $F_2 =$ `clashCas`$(jk,1) \wedge$ `murder`$(jk,1) : [0,3]$, $F_3 =$ `fedayeenAtt`$(jk,1) : [0,0.05]$.

Suppose we have action reward function R such that: $R(F_1) = 0.1$, $R(F_2) = 0.85$, and $R(F_3) = 0.7$. This function represents that subprograms that entail F_2 are much more preferable than those that entail F_1, and that F_3 is also a desirable formula to entail. ■

[4] We assume that ∞ represents a value for which, in finite-precision arithmetic, $\frac{1}{\infty} = 0$ and $x^\infty = \infty$ when $x > 1$. The IEEE 754 floating point standard satisfies these rules.

Definition 14. A *cost based query* is a four-tuple $\langle G : [\ell, u], s, cost_T, k\rangle$, where $G : [\ell, u]$ is an *ap*-formula, $s \in \mathcal{S}$, $cost_T$ is a cost function, and $k \in \mathbb{R}^+ \cup \{0\}$.

CBQA Problem. Given *ap*-program Π and cost-based query $\langle G : [\ell, u], s, cost_T, k\rangle$, return "Yes" if and only if there exists a state s' and sequence of states $seq = \langle s, s_1, \ldots, s'\rangle$ such that $cost^*_{seq}(s, s') \le k$, and $\Pi_{s'} \models G : [\ell, u]$; the answer is "No" otherwise.

The main difference between the BAQA problem presented above and CBQA is that in BAQA there is no notion of cost, and we are only interested in the *existence of some sequence* of states leading to a state that entails the *ap*-formula. Even though we are still interested in sequences, solutions now have associated values depending on the transitions they attempt and the desirability of the states they traverse. Since CBQA is a generalization of BAQA, the same intractability results hold here as well [28]. In the following, we investigate an algorithm for CBQA when the cost function is defined in terms of entailment-based reward functions; we will focus on a tractable approach to finding solutions, albeit not optimal ones.

A Heuristic Algorithm Based on Iterative Sampling

Given the exponential search space, we would like to find a tractable heuristic approach. We now show how this can be done by developing an algorithm in the class of *iterated density estimation* algorithms (IDEAs) [2]. The main idea behind these algorithms is to improve on other approaches such as Hill Climbing, Simulated Annealing, and Genetic Algorithms by maintaining a *probabilistic model* characterizing the best solutions found so far. An iteration then proceeds by (1) generating new candidate solutions using the current model, (2) singling out the best of the new samples, and (3) updating the model with the samples from Step 2. One of the main advantages of these algorithms over classical approaches is that the probabilistic model, a "byproduct" of the effort to find an optimum, contains a wealth of information about the problem at hand.

Algorithm DE_CBQA (Fig. 6) follows this approach to finding a solution to our problem. The algorithm begins by identifying certain *goal states*, which are states s' such that $\Pi_{s'} \models G : [\ell, u]$; these states are pivotal, since any sequence of states from s_0 to a goal state is a candidate solution. The algorithms in Sect. 3 can be used to compute a set of goal states. Continuing with the preparation phase, the algorithm then tests how good the direct transitions from the initial state s_0 to each of the goal states is; ϕ^* now represents the current best sequence (though it might not actually be a solution). The final step before the sampling begins occurs in line 5, where we initialize a probability distribution over all states,[5] starting out as the uniform distribution.

[5]In an actual implementation, the probability distribution should be represented implicitly, as storing a probability for an exponential number of states would be intractable.

Algorithm 4: DE_CBQA($\Pi, G : [\ell, u], s_0, T, h, k, numIter, giveUp$)

1. Initialize set of states $\mathcal{S}_G := getGoalStates(\Pi, G : [\ell, u])$;
2. test all transitions (s_0, s_G), for $s_G \in \mathcal{S}_G$; calculate $cost^*_{seq}(s_0, s_G)$ for each;
3. let ϕ_{best} be the two-state sequence that has the lowest cost, denoted c_{best};
4. let $\mathcal{S}' = \mathcal{S} - \mathcal{S}_G - \{s_0\}$; set $j := 2$;
5. initialize probability distribution P over $\mathcal{S}'$ s.t. $P(s) = \frac{1}{|\mathcal{S}'|}$ for each $s \in \mathcal{S}'$;
6. while $!giveUp$ do
7. $j := j + 1$;
8.   for $i = 1$ to $numIter$ do
9.   randomly sample (using P) a set H of h sequences of states of length j starting at s_0 and ending at some $s_G \in \mathcal{S}_G$;
10.   rank each sequence ϕ with $cost^*_{seq}(s_0, \phi(j))$;
11.   pick the sequence in H with the lowest cost c^*, call it ϕ^*;
12.   if $c^* < c_{best}$ then $\phi_{best} := \phi^*$; $c_{best} := c^*$;
13. $P :=$ generate new distribution based on H;
14. return ϕ_{best};

Fig. 6 An algorithm for CBQA based on probability density estimation

The *getGoalStates* function called in line 1 performs two tasks: first, it identifies subprograms Π' of Π such that $\Pi' \models G : [\ell, u]$; second, it identifies states s such that $\Pi_s = \Pi'$, for some Π' found in the first step. All such states are then labeled as *goal states*, since any sequence of states from s_0 to any goal state is a candidate solution.

The `while` loop in lines 6–13 then performs the main search; *giveUp* is a predicate given by parameter which simply tells us when the algorithm should stop (it can be based on total number of samples, time elapsed, etc.). The value j represents the length of the sequence of states currently considered, and *numIter* is a parameter indicating how many iterations we wish to perform for each length. Line 9 performs the sampling of sequences, while line 10 assigns a score to each based on the transition cost function. After updating the score of the best solution found up to now, line 13 updates the probabilistic model P being used by keeping only the best solutions found during the last sampling phase. The algorithm finally returns the best solution it found (if any). An attractive feature of DE_CBQA is that it is an anytime algorithm, i.e., once it finds a solution, given more time it may be able to refine it into a better one while always being able to return the best so far. We now discuss one way in which the probability distribution P in the DE_CBQA algorithm can be represented.

Representing the Probability Distribution via a Bayesian Network

It is reasonable to believe that, in real-world instances of CBQA, states and actions are not in general conditionally independent; as such, it is critical to explore an approach to maintaining our probability distribution that is capable of handling these cases. One such method is the Bayesian belief network [24], a directed acyclic graph modeling conditional dependencies among random variables. In our case, each node in the network structure represents a random variable covering all possible states for a single (ordered) position in the final sequence. For a given node, a state is assigned probability mass proportional to how likely it is to be included in a "good" sequence at the position associated with that node. These values are initially provided through uninformed sampling of the state space, while the structure of the final network is learned through standard machine learning techniques.

Since an exhaustive search for the optimal structure across all potential networks is superexponential in the number of variables—in our case, the length of the sequence—we can use a heuristic local search algorithm to perceive graph structure; for instance, a slightly modified K2 search algorithm with a fixed ordering based on the sampled sequences to emphasize speed of structure learning [6]. Our intuition is that neighboring nodes in the sequence are more likely to affect each other than those farther away. Many other heuristic search algorithms exist, but a discussion of their merits is outside the scope of this paper.

Sampling from the network is accomplished in two steps. First, recall that a state's probability mass at a root node in our Bayesian network is related only to the proportion of "good" training sequences containing that state at a specific location. With this in mind, for every root node, we take a weighted sample from its prior probability distribution table. Second, we sample the conditional probability table of each child node with respect to the partial assignment provided by sampling its immediate parents. In this way, we provide a method for sampling a full path through the state space that takes into account conditional dependencies (and, of course, *independencies*) between states, their ordering, and position.

In Sect. 6, we present the results of our experimental evaluation of the DE_CBQA algorithm using this approach, comparing to a baseline algorithm that uses a much simpler representation.

5 Parallel Solutions for Abductive Query Answering

In the previous sections, we presented algorithms for answering both basic and cost-based abductive queries, along with several heuristic approaches to improve the tractability of these computations. However, we can make further gains in scalability and computation time by identifying portions of these problems to compute in parallel. In this section, we present two explicitly parallel algorithms for solving CBQA problems. One algorithm will search for potential entailing states in parallel, allowing us to either examine more possible states, or to improve the running time of finding an entailing state. In addition, the iterative sampling for CBQA can be made more effective by parallelizing the sampling process, allowing for a more comprehensive search over the possible paths to goal states.

Algorithm 5: PAR_getGoalStates($\Pi, G : [\ell, u], N, giveUp$)

1. Initialize set of states $S_G := \emptyset$;
2. Initialize set of rules $HeurRules := \emptyset$;
3. Execute Steps 2, 3, and 4 of *simpleAnnBAQA*;
4. for $r_i \in candAct(\Pi, G : [\ell_G, u_G])$ with $Head(r_i) = F : [\ell_F, u_F]$ do
5.   if $[u_G = 1, F \models G, \text{and } \ell_G \leq \ell_F]$ or $[\ell_G = 0, G \models F, \text{and } u_G \geq u_F]$ then
6.   Add r_i to *HeurRules*;
7. $BatchSize := \left\lceil \frac{|2^{\mathcal{L}_{sta}}|}{N} \right\rceil$;
8. while $!giveUp$ do
9.   for parallel processes $n := 0$ to $N - 1$ do
10.   foreach $s_i \in 2^{\mathcal{L}_{sta}}$ where $i := (BatchSize * n)$ to $[(BatchSize * n) + BatchSize - 1]$ do
11.   if $\Pi_{s_i} \models HeurRules$ and $HeurRules \models G : [\ell_G, u_G]$ then
12.   Add s_i to S_G;
13. return S_G;

Fig. 7 A parallel algorithm for finding entailing states for the CBQA problem

5.1 Parallel Selection of Entailing States

Recall the DE_CBQA algorithm in Fig. 6 and the *getGoalStates* function invoked in line 1; this function returns entailing states, i.e., states s s.t. $\Pi_s \models \Pi'$. In practice, as we will see in Sect. 6, the large search space makes it intractable to find all such states, and so the number of goal states returned must be limited by the user. The implementation of *getGoalStates* that we developed for our experimental evaluation iteratively goes through potential goal states until one is found; the heuristic methods shown in Algorithm *simpleAnnBAQA-Heur-RC* (Fig. 4) are used to make quick (sound, but not complete) entailment checks.

Rather than looking at potential goal states in sequence, we can parallelize this procedure. Figure 7 contains a distributed version of *getGoalStates* called *PAR_getGoalStates* that will divide the state space and check for entailing states in parallel over N processors.

The DE_CBQA algorithm can now be run with *PAR_getGoalStates* in Line 1. With this method, the user can specify some termination condition *giveUp* (e.g., the number of goal states to find, the amount of search time, etc.) for the concurrent search for entailing states. In Lines 9 and 10, we divide the state space $2^{\mathcal{L}_{sta}}$ across N processors, and iterate through each batch in parallel to find entailing states until the *giveUp* condition is true. If the size of S_G is still limited to a single goal state, then *PAR_getGoalStates* can provide a direct speedup of the original method, using the distributed computation to more quickly identify an entailing state. However, we can also take advantage of the parallelization to find a larger number of goal states to test in the DE_CBQA algorithm, rather than simply looking at the first state found.

Algorithm 6: ParSampleAsynch_DE_CBQA($\Pi, G : [\ell, u], s_0, T, h, \boldsymbol{numIter}, \boldsymbol{giveUp}, N$)

1. Initialize set of states $\mathscr{S}_G$:= $getGoalStates(\Pi, G : [\ell, u])$;
2. test all transitions (s_0, s_G), for $s_G \in \mathscr{S}_G$; calculate $cost^*_{seq}(s_0, s_G)$ for each;
3. foreach parallel process $n := 0$ to $N-1$ do
4.   let ϕ_{best} be the two-state sequence that has the lowest cost, denoted c_{best};
5.   let $\mathscr{S}' = \mathscr{S} - \mathscr{S}_G - \{s_0\}$;   set $j := 2$;
6. $P :=$ new uniform probability distribution over $sequences(\mathscr{S}')$;
7.   while $!giveUp$ do
8. $j := j+1$;
9.   for $i = 1$ to $numIter$ do
10.   randomly sample (using P) a set H_n of h sequences of states of length j starting at s_0 and ending at some $s_G \in \mathscr{S}_G$;
11.   rank each sequence ϕ with $cost^*_{seq}(s_0, \phi(j))$;
12.   pick the sequence in H_n with the lowest cost c^*, call it ϕ^*;
13.   if $c^* < c_{best}$ then $\phi_{n-best} := \phi^*$; $c_{n-best} := c^*$;
14. P:= generate new distribution based on H_n;
15.   add ϕ_{n-best} to H_{total};
16. return sequence ϕ_{best} in H_{total} with lowest cost c_{best};

Fig. 8 An asynchronous parallel algorithm for CBQA using iterative distributive sampling

5.2 Parallel Sampling of State Paths

The sampling method in the DE_CBQA algorithm allows the user to specify the number of possible paths to examine to reach a particular goal state. In practice, the space of possible paths from the initial state to a goal state can be very large, and random sampling may not reliably be able to find a low-cost option within a tractable computation time. In Fig. 8 we present a distributed algorithm, *ParSampleAsynch_DE_CBQA*, that will divide the iterative sampling of state paths across n processors. Each of the N parallel nodes performs a separate round of iterative sampling, maintaining its own sequence probability distribution and returning the best sequence resulting from these samples. Then, in line 16, we return the overall ϕ_{best} sequence from each of the distributed samples. While this asynchronous computation is not the same as increasing the number of samples by a factor of N, as we are not using all samples to update the probability distribution, it does facilitate better coverage of the possible sequences. Because of this, we are more likely to find better sequences, and may be able to achieve this result with a fewer number of samples per iteration. We can of course also use the parallel version of *getGoalStates*, described above, along with either concurrent iterative sampling algorithm to further improve performance and results.

6 Experimental Results

In this section, we will report on a series of experimental evaluations that we carried out on the algorithms presented in Sects. 4 and 5; for reasons of space, we cannot include experimental results for BAQA (we refer the reader to [26] and [28] for a full set of empirical results). Also, due to the vast number of possible parameters in these algorithms, we chose to vary a subset of them for the purposes of this study.

We conducted experiments using a prototype JAVA implementation consisting of roughly 2,500 lines of code. Each *ap*-program used in the experiments consists of a set of randomly generated *ap*-rules, each with a randomly generated head and body. The head consists of either one or two clauses of length at most two variables each, with uniform randomly selected conjunction or disjunction connectors and random negation. The head is nontrivial; it is guaranteed to have at least one variable in at least one clause. Each *ap*-rule's body is generated by randomly selecting a conjunction of two atoms. The goal *ap*-formula is generated in a similar fashion, but with randomly generated upper and lower bounds. When experiments require a threshold goal, either the upper bound is set to 1 or the lower bound is set to 0.

6.1 Empirical Evaluation of Algorithms for CBQA

We carried out the following experiments on an Intel Core2 Q6600 processor running at 2.4 GHz with 8 GB of memory available; all runs were performed on Windows 7 Ultimate 64-bit OS, and made use of a single core.

For all experiments, we assume an instance of the CBQA problem with *ap*-program Π and cost-based query $Q = \langle G : [\ell, u], s, cost_T, k\rangle$. The required cost, transition, and reward values for both algorithms are assigned randomly in accordance with their definitions. We assume an infinite budget for our experiments, choosing instead to compare the numeric costs associated with the sequences returned by the algorithms.

A Baseline Algorithm. In the following, we will as a baseline a straightforward representation for the probability distribution: a mapping of states to the proportion of "good" sampled sequences that contain that state. We will refer to this method as the *naïve probability vector* approach. While this representation is neither memory nor computationally intensive, it ignores any subtle relationships that may exist between individual states or their ordering in the overall sequence. Intuitively, an informed sampling method should provide higher accuracy (i.e., lower sequence costs) at a greater computational cost, especially in instances when states and actions interact. To explore this intuition, we remove some of the randomness from our testing suite by seeding desirable paths through the state space. This is accomplished by manipulating the cost and transition functions between states, yielding low costs for specific sequences of states and high costs otherwise. In this way, obvious conditional dependencies are introduced into the world.

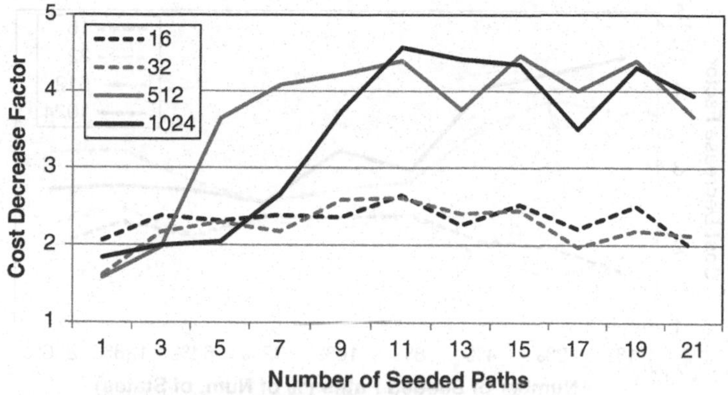

Fig. 9 Varying the number of seeded paths with a small (e.g., 16 or 32) number of states versus a larger (e.g., 512 or 1,024) state space

We compare the Bayesian method (implemented with WEKA [12]) against the naïve probability vector method. First, as a measure of result quality, we define the *cost decrease factor* to be the factor difference in the cost of the best sequence returned by the Bayesian method over that returned by the vector implementation. Higher cost decrease factors correspond to better relative Bayesian method performance. Figure 9 shows the cost decrease factor for very small amounts of seeded paths compared to different sizes of state spaces. For extremely small numbers of seeded paths, the Bayesian algorithm outperforms by roughly a factor of 2. This low number signifies similar performance to the vector method and is due to both DE_CBQA implementations missing the very few "carved" sequences in their initial sampling, before any probability distribution is constructed. The conditional network constructed from bad sampling is less useful; however, this problem can be easily solved by repetition of the algorithm.

Two trends, distinguished by the size of the state space, begin to form as we increase the number of seeded paths. When considering a larger number of seeded paths in larger state spaces, the Bayesian method shows its ability to discover dependencies in sampled sequences; however, when considering the same number of paths in a smaller state space, the Bayesian method continues to perform only slightly better than its vector counterpart. Carving too many (relative to the size of the state space) desirable paths essentially randomizes the transitions between states; for example, 20 paths through only 16 states alters overall dependencies far more than a similar number through 1,024 states. We explore this relationship further below.

Figure 10 shows the quality of results as the number of seeded paths is increased significantly. We see that the Bayesian network version performs admirably in large state spaces until roughly 8%, when its performance degrades to that of the Bayesian version in a smaller state space. As in Fig. 9, small instances of the problem stay roughly constant. Regardless of state space size, we see an increase in result quality of two to three over the naïve probability vector.

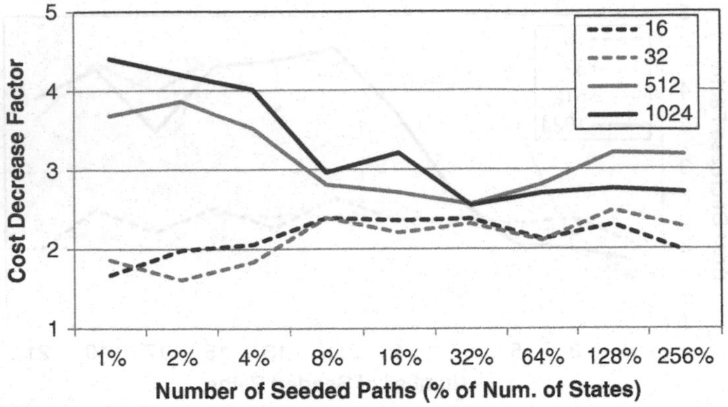

Fig. 10 Varying the number of seeded paths (and thus the level of conditional dependence in the world) as a percentage of the total number of states

We have seen that the more informed sampling method performs well, decreasing overall sequence cost. However, as our initial intuition suggested, the increased overhead of maintaining conditional dependencies slows the DE_CBQA algorithm significantly. Although the memory requirements of both algorithms increase linearly in the size of the number of states sampled, the Bayesian method is consistently slower than the vector method. This is due to a similar increase in the *runtime* complexity of the Bayesian method. The vector method represents probabilities as a simple mapping of states to real numbers; as such, an implementation with a constant lookup time data structure provides extremely fast sampling with a small memory footprint. For the more informed Bayesian variant, this relationship is based both on the number of initial iterations over the state space prior to the formation of the sampling structure and the maximum length of a sampled sequence. The Bayesian graph has as many nodes as there are states in a sampled sequence; furthermore, each of these nodes maintains knowledge of all unique states corresponding to a particular position in the sequence. Learning the structure of the network, storing the graph, and sampling from it are all dependent on the number of sampled states and sequence length. Thankfully, we can apply reasonable bounds to the number of samples, opting instead to instantiate multiple Bayesian networks over a smaller sample set.

When we include the additional cost of searching for entailing goal states (Line 1 of the DE_CBQA algorithm), both the naïve probability vector and informed Bayesian network methods scale similarly. We use the same fail-fast pessimistic approach to the heuristic goal search described earlier. Figure 11 shows how both algorithms scale with respect to an increase in number of states and number of rules. As before, the number of rules has a significantly higher effect on overall runtime than the number of states. We see that the algorithm scales gracefully to large state/action spaces. As we mentioned above, in our experience, real-world instances of CBQA tend to contain significantly *fewer* rules than states and actions [15]; as such, in these cases DE_CBQA scales quite well.

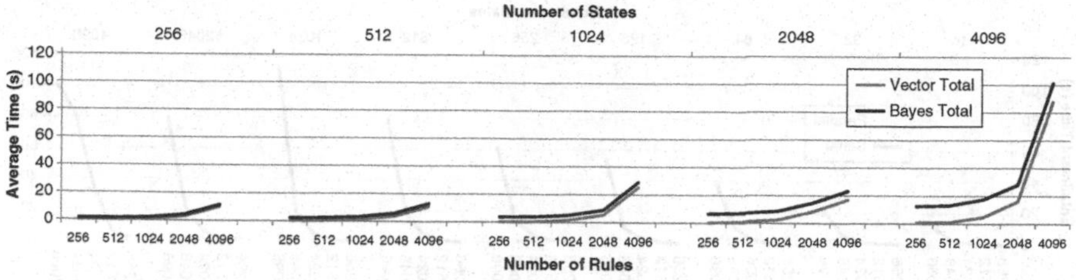

Fig. 11 Run time comparison as DE_CBQA scales with respect to number of states (*top axis*) and number of rules (*bottom axis*). Note the similarity in run time between the Bayesian and vector probability models

6.2 Empirical Evaluation of Parallel Algorithms for CBQA

We implemented the parallel algorithms using the Java Remote Method Invocation interface for distributed computation, and tested them on 15 nodes of a compute cluster, each with four 3.4 GHz cores and 8 GB RAM; the *ap*-programs and goals were generated in the same way as in the serial experiments (cf. Sect. 6, p. 532).

First, we compare the running time of the serial and parallel methods for finding entailing states to use in the DE_CBQA computation and demonstrate how *PAR_getGoalStates* can provide significant savings overall. We then compare the performance of the parallel algorithms for iterative sampling in DE_CBQA. Unfortunately, due to the synchronization and communication overhead associated with our particular implementation, the *ParSample_DE_CBQA* algorithm is quite intractable in practice, often taking 5 times the amount of running time for DE_CBQA using the naive vector distribution, and up to 35 times the running time for the Bayesian network distribution. The asynchronous version of this algorithm, *ParSampleAsynch_DE_CBQA*, is however able to concurrently run the DE_CBQA algorithm with only minimal impact from the communication required to initialize the problem and obtain the overall best sequence.

Second, we compare the quality of the sequences returned by the asynchronous parallel sampling algorithm and the serial DE_CBQA computation. Using 1,024 samples per iteration as a baseline for the serial algorithm, we run both algorithms over large rule and state spaces, varying the number of parallel samples per iteration. Because the parallel method takes distinct samples in parallel, it is able to explore more of the state space and find better sequences with a fewer number of samples.

Parallel methods for finding entailing states. Two experiments were performed to determine the effectiveness of *PAR_getGoalStates* as compared to the serial *getGoalStates* method. The default size of S_G (the set of goal states) in *getGoalStates* is either the total number of possible states or 50, whichever value is smaller. The first experiment uses this same cap of 50 entailing states, varying the number of states between 16 and 4,096 and the number of rules from 256 to 4,096. The parallel algorithm effectively divides the state space to find goal states concurrently, consistently running much more efficiently than the serial version

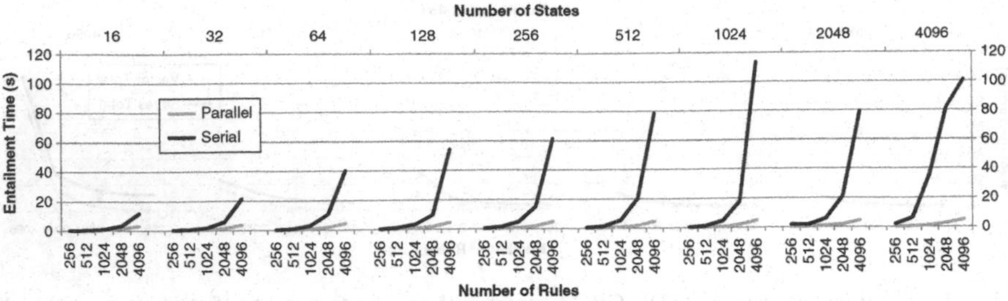

Fig. 12 Running time for both the parallel *PAR_getGoalStates* and serial *getGoalStates* methods to find up to 50 entailing states

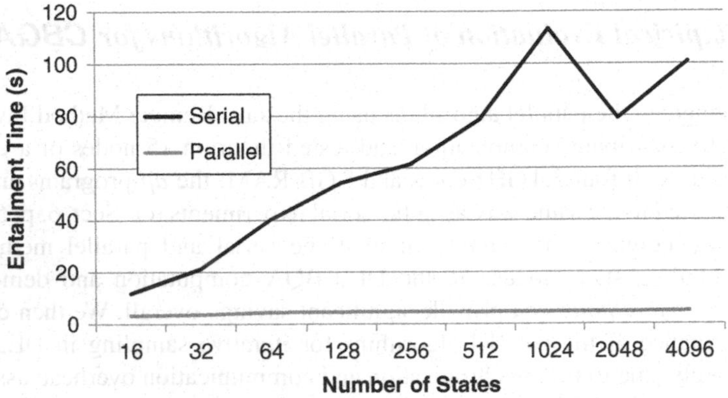

Fig. 13 Running time for both the parallel *PAR_getGoalStates* and serial *getGoalStates* methods to find entailing states. The number of states was varied between 16 and 4,096, and the number of rules held constant at 4,096

(Fig. 12). For 4,096 states and rules, the parallel entailment method requires 4.96 s, whereas serial selection is 20 times slower, taking 100.8 s. Furthermore, as shown in Figs. 13 and 14, the computation time required by the parallel algorithm increases only very slowly as the number of states and rules increase, indicating that this method will scale to a much larger number of states and larger programs. Because the entailment time is often a significant portion of the DE_CBQA algorithm, especially for large state-spaces, the parallel method provides significant overall savings.

Parallel iterative sampling. As discussed above, the communication and synchronization overhead required for the ParSample_DE_CBQA algorithm is far too costly in practice to make this method useful. However, empirical tests showed that the performance of the asynchronous parallel algorithm is very good with respect to the serial DE_CBQA algorithm. In Fig. 15, the running time of ParSampleAsynch_DE_CBQA is compared with the serial version for both the vector and Bayesian distribution methods. For the parallel computations, we performed 60

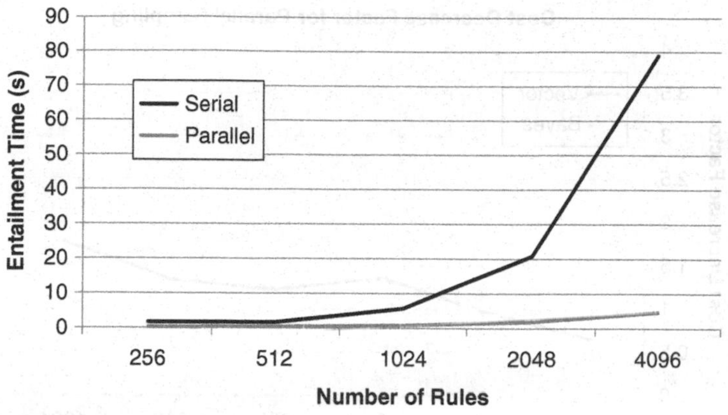

Fig. 14 Running time for both the parallel *PAR_getGoalStates* and serial *getGoalStates* methods to find entailing states. The number of rules was varied between 256 and 4,096, and the number of states held constant at 2,048

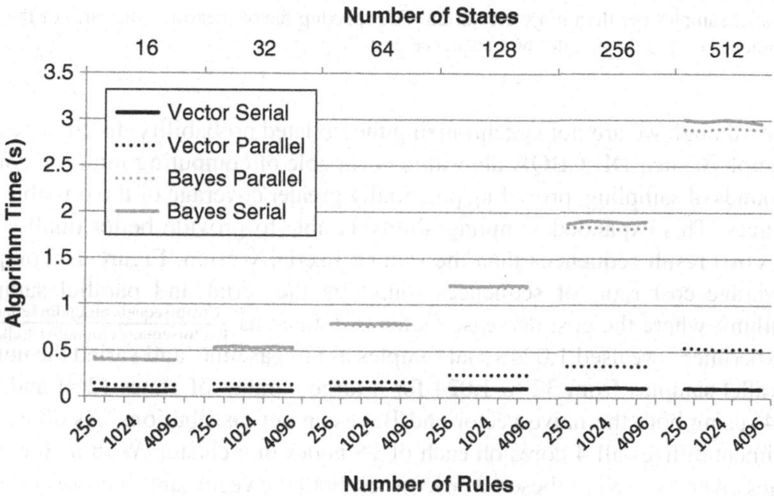

Fig. 15 Running time comparison of ParSampleAsynch_DE_CBQA and serial DE_CBQA using both the vector and Bayesian distributions

concurrent rounds of the DE_CBQA iterative sampling—using all 4 cores on each of 15 nodes of the compute cluster. When using the probability vector representation, the communication required to set up the remote computations and combine the final results still dominates the computation even in the asynchronous sampling—in many cases the parallel version takes at least twice as long. However, this difference is much smaller in the case of the Bayesian algorithm. A two-sample t-test at the 95% confidence level indicates with a very high p-value of 0.8881 that there is no significant difference between the running times of the parallel and serial algorithms with the Bayesian distribution.

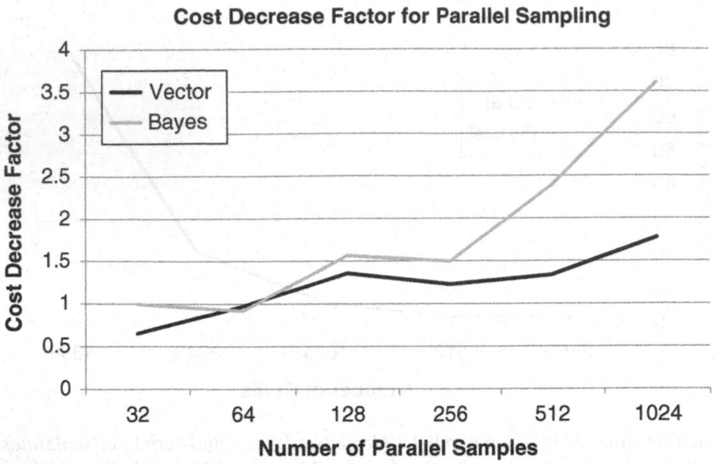

Fig. 16 Cost comparison of ParSampleAsynch_DE_CBQA and serial DE_CBQA using the vector and Bayesian distributions. The number of serial samples per iteration were held constant while the parallel samples per iteration were varied. The speedup factor measures the ratio of the serial best sequence cost to the parallel best sequence cost

Even though we are not synchronizing the updated probability distributions, the ParSampleAsynch_DE_CBQA algorithm is capable of computing multiple concurrent rounds of sampling, providing potentially greater coverage of the possible state sequences. This expanded sampling ability is able to provide better quality (i.e., lower cost) result sequences than the standard serial version. Figure 16 compares the average cost ratio of sequences found by the serial and parallel sampling algorithms, where the cost decrease factor is defined as $\frac{\text{Costofsequencefoundbyserial}}{\text{Costofsequencefoundbyparallel}}$. In this experiment, we used 1,024 serial samples as our baseline, and varied the number of parallel samples from 32 to 1,024 for a large number of states (2^{12}) and rules (1,024) using both the naive vector and Bayesian net distributions. As above, this experiment utilizes all 4 cores on each of 15 nodes in a cluster. With as few as 64 samples taken by each of these 60 processors, both the vector and Bayes versions of ParSampleAsynch_DE_CBQA achieve a quality almost at parity with 1,024 serial samples, with a cost decrease factor of 0.970690347 and 0.918439618 respectively. At 128 samples, both algorithms surpass the quality of the serial DE_CBQA and find sequences with much lower costs.

The overall efficiency gains of the ParSampleAsynch_DE_CBQA algorithm are illustrated in Fig. 17. Using the same parameters as the quality experiments just described, we also compared the running times of these algorithms. Not only does this method provide improved quality in the same amount of time for the default number of 1,024 samples—a 3.6 cost decrease factor for the Bayesian distribution (Fig. 16)—but we can achieve greater quality in a shorter period of time. For example, taking 128 parallel samples and using the Bayesian distribution,

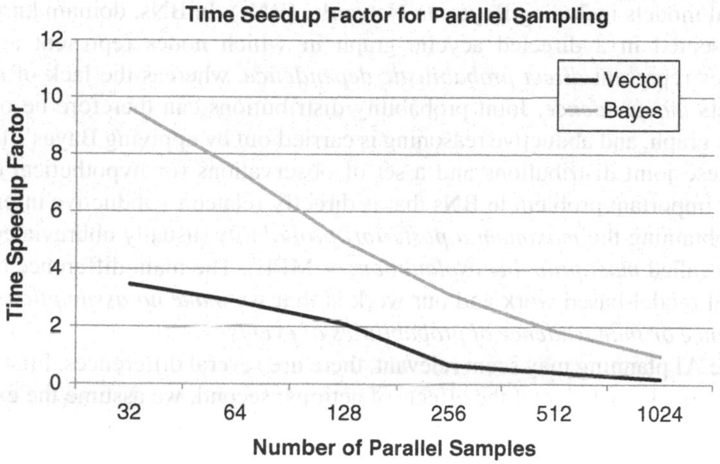

Fig. 17 Running time speedup of ParSampleAsynch_DE_CBQA versus serial DE_CBQA using both the vector and Bayesian distributions. The number of serial samples per iteration were held constant while the parallel samples per iteration were varied. The speedup factor measures the ratio of the serial running time to the parallel running time

ParSampleAsynch_DE_CBQA requires only about $\frac{1}{5}$ the time as DE_CBQA for 1,024 samples (16.26 and 3.29 s, respectively), but is able to find sequences with an average cost decrease factor of 1.57.

7 Related Work

Abduction has been extensively studied in diagnosis [5], reasoning with non-monotonic logics [7], probabilistic reasoning [13, 25], argumentation [16], planning [8], and temporal reasoning [8]; furthermore, it has been combined quite naturally with different variants of logic programs [1, 4]. An *abductive logic programming theory* is a triple (P, A, IC), where P is a logic program, A is a set of ground *abducible* atoms (that do not occur in the head of a rule in P), and IC is a set of classical logic formulas called *integrity constraints*. An explanation for a query Q is a set $\Delta \subseteq A$ such that $P \cup \Delta \models Q$, $P \cup \Delta \models IC$, and $P \cup \Delta$ is consistent. This is an abstract definition, independent of syntax and semantics; the variations in how such aspects are defined has lead to many different models.

David Poole et al. combined probabilistic and non-monotonic reasoning, leading to the development of Probabilistic Horn Abduction and eventually the Independent Choice Logic [25]. Christiansen [4] addresses probabilistic abduction with logic programs based on constraint handling rules. Though these models are related to our work, they either make general assumptions of pairwise independence of probabilities of events (such as in [25] or [4]) or are based on the class of graphical models including Bayesian Networks (BNs). In BNs, domain knowledge is represented in a directed acyclic graph in which nodes represent attributes and edges represent *direct probabilistic dependence*, whereas the lack of an edge represents *independence*. Joint probability distributions can therefore be obtained from the graph, and abductive reasoning is carried out by applying Bayes's theorem given these joint distributions and a set of observations (or hypothetical events). Another important problem in BNs that is directly related to abductive inference is that of obtaining the *maximum a posteriori probability* (usually abbreviated MAP, and also called *most probable explanation*, or MPE). The main difference between graphical model-based work and our work is that we *make no assumptions on the dependence or independence of probabilities of events.*

While AI planning may seem relevant, there are several differences. First, we are not assuming knowledge of the effects of actions; second, we assume the existence of a probabilistic model underlying the behavior of the entity being modeled. In this framework, we want to find a state such that *when the atoms in the state are added to the ap-program, the resulting combination* entails *the desired goal with a given probability.* While the italicized component of the previous sentence can be achieved within planning, it would require a state space that is exponentially larger than the one we use. In this space, the search space would be the set of all sets of atoms closed under consequence that are jointly entailed by any subprogram of the *ap*-program and any state (under the definition in this paper). This would cause states to be potentially exponentially bigger than those in this paper and would also exponentially increase their number.

8 Conclusions

There are many applications where we need to reason about the behaviors of actors about whom we can learn probabilistic rules of behavior. Examples of such applications include the modeling of terror groups [18, 19], the modeling of animal groups (e.g., groups of gorillas that exhibit behaviors such as *avoidance* of other gorilla groups, *attacks* on other gorilla groups, and so forth) [3]. The US Treasury, for instance, is interested in modeling behaviors of investor groups to learn their attitudes towards risk under different conditions; similarly, governments are interested in the impact of *policies* on groups (e.g., farmers). In many cases, we would like to *influence* these behaviors by understanding what actions we can take to ensure that the probability that a desired outcome occurs exceeds some threshold. This is further complicated by the fact that groups do not take actions "one at a time"; instead, these actions are often correlated and planned, and furthermore, the effects of these actions are not well understood.

We have formulated these problems via the Basic Abductive Query Answering (BAQA) and Cost-based Query Answering (CBQA) problems. We have presented heuristic algorithms that are relatively fast and sound, though incomplete, and developed innovative algorithms that maintain and update probability distributions as they run, allowing better estimation of solutions while reducing running times. A further important contribution is that of parallel algorithms for abduction in probabilistic logic.

Acknowledgements Some of the authors of this paper were funded in part by AFOSR grant FA95500610405, ARO grant W911NF0910206 and ONR grant N000140910685.

References

1. Baldoni M, Giordano L, Martelli A, Patti V (1997) An abductive proof procedure for reasoning about actions in modal logic programming. In: Selected papers from NMELP '96. Springer, London, pp 132–150
2. Bonet JSD, Isbell CL Jr, Viola PA (1996) MIMIC: finding optima by estimating probability densities. In: Proceedings of NIPS '96. MIT press, USA, pp 424–430
3. Bryson JJ, Ando Y, Lehmann H (2007) Agent-based modelling as scientific method: a case study analysing primate social behaviour. Philos Trans R Soc Lond B 362(1485):1685–1698
4. Christiansen H (2008) Implementing probabilistic abductive logic programming with constraint handling rules. In: Constraint handling rules. Springer, Berlin/New York, pp 85–118
5. Console L, Torasso P (1991) A spectrum of logical definitions of model-based diagnosis. Comput Intell 7(3):133–141
6. Cooper G, Herskovits E (1992) A Bayesian method for the induction of probabilistic networks from data. Mach Learn 9(4):309–347
7. Eiter T, Gottlob G (1995) The complexity of logic-based abduction. JACM 42(1):3–42
8. Eshghi K (1988) Abductive planning with event calculus. In: Proceedings of ICLP. MIT Press, USA, pp 562–579, ISBN 0-262-61056-6
9. Fagin R, Halpern JY, Megiddo N (1990) A logic for reasoning about probabilities. Inf Comput 87(1/2):78–128
10. Giles J (2008) Can conflict forecasts predict violence hotspots? New Sci (2647)
11. Hailperin T (1984) Probability logic. Notre Dame J Form Log 25(3):198–212
12. Hall M, Frank E, Holmes G, Pfahringer B, Reutemann P, Witten I (2009) The WEKA data mining software: an update. ACM SIGKDD Explor Newsl 11(1):10–18
13. Josang A (2008) Magdalena, L, Ojeda-Aciego, M, Verdegay, J.L. Abductive reasoning with uncertainty. In: Proceedings of the IPMU 2008, Torremolinos, Malaga, Spain. pp 9–16
14. Kern-Isberner G, Lukasiewicz T (2004) Combining probabilistic logic programming with the power of maximum entropy. Artif Intell 157(1–2):139–202
15. Khuller S, Martinez MV, Nau DS, Sliva A, Simari GI, Subrahmanian VS (2007) Computing most probable worlds of action probabilistic logic programs: scalable estimation for $10^{30,000}$ worlds. Ann Math Artif Intell 51(2–4):295–331
16. Kohlas J, Berzati D, Haenni R (2002) Probabilistic argumentation systems and abduction. Ann Math Artif Intell 34(1–3):177–195
17. Lloyd JW (1987) Foundations of logic programming, 2nd edn. Springer, Berlin/New York
18. Mannes A, Michael M, Pate A, Sliva A, Subrahmanian VS, Wilkenfeld J (2008) Stochastic opponent modeling agents: a case study with Hamas. In: Proceedings of ICCCD 2008, AAAI Press, USA, ISBN 978-1-57735-389-8
19. Mannes A, Michael M, Pate A, Sliva A, Subrahmanian VS, Wilkenfeld J (2008) Stochastic opponent modelling agents: a case study with Hezbollah. In: Liu H, Salerno J (eds) Proceedings of the first international workshop on social computing, behavioral modeling, and prediction, Springer, Germany, ISBN 978-0-387-77671-2

20. Mannes A, Shakarian J, Sliva A, Subrahmanian VS (2011) A computationally-enabled analysis of Lashkar-e-Taiba attacks in Jammu and Kashmir. In: Proceedings of EISIC. IEEE Computer Society, pp 224–229, ISBN 978-0-7695-4406-9
21. Ng RT, Subrahmanian VS (1992) Probabilistic logic programming. Inf Comput 101(2): 150–201
22. Ng RT, Subrahmanian VS (1993) A semantical framework for supporting subjective and conditional probabilities in deductive databases. J Autom Reason 10(2):191–235
23. Nilsson N (1986) Probabilistic logic. Artif Intell 28:71–87
24. Pearl J (1988) Probabilistic reasoning in intelligent systems: networks of plausible inference. Morgan Kaufmann, San Francisco
25. Poole D (1997) The independent choice logic for modelling multiple agents under uncertainty. Artif Intell 94(1–2):7–56
26. Simari GI, Subrahmanian VS (2010) Abductive inference in probabilistic logic programs. In: Technical communications of ICLP'10. LIPIcs, vol 7, Schloss Dagstuhl. Schloss Dagstuhl - Leibniz-Zentrum fuer Informatik 2010, ISBN 978-3-939897-17-0, pp 192–201
27. Simari GI, Dickerson JP, Subrahmanian VS (2010) Cost-based query answering in probabilistic logic programs. In: Proceedings of SUM 2010. LNCS. Springer, Berlin, Germany
28. Simari GI, Dickerson JP, Sliva A, Subrahmanian VS (2012) Parallel abductive query answering in probabilistic logic programs. Trans Comput Log

The Application of Search Games to Counter Terrorism Studies

Robbert Fokkink and Roy Lindelauf

The arrest of Saddam Hussein on 13th of December 2003 in a farmhouse near his hometown of Tikrit marked the end of military operation Red Dawn, a man hunt that had been planned by Major Brian J. Reed, who traced Saddam using social network analysis. Major Reed stated that: *the intelligence background and link diagrams that we built were rooted in the concepts of network analysis* [24]. The process of daily intelligence gathering led coalition forces to identify and locate more of the key players in the insurgent network [29]. This finally resulted in diagrams of Saddam's highly trusted relatives and clan members. A series of raids designed to capture some of those key individuals finally led to the information necessary to find Hussein. Operation Red Dawn took approximately half a year.

The arrests of the suspected war-criminals Radovan Karadžić and Ratko Mladić took much longer. The warrant for their arrest had been issued by the International Criminal Tribunal for the Former Yugoslavia in 1995. Karadžić was apprehended in July 2008, almost 3 years before Mladić was finally detained. Initially, the two had enjoyed political protection by the Serbian government, but when they were finally forced into hiding around 2002, they adopted different survival tactics. Karadžić changed his identity and practised as a psychologist in a private clinic in Belgrade. Mladić went into hiding with relatives in the countryside. Both men were tracked by surveillance of their family members.

None of these operations were on the same scale as the search for Osama bin Laden, who was finally captured and killed in Abbottabad, Pakistan, on May 2nd 2011. Bin Laden was initially thought to be hiding in the mountainous area on

R. Fokkink (✉)
Delft Institute of Applied Mathematics, TU Delft, P.O.Box 5031,
2600 GA, Delft, Netherlands
e-mail: r.j.fokkink@tudelft.nl

R. Lindelauf
Netherlands Defence Academy, De La Reijweg 120, 4818 BB, Breda, Netherlands
e-mail: rha.lindelauf.01@nlda.nl

V.S. Subrahmanian (ed.), *Handbook of Computational Approaches to Counterterrorism*,
DOI 10.1007/978-1-4614-5311-6_24,

the border between Afghanistan and Pakistan, and there were rumors that he had escaped to Saudi Arabia or Africa. He was finally located through his courier, hiding in a residential area near a military base. All men were caught from information that was obtained from their social networks.

In the search for a fugitive, governmental agencies have a broad array of resources at their disposal to generate information on the whereabouts of that individual. Think of human intelligence operatives who try to obtain information from human sources in the field. It is said for instance that the 300th Military Intelligence Brigade from Utah's Army National Guard developed intelligence that led to the capture of Saddam Hussein. In addition, geospatial intelligence can be used to create overlays of the terrain to indicate areas that are of higher interest because of features that make it easier to hide there. Communication between individuals that are known to be close or in the social circle of the fugitive can be intercepted using signals intelligence. Of course, if the fugitive is known to have medical problems this can also be used to the advantage of the searcher. It was rumored that Osama Bin Laden suffered from kidney related affliction. The presumption within the intelligence community was that he needed access to a dialysis machine. A fact that could be exploited in the determination of his whereabouts.

To analyze and develop all this information into useable intelligence, i.e., the most probable location of the fugitive and consequent search strategies, a plethora of analysis techniques have been developed and exist within the intelligence community. One of those techniques consists of mathematical modeling by the type of analysis with which we are concerned in this paper. This is because mathematically, the search for these fugitives can be described by the following zero-sum game. One player, a hider, wants to escape the other player, a searcher (or team of searchers). The hider seeks cover within a certain area that is known to the searcher, but the exact whereabouts of the hider are unknown. From time to time the hider may give up one hiding place and move to another hiding place within the same area. The hider wants to maximize the capture time τ while the searcher wants to minimize it. This is a *search game*.

The analysts in the team responsible for the search develop all the available information into several geographic overlays. Think for instance of an overlay representing the different languages in the respective region and another overlay depicting religious or ethnic boundaries [21, 23]. Furthermore a fugitive might want to have access to a highway or another easy escape route. Another overlay could target only those areas or cities with at least a certain amount of inhabitants. All this depends on the qualitative analysis of the fugitive's behavior and known habits. These overlays combined can then be transformed into a probability distribution over the area of interest representing the probability that a hider will dwell among that area. The theory behind search games has been developing for several decades, but the application of search games to terrorism computations has only just begun. So far the studies have remained theoretical as the models have not been sufficiently developed to be tested against practice. For instance, search games do no yet take the overlays into account that have been developed by qualitative analysis of fugitive behavior. In this chapter we will give an overview of theory of search games and the current state of search games that are of use in counterterrorism studies.

1 The Mathematics of Search Games

Modern textbooks on search games, containing much more information, are [2,20], a popular account is given in [13] and an older but very interesting reference is [31]. Search games are not to be confused with pursuit-evasion games. In a search game there is no direct visual contact between the players, while in pursuit-evasion games such a visual contact exists. A standard reference for dynamic pursuit-evasion games is [25], while combinatorial pursuit-evasion games can be found in [12,27].

1.1 A Brief History of Search Games

Search games were introduced by Rufus Isaacs in the final chapter of his monograph on differential games [25]. He defined search games as differential games with minimal information, stating that the value of these games for military and other applications is clear, but that the extension of the theory, from complete information to minimal information, is difficult. In the simplest search game, a female "hider" H secretes an object somewhere in a bounded region $\mathfrak{R}$ and her opponent S, a male "searcher", strives to find it in the least time. The searcher finds the secreted object if he arrives within a certain distance d. S can move about $\mathfrak{R}$ with almost complete freedom. The only limitation on the searcher's movement is a bounded speed $\leq w$. More explicitly, the only restriction on the continuous trajectory $f(t)$ that describes the searcher's movement is $|f(t_1) - f(t_2)| \leq |w(t_2 - t_1)|$. This is called a search game with an immobile hider. It is a zero-sum game. The payoff to the hider H is the capture time τ that it takes the searcher to find the secreted object.

If the space $\mathfrak{R}$ is homogeneous, i.e., there are no good or bad places to hide, then the hider will secrete the object uniformly randomly. The searcher will search the space exhaustively, starting the search from a random location and never searching the same place twice. A closed trajectory $f(t)$ is called a tour if it has minimal length and if each position in $\mathfrak{R}$ is within distance d of $f(t)$. If T is the total time that it takes S to tour $\mathfrak{R}$, then the value of the game is $\frac{T}{2}$, see [25, Theorem 12.3.1]. Indeed, if H hides uniformly, or if S tours the space at maximum speed starting from a uniformly random location, then the capture time τ is uniformly distributed on $[0, T]$.

The analysis of the game remains the same if S gets more agents. If S leads a team of s identical and independent agents, then $\mathfrak{R}$ is divided into s equal parts, to be searched by one agent. The value of the game thus reduces to $\frac{T}{2s}$, so this is equivalent to increasing the speed of a single searcher by a factor s. The analysis of the game becomes more involved if H gets h identical objects and the searcher has to find them all. If H places the objects uniformly randomly and independently, then τ is the maximum of h independent random variables that are uniformly distributed on $[0, T]$. So, on average S will find the objects in time $\frac{hT}{h+1}$, which puts a lower bound on the value of the game [25, Lemma 12.3.1]. However, this hider strategy may not be optimal. If $\mathfrak{R}$ is a circular region and if $h = 2$, then it is optimal to hide the objects at diametrically opposite locations. In that case, S will find the objects in average time $\frac{3T}{4}$. It is not completely straightforward to compute the value of the search game with h objects, not even in the relatively simple case in which the region $\mathfrak{R}$ is a circle. In general, search games get much more difficult to solve if there are multiple hiders.

The same search game with a mobile hider is called the game between the Princess and the Monster by Isaacs. In this game, the hider gets full mobility, not even bounded by a maximum speed. Isaacs conjectured that the sole decision of importance rests with H: how fast should she run? If she remains immobile, she is found at average time $\frac{T}{2}$. If she runs at unlimited speed, then she will run into S within no time. Somewhere between these extremes, there must be an optimal speed, but where? This question was settled by Shmuel Gal in a groundbreaking paper [16]. He showed that if $d \to 0$ then the value of the game converges to T. Gal constructs an optimal hider strategy in which H rests at a point $\mathbf{x}_i$ for a fixed amount of time h before moving to a uniformly random $\mathbf{x}_{i+1}$ in a straight line with a speed that is bounded by 1 and the searcher's speed w. If $w > 1$ and if $\mathfrak{R}$ is a two-dimensional region of area μ and diameter R, then in Gal's solution the optimal resting time h is given by

$$h = \sqrt[4]{\frac{37\pi R^3\mu^2}{16d^3}} \tag{1}$$

T is proportional to μ/d and R is the maximum time it takes H to move between consecutive rest points. Furthermore μ is proportional to R^2 with a proportional constant that depends on the shape of the domain. If $\mathfrak{R}$ is circular, then the proportional constant is close to one, at $\pi/4$, but if $\mathfrak{R}$ is long and narrow, then the constant is close to zero. Rewriting the resting time h in these terms, then we get that $h = c\sqrt[4]{T^3}$ for a constant c that depends on the shape of the domain and on the root $\sqrt[4]{R}$ of its diameter. Under this hider strategy, the searcher never gets the time to search a significant proportion of $\mathfrak{R}$ before the hider moves to a new position. The probability of capture therefore does not increase over time, which implies that τ is exponentially distributed if d is arbitrarily small. In general, against an immobile hider τ is uniformly distributed and against a mobile hider τ is exponentially distributed, see also [2, p. 47].

Gal's original solution of the princess and monster game applies to convex two-dimensional domains only. Extensions of Gal's solution were later given by Lalley and Robbins [26]. Gal himself showed how to extend his solution to non-convex domains [17] and Garnaev [19] showed how to solve the game for non-homogeneous domains. The solution of Isaacs' princess and monster game applies only if $d \to 0$ and it depends on the ability of the hider to move without running a

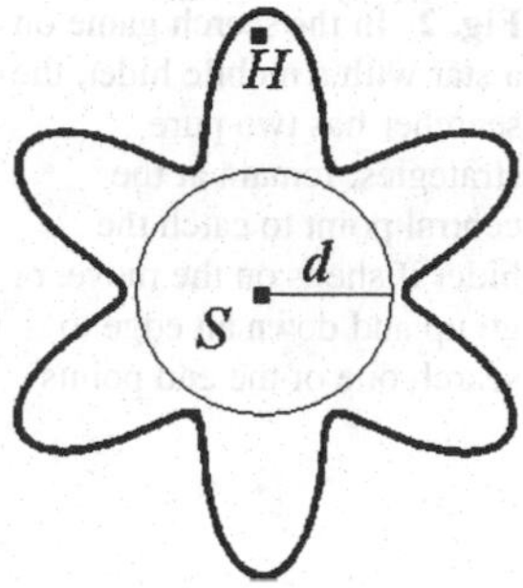

Fig. 1 In a star shaped domain, the searcher may decide not to tour the space at full speed, but to stay near the center of the space where there is a high probability of detection if the hider moves. The advantage of staying in the center decays only if $d \to 0$

large risk of detection. If a moving hider runs a large risk of detection, then the game is much more difficult to solve. Such a game is illustrated in Fig. 1. In that case, the searcher has the opportunity of remaining in the center, in ambush, so the hider will no longer move at a fixed time but at a random time. This significantly complicates the analysis and momentarily, such games are not completely understood. A first study on the topic is [6].

1.2 Search Games on Networks

To avoid technicalities, it is convenient to study search games on networks instead of spatial domains. The maximum speed of the searcher is normalized to one. The maximum speed of the hider is unbounded. The radius of detection d is taken equal to zero, so the hider is found once S and H are in the exact same location. Gal [16] already solved the game with an immobile hider on a tree. He showed that if S starts the search at the root of the tree, then an immobile H will hide in one of the leaves with a probability that is proportional to the distance from the root of the tree. S uses a Chinese postman tour of the tree, i.e., a closed path that traverses every edge at least once and is of minimal length. S flips a coin and follows the tour equiprobably in either direction. This result of Gal initiated the study of search games on networks with an immobile hider and Gal's solution was extended from the tree to much more general graphs and more general starting positions, see for instance [3, 18, 30] or [2, Chap. 3]. An algorithmic point of view was taken by Anderson and Aramendia [9] and by Von Stengel and Werchner [34], who showed that solving the game is NP hard [9, 10].

The literature on network search for an immobile hider is extensive. Much less is known about search games on networks with an mobile hider. The game has only been solved if the network is a circle [2] or if the network consists of two nodes that are connected by three edges [1]. It has not even been solved for the simplest possible network: the interval [4]. To illustrate the problem in solving these games, consider the search game on a star such as depicted in Fig. 2. H will hide at one of the end points of the star, but will occasionally move to another end point at

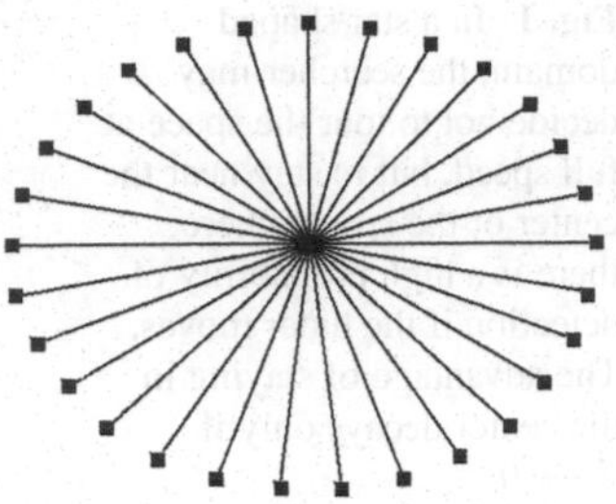

Fig. 2 In the search game on a star with a mobile hider, the searcher has two pure strategies: remain at the central point to catch the hider if she is on the move, or go up and down an edge to search one of the end points

maximum speed. S will search edges of the star, but will occasionally stand still in the central vertex in ambush of a moving hider. This game is far from solved, but recent progress has been made in [33] under the assumption that the players are noisy. If both S and H are aware of the number of nodes n that have been searched since the last time that H moved, then the value of the game depends on the state variable n only. The approach in [7] is analogous but slightly different. There a continuous game is considered in which S can either search a part of the space X or wait in ambush. H can decide to remain still or move to another random location in X. If S searches and H stands still, then S may capture H with a probability that is proportional to the amount of space that has been searched. If S ambushed and H moves, then H is captured. If $s(t)$ is the proportion of the space that has been searched at time t, then S controls $s'(t)$. If τ denotes the *expected* capture time and h is the time that the hider moves, then it can be shown that

$$\tau = \int_0^h ts'(t)dt + (1 - s(h))s'(h)(\tau + h) \tag{2}$$

The hider is indifferent under optimal play, in other words, $\frac{d\tau}{dh} = 0$ and differentiating the integral equation gives

$$-(s'(h))^2\tau + (1 - s(h))(1 + s''(h)\tau) = 0 \tag{3}$$

The solution of this differential equation is subtle, since τ is not yet known. For any value of τ one can compute the solution from the initial condition is $s(0) = 0$. This solution has to satisfy that the expected capture time against an immobile hider is equal to τ, since the hider is indifferent against an optimal searcher strategy. It is shown in [7] that optimally $s(t) = t - \frac{1}{4}t^2$ and that $\tau = \frac{2}{3}$.

One game that has been extensively studied is the search game on the network with two nodes that are connected by multiple edges, as in Fig. 3. It has been solved under the 'non-loitering restriction', which means that the searcher is not allowed to remain at a node in ambush. If S has initial position at node O and H has initial position at node A, then it is shown in [2, chap. 4] that it is optimal for the searcher to move from O to A at maximum speed along a random edge. The hider waits at A and only moves along a random edge to O just before the searcher reaches A, at the same speed. Both players keep moving up and down along random edges. The capture

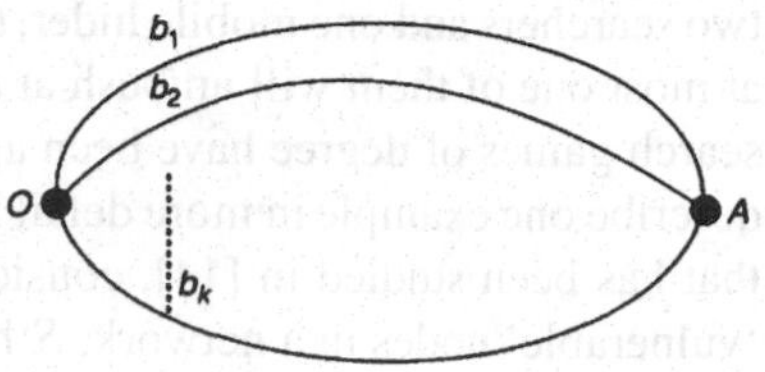

Fig. 3 The search game on a network with two nodes and k arcs of equal length is easy to solve under the non-loitering restriction, but remains unsolved for $k > 3$ if this restriction is dropped

time τ is geometrically distributed with parameter $\frac{1}{k}$. If the non-loitering restriction is dropped, then these strategies are no longer optimal and it is known that the searcher ambushes at the nodes with positive probability [1]. The solution of the game is unknown for $k > 3$. In general, search games are hard to solve. Dresher's guessing game, which is one of the first search games that has been studied, was solved only recently [15].

1.3 Games of Degree and Multi-agent Games

Isaacs distinguishes between games of kind and games of degree. In a *game of kind* one considers the average time τ: the searcher's objective is to find the hider in least expected time. In a *game of degree*, the searcher has to find the hider before a given time. For instance, if H is a suicide-bomber then S should catch H before she reaches her destination. For applications to counter-terrorism studies, search games of degree are more important than search games of kind. Despite this fact, they are less well studied.

The simplest model of a search game of degree with an immobile hider is the hypergraph incidence game [32, p. 5]. This is a simple game that is played on a finite set X. The searcher chooses one element $F \in 2^X$ from some given family of subsets and the hider chooses an element $x \in X$. The searcher wins iff $x \in F$. To see why this game is relevant, consider the search game on a tree where S starts the search from the root of the tree and has to find H within a certain time t_0. In other words, the searcher chooses a path F of length t_0 and the hider chooses an element x. The searcher wins if $x \in F$. The dominant strategies of the searcher correspond to end points of the tree. So H chooses an element from the end points E of the tree. Any path of length t_0 intersects E in a subset. So S chooses some subset of E. The search game of degree on a tree reduces to the hypergraph incidence game. One application of the hypergraph incidence game to network design is given in [22]. A protector chooses a spanning tree $\mathscr{T}$ in a network and an attacker chooses an edge E. The protector wins if $E \in \mathscr{T}$. The value of the game can be interpreted as a level of safety for the network, which can be used in its design against intrusion.

Another class of search games that deserve more attention are multi-agent search games. Isaacs pointed out that in a homogeneous region, two searchers simply divide $\mathfrak{R}$ into two halves and search them separately and independently. In a non-homogeneous region, this is no longer true. For instance, in the game on a star with two searchers and one mobile hider, the searchers will coordinate their search so that at most one of them will ambush at the central node at any given time. Multi-agent search games of degree have been applied to counter-terrorism studies and we will describe one example in more detail in the next section. To mention another example that has been studied in [14], consider a game in which a H tries to reach certain 'vulnerable' nodes in a network. S has a number of agents that can put road blocks on certain edges of the network. This is a variation on the hypergraph incidence game, in which both S and H choose a subset of 2^E, where E denotes the set of edges of the network. These subsets being the set of road blocks and the path. S wins iff the two sets have non-empty intersection.

2 Some Counter-Terrorism Search Games

Search games that can be applied to counter-terrorism studies have been developed only recently, and as of yet remain theoretical tools that have not been tested against practice yet. We exhibit some of these games here: the patrolling game and the disperse-or-unite game, introduced by Alpern and his co-workers, and the fugitive game, introduced by Owen and McCormick. We also include a new analysis on multi-agent search on the star network, as a first step towards the analysis of the multi-agent search games that are of interest to counterterrorism studies.

2.1 The Patrolling Game

The patrolling game [8], which is a game of degree, models the scheduling and deployment of patrols that safeguard vulnerable facilities. A searcher moves around a network G to protect it against a hider that attacks one of its nodes. Time is discrete and S moves from one node to the next in one unit of time. A patrol is a walk w around G during some time interval $[0, T]$, where $w(i)$ and $w(i+1)$ are either the same or adjacent. H selects a time h and a node k to carry out the attack, she needs m time units. If $k \in w[h, h+m-1]$, then S wins and otherwise he loses. The value of the game, i.e., the probability of a searcher win, represents the level of safety of the network against intrusion and the optimal hider strategy represents weaknesses in the network.

The patrolling game is a variation on the search game with an immobile hider. It is a game of degree rather than a game of kind, and it can also be considered as a variation on the hypergraph incidence game. As an illustration, consider the patrolling on the line graph with six nodes with $T = 4$ and $m = 3$. This game can be represented by the time-space domain $\{0, 1, 2, 3, 4\} \times \{0, 1, 2, 3, 4, 5\}$. A patrol is a path $(t, w(t))$ and an attack is a subset $\{(i, j-1), (i, j), (i, j+1)\}$. The solution of this game is illustrated in Fig. 4. An optimal mixed hider strategy is illustrated by representing only the central time of the attack (i, j) with its probability. The hider

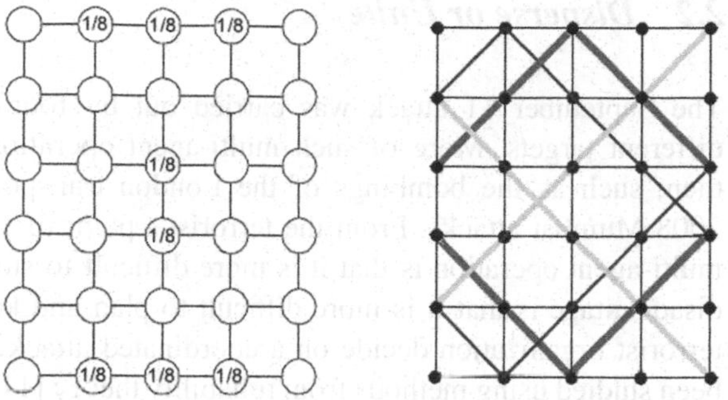

Fig. 4 The solution of the patrolling game on the line graph with six nodes with $T = 5$ and $m = 3$ following [8]. The horizontal axis represents time and the vertical axis represents space. The diagram on the *left* illustrates the hider strategy and the diagram on the *right* illustrates the searcher strategy

attacks nodes 0 and 5 with probability $\frac{1}{8}$ at each possible time moment. The hider attacks nodes 2 and 3 only during the time interval $[1, 3]$ with probability $\frac{1}{8}$. No patrol can prevent more than three of these possible attacks, so the hider wins with probability $\frac{5}{8}$ under this strategy. The optimal mixed searcher strategy is illustrated by paths in the time-space domain. The patrol $\{(0, 2), (1, 1), (2, 0), (3, 1), (3, 2)\}$ and its reflection $\{(0, 3), (1, 4), (2, 5), (3, 4), (3, 3)\}$ are chosen with probability $\frac{1}{4}$. The other four patrols that start from nodes 1 and 4 are chosen with probability $\frac{1}{8}$ each. It is not hard to verify that each attack is intercepted with probability of at least $\frac{3}{8}$. An attack at nodes 1 or 4 is intercepted with a probability $\frac{1}{2}$, which is why the hider does not attack one of these nodes. It is shown in [8] that the hider will never attack such a penultimate node. Other results from that paper include a method to decompose a network and other reduction techniques, to facilitate the computation of optimal strategies. Below, we suggest possible extensions of the patrolling games that deserve further study.

If we reverse the time and make the searcher patrol in the opposite direction, then we get another optimal strategy that can be combined with the original patrol. A searcher then walks up and down the same patrol, and we can consider a game that is played over indefinite time, rather than over a fixed time interval, which is more realistic. Suppose that the hider is able to observe the searcher's movement and can carry out the attack at any given time. The searcher can then no longer carry out a deterministic patrol, otherwise the hider could predict the movement of the searcher. So the patrol has to be a random walk rather than a deterministic walk, i.e, S moves from a node i to a neighboring node j with probability p_{ij}. In this game, S has to optimize the p_{ij} to optimize the capture probability, which is a Markov decision problem. If $m = 1$ then the capture probability at node i is equal to the probability p_i in the limiting distribution of the random walk. In this case S has to maximize the minimum p_i, which is a standard problem. If $m > 1$ the computations get more involved.

A multi-agent patrolling game with multiple hiders has not yet been studied. If two or more agents patrol the network, then they have to find an optimal way to coordinate their patrols. We will study this problem in more detail below. If two or more agents plan an attack, then they have to coordinate that attack and carry it out simultaneously. Such a game is the topic of the following section.

2.2 Disperse or Unite

The September 11 attack was carried out by four different groups with three different targets. More of such multi-agent operations have been planned since then, such as the bombings of the London transport network in 2005 and the 2008 Mumbai attacks. From the terrorist's point of view, an advantage of such a multi-agent operation is that it is more difficult to stop once it is underway, but a disadvantage is that it is more difficult to plan and keep secret. So what makes a terrorist organization decide on a coordinated attack? This is a dilemma that has been studied using methods from reliability theory [11], but it has also been studied by means of a search game of degree [5].

Consider a search game in which the hider can secrete various objects, but placing these objects requires an effort. The more time the hider invests in placing an object, the more difficult it becomes for the searcher to retrieve it. H has k objects that can be placed in n locations. In each location, H can invest a time t_i to place one or more objects. The total time that is available to H is normalized to $t_1 + \cdots + t_n = 1$. The more time the hider invests, the more time it takes the searcher to retrieve the hidden object from that location. If H has invested time t_i and if S searches location i, then it takes the searcher time t_i to uncover the hidden object. The searcher invests time s_i at location i and has a total time $s = s_1 + \cdots + c_n > 1$ to uncover the objects. S may choose to search a location exhaustively, i.e., $s_i = 1$, but may also invest less time. S knows how many objects have been hidden and he does not have to decide beforehand how much time s_i is invested in location i. If S uncovers all objects, then the attack is stopped and the searcher wins. Otherwise the hider wins.

It turns out that this game is not easy to solve and an efficient algorithm to compute the optimal strategies has not yet been found. To illustrate the game, consider the case of $k = n = 2$, i.e., two locations and two objects. If $s > 2$ then S has time to exhaustively search both locations and he always wins. If $\frac{3}{2} \leq s < 2$ then the searcher randomly chooses one location for an exhaustive search and spends the remaining $s - 1$ in the other location. So S either distributes the search time as $(1, s-1)$ or as $(s-1, 1)$. Against any hider strategy (t_1, t_2) the searcher wins if he exhaustively searches the location that has the largest t_i. So S wins with probability at least $\frac{1}{2}$. On the other hand, the hider may choose one random location and hide both objects there. In other words, the hider distributes the time like $(1, 0)$ or $(0, 1)$ and since the searcher can never win against both strategies, the hider wins with probability at least $\frac{1}{2}$ as well. The game gets interesting for $1 < s < \frac{3}{2}$, when the hider uses decoy objects. The hider either places both objects in both locations with an equal time investment of $(\frac{1}{2}, \frac{1}{2})$, or the hider places both objects in one location, but invests time $\frac{1}{2}$ in the first object and invests the remaining time in the other object. So, the placement of the objects is equiprobably $(\{\frac{1}{2}, 1\}, 0)$ or $(0, \{\frac{1}{2}, 1\})$. If the searcher uncovers an object after time $\frac{1}{2}$, then the remaining object is either at the same location and requires an exhaustive search, or it is hidden in the other location and requires a search of time $\frac{1}{2}$. Since $s < \frac{3}{2}$, the remaining time is not enough to

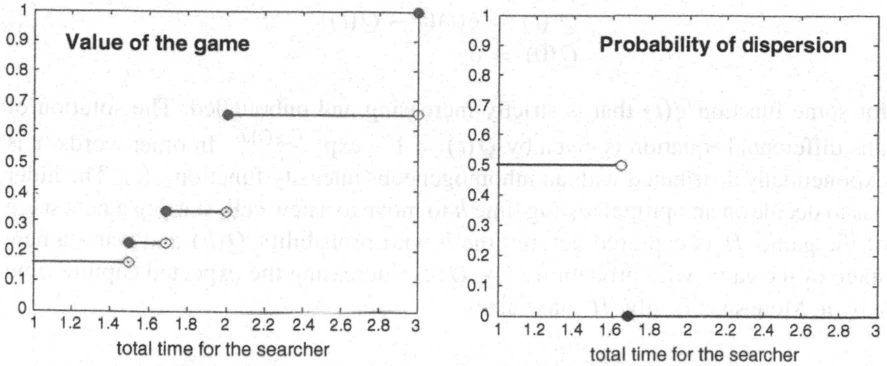

Fig. 5 The value of the dispersion game is defined as the probability of a searcher win. The diagram on the *left* gives the value of the game, which increases with s; the diagram on the *right* shows that the hider will attack only one location as soon as s passes a threshold value

cover both possibilities and the searcher has to decide. The optimal searcher strategy for $s < \frac{3}{2}$ turns out to be as follows. Search until the first object is found. With probability $\frac{2}{3}$ carry on the search in the same location and with probability $\frac{1}{3}$ start searching the other location. The searcher wins with probability $\frac{1}{3}$.

The analysis of the game quickly gets more complicated for larger values of k and n. The only game that has been fully solved so far is $k = 2$ and $n = 3$. The value of the game increases with s and the searcher surely wins if $s \geq 3$. As s decreases, the probability of a hider win increases and at a certain point the hider will start placing the two objects in different locations. In other words, if the searcher has not much time, then the hider disperses. Otherwise, the hider "unites", hiding the objects at the same place in the same location (Fig. 5).

The solution of the game for $k = 2$ and $n = 3$ indicates that the hider places the objects in such a way that the searcher does not gather any information during the search. If the searcher uncovers an object say after time r, then the game reduces to finding $k - 1$ objects in n remaining locations that require an exhaustive search of time $1 - r$. It must be possible to substantially extend the mathematical analysis in [5] by analyzing the recursion that underlies the game.

2.3 *Finding a Moving Fugitive*

The fugitive game [28] is a search game of kind with a mobile hider who may hide in a finite number of locations. If H moves from one location to another, then S knows that H has moved but does not know to which location, so then a new stage of the game begins. Let $Q(t)$ be the probability distribution function of the capture time τ. As time passes during a stage of the game, the probability of capture increases. This is described by the linear differential equation

$$\begin{aligned} Q'(t) &= g(t)(1 - Q(t)) \\ Q(0) &= 0 \end{aligned} \tag{4}$$

for some function $g(t)$ that is strictly increasing and unbounded. The solution of this differential equation is given by $Q(t) = 1 - \exp^{-\int_0^t g(s)ds}$. In other words, τ is exponentially distributed with an inhomogeneous intensity function $g(t)$. The hider has to decide on an optimal resting time h to move to a new cell, starting a new stage of the game. H is captured before time h with probability $Q(h)$ and starts a new stage of the game with probability $1 - Q(h)$, increasing the expected capture time τ by u. More specifically, H maximizes

$$\int_0^h tQ'(t)dt + (1 - Q(h))(h + \tau) \tag{5}$$

where τ denotes the expected capture time at the start of the game. To find the optimal value of h we differentiate to find that $\tau = (1 - Q(h))/Q'(h) = 1/g(h)$ and thus the optimal resting time satisfies

$$h = g^{-1}(1/\tau) \tag{6}$$

This is an implicit equation, since τ can only be computed from h. To solve h, notice that the expression in (5) is equal to τ, so this reduces to an equation with only one unknown. Interestingly, and somewhat counter-intuitively, if τ is large then the resting time is small, but if τ is small, then the resting time is large. In other words, the hider moves around a lot if the expected capture time is long. If the expected capture time is short, then the hider sits still.

Up to this point, the fugitive game is similar to the ambush game in [7] that we discussed above. Equations 4 and 5 are analogous to Eqs. 2 and 3. The hider moves after a fixed resting time h, as in Gal's solution of the princess and monster game, given in Eq. 1. A new element in the analysis of [28] is that it takes betrayal into account. Equation 4 can be interpreted as a probability of finding the hider by betrayal. The probability of betrayal increases with time, i.e., $Q'(t)$ is increasing, and this is one incentive for the hider to move. Another incentive to move is a search in the location where H is hiding. This is modeled by increasing the function $g(t)$ in Eq. 4 to another function $f(t) > g(t)$. The optimal resting time if S is searching the right location is then computed for $f(t)$ instead of $g(t)$.

A new stage of the game begins each time the hider moves to a new location and, since $f(t)$ and $g(t)$ depend on the hider's location, the expected capture time will depend on choice of the hider and the searcher at the start of a new stage. This is a stochastic game. This game has a well defined value, if the hider is only allowed to move a finite number m of times and then the game can then be solved by backward induction. It is proved in [28] that the expected capture time converges as m goes to infinity. Thus, the game can be described by a Markov chain. The searcher can then base his strategy on the transition probabilities and the equilibrium distribution of this Markov chain.

2.4 Some Remarks on Multi-agent Search Games

The fugitive game is the most intricate model of a counterterrorism game that is available at the moment. However, it is a single agent game that does not yet incorporate the possibility of the searcher laying an ambush. Such search games have not yet been reported in the literature and that is why we consider such games here, making some preliminary observations. Consider a game on a star, as in Fig. 2, with two searchers and one hider and a very large number of nodes. This game can be described by an integral equation that is similar to Eq. 2 but now $s'(t)$ is bounded from above by 2 instead of 1, since two searchers double the speed. If one searcher ambushes, then the other will continue the search, so $s'(t)$ is bounded from below by 1 instead of 0. The probability that H is caught in ambush at time h is equal to $s'(h) - 1$, so the expected capture time is given by

$$\tau = \int_0^h ts'(t)dt + (1 - s(h))(s'(h) - 1)(\tau + h) \tag{7}$$

This is only a slight variation on Eq. 2 and it leads to the following differential equation for $s(t)$

$$2s(h) - (\tau + h)s'(h) + \tau s''(h) - hs(h)s''(h) - \tau s'(h)^2 - 1 = 0 \tag{8}$$

with initial condition $s(0) = 0$. This differential equation can only be solved numerically and as we already indicated in the discussion of the ambush game with one searcher, the solution of this equation is subtle since τ is a priori unknown.

One can extend this analysis to other networks, but the solution of the differential equation will quickly get more complicated. For instance, the search game on the network with k arcs and two nodes as in Fig. 2 can be modeled by letting H be captured with probability $\frac{1}{2}$ if H moves and S ambushes. Equation 7 can easily be adapted to take this into account. In the same way one can extend the analysis to consider multiple searchers. The resulting differential equation requires some numerical work, but the method to solve this game is clear. A search game with multiple hiders, however, will be much more difficult to solve.

3 Summary

We have exhibited several search games that have been developed for counter terrorism studies. Most of these games are single agent. A more realistic approach requires the study of multi-agent games. Such games are very hard to solve, efficient algorithms are still to be developed, and the connection with overlay techniques as of yet remains absent. We have indicated how the differential equations approach in [28] can be adapted to study a multi-agent search game of kind on the star network, against a single hider. The extension of this analysis to arbitrary networks and multiple hiders remains a challenging task.

References

1. Alpern S, Asic M (1986) Ambush strategies in search games on graphs. SIAM J Control Optim 24(1):66–75
2. Alpern S, Gal S (2003) The theory of search games and rendezvous. International series in management science, vol 55. Kluwer, New York
3. Alpern S, Baston V, Gal S (2008) Network search games with immobile hider without a designated starting point. Int J Game Theory 37:281–302
4. Alpern S, Fokkink R, Lindelauf R, Olsder GJ (2008) The princess and monster game on an interval. SIAM J Control Optim 47(3):1178–1190
5. Alpern S, Fokkink R, Op den Kelder J, Lidbetter T (2010) Disperse or unite, a model of coordinated attack. In: Proceedings of the first international conference on decisions and games for security (GameSec), Berlin. pp 220–233
6. Alpern S, Fokkink R, Gal S, Timmer M (2011) On search games that involve ambush (reprint, 2012)
7. Alpern S, Fokkink R, Timmer M, Casas J (2011) Ambush frequency should increase over time during optimal predator search for prey. J R Soc Interface 8(64):1665–1672. doi: 10.1098/rsif.2011.0154
8. Alpern S, Papadaki K, Morton A (2011) Patrolling games. Oper Res 59(5):1246–1257
9. Anderson EJ, Aramendia MA (1990) The search game on a network with immobile hider. Networks 20(7):817–844
10. Anderson EJ, Aramendia MA (1992) A linear programming approach to the search game on a network with mobile hider. SIAM J Control Optim 30(3):675–694
11. Bier V, Olivers S, Samuelson L (2007) Choosing what to protect. Defensive allocation against an unknown attacker. J Publ Econ Theory 9(4):563–587
12. Bonato A, Nowakowski RJ (2011) The game of cops and robbers on graphs. AMS Student Mathematical Library 61, 2011
13. Chrobak M (2004) A princess swimming in the fog looking for a monster cow. ACM SIGACT News 35(2):74–78. doi:10.1145/992287.992304
14. Dickerson JP, Simari G, Subrahmanian VS, Kraus S (2010) A graph-theoretic approach to protect static and moving targets from adversaries. In: Proceeding of the ninth international conferences on autonomous agents and multiagent systems (AAMAS-2010), Toronto. pp 299–306
15. Fokkink R, Stassen M (2011) An asymptotic solution of Dresher's Guessing game, second international conferences on decisions and games for security (GameSec), Maryland. pp 104–116
16. Gal S (1979) Search games with mobile and immobile hider. SIAM J Control Optim 17(1): 99–122. doi:10.1137/0317009
17. Gal S (1980) Search games. Academic, New York
18. Gal S (2000) On the optimality of a simple strategy for searching graphs. Int J Game Theory 29:533–542
19. Garnaev AY (1992) A remark on the Princess and Monster search game. Int J Game Theory 20(3):269–276. doi:10.1007/BF01253781
20. Garnaev AY (2000) Search games and other applications of game theory. Springer, Berlin
21. Gill P, Phythian M (2006) Intelligence in an insecure world. Polity Press, Cambridge
22. Gueye A, Walrand JC, Anantharam V (2010) Design of network topology in an adverserial environment. In: Proceedings of the first international conference on decisions and games for security (GameSec), Berlin. pp 1–20
23. Herman M (1996) Intelligence power in peace and war. Cambridge University Press, Cambridge
24. Hougham V (2008) Sociological skills used in the capture of Saddam Hussein. Footnotes 33(6). http://wwww.asanet.org/footnote/julyaugust05/index.html

25. Isaacs R (1965) Differential games, a mathematical theory with applications to warfare and pursuit, control and optimization. Wiley, New York
26. Lalley S, Robbins H (1987) Asymptotically minimax stochastic search strategies in the plane. Proc Natl Acad Sci 84(8):2111–2112
27. Megiddo N, Hakimi SL, Garey MR, Johnson DS, Papadimitriou C (1988) The complexity of searching a graph. J ACM 35(1):18–44
28. Owen G, McCormick GH (2008) Finding a moving fugitive. A game theoretic representation of search. Comput Oper Res 35:1944–1962
29. Petraeus H, Amos JF (2007) Counterinsurgency field manual, U.S. Army Field Manual on Tactics, Intelligence, Host Nation Forces, Airpower, University of Chicago Press, Chicago
30. Potters JAM, Reijnierse JH (1993) Search games with immobile hider. Int J Game Theory 21:385–394
31. Ruckle W (1983) Geometric games and their applications. Pitman, London
32. Scheinerman ER, Ullman DH (2008) Fractional graph theory. http://www.ams.jhu.edu/ers/fgt
33. Timmer M (2008) Rendezvous on an interval and a search game on a star. MSc thesis, TU Delft. repository.tudelft.nl
34. Von Stengel B, Werchner R (1997) Complexity of searching an immobile hider in a graph. Discret Appl Math 78:235–249

Temporal and Spatial Analyses for Large-Scale Cyber Attacks

Haitao Du and Shanchieh Jay Yang

1 Introduction

Prevalent computing devices with networking capabilities have become critical cyber infrastructure for government, industry, academia and every-day life. As their value rises, the motivation driving cyber attacks on this infrastructure has shifted from the pursuit of notoriety to the pursuit of profit [1, 2] or political gains, leading to cyber terrorism on various scales. Cyber terrorism has had its share of case studies and definitions since late 1990s and early 2000s [3–5]. A common denominator of the definition of cyber terrorism is the threat posed through the use of cyber infrastructure, especially the Internet. Stuxnet, a malware discovered in June 2010, which was a directed attack against the Iranian nuclear program [6], represented a milestone on cyber warfare and posed a new challenge to analyze and understand cyber attacks due to its complexity in attack strategy. While cyber terrorism can have many elements beyond exploiting cyber vulnerabilities, this chapter focuses on analyzing techniques that process observables of malicious activities in the cyberspace.

As new cyber vulnerabilities are discovered by few elite attackers, they are routinely bought and sold by underground organizations. Corresponding attack tools take advantage of the novel vulnerabilities and recruit new zombie (compromised) hosts for the attacker. According to the Symantec Internet security threat report [1], the largest botnet observed in 2010 had over one million bots under control, and underground economy advertisements promote 10,000 bots for $15. Large-scale cyber attacks can take the traditional form of a botnet, from which a large number of hosts perform similar actions, e.g., Distributed Denial-of-Service (DDoS) or distributed stealthy scans [2]; they can also consist of a team of colluding sources

H. Du (✉) • S.J. Yang
Rochester Institute of Technology, One Lomb Memorial Drive, Rochester, NY, USA
e-mail: hxd1011@rit.edu; jay.yang@rit.edu

V.S. Subrahmanian (ed.), *Handbook of Computational Approaches to Counterterrorism*, DOI 10.1007/978-1-4614-5311-6_25,

dividing up tasks, interleaving actions over time and dispersing over the IP and port spaces to conceal their overall strategy. It is not uncommon for an enterprise or global network to face multiple coordinated attack teams simultaneously along with other large-scale malicious activities. This overwhelming data has led to challenges to extract, comprehend and predict the diverse attack strategies and goals within the mixture of various coordinate cyber attacks.

Combating against large-scale, coordinated attacks requires advances on various fronts, including intrusion detection, alert correlation, attack characterization, attack prediction, and host clustering. Advances in intrusion detection [7, 8], though not perfect, have provided significant observables or alerts that contain attributes of individual malicious actions. Alert correlation, e.g., [9–17], processes and groups the observed alerts based on their similarity or a pre-defined attack scenarios. Attack characterization and prediction, e.g., [16, 18–23], aims to analyze the *temporal* or sequential characteristics of alert sequences of individual attack sources, aiming at predicting behaviors of future actions. Host clustering, e.g., [24–28], explores the *spatial* characteristics as well as packet and flow level anomalies among Internet hosts, to group them into clusters of normal, infected or botnet hosts.

While the computational techniques used to analyze cyber attacks are advancing, the attack tools, hacker skills and the coordination strategies are also becoming more sophisticated. This calls for new techniques to recognize coordinated attack teams and their strategies when the critical observables are embedded in the large volume and diverse malicious activities. This chapter will first summarize the advances in alert correlation, attack characterization and prediction, and host clustering. From there, we will discuss our recent and ongoing works that investigate how temporal and spatial analyses can be used to discover attack sources that play different roles in a cyber attack and to discover different coordinated attack strategies. The task of simultaneously grouping attack sources into coordinated teams and recognizing the coordination strategies is a daunting challenge. The chapter will end by discussing the open research problems in recognizing coordinated attack teams and strategies in large-scale cyber attack environment.

2 Intrusion Detection and Alert Correlation

The essential goal of a Intrusion Detection System (IDS) is to differentiate malicious activities from the normal ones. For more than two decades, significant effort has been put into advancing intrusion detection via anomaly-based and signature-based systems. Anomaly-based detection techniques model the users or systems normal behavior, and report outliers as suspicious activities. Signature-based systems usually maintain a database of malicious behavior signatures and use pattern matching techniques to detect and report suspicious actions. Details of intrusion detection techniques can be found in several survey papers, e.g., [8, 29–31].

Intrusion detection is challenging because of the variety of normal behaviors and fast changing cyber environment with new vulnerabilities and attacks.

Table 1 An example of alert correlation scenario (The example is extracted from [14])

Alert ID	Description	Sensor	Start/end time	Source	Target
1	IIS exploit	N1	12.0/12.0	80.0.0.1	10.0.0.1, port: 80
2	Scanning	N2	10.1/14.8	31.3.3.7	10.0.0.1
3	Port scan	N1	10.0/15.0	31.3.3.7	10.0.0.1
4	Apache exploit	N1	22.0/22.0	31.3.3.7	10.0.0.1, port: 80
5	Bad request	A	22.0/22.1		localhost, Apache
6	Local exploit	H	24.6/24.6		linuxconf
7	Local exploit	H	24.7/24.7		linuxconf

While intrusion detection techniques continues to evolve and improve, the overwhelming and heterogeneous IDS alerts has made the analysis difficult and unable to provide an effective situation assessment. As a result, *alert correlation* has become a popular topic in the past decade [32]. Ideally, the goal of alert correlation is to determine collections of IDS alerts, where each collection corresponds to a high level description of the attack.

The tasks of alert correlation can be categorized into Normalization (Norm.), Aggregation (Agg.), Correlation (Corr.) and Strategy Analysis (SA) [33]. The normalization organizes the format of alerts from heterogeneous IDS sensors. Aggregation combines alerts that share the same root causes, e.g., originated from same source IP or attack the same target. In the correlation step, aggregated alerts are mapped into an attack scenario template. Further, the causal relationship, i.e., pre-condition and post-condition, can be used in strategy analysis to infer attack intention and strategies.

The details of alert correlation and a performance evaluation can be found in [33, 34]. Table 1 uses an example to illustrate an ordered sequence of seven alerts for alert correlation. The example is extracted from [14] which provides a comprehensive framework for alert correlation.

The monitored network has four heterogeneous IDS sensors, i.e., network based IDS 1 and 2, (N1,N2), host based and application based IDS (H, A). The attacker (31.3.3.7) first launches a port scan against 10.0.0.1 and discovers the vulnerability of Apache server. After scanning, the attacker performs a successful Apache buffer overflow exploit on the target and obtains user privilege on the server. Finally, the attacker launches privilege escalation by using a local exploit *linuxconf*. In addition, there is one noisy alert triggered by a worm that probes the same target while the attack is in progress. The ideal output of alert correlation would successfully group Alerts #2 and #3 as malicious scanning, Alerts #4 and #5 as vulnerability attempts, Alerts #6 and #7 as privilege escalation, and the noisy Alert #1 will be marked as irrelevant. After the aggregation of alerts, the scanning, vulnerability attempts and privilege escalation should be correlated and reported to security analysts. This example shows that once correlated, an analyst can be more effective to process the high-level attack descriptions instead of individual alerts.

The methodologies widely used in correlation engines include similarity based clustering and causal relationship based (pre/post condition) reasoning. The attack

Table 2 Review of alert correlation work

Ref.	Framework	Methodology
[35]	Agg.	Similarity based clustering
[12]	Corr., SA	Pre/post condition based correlation
[11]	Norm., Agg., Corr.	Exam source-target relationship
[36]	Agg., Corr.	Pre/post condition based correlation
[15]	Corr.	Similarity based, Pre/post condition based correlation
[37]	Agg., Corr.	Bayesian networks based inference
[38]	Corr.	Bayesian networks, Pre/post condition based correlation
[18]	Corr.	Bayesian networks based inference
[9]	Agg., Corr.	Similarity based clustering
[14]	Agg., Corr.	Complete system, attack scenario based correlation
[39]	Corr.	Data mining for frequent structured patterns

scenario templates can be pre-defined or automatically learned from data. Uncertainties usually are captured with Bayesian networks, which will be reviewed in next section. Table 2 is a summary of representative alert correlation approaches.

Alert correlation synthesizes the raw IDS alerts into attack scenarios and provides better situation awareness to security analysts. However, the ever-changing environment, e.g., software patches, new IDSs, customized alerts, and new exploits, make alert correlation challenging. On the other hand, the attack-scenario (pre/post-condition) based approaches are accurate but will not scale for a large number of diverse and unknown attacks, similar to the limitation of signature-based intrusion detection. The existence of large-scale sophisticated coordinated attacks calls for further advances in computationally effective attack characterization, prediction and host clustering.

3 Attack Characterization and Prediction

Bayesian networks are widely used to model uncertainty in the security context because the conditional dependency fits perfectly with pre/post-condition and attack scenarios. Several work [18, 40, 41] utilizes Bayesian networks for predicting high-level goal of an attack. Figure 1 gives an example of Bayesian networks modeling applied to the security context.

Consider a simplified attack scenario with four random variables representing the stages of an attack. Let B denote *install back door on the system*, C as *compromise application account and password*, M as *Monitor confidential transactions*, and S denote *Successfully obtain the confidential data*. The conditional dependencies of the random variables can be described in Fig. 1. It indicates that the success of C (compromise application account and password) and M (monitor confidential transactions) are independent given their parent B (install back door on system). Meanwhile, S (Successfully obtain the confidential data) depends on C and M.

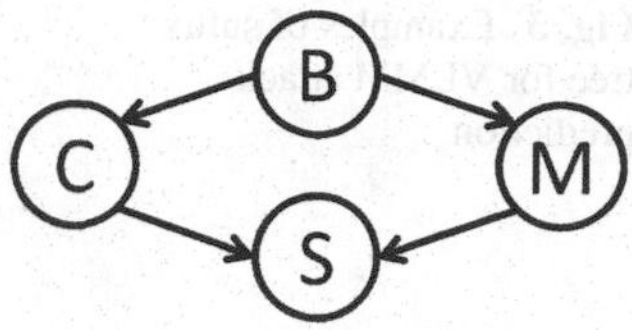

Fig. 1 Example of Bayesian networks for attack prediction

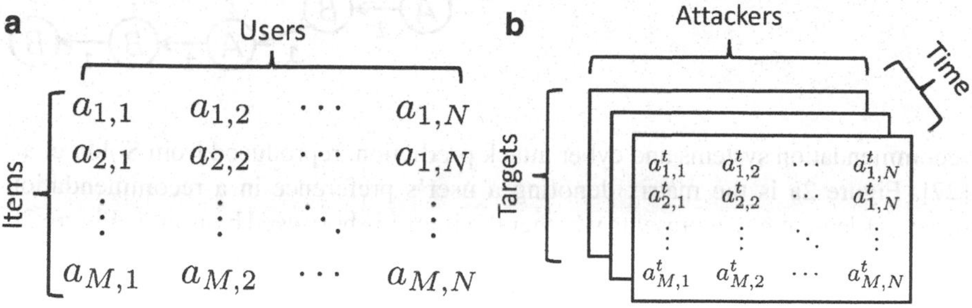

Fig. 2 Attack prediction using recommendation system. (**a**) Recommendation system. (**b**) Network attacks (Figure is reproduced from [27])

In addition to the structure of the model, the parameters i.e., the conditional probabilities, of the model could potentially be derived from security knowledge. In particular, a complete model will specify $P(B)$, $P(C|B)$, $P(M|B)$ and $P(S|C,M)$. Therefore, according to the conditional dependencies, the joint distribution $P(B,C,M,S)$ can be decomposed by $P(B)P(C|B)P(M|B)P(S|C,M)$. With simplified joint probability, one can perform any inference on interested events.

There are many challenges when using Bayesian networks for cyber attack modeling and prediction. The key problem for Bayesian networks is the assumption of the model. Unlike other fields, the model structure and parameters have very high uncertainty for multistage cyber attacks. Furthermore, training using up-to-date multistage data is almost impossible, as little ground truth exists for the stages a cyber attack goes through. Manually specifying both the structure and the parameters may be error-prone for a large network.

Because of its limitations on scalability and the requirement of domain knowledge, Bayesian networks are typically used to predict the high-level goal of an attack, for example, whether the attacker will compromise an application account and password. On the other hand, sequence modeling, e.g., [21, 22, 27], has been utilized to predict more detailed attack actions, e.g., the next attack target or service.

Generally speaking, sequence modeling techniques learn attack patterns from observed attack sequences and predict the future actions of a given sequence based on the aggregate likelihood of similar attack patterns. Soldo et al. [27] developed a cyber attack prediction system by drawing an analogy from the context of recommendation systems [42], which has been used to recommend movies based on similar users' preferences. Figure 2 shows an example of this analogy between

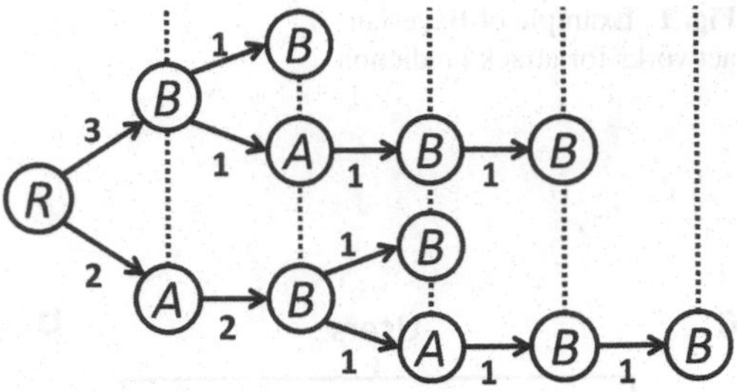

Fig. 3 Examples of suffix tree for VLMM attack prediction

recommendation systems and cyber attack prediction, reproduced from Soldo et al. [27]. Figure 2a is the matrix denoting a user's preference in a recommendation system. Element $a_{i,j}$ represents whether the item i is borrowed by user j. Figure 2b presents a similar idea in the context of cyber attacks. Similar preference on the choice of targets from similarly behaving attacks are used to 'recommend/predict' the targets of a given attack.

A more explicit approach to extract the sequential dependencies between attack action attributes within each attack sequence is to use a Variable Length Markov Model (VLMM [21]). Consider an ongoing attack with N observed actions $\{X_1, X_2, \cdots, X_N\}$. A model of order o assumes the current observed event is conditional depend on previous o events. The probabilities are obtained from sample counts in historical and ongoing attack sequences. A sequence of length N will contribute to the building of oth order models for $1 \leq o \leq N$. More specifically, a sequence of length N will provide one sample to the nth order model, two samples to the $(N-1)$th order model, ..., and N samples to the 1st order model. For implementation, a suffix tree can be used to record the samples and to store models of different orders, which allows making predictions based on observed context in $O(N)$ time given a sequence of length N. Figure 3 shows the suffix tree corresponding to a single sequence of 'A, B, A, B, B', where the edge weights indicate the number of times the corresponding transition has occurred.

In reality, the suffix tree will be built with many attack sequences and continuously updated with incoming alerts. The overall suffix tree represents the various possible Markov relationships of different orders. For a given ongoing sequence with length K, the for oth order model $P_o(X_K)$, $\forall -1 \leq o \leq K$ can be found from the suffix tree in $O(K^2)$ time, which may be further simplified. The $P_o(X_{K+1})$ can then be blended to make the prediction for next event as

$$P(X_{K+1}) = \sum_{o=-1}^{K} w_o \cdot P(X_{K+1}|X_{K-o+1}, \cdots, X_K)$$

where w_o is the weight associated with the oth order model, and $\sum_{o=-1}^{K} w_o = 1$. Note that finite sequences should be penalized by their rarity and rewarded by their specificity. The weights are designed to be adaptive to take into account the Bias-Variance trade off [43]. Examples of the weight functions can be found in [44].

Fava [21] shows that there is no significant performance difference when using different weight functions, as long as the weights satisfy the properties described above. Notice that the summation starts at -1. The minus-one order model assigns all characters a probability of $1/|\Omega|$ to prevent the *zero frequency problem* [44]. The zero order model assumes all observations are independent and holds the frequency count of all $X \in \Omega$. For the given example in Fig. 3, after observing 'A, B, A, B, B', we predict next event by considering the minus-one order model ($P_0(A) = P_0(B) = 0.5$), zero order model ($P_1(A) = 0.4$, $P_1(B) = 0.6$), first order model ($P_2(A|A) = 0$, $P_2(B|A) = 0.5$, $P_2(B|B) = 0.25$, $P_2(B|B) = 0.25$), all the way to the fifth order model. The predictions in different models will be blended using different weights.

The VLMM model allows us to discover patterns within attack sequences without explicitly defining attack plans [21,22]. In fact, an ongoing attack sequence can match patterns from numerous different types of preceding attack sequences. A VLMM combines the probabilities associated with all matched patterns and produce a better guess.

4 Host Clustering and Botnet Detection

Bayesian networks and sequence modeling enable us to predict future events of ongoing attacks. Among all attack prediction work, selecting the right granularity for modeling is crucial, which needs to take all factors into consideration, including the usage (training and testing) of the data, scalability, etc.

On the other hand, in addition to focusing on examining the detailed action, in a large scale, host clustering work allows us to group a large number of (normal or malicious) hosts that share similar behavior into clusters. The input for host clustering usually is passive Internet backbone traffic [24–26, 28] or malicious traffic [27, 45] and the outputs are host clusters with behavior pattern descriptions, which could be useful for anomaly detection and attack profiling. Some malicious activities, such as worms, scanning and DDoS attacks, can be detected behavior by host clustering work [26, 28]. An example of host clustering for botnet detection can be found in [26]. Scan activity, spam activity, binary downloading and exploit activity are classical activities that botnet hosts would behave. By looking at the host clusters that exhibit these different activities, one can infer the structure of botnet zombie hosts.

The key differences between various types of host clustering work can be understood as selecting different *features* to form a multi-dimensional data point and *objective function* for clustering them. The data point $\mathbf{x}$ represents an attacking host that contains several features on security context. Among all the basic attributes in the communications, the widely used features are statistics from flow information (source IP, destination IP) and protocol information (TCP/IP protocol, source port, definition port). Figure 4 is an example of data points from host clustering work [25]. Every point in the figure represents one attacking host. Three features, Relative

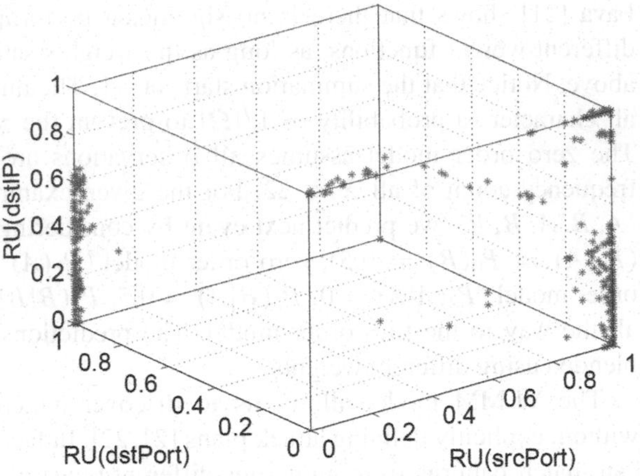

Fig. 4 An example of data points (The figure is extracted from [25])

Table 3 Review of host clustering work

Ref.	Features	Methodology
[26]	Flows and packets statistics	X-means clustering
[24]	Flows and protocol statistics	Hierarchical clustering
[25]	Relative uncertainty on flows	Customized algorithm
[28]	Relative uncertainty on flows	Spectral clustering
[45]	Statistics on flows	Hierarchical clustering
[27]	Statistics on flows	Customized algorithm

Uncertainty (RU) on source port, destination port, and destination IP are taken into account to characterize an attacker's behavior. The data points in Fig. 4 exhibit cluster patterns,. After selecting the features, the next step is clustering the data points, which optimizes on specific objective function. Different objective functions group data points based on different principles. Some of them minimize the distance between data points and cluster center (e.g., K-means clustering) and some utilizing the graph cut notion (e.g., spectral clustering) to separate points are distinct from each other. Table 3 is a summary review of host clustering work.

Host clustering analyzes a large number of attacking sources by grouping similar attack sources and extracting patterns of clusters. It may be ineffective to discover sophisticated attacks, which take slow actions and utilize multiple attack sources. Nevertheless, the features and the clustering analysis methodology can be very helpful for treating overwhelming numbers of alerts.

5 Coordinated Attacks

The mixture of organized cyber crimes, terrorism and random attacks against enterprise and government networks has led to asymmetric cyber battlefields filled with large-scale cyber attacks. In this chapter, the term *Coordinated Attack* is used to

Table 4 An example of coordinated attack from real-world data

No.	Time	Src. IP	Dest. IP	Snort alert description
1	11:02:07	10.13.148.213	10.14.0.100	(portscan) TCP Portscan
2	12:52:02	10.13.148.223	10.14.0.1	ICMP PING NMAP
3	12:52:02	10.13.148.223	10.14.0.2	ICMP PING NMAP
4	12:52:02	10.13.148.223	10.14.0.3	ICMP PING NMAP
5	12:52:02	10.13.148.223	10.14.0.4	ICMP PING NMAP
6	12:52:02	10.13.148.223	10.14.0.5	ICMP PING NMAP
7	12:52:02	10.13.148.223	10.14.0.6	ICMP PING NMAP
8	12:52:02	10.13.148.223	10.14.0.7	ICMP PING NMAP
9	12:52:02	10.13.148.223	10.14.0.8	ICMP PING NMAP
10	12:52:02	10.13.148.223	10.14.0.9	ICMP PING NMAP
11	12:52:02	10.13.148.223	10.14.0.10	ICMP PING NMAP
12	14:36:15	10.13.148.217	10.14.1.9	(portscan) TCP Portscan
13	14:51:26	10.13.148.210	10.14.1.17	WEB-MISC cat%20 access
14	15:02:30	10.13.148.210	10.14.1.17	WEB-MISC cat%20 access
15	16:15:37	10.13.148.150	10.14.1.17	WEB-MISC http directory traversal
16	16:25:35	10.13.148.50	10.14.1.17	WEB-MISC /etc/passwd
17	17:00:22	10.13.148.150	10.14.1.17	WEB-MISC http directory traversal
18	17:11:03	10.13.148.150	10.14.1.17	WEB-MISC http directory traversal

describe sophisticated, stealthy attacks which take advantage of multiple host level attack sources and are conducted by an attacker or a group of attackers.[1]

Table 4 gives an example coordinated attack by listing a sequence of Snort [7] IDS alerts from real-world data. Basic alert attributes are listed, which include time, source IP, target IP and attack signature. The attack signature is the Snort description of alert which can be looked up in an IDS database to get more information about the attack action.

There is interesting coordination among attack sources described in Table 4. According to the alert description, Alerts #1–12 are reconnaissance actions, and Alerts #13–18 are web server vulnerability attempt actions. Although these two steps are closely related, i.e., vulnerability attempts depend on reconnaissance, but they originated from different sources. Attack source 10.13.148.218 is utilized for probing across the target space within the subnet 10.14.0.x. After the reconnaissance, 10.13.148.210 and 10.13.148.150 are utilized to try different vulnerabilities on web services.[2]

In fact, this is a simple example where source IP subnet alone can be used to identify the corresponding team. This property occurs because the example is

[1]In general, it is hard to infer about the attacker(s), because the basic unit of most observable is at host level (e.g., IP address) but not person level.

[2]We also noticed there are missing steps between Alerts #12 and #13. Vulnerability attempts are conducted against 10.14.1.17, but no alerts indicates the target has been probed and discovered. This is because Snort alert is one type of observed evidence, and it may not be a comprehensive since current IDS cannot perform the perfect detection.

Table 5 Examples of coordinated attacks with different strategies

Src. IP	Dest. IP	Snort alert description	Team
10.13.1.86	10.100.113.48	(portscan) TCP Filtered Portsweep	EF
10.13.1.86	10.120.113.42	(portscan) TCP Filtered Portsweep	EF
10.13.1.86	10.199.113.36	(portscan) TCP Filtered Portsweep	EF
10.13.1.32	10.100.113.24	(portscan) TCP Filtered Portsweep	EF
10.13.1.32	10.100.113.9	(portscan) TCP Filtered Portscan	EF
10.13.1.32	10.100.113.3	(portscan) TCP Filtered Portscan	EF
10.13.1.32	10.100.113.8	(portscan) TCP Filtered Portscan	EF
10.13.1.32	10.100.113.7	(portscan) TCP Filtered Portscan	EF
10.33.1.13	10.100.133.1	ICMP PING NMAP	CM
10.33.1.12	10.100.133.2	ICMP PING NMAP	CM
10.33.1.16	10.100.133.3	ICMP PING NMAP	CM
10.33.1.16	10.100.133.4	ICMP PING NMAP	CM
10.33.1.17	10.100.133.5	ICMP PING NMAP	CM
10.33.1.12	10.100.133.6	ICMP PING NMAP	CM
10.33.1.20	10.100.133.7	ICMP PING NMAP	CM
10.33.1.16	10.100.133.8	ICMP PING NMAP	CM

extracted from International Capture the Flag (ICTF) hacking competition [46, 47] and participating teams are assigned specific range of source IP. In the real world where IP spoofing and zombie machines are common, inferring coordinated attack sources by source IP will not be applicable.

Because multiple attack sources are involved, coordinated attacks make detection more difficult on the defense side. Table 5 includes a sequence of alert data from the ICTF hacking competition [46]. Alerts triggered by two teams, *ENOFLAG* (EF) and *Chocolate Makers* (CM), are listed in the table. Both teams utilize multiple attacking sources to work together for achieving a certain goal. Basic alert attributes are listed, and the last column shows the team assignment for each attack source.

The two teams shown in Table 5 have different strategies, i.e., the attacks performed by two teams are different in both spatial and temporal domains. For the reconnaissance stage, team EF is more *centralized*. EF mainly uses three hosts for comprehensive reconnaissance. On the other hand, team CM is more *distributed*. Eleven sources are used for probing the whole target space. If we use vertices to represent hosts and edges represent attacks, the graphical representations [3] are very different for two teams, as shown in Fig. 5.

From the temporal domain, the sequences of attack sources by the two teams are also different from each other. Team EF's members perform TCP port sweep and port scan against the target space. For team CM, six attack sources collaboratively use NMAP to probe the target space. Therefore, for all alerts listed in the table, building the action sequence for each attack source will lead to following results:

[3]Table 5 only listed eight alerts for each team. Figure 5 represents two teams' all observed alerts.

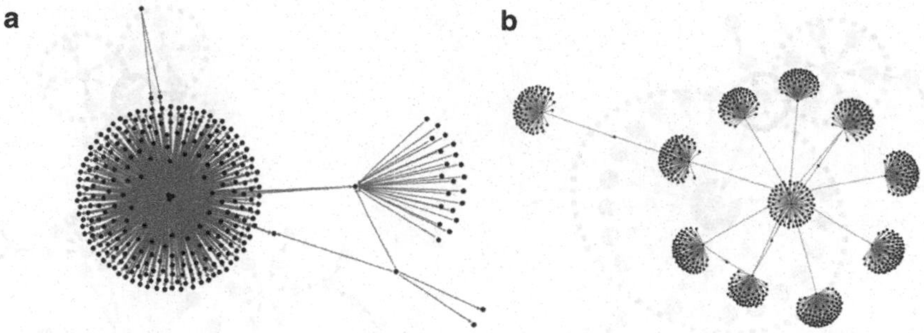

Fig. 5 Example of spatial feature for coordinated attacks. (**a**) Spatial relationships of attacks by EF. (**b**) Spatial relationships of attacks by CM

team EF will have two sequences with length 4 and 6 respectively. Team CM will have ten sequences, which with only one action.

Classical work may not be effective for dealing coordinated attacks because the actions are distributed into different attack sources. For alert correlation, with missing observables, it is very challenging to perform correlation by examining the pre-condition and post-condition. For example, consider the attack conducted by team CM given in Table 5, coordinated reconnaissance, the pre/post-condition and attack scenario based approach will fail to group the coordinating attackers. For attack characterization and prediction, Bayesian networks and sequence modeling may be also inefficient in the presence of missing actions, because coordinated attack will break the dependence structure of the model (for both Bayesian networks and Markov model). Host clustering may work for same cases where attack sources behave similarly, such as botnet DDoS attack, but may not work for the general case of coordinated attack.

6 Spatial and Temporal Analyses for Coordinated Attacks

As discussed earlier, classical work may not be effective on coordinated attacks because it mostly focuses on examining individual attack source's actions, which will only give *local information* and ignores the *global information* on the whole network. On the other hand, classical work suggests Markov models and graph-based analysis can greatly benefit the analysis for large-scale attacks. Using spatial features can provide additional insights about the relationship of the attack sources and the use of sequence modeling can help for characterize the attack behavior.

To analyze the spatial feature, an Attack Social Graph (ASG) can be defined as follows to represent attacks in a given network [45] as follows:

An Attack Social Graph $ASG_T(V, E)$ is a directed graph representing the malicious traffic within a time interval $[0, T]$, where a vertex $v \in V$ is a host, and an

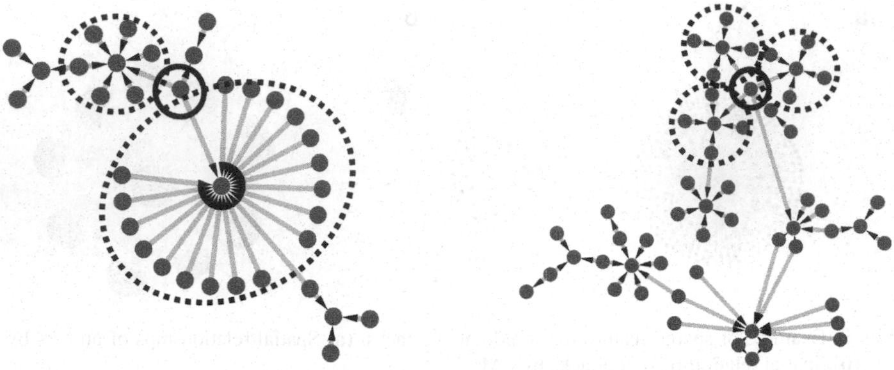

Fig. 6 Example of ASG analysis

edge $e_{(u,v)} \in E$ exists if attacks are observed from u to v. Edge direction is from the attacking host to its target.

In addition to the hacking competition data, ASGs exhibit many interesting patterns on other real-world data, which can provide us important insights on understanding coordinated attacks. For example, consider two ASG subgraphs shown in Fig. 6 which are extracted from CAIDA Network Telescope data [48, 49]. In Fig. 6a, b, the attack sources circled with the dotted line act similarly to attack one and only one target. This is unlikely to happen by chance due to the large target space. On the other hand, the two attack sources circled with the solid line in Fig. 6 are suspicious. They attack several *heavily attacked targets* and are the common denominator among the all attacking sources. One possible interpretation of such a situation is that this is a coordinated attack. The sources in dotted circles are zombie machines controlled by the two sources in the solid circles.[4]

Based on the intuition given in Fig. 6, two approaches can be applied to analyze ASG. The first approach is to calculate certain graph properties, such as centrality distribution [50]. The second approach is to define the labels that have specific meanings, and analyze attack sources with label sequences. For the spatial labeling approach, *Attack Conspirator*, *Heavily Attacked Target (HAT)* can be defined as follows:

Let $T_u \doteq \{v \mid e_{(u,v)} \in E, v \in V_t\}$ be the targets attacked by $u \in V_s$, the attack conspirators of $u \in V_s$, denoted as C_u, is a set of vertices: $C_u \doteq \{v \mid T_u \cap T_v \neq \emptyset,\ v \in V_s\}$, where V_t and V_s represent the set of targets and attacking sources respectively.

A heavily attacked target is a vertex $v \in V_t,\ \ s.t.\ d_{\text{in}}(v) \geq H$, where H is a pre-selected threshold.

[4]In the real-world data set, there is no ground truth suggesting how attack sources are coordinated. We collect evidence to support our assumption. In this two example, inter-arrival time and geographical information are consistent with the assumption of leader and zombie hosts.

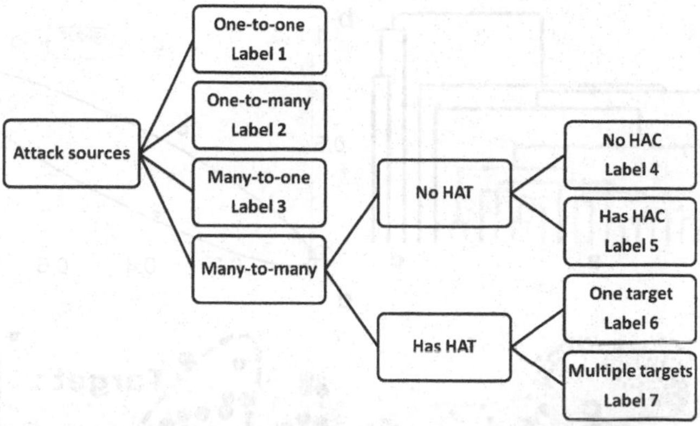

Fig. 7 Attack source labeling

Using these definitions, each source can be characterized by its spatial pattern. In particular, we can label the attack sources according to their behavior, and examine their roles in coordinated attack. There are many disjoint subgraphs in each ASG, the first level of labeling is to differentiate the sources based on the type of subgraph they reside: *one-to-one*, *one-to-many*, *many-to-one* and *many-to-many* relationships. The first three ASG subgraph types are relatively easy to analyze in the security context. *One-to-one* relationships could indicate the sensor was triggered by chance[5] or focused attack on specific target. *One-to-many* relationships represent service scanning on a set of targets. *Many-to-one* relationships represent DDoS or coordinated attacks. In the case of *many-to-many* subgraphs, additional factors are needed to differentiate the attack sources. Specifically, we use the HAT to differentiate whether a source is part of a potentially coordinated attack, i.e., unlikely to happen by chance. The attack sources that do not have any HAT are further differentiated depending on whether they have any Heavily Attacking Conspirator (HAC).[6] The idea here is to examine whether the source, which can be attacking because of mis-configuration or it has a specific target, has a conspirator that is part of a potentially coordinated attack. Note that, the sources that have at least one HAT must have HACs. We further differentiate sources with only one HAT or multiple targets. Figure 7 summarizes our labeling approach.

Spatial relationships among attack sources and targets are important features of coordinated attacks. With the degree centrality measure and labeling scheme, one can characterize the possible role for given attacking sources. Figure 8 gives the result of degree centrality based clustering results. In Du's work [45], connectivity

[5]The spatial pattern example is extracted from UCSD data set, in which the alert can be triggered by chance such as mis-configuration. In hacking competition data, we do not have such a case.

[6]Conspirator to be a Heavy Attacking Conspirator, if it has at least one HAT

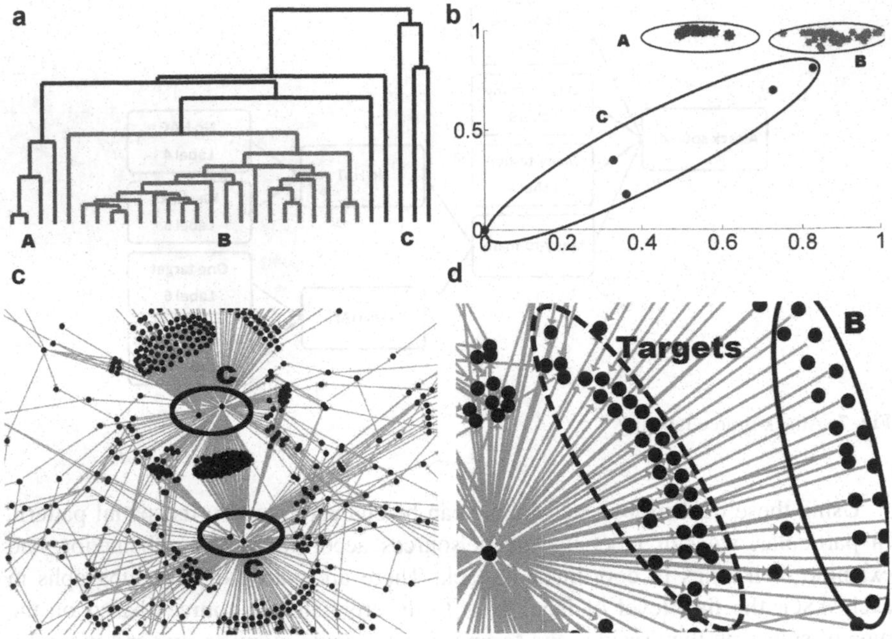

Fig. 8 Degree centrality based attack sources clustering. (**a**) Dendrogram of attack sources. (**b**) Cluster of attack sources. (**c**) Cluster C in ASG. (**d**) Cluster B in ASG

attributes and hierarchical clustering are used for categorize the coordinated attack. Figure 8a, b are the clustering results at the top level and Fig. 8c, d show the corresponding attack sources in the ASG.

By analyzing the clusters from Fig. 8a, we find several sets of interesting collaborative attack patterns. First, the five feature points within cluster C are actually five distinct attack sources that attack a large number of targets within the network monitored by the Network Telescope. These five sources are outliers on the 2D plane (see cluster C in Fig. 8b), it is shown in Fig. 8c to identify these sources in the ASG. Second, some attack sources form cluster B since these sources all have a "special" conspirator 0.211.214.160,[7] which is an attack sources in Cluster C and has out-degree $18,920$. Being a conspirator of such hosts makes their features significantly different from others and thus forms a cluster. Figure 8d identifies some of the Cluster B sources in a zoom-in view of the corresponding ASG.

For the spatial labeling approach, by analyzing the joint probabilistic distribution over time of the spatial labels, the patterns can be extracted from a very large training set (2,322,134 attack sources). For convenience, let $d(p)$ denote the number of unique values in a label pattern p. We define this value as the *diversity* of an

[7] For anonymity reason, in this chapter, the first byte of IP address from real-world data is masked with 0.

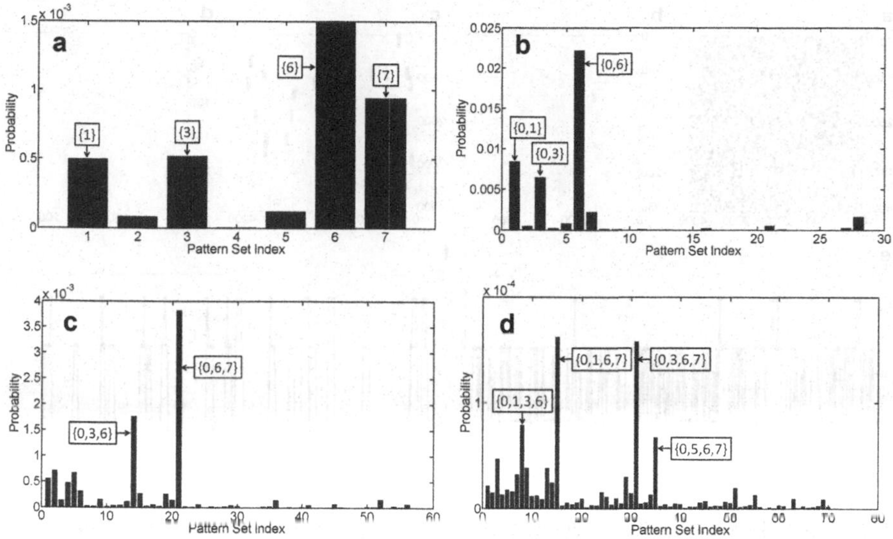

Fig. 9 Spatial pattern sets probabilities. (**a**) Pattern sets for $d(p) = 1$. (**b**) Pattern sets for $d(p) = 2$. (**c**) Pattern sets for $d(p) = 3$. (**d**) Pattern sets for $d(p) = 4$

attack sequence. Furthermore, let $\{a, b\}$ denote the label patterns with $d(p) = 2$ that contain x and y regardless of the order over which they occur. Similarly, we can define $\{a, b, c\}$ and $\{a, b, c, d\}$ for patterns with $d(p) = 3$ and $d(p) = 4$, respectively. Figure 9 shows the probabilities of occurrence for label patterns with $d(p) = 1, 2, 3$ and 4., for which there are a total of 8, 28, 56, 78 patterns in each set, respectively. The most popular label pattern sets are highlighted in the subfigures. These patterns represent the cases where DDoS and distributed scanning occurred and sometimes switched targets.

The connectivity features are also effective for differentiating different attack behavior. Consider two different probing behaviors: *web probing* and *share probing*. The *web probing* attacks ports 80, 8000 and 8080, and it is widely used to identify live targets at the beginning of the reconnaissance. The destination ports of *share probing* include ports 139 and 445, and it is also used for host discovery and OS fingerprinting. The spatial label can effectively differentiate these two different behaviors.

Figure 10 is an example to compare the target IPs, target ports, and label sequences of two attack sources from each group. The subfigures (a), (b) and (e) on the left show the behavior of *web probe* and those on the right (c), (d) and (f) show the share probe for comparison. Note that target IP and target ports are two key factors to describe an attacking behavior. The target IPs and ports are shown in a 2^{12} by 2^{12} IP space and a 2^8 by 2^8 port space. Subfigure (e), (f) provide the corresponding label sequence.

Comparing *web probe* with *share probe*, their target port and IP selections are distinct. In addition to ports 80, 8080 and 8000, *web probe* also attacks other

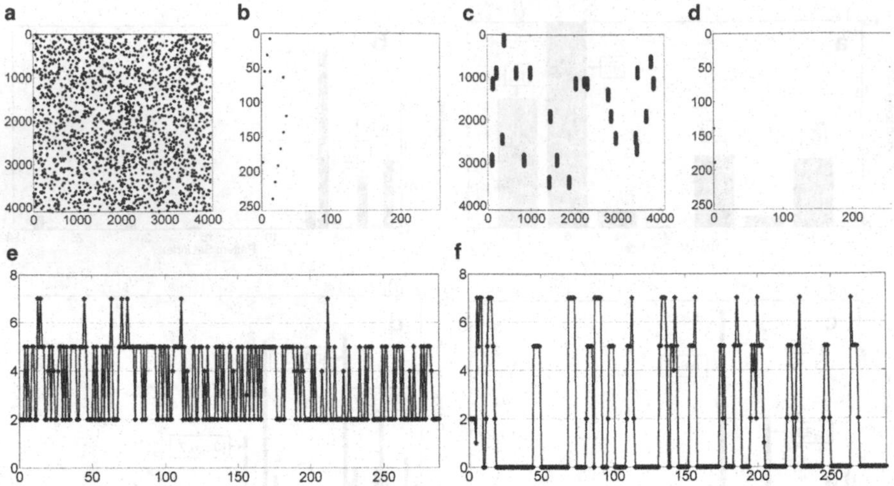

Fig. 10 Label sequence comparison for different attacking strategies. (a) Target IP. (b) Target port. (c) Target IP. (d) Target port. (e) Label sequence. (f) Label sequence

variations, such as ports 808, 1080 and 2080. *share probe* only attacks ports 139 and 445. In terms of target IP, *web probe* often explores randomly over the IP space. *share probe* focuses on scanning individual subnets – the strips in (c) represent the continuous target IPs.

The label sequences of different strategies are very different from one to another. For *web probe*, the majority of the labels are Label-2s and Label-5s, along with other non-zero labels spreading over 24 h. In such case, it is more likely to be an automatic script attack, and have few HATs (Label-6 and Label-7). For *share probe*, the majority of the labels are Label-0s with occasional occurrences of Label-2s, Label-5s and Label-7s. This suggests *share probe* is somewhat sporadic with short breaks, which is consistent with it scanning on a subnet basis. Furthermore, Label-7 will also occur sometimes because the concentration of target IPs is likely to hit some HATs.

As discussed earlier, the challenge of discovering coordinated attacks is due to the stealthy actions distributed across multiple attack sources. The spatial features can effectively discover such sophisticated attacks buried in the overwhelming data. For example, consider an attack source 0.86.249.218, which sent 59 malicious packets over 24 h in the Network Telescope data set [49] (the whole data set contains over 10^9 packets).

Table 6 gives the details of the first ten packets if only traffic volume and target range are considered. There are 21 distinct targets, seemingly randomly selected, and 4 distinct ports (ports 80, 8080, 808 and 8000). There is no evidence suggesting this attack source is important and worthy of further investigation. However, the label sequence for this attack source suggests it could be indicative of advanced attack, where the hacker is switching between compromised hosts,

Table 6 Packet level details for attack source 0.86.249.218

Time	Source IP	Target IP	Protocol	Target port
19:11:08	0.86.249.218	0.145.245.69	TCP	80
19:11:21	0.86.249.218	0.145.245.69	TCP	80
19:27:39	0.86.249.218	0.111.153.210	TCP	80
19:28:53	0.86.249.218	0.140.180.153	TCP	8080
19:27:40	0.86.249.218	0.111.153.210	TCP	80
19:28:52	0.86.249.218	0.140.180.153	TCP	8080
19:30:11	0.86.249.218	0.141.139.196	TCP	80
19:29:48	0.86.249.218	0.141.139.196	TCP	80
19:30:22	0.86.249.218	0.141.68.93	TCP	80
19:27:36	0.86.249.218	0.111.153.210	TCP	80

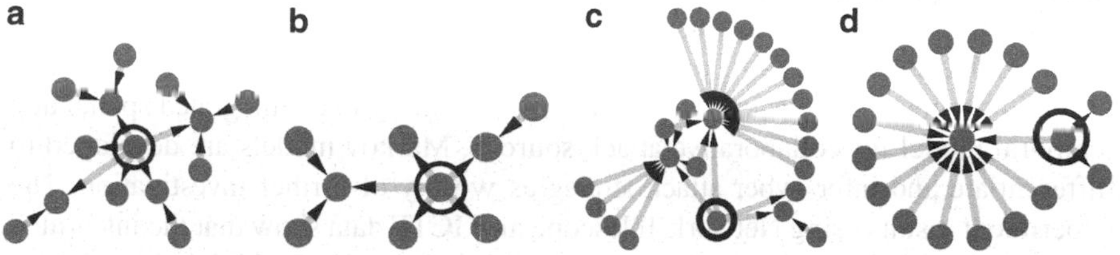

Fig. 11 Label Sequence and ASG Subgraphs for Attack Source 0.86.249.218. (**a**) Attacks in T_6. (**b**) Attacks in T_7. (**c**) Attacks in T_8. (**d**) Attacks in T_9

which can be exhibited by the ASG subgraphs. Therefore we further verify our discoveries by examining the corresponding ASG over time. Figure 11a–d give the ASG subgraphs in four consecutive time frames T_6 to T_9. The attack source 0.86.249.218 is highlighted with the solid circle. For better visualization, only the attack source, its targets and its conspirators are shown. The extracted ASGs show that the attack source is not a inconsequential attacker. Although the malicious traffic volume and the out-degree of the attack source are both small, most of its targets are heavily attacked. During the entire 20 min, the attack methodically narrowed down the point-of-interest and increased the volume of attacks on specific targets. At T_6, it attacked six targets, and four of them were also attacked by other sources. At T_7, it reduced the range of targets by 1. Among the remaining five targets, two were attacked by others. For T_8, the target declined to 4, but 2 of them were heavily attacked by 7 and 14 other sources, respectively. In addition, there were three suspicious sources that attacked both HATs. Note that, in a large targets space, it is unlikely to have four hosts simultaneously select the same two targets. One can hypothesize that these attack sources are controlled as zombie machines or collaborated together for the attack. Finally, at T_9 the source reduced the attack to only three targets and focused on one to perform comprehensive attacks using multiple hosts. This example illustrates that an attack source with transitions between Label-6 and Label-7 can be critical and worth further investigation.

7 Conclusion

The mixture of organized cyber crimes, terrorism and random attacks against enterprise and government networks has led to asymmetric cyber battlefields filled with large-scale cyber attacks. To obtain a timely situation awareness from overwhelming, diverse and evolving data, analysts can benefit from effective computational techniques performing intrusion detection, alert correlation, attack characterization and prediction, host clustering, and coordinated attack analysis. This chapter reviews key contributions in host clustering, attack characterization and prediction, and discusses our integrated temporal and spatial analyses for analyzing large-scale coordinated attacks.

Drawing the analogy from social network analysis, an Attack Social Graph is defined to represent the relationship between attack sources. Applying the notion of degree centrality and agglomerative hierarchical clustering, various types of collaborative attack, or spatial patterns are discovered. These spatial patterns enable a labeling scheme for attack sources over time, resulting in an integrated spatial and temporal model for collaborative attack sources. Markov models are developed to differentiate and infer cyber attack strategies worthy of further investigation. The experiment results using Network Telescope and ICTF data show that the integrated spatial and temporal analyses can provide additional insights for high impact attacks that are not trivial by applying traditional statistical or anomaly analyses.

Simultaneously identifying attack sources belonging to the same coordinated attack groups and the overall attack strategy remains an open challenge. This chapter offers a viable approach to analyze attack strategies by exploring not only the sequential relationship between the attack actions performed by an individual attack source, but also the relationships exhibited in attack actions among attack sources. Expanding from this research, new approaches may be developed to identify coordinated attack *group* behavior.

References

1. Fossl M et al (2010) Symantec internet security threat report for 2010. Technical Report
2. Zhou C, Leckie C, Karunasekera S (2010) A survey of coordinated attacks and collaborative intrusion detection. Comput Secur 29(1):124–140
3. Denning DE (2000) Cyberterrorism: testimony before the special oversight panel on terrorism committee on armed services US house of representatives. Nova Science Pub. Inc, New York
4. Flemming P, Stohl M (2001) Myths and realities of cyberterrorism. In: Proceedings of the international conference on countering terrorism through enhanced international cooperation. ISPAC, pp 70–108
5. Gordon S, Ford R (2002) Cyberterrorism? Comput Secur 21(7):636–647
6. Langner R (2011) Stuxnet: dissecting a cyberwarfare weapon. IEEE Secur Priv 9(3):49–51
7. Roesch M et al (1999) Snort-lightweight intrusion detection for networks. In: Proceedings of the 13th USENIX conference on system administration. USENIX, Berkeley, CA pp 229–238

8. Fuchsberger A (2005) Intrusion detection systems and intrusion prevention systems. Inf Secur Tech Rep 10(3):134–139
9. Valdes A, Skinner K (2001) Probabilistic alert correlation. In: Proceedings of the international symposium of the recent advances in intrusion detection (RAID'01). Springer, Berlin, pp 54–68
10. Dain O, Cunningham RK (2001) Fusing a heterogeneous alert stream into scenarios. In: Proceedings of ACM workshop on data mining and security ACM, New York,
11. Debar H, Wespi A (2001) Aggregation and correlation of intrusion-detection alerts. In: Proceedings of the international symposium of the recent advances in intrusion detection (RAID'01). Springer, Berlin, pp 85–103
12. Cuppens F, Miège A (2002) Alert correlation in a cooperative intrusion detection framework. In: Proceedings of IEEE symposium on security and privacy: IEEE, New York, pp 202–215
13. Cheung S, Lindqvist U, Fong MW (2003) Modeling multistep cyber attacks for scenario recognition. In: Proceedings of DARPA information survivability conference and exposition, IEEE, New York, vol 1. pp 284–292
14. Valeur F, Vigna G, Kruegel C, Kemmerer R (2004) A comprehensive approach to intrusion detection alert correlation. IEEE Trans Dependable Secur Comput 1(3):46–169
15. Ning P, Xu D, Healey CG, Amant RS (2004) Building attack scenarios through integration of complementary alert correlation methods. In: Proceedings of the 11th annual network and distributed system security symposium (NDSS'04). pp 97–111
16. Arnes A, Valeur F, Kemmerer R (2006) Using hidden markov models to evaluate the risk of intrusions. In: Proceedings of the international symposium of the recent advances in intrusion detection (RAID'06), Hamburg, Germany, Springer, Berlin
17. Stotz A, Sudit M (2007) INformation fusion engine for real-time decision-making (INFERD): a perceptual system for cyber attack tracking. In: Proceedings of 10th IEEE international conference on information fusion, IEEE, New York
18. Qin X, Lee W (2004) Attack plan recognition and prediction using causal networks. In: Proceedings of the 20th ACM annual computer security applications conference. ACM, New York, pp 370–379
19. Wang L, Liu A, Jajodia S (2006) Using attack graphs for correlating, hypothesizing, and predicting intrusion alerts. Comput Commun 29(15):2917–2933
20. Holsopple J, Yang SJ (2008) FuSIA: future situation and impact awareness. In: Proceedings of the 11th ISIF/IEEE international conference on information fusion, IEEE, New York
21. Fava D, Byers S, Yang S (2008) Projecting cyberattacks through variable-length markov models. IEEE Trans Inf Forensics Secur 3(3):359–369
22. Du H, Liu D, Holsopple J, Yang S (2010) Toward ensemble characterization and projection of multistage cyber attacks. In: Proceedings of the 19th IEEE international conference on computer communications and networks (ICCCN'10). EEE, New York pp 1–8
23. Soldo F, Le A, Markopoulou A (2011) Blacklisting recommendation system: using spatio-temporal patterns to predict future attacks. IEEE J Sel Areas Commun 29(7):1423–1437
24. Wei S, Mirkovic J, Kissel E (2006) Profiling and clustering internet hosts. In: Proceedings of the 6th IEEE international conference on data mining (ICDM'06). IEEE, New York, pp 269–275
25. Xu K, Zhang Z, Bhattacharyya S (2005) Profiling internet backbone traffic: behavior models and applications. ACM SIGCOMM Comput Commun Rev. USENIX, Berkeley, CA 35(4): 69–180
26. Gu G, Perdisci R, Zhang J, Lee W (2008) BotMiner: clustering analysis of network traffic for protocol-and structure-independent botnet detection. In: Proceedings of the 17th conference on security symposium. USENIX Association. USENIX, Berkeley, CA pp 139–154
27. Soldo F, Le A, Markopoulou A (2010) Predictive blacklisting as an implicit recommendation system. In: Proceedings of IEEE INFOCOM'10. IEEE, New York, pp 1–9
28. Xu K, Wang F, Gu L (2011) Network-aware behavior clustering of Internet end hosts. In: Proceedings of IEEE INFOCOM'11. IEEE, New York, pp 2078–2086

29. Debar H, Dacier M (1999) Towards a taxonomy of intrusion-detection systems. Comput Netw 31(8):805–822
30. Tsai C, Hsu Y, Lin C, Lin W (2009) Intrusion detection by machine learning: a review. Expert Syst Appl 36(10):11994–12000
31. Wu S, Banzhaf W (2010) The use of computational intelligence in intrusion detection systems: a review. Appl Soft Comput 10(1):1–35
32. Bass T (2000) Intrusion detection systems and multisensor data fusion. Commun ACM 43(4):99–105
33. Sadoddin R, Ghorbani A (2006) Alert correlation survey: framework and techniques. In: Proceedings of the ACM international conference on privacy, security and trust. ACM, New York, pp 1–10
34. Haines J, Ryder D, Tinnel L, Taylor S, Kewley Ryder D (2003) Validation of sensor alert correlators. IEEE Secur Priv 1(1):46–56
35. Cuppens F (2001) Managing alerts in a multi-intrusion detection environment. In: Proceedings of the 17th ACM annual computer security applications conference. ACM, New York, 32
36. Ning P, Cui Y (2002) Constructing attack scenarios through correlation of intrusion alerts. In: Proceedings of the 9th ACM conference on computer and communications security. pp 245–254
37. Iyer P, Reeves D et al (2004) Reasoning about complementary intrusion evidence. In: Proceedings of the 20th ACM annual computer security applications conference. ACM, New York, pp 39–48
38. Qin X (2005) A probabilistic-based framework for INFOSEC alert correlation. Ph.D. dissertation
39. Sadoddin R, Ghorbani AA (2009) An incremental frequent structure mining framework for real-time alert correlation. Comput Secur 28(3–4):153–173
40. Li JH, Levy R (2010) Using Bayesian networks for cyber security analysis. In: Proceedings of the 40th IEEE/IFIP international conference on dependable systems & networks (DSN'10). pp 211–220
41. Li J, Ou X (2010) Uncertainty and risk management in cyber situational awareness. Adv Inf Secur 46:51–68
42. Adomavicius G, Tuzhilin A (2005) Toward the next generation of recommender systems: a survey of the state-of-the-art and possible extensions. IEEE Trans Knowl Data Eng 17(6): 734–749
43. Hastie T, Tibshirani R et al (2001) The elements of statistical learning: data mining, inference and prediction. Springer, Berlin/New York
44. Bell T, Cleary J, Witten I (1990) Text compression. Prentice-Hall, Englewood
45. Du H, Yang S (2011) Discovering collaborative cyber attack patterns using social network analysis. In: Proceedings of social computing, behavioral-cultural modeling and prediction (SBP'10). Springer, Berlin/Heidelberg, pp 129–136
46. Childers N, Vigna G et al (2010) Organizing large scale hacking competitions. In: Proceedings of detection of intrusions and malware, and vulnerability assessment (DIMVA'10), vol 6201. Springer, Berlin/Heidelberg, pp 132–152
47. ICTF Data set [Online]. Available: http://ictf.cs.ucsb.edu/data.php. Accessed Jan 2012
48. Moore D, Shannon C, Voelker G, Savage S (2004) Network telescopes: technical report. Technical Report
49. Aben E et al The CAIDA UCSD network telescope two days in November 2008 dataset [Online]. Available: http://www.caida.org/data/passive/telescope-2days-2008_dataset.xml. Accessed Jan 2012
50. Bonacich P (1987) Power and centrality: a family of measures. Am J Sociol 92:1170–1182